High-Intensity Ultrasonics

High-Intensity Ultrasonics
Theory and Industrial Applications

Oleg V. Abramov

Kurnakov Institute of General and Inorganic Chemistry
Moscow, Russia

CRC Press
Taylor & Francis Group
Boca Raton London New York

CRC Press is an imprint of the
Taylor & Francis Group, an **informa** business

CRC Press
Taylor & Francis Group
6000 Broken Sound Parkway NW, Suite 300
Boca Raton, FL 33487-2742

First issued in paperback 2020

ISBN-13: 978-0-367-45570-5 (pbk)
ISBN-13: 978-90-5699-041-1 (hbk)

Visit the Taylor & Francis Web site at
http://www.taylorandfrancis.com

and the CRC Press Web site at
http://www.crcpress.com

British Library Cataloguing in Publication Data

Abramov, O.V. (Oleg Vladimirovich)
High-intensity ultrasonics : theory and industrial
applications
1. Ultrasonic waves - Industrial applications
I. Title
620.2'8

CONTENTS

Preface

This book is concerned with the physical and technical aspects of high-intensity ultrasound and its applications in industry.

There is already a lot of relevant literature describing the potential benefits of innovative ultrasonic technological processes. However, most of the published information deals with particular problems of ultrasound and is scattered throughout highly specialized journals, therefore it does not often come into the view of specialists.

The monographs available either treat certain problems of the physics of ultrasound or concentrate on its specific applications. To our knowledge, there have been no attempts, either in Russia or abroad, to generalize the physical, technological, and practical aspects of ultrasound in the scope of one monograph.

In fact, the only book where the problems of physics and technology of ultrasonics were comprehensively approached was a three-volume work edited by L. B. Rozenberg: *Physical Processes of Ultrasonic Technology*, Nauka Publishers (1967–1970). The present monograph is an attempt to update the material included in Rozenberg's book.

This book mainly includes the results of Russian authors, which were published in Soviet and Russian journals and are thus unfamiliar to foreign readers. Parts II and III of this book are mainly based on the author's research at the Institute of General and Inorganic Chemistry, Institute of Solid Physics, and in the Bardin Research Institute of Ferrous Metallurgy.

Many of the problems treated in this book were fruitfully discussed with my collaborators and my foreign colleagues, recognized experts in the field of ultrasonics. For this I am deeply grateful to Dr V. O. Abramov, Dr M. Arsanov, Professor J. Berlan (France), Dr Yu. Ya. Borisov, Dr A. De Paoli (Germany), Dr O. N. Gradov, Dr K. Graf (United States), Professor V. P. Zubov, Mr S. Jacke (United States), Mr G. Kaess (United Kingdom), Mr J. Keller (Switzerland), Mr R. Hunicke (United States), Mr D. Hunicke (United States), Dr A. I. Kirshin,

Dr L. O. Makarov, Professor T. Mason (United Kingdom), Academician Yu. A. Osip'yan, Professor E. Rivin (United States), Dr A. A. Rukhman, Dr R. G. Sarukhanov, Dr S. Svegla (Slovakia), Dr U. Stolz (Germany), Professor K. Suslick (United States), Professor J. Wissler (Germany), Mr M. Walter (Germany), Professor G. I. Eskin, Dr D. Zimmermann (Belgium), and others.

I also wish to thank N. B. Abramova, T. O. Abramova, N. N. Manokhina, E. S. Prosina, A. I. Teplyakova, L. T. Frolova, and Zh. Yu. Chashechkina for their help in preparing this manuscript.

Introduction

There is currently an increasing understanding, at the community level, of the benefits of the latest scientific and technological advances. Among these, high-intensity ultrasound deserves particular attention.

In the final analysis, the course of any physicochemical reaction in a given system is determined by temperature, pressure, their rates, and the velocity of movement of particular parts of the system. By varying these parameters, one can control physicochemical reactions and thereby the overall technological process(es) of the manufacturing and processing of materials, metabolism of micro- and macroorganisms, the modes of equipment operation, and so on.

Innovative technological tools, such as lasers, plasmas, electromagnetic fields, and high-energy particle fluxes, appreciably widen the range of means to control physicochemical processes. In particular, they may substantially raise attainable temperature, pressure, and their rates, as well as the velocity of convective flows in liquids and structural imperfection movement in solids, as compared to those occurring in natural processes and present-day technologies.

In terms of the foregoing, industrial implementation of the latest scientific advances in the processes of physicochemical transformations could appreciably contribute to the development of efficient technologies and to the national economy as a whole.

Among recent technological developments, high-power ultrasound occupies a special position as a powerful tool to affect a substance.

At the community level, the role of ultrasound, as part of acoustics, is manifold. The diagram given below illustrates the spheres of potential ultrasonic applications. Those representing a particular interest in the scope of this book are highlighted.

Relationship of Acoustics with Fundamental
and Applied Sciences and the Arts

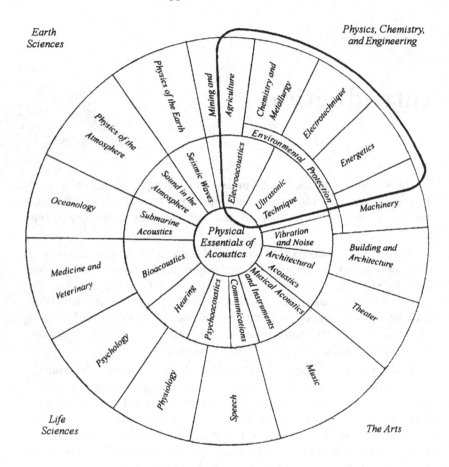

*Earth
Sciences*

*Physics, Chemistry,
and Engineering*

*Life
Sciences*

The Arts

The most promising fields of ultrasonic application involve

(1) Mineral processing;

(2) Metallurgy, machinery, electronics, and electroengineering;

(3) Chemical technology and biotechnology;

(4) Agriculture, food and light industries;

(5) Goods manufacturing;

(6) Energetics;

(7) Environmental protection;

(8) Medicine and veterinary.

Ultrasonic vibrations of kilo and megahertz frequencies and intensities of more than 0.1 W/cm^2 usually used in practice may cause irreversible transformations in processed media.

The first relevant works date back to the end of the 1920s, when Wood and Loomis observed ultrasonically driven atomization and emulsification of liquids as well as structural alterations in solidified organic substances and living tissues.

The research along this line was extended by Soviet scientists. In the 1930s–1940s, V. I. Danilov and S. Ya. Sokolov undertook studies to elucidate the mechanism of crystallization of organic substances and metals in an ultrasonic field.

Considerable progress in ultrasonic studies began from the mid 1950s, when L. B. Rozenberg and his collaborators at the Acoustic Institute, started their research into the propagation of high-power ultrasound in liquids and solids with an emphasis on ultrasonic cutting and welding.

Ultrasonic welding was also studied by a group headed by Academician G. A. Nikolaev, Moscow State Technical University.

In Minsk, a research group headed by Belorussian Academicians E. G. Konovalov and V. P. Severdenko greatly enhanced ultrasonic methods for metal shaping.

Many foreign scientists, among them Polman (Germany), Crowford (United States), Noltingk and Neppiras (England), Kikuchi (Japan), Langenecker, Weiss (Austria), Greguch (Hungary), and others extensively studied ultrasonic refining, cutting, welding, emulsification, and other applications of ultrasound.

The number of laboratories engaged in high-power ultrasonic research is gradually increasing. In addition to those mentioned above similar investigations are being successfully performed in the laboratories of some academic institutes, such as the Moscow Institute of Solid State Physics (Russian Academy of Sciences), the Institute of Strength Problems, and the Institute of Metal Physics (both of the National Academy of Sciences, Ukraine) and in many educational institutes, such as the Moscow Institute of Steels and Alloys, Moscow Highway Institute, Moscow Institute of Aviation; and in specialized institutes, such as the All-Russia Institute of Light Alloys and Central Research

Institute of Ferrous Metallurgy. The Moscow Institute of Radioelectronics and Automatics and the Moscow Institute of Steels and Alloys provide graduates trained in ultrasonic engineering and technology.

Research efforts in studying high-power ultrasound propagation in liquids and solids led to the understanding of some effects occurring at interfaces and thereby contributed to the development of various ultrasonic technologies.

The benefits of using ultrasound include an appreciable shortening of production processes, the reduction of power consumption, and improvement of product quality.

As regards the prerequisites of the employment of high-power ultrasound as a potent technological tool, it should be emphasized that ultrasonic vibrations may significantly influence heat and mass transfer in liquids, modify the structure and properties of solids and thereby interfere with their interactions. This is primarily due to nonlinear effects (cavitation, acoustic streaming, and movement of dislocations) that accompany the propagation of high-power ultrasound in media.

This monograph covers a wide range of problems associated with industrial applications of ultrasound. It gives a unified consideration of the physical background for the ultrasonic technology of materials and principles underlying specific approaches and methods for feeding vibrations into a load. The appropriateness of such consideration is readily apparent from the fact that the applicability of ultrasound in particular production processes is determined by common phenomena arising in response to ultrasonic irradiation – cavitation and acoustic streaming in melts and increased density of structural imperfections in solids. This dictates the structure of the monograph which is in three parts.

Part I introduces the reader to the physical fundamentals of high-power ultrasound. Some chapters treat the regularities of the propagation of high-power ultrasound in liquid media, as well as the related nonlinear phenomena (cavitation, acoustic streaming, and dispersion), and give the theoretical analysis of metal crystallization in an ultrasonic field. Other chapters deal with the propagation of ultrasound in solids, alterations in dislocational structure caused by alternating stresses, as well as with the analysis of the ultrasonic effects on a solidified metal subjected to heating and deformation.

Part II of the book is concerned with the design principles of ultrasonic generators, mechanoacoustic radiators, and other vibrational systems, as well as the control of acoustic parameters during the feeding of vibrations into a processed medium.

Part III envisages problems associated with the employment of high-power ultrasound in metallurgy for ore dressing, degassing of metal melts during their crystallization, production of powders and cast composites, coating, and zonal crystallization. The applications of high-intensity ultrasound for metal shaping, thermal and thermochemical treatment, welding, cutting, refining, and surface hardening are also discussed.

Part III considers the mechanism of ultrasonic action on particular production processes, the feasible methods of ultrasound feeding into a load, the concrete diagrams of ultrasonic processing of materials, accompanying effects, and the technological benefits of using ultrasound.

Part I
Physical Aspects of Ultrasonics

An understanding of the specificities of high-power ultrasonic effects on technological processes is possible if the reader has a certain knowledge of engineering sciences and ultrasonics. As the book is primarily for those who are concerned, in one way or another, with material sciences and engineering, it seems reasonable in the first part of the book to introduce the reader to the physical and technical aspects of ultrasonics which is a branch of the wave motion science covering a diversity of phenomena, such as the propagation of elastic waves in water, earth crust, air, electromagnetic waves of a radio-frequency range, light waves, etc.

The first part of the book deals with the problems that will be considered to some extent in its subsequent parts. These involve the basic principles of ultrasonics: the regularities of vibrational motion and propagation of low- and finite-amplitude waves in fluids and solids, as well as nonlinear effects at interfaces.

Chapter 1

Low-Amplitude Vibrations and Waves

This chapter gives a brief description of the regularities of oscillatory and wave motion as well as the principles of the propagation of low-amplitude waves in fluids and solids.

The fundamentals of classical acoustics were formulated in the works of G. Galilei, M. Mersenne, L. Euler, V. Weber, and G. Helmholtz. The book by J. Rayleigh "Theory of sound" [1] completes the main stage of development of classical acoustics. For a more detailed information on acoustic fundamentals, the reader is referred to [2–10].

1.1 Harmonic Oscillator

Vibrations, or periodical reciprocating motion, is a widespread form of motion. Mechanical oscillations (e.g. of a pendulum, tuning fork, etc.) can be considered as repetitive alterations of position and velocity of a body or its parts. A change in the current and voltage in a circuit or movement of electrons in atoms, as well as other relevant processes, may give rise to electrical oscillations. Although different, various oscillations obey the same laws and can be described by the same equations.

Among a diversity of oscillations, those of a linear harmonic oscillator are considered to be simplest (Figure 1.1).

If mass m is disturbed from equilibrium by stretching or compressing a spring and then letting it free, the mass begins to oscillate around its equilibrium position*. Such oscillations are known as free or nat-

---

*It is assumed that the system possesses one degree of freedom (i.e. its position at any time can be described by a single parameter), the spring is ideally elastic

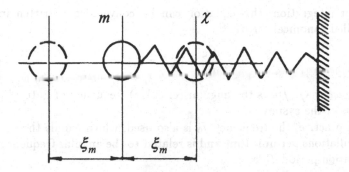

Figure 1.1. Vibrations in the simplest mechanical system.

ural, as opposed to forced oscillations, when the system is subject to external force. Oscillations would not be damped if energy loss is absent, i.e. if the system is conservative. Let us consider the simplest case of undamped natural oscillations, which makes it possible to elucidate on the contribution from various parameters of real systems to the oscillatory process.

The system under study is acted upon by two forces, inertial F_i and elastic F_e. The law of motion can be written as

$$F_i + F_e = 0. \tag{1.1}$$

According to the force law

$$F_i = m\frac{d^2\xi}{dt^2}, \tag{1.2a}$$

where ξ is the displacement, $d^2\xi/dt^2$ is the acceleration equal to the second derivative of displacement with respect to time t.

For an ideal spring, the elastic force F_e counteracting the extension is given by

$$F_e = \chi\xi, \tag{1.2b}$$

where χ is the coefficient of elasticity which is equal to the ratio of the acting force to the mass displacement (i.e. to the extension or compression of the spring).

In view of (1.2a) and (1.2b), the equation of motion (1.1) can be rewritten as

$$m\frac{d^2\xi}{dt^2} + \chi\xi = 0. \tag{1.3a}$$

and has no mass, while the displacement from equilibrium is small.

For integration, this equation can be conveniently written in the so-called canonical form

$$\frac{d^2\xi}{dt^2} + \omega_0^2\xi = 0, \qquad (1.3b)$$

where $\omega_0 = \sqrt{\chi/m}$ is the angular (circular) frequency of natural oscillations of the system.

In practice, the frequency f_0 is also used, which equals the number of oscillations per unit time and is related to the angular frequency and oscillation period T_0 as

$$f_0 = \frac{1}{2\pi}\omega_0 = \frac{1}{T_0}. \qquad (1.4)$$

The general solution to equation (1.3b) has the form

$$\xi = A\cos\omega_0 t + B\sin\omega_0 t = \xi_m\cos(\omega_0 t + \theta), \qquad (1.5a)$$

where $A = \xi_m\cos\theta$; $B = -\xi_m\sin\theta$; ξ_m is the amplitude of oscillations (i.e. the maximum displacement of the mass from its equilibrium position); θ is the oscillation phase (a phase shift with respect to some phase taken as zero).

Instead of the real representation of equation (1.5a) for harmonic oscillations, its complex representation can be used:

$$\xi = \xi_m e^{j(\omega_0 t + \theta)}, \qquad (1.5b)$$

where $j = \sqrt{-1}$.

Real physical oscillations are described by the real part of equation (1.5b). The convenience of complex representation is determined by the ease and clearness of operations with complex quantities.

The quantities A and B, or ξ_m and θ, are defined by initial conditions.

If the mass is displaced to a distance $\xi = \xi_m$ and then set free without a push, then the initial conditions for the system are as follows

$$\xi = \xi_m, \quad \frac{d\xi}{dt} = 0 \quad \text{at} \quad t = 0. \qquad (1.6)$$

The solution to equation (1.3b) under initial conditions (1.6) can be written as

$$\xi = \xi_m\cos\omega_0 t, \qquad (1.7a)$$

which represents an analytical expression for the displacement at an arbitrary moment in the case of free undamped oscillations of the system (Figure 1.2).

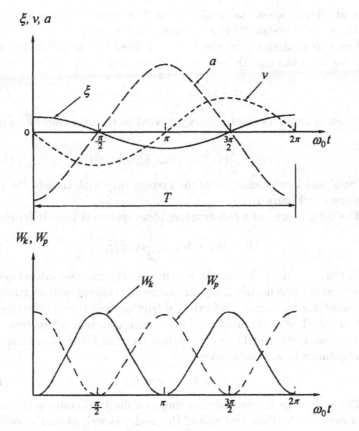

Figure 1.2. Time cources of displacement ξ, vibrational velocity v, acceleration a, kinetic (W_k) and potential (W_p) energy.

Expressions for the velocity and acceleration of the mass can be derived from (1.7a) by differentiating it with respect to time

$$v = \dot{\xi} = \frac{d\xi}{dt} = -\omega_0 \xi_m \sin \omega_0 t = -v_m \sin \omega_0 t \qquad (1.7b)$$

$$a = \ddot{\xi} = \frac{d^2\xi}{dt^2} = -\omega_0^2 \xi_m \cos \omega_0 t = -a_m \cos \omega_0 t, \qquad (1.7c)$$

where $v_m = \omega_0 \xi_m$ and $a_m = \omega_0^2 \xi_m$ are the amplitudes of vibrational velocity and acceleration, respectively.

It is seen from (1.7a)–(1.7c) that displacement, velocity, and acceleration change with time with the same frequency, their phases being different. Thus, when the displacement ξ is maximum, the velocity v is zero, and vice versa. This is apparent in Figure 1.2.

The potential energy of the system defined by the displacement of the mass from the equilibrium can be described by the expression

$$W_p = \frac{1}{2}\chi\xi^2 = \frac{1}{2}\chi\xi_m^2 \cos^2 \omega_0 t. \tag{1.8a}$$

At the same time, its kinetic energy related to the mass motion is given by

$$W_k = \frac{1}{2}mv^2 = \frac{1}{2}m\omega_0^2\xi_m^2 \sin^2 \omega_0 t. \tag{1.8b}.$$

Potential and kinetic energies of the system vary with time in the manner shown in Figure 1.2.

The total energy of a conservative ideal system is time-independent

$$W = W_p + W_k = \frac{1}{2}m\omega_0^2\xi_m^2. \tag{1.9}$$

At the same time, free oscillations in a real (nonconservative) system cannot persist indefinitely long. Its mechanical energy will be gradually dissipated due to internal and external friction, and free oscillations will be damped. The equation of motion for such a system differs from that for an ideal system (1.1) in that it involves the term F_l representing the resistance to a moving body

$$F_i + F_e + F_l = 0. \tag{1.10}$$

The resistance to the body moving in a fluid depends on the velocity of motion, the form and size of the body, as well as on the medium properties. In the range of small velocities, the resistance force is proportional to oscillatory velocity:

$$F_l = b\frac{d\xi}{dt}, \tag{1.11}$$

where b is the coefficient of proportionality known also as the friction coefficient. In view of the foregoing, the equation of motion with losses (1.10) can be rewritten as

$$\frac{d^2\xi}{dt^2} + 2\beta\frac{d\xi}{dt} + \omega_0^2\xi = 0, \tag{1.12}$$

where $\beta = b/(2m)$ is the damping coefficient proportional to losses. Sometimes it appears expedient to introduce the constant τ called the relaxation time and related to the damping coefficient by formula

$$\tau = \frac{1}{2\beta}. \tag{1.13}$$

The solution to equation (1.12) under the initial conditions (1.6) has the form:

$$\xi_l = \xi_m e^{-\beta t} \cos \omega_l t = \xi_m e^{-\beta t} e^{j \omega_l t}, \qquad (1.14)$$

where

$$\omega_l = \sqrt{\omega_0^2 - \beta^2} = \sqrt{\omega_0^2 - \left(\frac{1}{2\tau}\right)^2} = \omega_0 \sqrt{1 - \left(\frac{1}{2\omega_0 \tau}\right)^2}. \qquad (1.15)$$

Analysis of (1.14) shows that the amplitude of oscillations, $\xi_m e^{-\beta t}$, decreases with time, while the frequency remains constant, although being smaller than in the absence of friction in the system (Figure 1.3).

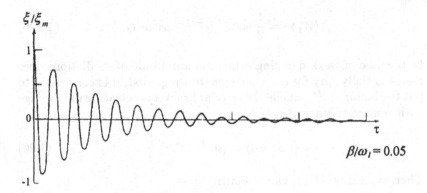

Figure 1.3. Time course of displacement in damped system.

By the period is meant the duration of one complete to and fro movement of an oscillating particle. After $t_l = \frac{1}{\beta} = 2\tau$, the displacement amplitude decreases by a factor of e. If $\omega_0 \tau \gg 1$, then (1.14) can be rewritten as

$$\xi_l = \xi_m e^{-\beta t} \cos \omega_0 t = \xi_m e^{-\frac{t}{2\tau}} \cos \omega_0 t. \qquad (1.16)$$

This case corresponds to the lowest damping of oscillations.

To estimate the extent of damping in the system one can relate it to a natural time scale, i.e. to the oscillation period. Damping can be characterized by a decrease in the amplitude for the period, then, instead of the damping factor β, one can conveniently use a logarithmic decrement of damping $\beta_e = \ln(\xi_{mn}/\xi_{m(n+1)})$, where ξ_{mn} and $\xi_{m(n+1)}$ are the amplitudes of nth and $(n+1)$th oscillations, respectively.

The logarithmic damping decrement β_e is related to the damping factor β by the formula

$$\beta_e = \beta T. \qquad (1.17)$$

Taking into account the relations (1.8) and (1.16), one can obtain expressions for the kinetic and potential energies of the harmonic oscillator for the extreme case of weak damping:

$$W_k = \frac{1}{2} m \omega_0^2 \xi_m^2 e^{-t/\tau} \sin^2 \omega_0 t, \tag{1.18a}$$

$$W_p = \frac{1}{2} m \omega_0^2 \xi_m^2 e^{-t/\tau} \cos^2 \omega_0 t. \tag{1.18b}$$

The mean kinetic and potential energies for one period comprise

$$\langle W_k \rangle_T = \frac{1}{2} m \omega_0^2 \xi_m^2 \langle e^{-t/\tau} \sin^2 \omega_0 t \rangle, \tag{1.19a}$$

$$\langle W_p \rangle_T = \frac{1}{2} m \omega_0^2 \xi_m^2 \langle e^{-t/\tau} \cos^2 \omega_0 t \rangle. \tag{1.19b}$$

In the case of weak damping, when the amplitude of oscillations does not essentially vary for the span equal to the period, it is convenient to put the factor $e^{-t/\tau}$ outside the angular brackets indicating the average with respect to time. Besides,

$$\langle \cos^2 \omega_0 t \rangle = \langle \sin^2 \omega_0 t \rangle = \frac{1}{2}. \tag{1.20}$$

Then expression (1.19) can be rewritten as

$$\langle W_p \rangle_T = \langle W_k \rangle_T \cong \frac{1}{4} m \omega_0^2 \xi_m^2 e^{-t/\tau}. \tag{1.21}$$

The mean energy dissipated per unit time is the time derivative of energy taken with the opposite sign

$$\langle \Delta W \rangle = -\frac{d}{dt} \langle W \rangle \cong -\frac{d}{dt} (\langle W_k + W_p \rangle_T) \cong \frac{1}{\tau} \left(\frac{1}{2} m \omega_0^2 \xi_m^2 e^{\frac{-t}{\tau}} \right), \tag{1.22a}$$

while that dissipated for the period is

$$\langle \Delta W \rangle_T = \frac{\langle W \rangle_T}{\tau}. \tag{1.22b}$$

Thus, it can be inferred from the foregoing that the conservative system is characterized only by two parameters, the mass m and elasticity χ, while the properties of a real system are determined by three parameters (additionally to the above two, by the friction (or damping) factor characterizing energy loss in the medium).

To make oscillations in the system with losses undamped, the system must be applied with an external periodic (e.g., harmonic) force:

$$F = F_m \cos \omega t. \tag{1.23}$$

Here, F_m is the force amplitude, ω is its frequency. Such oscillations are called the forced oscillations. The equation of motion in this case can be written as

$$\frac{d^2\xi}{dt^2} + 2\beta\frac{d\xi}{dt} + \omega_0^2\xi = \frac{F_m}{m}\cos\omega t = \frac{F_m}{m}e^{j\omega t}. \tag{1.24}$$

This equation is derived from (1.12), to the right-hand side of which expression (1.23) for the external force is added. For convenience, expression (1.23) is written in the complex form.

The solution to (1.24) can be written as

$$\xi(t) = \xi_l(t) + \xi_F(t). \tag{1.25}$$

The first term on the right-hand side of (1.25), $\xi_l(t) = \xi_m e^{-\beta t} \times e^{j(\omega t + \theta)}$, ($\theta$ is the initial phase of oscillations) describes free damped oscillations, while the second term describes forced oscillations. After some time, because of free oscillations damping, the system will display only forced oscillations which can be described by the expression

$$\xi_F(t) = \xi_{Fm}e^{j(\omega t + \psi)}, \tag{1.26}$$

where ξ_{Fm} is the amplitude of forced oscillations equal to

$$\xi_{Fm} = \frac{F_m}{m\sqrt{(\omega_0^2 + \omega^2)^2 + 4\omega^2\beta^2}}. \tag{1.27}$$

ψ is the phase difference between the applied force and the displacement it produces, so that

$$\text{tg}\,\psi = -\frac{2\beta\omega}{\omega_0^2 - \omega^2}. \tag{1.28}$$

It is seen that ψ is negative at all ω, i.e., the displacement lags behind the driving force.

Analysis of (1.27) indicates that when the frequency of external force approaches the natural frequency of the system, the amplitude of forced oscillations drastically increases. This phenomenon is referred to as resonance. Figure 1.4,a shows a set of curves illustrating the

ω-dependence of the amplitude ξ_{Fm} for various damping factors. At resonance ($\omega = \omega_0 = \omega_r$), the amplitude becomes maximum

$$\xi_{Fm,\max} = \frac{F_m}{2m\beta\omega_r}. \qquad (1.29)$$

An ideal system without damping would have an infinitely great amplitude of resonant oscillations.

One of the important features of a harmonic oscillator is its increased displacement amplitude at resonance, as compared with the displacement produced by a constant force ($\omega = 0$)

$$\xi_{F,0} = \frac{F_m}{m\omega_0^2}. \qquad (1.30)$$

By analogy with electrotechnical terminology, we use here a Q-factor to characterize the resonant properties of the system:

$$Q = \frac{\xi_{Fm,\max}}{\xi_{F0}} = \frac{\omega_r}{2\beta} = \frac{\pi}{\beta_e}. \qquad (1.31)$$

Thus, the Q-factor of the vibrating system is inversely related to the logarithmic damping decrement and can be used to characterize losses in the system.

Physically, Q-factor represents the ratio of the total energy W accumulated in a vibrating system for the unit time (period) to the energy loss ΔW. With allowance for expressions (1.22), one can write:

$$Q = \frac{2\pi\langle W\rangle_T}{\langle\Delta W\rangle_T} = \omega_r\tau. \qquad (1.32)$$

The important characteristic of a resonant system is the extent of the amplitude decrease caused by deviation of its frequency from the resonant value.

This characteristic is accounted for by the resonance half-width $\Delta\omega = \omega_2 - \omega_1$, where ω_1 and ω_2 are the frequencies at which the square of their displacement amplitude ξ'_{Fm} is equal to $\frac{1}{2}\xi^2_{Fm,\max}$; therefore, $\xi'_m = \xi_{Fm,\max}/\sqrt{2}$. The square in this expression is due to the fact that the energy of the harmonic oscillator is proportional to the square of the displacement amplitude.

One can show that the resonance half-width is related to the Q-factor by expression:

$$\Delta\omega = \frac{\omega_r}{2Q}. \qquad (1.33)$$

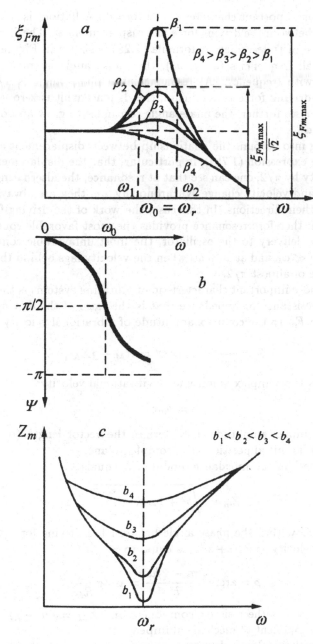

Figure 1.4. Frequency dependence of (*a*) displacement amplitude, (*b*) the phase angle between the force and displacement, and (*c*) the mechanical impedance modulus of forced vibrations.

Another important characteristic of forced oscillations is the phase difference between the driving force and displacement ψ. The frequency dependence of ψ described by formula (1.28) is shown in Figure 1.4,b.

At small frequencies ($\omega \ll \omega_0$), the phase angle is small, but it increases with frequency. At resonance, the phase angle is $\pi/2$, i.e. when the driving force is maximum, the displacement is zero. As the frequency rises further, the phase angle continues to grow approaching π at $\omega \gg \omega_0$.

Taking into account the relationship between displacement and velocity (see expression (1.7b)), in particular, that the displacement lags the velocity by $\pi/2$, one can see that at resonance, the alternating driving force and velocity change synchronously, i.e. they are always coincident in their directions. In this case, the work of the driving force is maximum; therefore resonance provides the most favorable conditions for energy delivery to the oscillator, the most unfavorable conditions being at $\omega \ll \omega_0$ and at $\omega \gg \omega_0$ when the velocity lags behind the force by a value of almost $\pi/2$.

One more important characteristic of vibrating systems is their mechanical resistance or impedance that is the ratio of the driving force amplitude F_m to the complex amplitude of vibrational velocity v

$$Z = \frac{F_m}{v} = Z_m e^{-j\varphi} = R_M + jX_M, \qquad (1.34)$$

where v is the complex amplitude of vibrational velocity:

$$v = v_m e^{j\varphi}.$$

The complex amplitude characterizes the vector length (or amplitude) and its initial position on a complex plane.

The mechanical impedance module Z_m equals

$$Z_m = \sqrt{b^2 + \left(\omega m - \frac{\chi}{\omega}\right)^2}. \qquad (1.35)$$

One can show that the phase angle between the driving force and oscillatory velocity, $\varphi = \psi + \pi/2$, is equal to

$$\varphi = \operatorname{arctg} \frac{\omega m - \chi/\omega}{b} = \operatorname{arctg} \frac{X_M}{R_M}, \qquad (1.36)$$

where $R_M = b$ is the resistive component and $X_M = \omega m - \chi/\omega$ is the reactive component of mechanical impedance.

The mechanical impedance can be graphically represented in a complex system of coordinates, where the abscissa is R_M, and the ordinate is X_M.

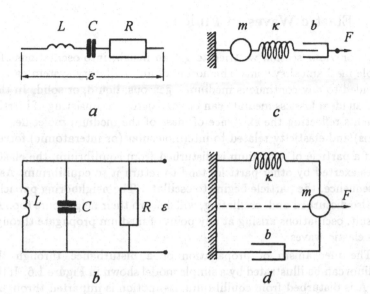

Figure 1.5. An electric vibrational circuit with (a) series and (b) parallel elements and a corresponding mechanical vibrational system with (c) series and (d) parallel elements.

Impedance characterizes the response of the system to a driving force. The impedance of a given mechanical system at a given frequency is constant. Figure 1.4 presents the frequency dependence of impedance module for different b. Being subjected to equal forces, the systems with a higher impedance oscillates at a smaller frequency than the systems with a lower impedance. As seen from expression (1.35), the impedance is minimal at resonance when it is equal to resistance.

Mechanical impedance can be also referred to more complex mechanical oscillating systems, including those with distributed parameters. Since various particles of a complex mechanical oscillating system subject to a force can oscillate with different velocities, the impedance is defined as the ratio of the applied force to the vibrational velocity of that point to which this force is applied.

Analysis of electromechanical transducers and waveguides is often performed with the use of electromechanical analogies, which allows the employment of well-developed methods of the theory of electric circuits.

Indeed, one can show that equation describing forced oscillations in an electric circuit with electron driving force ϵ, self-inductance L, capacitance C, and resistance R (Figure 1.5) is analogous to equation (1.24) characterizing oscillations in a mechanical system.

1.2   Elastic Waves in Fluids

In the previous section we tried to get an insight into oscillations of a simple mechanical system. The notion of an oscillatory system can be extended to any continuous medium – gaseous, liquid, or solid. In this case, an ideal lossless medium can be considered as consisting of inertial elements reflecting the existence of mass of the medium molecules (or atoms) and elasticity related to intermolecular (or interatomic) forces.

If a particle of a medium is disturbed from equilibrium, the elastic forces exerted by other particles tend to return it to equilibrium. As a consequence, the particle begins to oscillate. The neighboring particles are also disturbed and, oscillating, will disturb their neighbours too. As a result, oscillations arising at any point of medium propagate through it as elastic waves.

The mechanism of propagation of a disturbance through the medium can be illustrated by a simple model shown in Figure 1.6. If the ball A is disturbed from equilibrium, its motion is imparted through a spring to the ball B that is displaced from an equilibrium position with a delay due to inertia. In turn, the ball B transmits its motion through

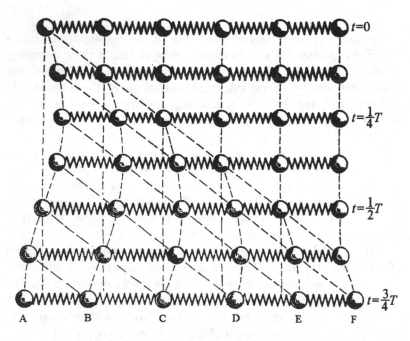

Figure 1.6.   Mechanical model of wave propagation in a medium.

the next spring to the ball C. As a result, the disturbance propagates over the system at some finite velocity.

It is apparent from this model that elastic waves represent propagating alternating compressions and rarefactions.

To describe the wave motion, one has to establish the relationship between the disturbance (i.e. the displacement of the medium particles from their equilibrium positions), time, and distance from the source of oscillations. This can be easily done for the case of elastic waves propagating in liquids and gases.

The elastic wave travelling in fluids represents an alternating flow and obeys the laws of hydrodynamics. A complete set of hydrodynamic equations can be written as follows*:

$$\rho \frac{dv}{dt} = -\operatorname{grad} P - \rho (\operatorname{grad} \mathfrak{f}) + F_v, \tag{1.37}$$

$$\frac{\partial \rho}{\partial t} + \operatorname{div} (\rho v) = 0, \tag{1.38}$$

$$\mathfrak{f}(P, \rho, T) = 0, \tag{1.39}$$

where ρ is the medium density; P is pressure; v is the velocity vector; T is absolute temperature; F_v are viscous forces; $\mathfrak{f}$ is the potential of external forces.

The Euler equation (1.37) describes the motion of particles subjected to elastic forces, i.e. external forces, such as gravity or electricity, and internal forces, such as shear stress in a flowing fluid.

The elastic forces are defined as the pressure per unit volume $(\operatorname{grad} P)$. In the case of conservative external forces with the potential reduced to unit mass, $\mathfrak{f}$, they can be written as $\rho(\operatorname{grad} \mathfrak{f})$. The internal forces are those due to viscosity, F_v. The equation (1.38) known as the equation of continuity is valid if there are no discontinuities in the medium, as in the absence of cavitation.

The equation of state (1.39) has no standard form.

Equations (1.37–1.39) are nonlinear, but they can be linearly approximated for the case of low-amplitude acoustic waves.

Consider the approximated equations of motion and continuity for a homogeneous medium possessing an ideal fluidity (i.e., without a shear stress directed tangentially to the medium motion), absolute elasticity, and zero heat conductivity.

---

*Hereinafter, we use in representations the Hamiltonian $\nabla = i\frac{\partial}{\partial x} + j\frac{\partial}{\partial y} + k\frac{\partial}{\partial z}$ and Laplacian $\Delta = \nabla^2 = \frac{\partial}{\partial x^2} + \frac{\partial}{\partial y^2} + \frac{\partial}{\partial z^2}$. In this case, $\nabla a = \operatorname{grad} a$, $\nabla a = \operatorname{div} a$, and $\nabla \times a = \operatorname{rot} a$, where $a = a(x, y, z)$; a is vector; i, j, k are unit vectors.

Such medium in the state of mechanical and thermal equilibrium is completely described by three parameters – pressure P_0, temperature T_0, and density ρ_0 which vary when elastic oscillations propagate through the medium. To describe elastic waves, one has to relate the rate of changing of the above parameters to the amplitude or velocity of oscillations.

Consider the volume element dx, dy, dz (Figure 1.7,a) whose velocity v can be decomposed into three components v_x, v_y, and v_z parallel to the element faces. The motion of the element is accompanied by changes in pressure. For example, if pressures exerted on the faces A_1 and A_2 are P and $P + (\partial P/\partial x)\,dx$, respectively, then the difference in the pressures exerted on the faces A_1 and A_2 is $(\partial P/\partial x)\,dx$.

The resultant force acting on the volume element is

$$F_x = -\frac{\partial P}{\partial x}\,dx\,dy\,dz. \tag{1.40}$$

The minus sign on the right-hand side of this equation is due to the choice of a positive force as acting in the x direction.

In accordance with the force law

$$\frac{\partial P}{\partial x}\,dx\,dy\,dz = -\rho\frac{dv_x}{dt}\,dx\,dy\,dz \tag{1.41a}$$

and, respectively, for other directions

$$\frac{\partial P}{\partial y} = \rho\frac{dv_y}{dt}, \tag{1.41b}$$

$$\frac{\partial P}{\partial z} = -\rho\frac{dv_z}{dt}. \tag{1.41c}$$

In the case of low-amplitude acoustic waves, changes in density ρ' and pressure p caused by the medium disturbance are small as compared with ρ_0 and P_0. In other words,

$$\left|\frac{P - P_0}{P_0}\right| = \left|\frac{p}{P_0}\right| \ll 1, \tag{1.42a}$$

$$\left|\frac{\rho - \rho_0}{\rho_0}\right| = \left|\frac{\rho'}{\rho_0}\right| \ll 1. \tag{1.42b}$$

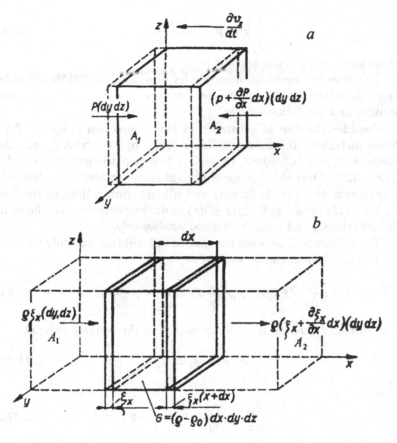

Figure 1.7. Illustrations to the derivation of equations of (*a*) motion and (*b*) medium continuity.

In the general case, the velocity of the medium particles is a function of time and coordinates; therefore, time derivatives should be taken with allowance for the time-dependence of coordinates

$$\frac{dv}{dt} = \frac{dv}{\partial t} + v_x \frac{\partial v}{\partial x} + v_y \frac{\partial v}{\partial y} + v_z \frac{\partial v}{\partial z} = \frac{\partial v}{\partial t} + v\,\mathrm{grad}\,v. \qquad (1.43)$$

For low-amplitude waves, the first term on the right-hand side of expression (1.43) (local acceleration) is significantly higher than the second term (convective acceleration). With allowance for this and the condition (1.42), the expressions (1.41) can be rewritten as

$$\operatorname{grad} p = -\rho_0 \frac{\partial v}{\partial t}, \qquad (1.44)$$

where $p = P - P_0$ is the acoustic pressure*.

To derive the equation of continuity, one has to consider the substance flow through the volume element in which there are no discontinuities and oscillation sources.

Consider the flow of substance in the x direction (Figure 1.7,b). When disturbed, the substance layer shifts by the value $\xi_x(x)$. At points with the coordinate $(x + dx)$, the displacement is given by $\xi_x(x + dx)$. Then the masses of the substance influent into the volume element through the face A_1 and effluent from it through the face A_2 are $\rho \xi_x(x)\, dy\, dz$ and $\rho \xi_x(x + dx)\, dy\, dz$, respectively. The flows in the directions y and z can be defined analogously.

The difference G between the masses influent into and effluent from the volume element can be written as

$$-G = \rho \frac{\partial \xi_x}{\partial x}\, dx\, dy\, dz + \rho \frac{\partial \xi_y}{\partial y}\, dy\, dz\, dx + \rho \frac{\partial \xi_z}{\partial z}\, dz\, dx\, dy. \qquad (1.45)$$

and expressed through the density change in this volume element

$$G = (\rho - \rho_0)\, dx\, dy\, dz. \qquad (1.46)$$

Then,

$$\frac{\partial \xi_x}{\partial x} + \frac{\partial \xi_y}{\partial y} + \frac{\partial \xi_z}{\partial z} = -\eth \qquad (1.47a)$$

or in the vector representation

$$\operatorname{div} \boldsymbol{\xi} = -\eth, \qquad (1.47b)$$

where

$$\eth = \frac{\rho - \rho_0}{\rho} \cong \frac{\rho - \rho_0}{\rho_0}.$$

It should be noted that expression (1.47b) can be written in the form

$$\operatorname{div} \boldsymbol{\xi} = \Omega, \qquad (1.47c)$$

where $\Omega = dV/V$ is the relative change of volume.

---

*The SI-unit of acoustic pressure is 1 Pa = 1 N/m²; its other units are 1 bar = 10^6 dynes/cm² = 10^5 Pa and 1 atm = 10^6 dynes/cm² = 10^5 Pa.

By differentiating expression (1.47a) with respect to time one can obtain the equation of continuity (1.38) which, if condition (1.42b) is fulfilled, takes the form:

$$\frac{\partial \rho}{dt} + \rho_0 \operatorname{div} v = 0. \tag{1.48a}$$

For a noncompressible fluid ($\rho = \text{const}$), the equation of continuity has the form

$$\operatorname{div} v = 0. \tag{1.48b}$$

The equation of state (1.39) relates the pressure, density, and temperature of the medium. For an ideal gas and adiabatic process of wave propagation, the equation of state can be written as

$$PV_M^\gamma = RT, \tag{1.49}$$

where $\gamma = C_p/C_v$ is the ratio of heat capacities at constant pressure, C_p, and constant volume, C_v; R is the gas constant; V_M is the molecular volume.

The empirical equation of state for a liquid can be given by

$$P = A \left(\frac{\rho}{\rho_0} \right)^n - B, \tag{1.50}$$

where A, B, n are constants.

The equation of state can be linearized by introducing the modulus of uniform compression (modulus of elasticity) as the ratio of acoustic pressure to the relative change in the element volume caused by this pressure

$$K = \frac{p}{\eth}. \tag{1.51a}$$

By going on to density and sound velocity in the medium, c_0, one can obtain

$$p = c_0^2 \rho'. \tag{1.51b}$$

Equations (1.44), (1.48a), and (1.51b) constitute a set of approximated linearized equations describing, to a sufficient accuracy, the propagation of waves in an ideal fluid.

By means of respective transformations of these equations, one can obtain the so-called wave equation. In particular, eliminating p and ρ, we arrive at

$$\operatorname{div} \operatorname{grad} v = \frac{\rho_0}{K} \left(\frac{\partial^2 v}{\partial t^2} \right). \tag{1.52}$$

Analogously, one can get expressions in which the only variable will be ξ, p, or $\eth$.

The wave equation (1.52) describes acoustic parameters (pressure, velocity, etc.) at any instant of time at any point x, y, z of acoustic wave.

The parameter $K/\rho_0 = (\partial p/\partial \rho)_S = c_0^2$ in equation (1.52) represents the square of the sound velocity in the medium (S is entropy).

In some cases (especially in molecular acoustics of fluids), it is more convenient to use, instead of volume elasticity, the compressibility β

$$\beta = \frac{1}{K} = \frac{\eth}{p}. \qquad (1.53)$$

The velocity plane waves depends on the density and compressibility of a medium. In turn, the compressibility depends on the extent to which temperature is levelled off during the phases of compression and rarefaction in the process of wave propagation. If temperature does not have time to be levelled off over a half-period, then the elastic properties of the medium are characterized by the adiabatic compressibility β_S, but if temperature is levelled off over the half-period (in this case, temperature is equal at all points of the medium in spite of pressure difference), the elastic properties are characterized by the isothermal compressibility β_T.

Adiabatic and isothermal compressibilities are related as follows

$$\frac{\beta_T}{\beta_S} = \frac{C_p}{C_v} = \gamma. \qquad (1.54)$$

One can show that for an ideal gas

$$\beta_T = \frac{1}{P_0}, \qquad (1.55)$$

which allows the sound velocity in the gas to be determined from the expression

$$c_0 = \sqrt{\frac{\gamma P_0}{\rho_0}}. \qquad (1.56)$$

The sound velocity in simple liquids can be determined by formula from [11]

$$c_0 = \sqrt{\frac{\gamma}{\rho_0 \beta_T}}. \qquad (1.57)$$

Knowing the sound velocity, density, and heat capacity of a fluid at constant pressure, one can calculate such parameters as β_S, C_v, and γ using expressions (1.54) and (1.57).

Table 1.1 Elastic wave velocities in some gases at 0°C [12].

Gas	Chemical formula	Elastic wave velocity C, m/s	Specific heat capacities ratio, γ
Hydrogen	H_2	1284	1.40
Helium	He	965	1.66
Nitrogen	N_2	334	1.40
Oxygen	O_2	316	1.39
Neon	Ne	435	–
Argon	Ar	319	1.67
Carbon oxide	CO	338	1.40
Carbon dioxide	CO_2	259	1.30
Sulfur dioxide	SO_2	213	1.29
Air	–	331	1.40
City gas	–	453	–

Tables 1.1 and 1.2 summarize the sound velocity data for some gases and liquids (in particular, metal melts), as well as the specific heat capacities ratios, temperature coefficients, and conditions of measurements.

As follows from expressions (1.56) and (1.57), sound velocity in ideal liquids and gases is generally independent of frequency, although in some cases of abnormal ultrasonic wave attenuation, sound velocity is frequency-independent. By analogy with optics, this effect is referred to as a sound dispersion.

In gases, sound dispersion takes place at relatively low frequencies, for instance, in carbon dioxide it is observed in a range of 100–1000 kHz. At the same time, sound dispersion in liquids occurs only at very high frequencies which are virtually unattainable.

For analysis of wave processes it is often convenient to introduce the velocity potential Φ, which is a scalar characteristic of wave processes possessing no physical sense. The relation of this potential to the sound wave parameters is given by expressions

$$v = -\operatorname{grad}\Phi, \quad p = \rho_0 \cdot \frac{\partial \Phi}{\partial t}, \quad \rho = -\frac{\rho_0}{c_0^2}\frac{\partial \Phi}{\partial t}. \tag{1.58}$$

In terms of the velocity potential, the wave equation (1.52) can be rewritten as

$$\operatorname{div}\operatorname{grad}\Phi = \frac{1}{c_0^2}\frac{\partial^2 \Phi}{\partial t^2} \tag{1.59a}$$

Table 1.2   Elastic wave velocities in some liquids and melts [12–14].

Liquid (melt)	Temperature, °C	Density ρ_L, g/cm³	Elastic wave velocity c, m/s	Temperature coefficient α_T, m/s grad	Wave resistance $w_0 = \rho_0 c_0 \times 10^3$, g/cm² s
Acetone	20	0.792	1192	–	94
Ethanol	20	0.789	1180	–	93
Gasoline	20	–	1162	–	–
Glycerol	20	1.261	1923	–	242
Olive oil	20	0.905	1405	–	127
Water	20	0.997	1483	–	148
Lithium	190	0.523	5000	–	261
Sodium	100	0.946	2395 ± 25	0.3	226
Potassium	65	0.840	1820 ± 20	0.5	152
Rubidium	40	1.491	1260 ± 10	0.4	187
Cesium	30	1.850	967 ± 10	0.3	178
Copper	1100	8.554	3270	1.0	2797
Silver	970	10.017	2770	0.5	2775
Zinc	425	6.676	2790 ± 60	–	1862
Cadmium	325	8.243	2200 ± 20	0.5	1813
Mercury	50	13.889	1440 ± 10	0.7	2000
Gallium	35	6.111	2740 ± 50	–	1674
Indium	160	7.117	2215 ± 20	0.5	1578
Tallium	310	11.467	1625 ± 15	–	1863
Tin	240	5.595	2270 ± 20	0.7	1270
Lead	335	10.950	1790 ± 15	0.5	1960
Bismuth	280	10.164	1635 ± 05	0.5	1662

or

$$\frac{\partial^2 \Phi}{\partial t^2} - c_0^2 \Delta \Phi = 0. \tag{1.59b}$$

In the case of plane waves ($\Phi_x = \Phi$, $\Phi_y = \Phi_z = 0$), equation (1.59b) takes the form

$$\frac{\partial^2 \Phi}{\partial t^2} - c_0^2 \frac{\partial^2 \Phi}{\partial x^2} = 0. \tag{1.60}$$

The solution to this equation is given by

$$\Phi = f\left(t - \frac{x}{c_0}\right) \tag{1.61a}$$

or

$$\Phi = f\left(t + \frac{x}{c_0}\right).$$ (1.61b)

These expressions describe the wave propagating without distortion of its form in the direction of increasing (1.61a) or decreasing (1.61b) x with the wave velocity c_0.

Such waves are known as the traveling plane waves.

If the function f is harmonic, expressions (1.61) can be written as

$$\Phi = \Phi_m \cos \omega \left(t \mp \frac{x}{c_0}\right)$$ (1.62a)

or in the complex form

$$\Phi = \Phi_m \cos\left(\omega t \mp kx\right) = \Phi_m e^{j(\omega t \mp kx)},$$ (1.62b)

where k is the coefficient of proportionality.

Based on expressions (1.58) and (1.62), one can find the values for acoustic pressure, oscillatory velocity, and displacement in the sound wave

$$p = -\Phi_m \rho_0 \omega \sin\left(\omega t - kx\right) = p_m \sin\left(\omega t - kx\right)$$ (1.63a)

$$v = -\Phi_m \frac{\omega}{c_0} \sin\left(\omega t - kx\right) = v_m \sin\left(\omega t - kx\right)$$ (1.63b)

$$\xi = \frac{\Phi_m}{c_0} \cos\left(\omega t - kx\right) = \xi_m \cos\left(\omega t - kx\right).$$ (1.63c)

At each point ($x = $ const) of the medium, Φ oscillates harmonically with the amplitude Φ_m and circular frequency ω, which are the same for all x. The phase of oscillations ($\theta = kx$) depends on the distance from the plane $x = 0$. Such waves are known as the traveling sinusoidal waves. The coefficient of proportionality k is called the wave number. The spatial period or distance λ, such that for any x is valid

$$\Phi(x + \lambda) = \Phi(x),$$

is called the wavelength physically representing the distance between two successive compressions or rarefactions. Wavelength, wave number, wave velocity, and frequency are related as follows

$$\lambda = \frac{2\pi}{k} = c_0 T = \frac{2\pi c_0}{\omega} = \frac{c_0}{f}.$$ (1.64)

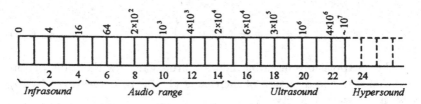

Figure 1.8. Logarithmic frequency scale for elastic vibrations.

These relations are valid for waves of any origin and frequency. The frequency range of elastic oscillations is very broad: from fractions to 10^{13} Hz when ultrasonic wavelengths become comparable with interatomic distances in liquids and solids. In gases, the upper limit of frequencies is restricted by a free path length of molecules and is much smaller ($\sim 10^9$ Hz). Elastic oscillations can be conventionally classified into four groups according to the following frequency ranges

  – infrasound, below 20 Hz;

  – audible sound, $20 - 20 \times 10^3$ Hz;

  – ultrasound, $20 \times 10^3 - 10^9$ Hz,

  – hypersound, above 10^9 Hz.

The human ear perceives frequencies ranging from 20 Hz to $7 \div 25$ kHz. Taking into account a significant spread in the upper limit of audible frequencies, it has been conventionally defined to be equal to 20×10^3 Hz.

Figure 1.8 shows the logarithmic frequency scale $\lg_2 f = 1, 2, 3, \ldots,$ n, where integers $1, 2, 3, \ldots, n$ are the octave numbers. The first four octaves correspond to infrasound and the next ten octaves to audible sound. The ultrasonic range begins from the fifteenth octave.

In addition to traveling plane waves whose amplitude and phase are the same at all points of the plane, there are some other types of waves, e.g. spherical and cylindrical waves, whose amplitudes depend on the distance from the source*. Some formulas relating acoustic parameters of plane, spherical, and cylindrical waves are summarized in Table 1.3.

---

*Spherical and cylindrical waves are not detailed in this book. For this the reader is referred to [4, 6].

Table 1.3   Relationship between acoustic parameters of plane, spherical, and cylindrical waves.

Plane waves	Spherical waves	Cylindrical waves
$\Phi = \Psi_m \cos(\omega t - kx)$	$\Phi_{sp} = \dfrac{\Phi_{spm}}{r}\cos(\omega t - kr)^*$	$\Phi_c = \dfrac{\Phi_{cm}}{\sqrt{r}}\cos(\omega t - kr)^*$
$\xi = \dfrac{\Phi_m}{c_0}\cos(\omega t - kx)$	$\xi_{sp} = -\dfrac{\Phi_{spm}}{c_0 r}\left(1+\dfrac{1}{k^2 r^2}\right)^{1/2}$ $\times \cos(\omega t - kr - \varphi)$	$\xi_c = \dfrac{\Phi_{cm}}{c_0 r}\left(r+\dfrac{1}{4rk^2}\right)^{1/2}$ $\times \cos(\omega t - kr - \varphi)$
$v = -k\Phi_m \sin(\omega t - kx)$	$v_{sp} = -\dfrac{k\Phi_{spm}}{r}\left(1+\dfrac{1}{k^2 r^2}\right)^{1/2}$ $\times \sin(\omega t - kr - \varphi)$	$v_c = -\dfrac{k\Phi_{cm}}{r}\left(r+\dfrac{1}{4rk^2}\right)^{1/2}$ $\times \sin(\omega t - kr - \varphi)$
$p = -kw_0\Phi_m \sin(\omega t - kx)$	$p_{sp} = -\dfrac{kw_0\Phi_{spm}}{r}$ $\times \sin(\omega t - kr)$	$p_c = -\dfrac{kw_0\Phi_{cm}}{\sqrt{r}}$ $\times \sin(\omega t - kr)$
$\vartheta = \dfrac{k}{c_0}\Phi_m \sin(\omega t - kx)$	$\vartheta_{sp} = -\dfrac{k\Phi_{spm}}{c_0 r}$ $\times \sin(\omega t - kr)$	$\vartheta_c = -\dfrac{k\Phi_{cm}}{c_0 \sqrt{r}}$ $\times \sin(\omega t - kr)$
$w_0 = \rho_0 c_0$	$w_{sp} = \dfrac{\rho_0 c_0}{1+\dfrac{1}{k^2 r^2}}\left(1+\dfrac{j}{kr}\right)$	$w_c = \dfrac{\rho_0 c_0}{1+\dfrac{1}{4k^2 r^2}}\left(1+\dfrac{j}{2kr}\right)$
$\operatorname{tg}\varphi = 0$	$\tan\varphi = \dfrac{1}{kr}$	$\tan\varphi = \dfrac{1}{2kr}$

* φ is the phase angle between vibrational velocity and pressure. The subscripts 'sp' and 'c' denoting the distance from the source for spherical and cylindrical waves are omitted.

The notion of mechanical resistance (see (1.34)) can be applied to media with distributed parameters. Taking into account the relation (1.58), one can write the general expression

$$Z = \frac{p}{v} = \rho_0 \frac{\partial\Phi/\partial t}{-\operatorname{grad}\Phi} \tag{1.65a}$$

and for the plane wave

$$Z = \rho_0 c_0 = w_0. \tag{1.65b}$$

The quantity w_0 known as the wave resistance of medium is an important parameter (such as sound velocity) characterizing elastic waves. The wave resistance depends neither on amplitude nor frequency.

The propagation of elastic waves is accompanied with the transfer of energy from one point of medium to another. Back to the model in Figure 1.6, one can note that energy flow in the direction of wave propagation is associated with a periodical conversion of potential energy into kinetic and vice versa. A compressed spring possesses some amount of potential energy, while moving balls – some amount of kinetic energy.

For a system with distributed parameters, one can write the expression for the kinetic energy of a medium element with volume dV and density ρ_0

$$dW_k = \frac{1}{2}\rho_0 v^2 dV \qquad (1.66a)$$

and expression for the potential energy

$$dW_p = \left(\int_0^\vartheta p\, d\vartheta\right) dV = \frac{1}{2}\frac{p^2}{\rho_0 c_0}\, dV. \qquad (1.66b)$$

The total energy is

$$W = \frac{1}{2}\int_V \left(\rho_0 v^2 + \frac{p^2}{\rho_0 c_0^2}\right) dV. \qquad (1.67)$$

The energy density can be given by

$$\mathfrak{w} = \frac{W}{V} = \frac{1}{2}\left(\rho_0 v^2 + \frac{p^2}{\rho_0 c_0^2}\right). \qquad (1.68)$$

The transfer of energy is associated with the medium disturbance. The energy flow $\mathfrak{q}$ (or power N) from the volume element dV equals the energy change in this element per unit time. Based on (1.67), one can write the vector expression

$$\mathfrak{q} = N = \frac{\partial W}{\partial t} = -\int_V (v\,\mathrm{grad}\,p + p\,\mathrm{div}\,v)\,dV = -\int_V \mathrm{div}\,(pv)\,dV. \quad (1.69)$$

This energy flow passes through the surface S surrounding the volume element

$$\mathfrak{q} = \int_S (pv)_n\, dS = \int_S q\, dS. \qquad (1.70)$$

The quantity $q = pv$ called the energy flux density has the physical sense analogous to that of Poynting's vector in the theory of electromagnetic fields.

It is seen from expression (1.70) that the directions of energy flow and wave velocity coincide, which indicates that only longitudinal waves can propagate through liquids and gases.

Let us write the equation for the energy flux of a plane harmonic traveling wave. Using expressions for pressure and velocity from (1.63), one can obtain

$$q = k\rho_0\omega\Phi_m^2 \sin^2(\omega t - kx) = \frac{k\rho_0\omega\Phi_m^2}{2}\left[1 + \sin 2(\omega t - kx)\right]. \quad (1.71)$$

It is seen from (1.71) that energy flow is at a maximum two times over the period, the frequency of its change at any point being 2ω. Based on expression (1.71), one can average the energy flux over the period

$$q = \frac{1}{T}\int\limits_0^T q(t)\,dt = \frac{1}{2}k\rho_0\omega\Phi_m^2 = \frac{1}{2}\frac{\rho_0}{c_0}\omega^2 p_m^2 = \frac{1}{2}\rho_0 c_0\omega^2\xi_m^2. \quad (1.72a)$$

Thus, the mean energy flux in the traveling wave is proportional to the square of the oscillation frequency times the square of the amplitude. This quantity is called the sound power or intensity. Formula (1.72a) can be transformed into the following convenient expressions for the traveling wave intensity I

$$I = q = \frac{1}{2}\rho_0 c_0\omega^2\xi_m^2 = \frac{1}{2}w_0 v_m^2 = \frac{1}{2}\frac{p_m^2}{w_0} = \frac{1}{2}p_m v_m = wc_0. \quad (1.72b)$$

Sound intensity is usually measured in W/cm^2.

In acoustics, it is a common practice to use relative units expressing the value of some quantity relative to its certain arbitrarily chosen reference value. This is because the range of sound intensities can be very large. Thus, the ratio of the loud noise intensity to the audibility threshold can be as high as 10^{14}. In such a situation, the logarithmic scale can be conveniently used. In acoustics, the relative values of energy, power, sound intensity, and acoustic pressure are typically measured in decibels (dB)

$$L_W = 10\lg\frac{W}{W_0}, \quad (1.73a)$$

$$L_I = 10\lg\frac{I}{I_0}, \quad (1.73b)$$

$$L_p = 10\lg\left(\frac{p}{p_0}\right)^2 = 20\lg\frac{p}{p_0}, \quad (1.73c)$$

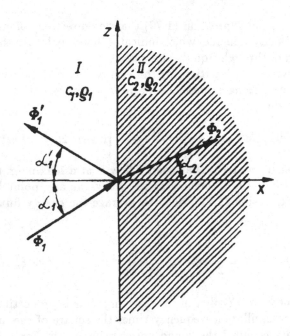

Figure 1.9. Propagation of a plane wave through the two liquid media interface.

where W_0, I_0, and p_0 are some reference values of energy, sound intensity, and acoustic pressure, respectively.

By I_0 is taken the value 10^{-16} W/cm^2, while the reference acoustic pressure in air, $p_0 = 2 \times 10^{-5}$ Pa.

It follows from (1.73) that acoustic pressure increases by 3 dB when sound intensity doubles.

In electroacoustics, sound intensity is sometimes expressed as the natural logarithm of the ratio of voltages

$$L_U = \ln \frac{U}{U_0}. \tag{1.74}$$

The respective values are expressed in Napiers (Np), the following relation being valid

$$1 \text{ Np} = 8.686 \text{ dB}.$$

The elastic wave incident on the interface between two media with different wave resistances is partially reflected from and partially refracted into the second medium. As an example, let us consider some regularities of the acoustic wave behaviour illustrated in Figure 1.9. The elastic wave with the velocity potential Φ_1 propagating through liquid I ($w_{01} = \rho_1 c_1$) is incident on the interface with liquid II ($w_{02} = \rho_2 c_2$) at an angle of α_1. The reflected and refracted waves possess the velocity potentials Φ_1' and Φ_2', respectively

$$\Psi_1 = \Phi_{m1} \exp j\omega t \exp \left[-j\frac{\omega}{c_1}(x\cos\alpha_1 + z\sin\alpha_1) \right], \qquad (1.75a)$$

$$\Phi_1' = \Phi_{m1} \exp j\omega t \exp \left[-j\frac{\omega}{c_1}(-x\cos\alpha_1' + z\sin\alpha_1') \right], \qquad (1.75b)$$

$$\Phi_2 = \Phi_{m2} \exp j\omega t \exp \left[-j\frac{\omega}{c_2}(x\cos\alpha_2 + z\sin\alpha_2) \right]. \qquad (1.75c)$$

With allowance for the wave continuity at the interface ($x = 0$), one can write the following boundary conditions

$$p_1 + p_1' = p_2, \qquad (1.76a)$$

$$v_{1x} + v_{1x}' = v_{2x}, \qquad (1.76b)$$

$$v_{1z} = v_{1z}'. \qquad (1.76c)$$

As follows from analysis of expressions (1.75), the boundary conditions (1.76) are fulfilled when the following relations are valid

$$\alpha_1 = \alpha_1' \qquad (1.77)$$

and

$$n = \frac{\sin\alpha_2}{\sin\alpha_1} = \frac{c_2}{c_1}, \qquad (1.78)$$

where n is the index of refraction. The expression (1.78) represents Snell's law.

The relationship between the velocity potentials of the incident and reflected waves is as follows

$$\frac{\Phi_1'}{\Phi_1} = \frac{w_2\cos\alpha_1 - w_1\cos\alpha_2}{w_2\cos\alpha_1 + w_1\cos\alpha_2}, \qquad (1.79)$$

while for the refracted wave one can write

$$\frac{\Phi_2}{\Phi_1} = \frac{2\rho_1 c_2 \cos\alpha_1}{w_2\cos\alpha_1 + w_1\cos\alpha_2}. \qquad (1.80)$$

Sometimes it is convenient to introduce the coefficient of reflection

$$b_r = \frac{p_1'}{p_1} = -\frac{v_1'}{v_1} = \frac{\Phi_1'}{\Phi_1} \qquad (1.81a)$$

and the coefficient of refraction

$$b_p = \frac{w_1 p_2}{w_2 p_1} = \frac{v_2}{v_1} = \frac{c_1 \Phi_2}{c_2 \Phi_1}. \tag{1.81b}$$

Analysis of expressions relating the energy and intensity of incident, reflected, and refracted waves indicates that

$$W_1 = W_1' + W_2; \quad I_1 = I_1' + I_2 \frac{\cos \alpha_2}{\cos \alpha_1}. \tag{1.82}$$

The intensity of reflected (I_1') and refracted (I_2) waves can be expressed through the intensity I_1 of the incident wave

$$I_1' = I_1 b_r^2 = I_1 \frac{(w_2 \cos \alpha_1 - w_1 \cos \alpha_2)^2}{(w_2 \cos \alpha_1 + w_1 \cos \alpha_2)^2}, \tag{1.83a}$$

$$I_2 = I_1 b_p^2 = I_1 \frac{4\rho_1^2 c_2^2 \cos^2 \alpha_1}{(w_2 \cos \alpha_1 + w_1 \cos \alpha_2)^2}. \tag{1.83b}$$

In the case of normal incidence ($\alpha_1 = \alpha_2 = 0$), expressions (1.83) are simplified

$$I_1' = I_1 \frac{(w_2 - w_1)^2}{(w_2 + w_1)^2}, \tag{1.84a}$$

$$I_2 = I_1 \frac{4\rho_1^2 c_2^2}{(w_2 + w_1)^2}. \tag{1.84b}$$

It is seen from (1.83) and (1.84) that when the wave resistances of media are equal and the primary wave is incident normally to their interface, acoustic energy is not reflected ($b_r = 0$). The interface between water and carbon tetrachloride can serve as an example. Indeed, in spite of different densities of these liquids (1.0 and 1.5 g/cm^3, respectively) and different sound velocities (1500 and 1000 m/s, respectively), their wave resistances are equal; therefore, $b_r = 0$. As a result, $b_p = 1$ and acoustic energy passes from one medium into the other without loss.

The greater the difference between the wave resistances of media, the higher the coefficient of reflection from their interface. Thus, wave energy is almost totally reflected from the water-air interface, as the wave resistances of air and water are 44.2 and 1.5×10^5 g/cm s, respectively.

If the coefficient of reflection $b_r = 1$ at $\alpha_1 = 0$, then the interference of the incident and reflected waves gives rise to a standing wave. This can be the case for the wave reflected from the medium with $w_2 = \infty$ or $w_2 = 0$. If $w_2 = \infty$ (reflection from a perfectly rigid wall), one can write

$$\Phi_s = \Phi_1 + \Phi_1' = 2\Phi_m \cos kx \, \cos \omega t, \qquad (1.85a)$$

$$p_s = 2p_{m1} \cos kx \, \cos \omega t, \qquad (1.85b)$$

$$v_s = 2v_{m1} \sin kx \, \sin \omega t. \qquad (1.85c)$$

It can be inferred from (1.85) that the variables x and t are functionally separated, which means that the oscillatory parameters (velocity potential Φ_s, acoustic pressure p_s, oscillatory velocity v_s, etc.) of the standing wave change with equal frequency, although their amplitudes are different at various points. Thus, acoustic pressure amplitude changes from the value $p_{ms} = 0$ for the planes lying at the distance

$$x = \left(n + \frac{1}{2}\right)\frac{\lambda}{2}, \quad n = 1, 2, 3 \qquad (1.86a)$$

from the interface ($x = 0$) to the value $p_{ms} = 2p_{m1}$ for the planes lying from it at the distance

$$x = n \cdot \frac{\lambda}{2}, \quad n = 0, 1, 2. \qquad (1.86b)$$

The points for which the oscillatory parameters (Φ_s, p_s, v_s) equal zero at any time are called the standing wave nodes, whereas those for which these parameters are maximum and equal to double amplitude values of respective parameters in the incident wave are called the loops or antinodes.

Analysis of expressions (1.86) shows that oscillatory velocity and acoustic pressure are out of phase by $\pi/2$ and their nodes and antinodes are spatially displaced by $\lambda/4$ (Figure 1.10,a).

The specific wave resistance of standing wave is a wholly imaginary quantity

$$w_s = \frac{p_{ms}}{v_s} = jw_1 \mathrm{ctg}\, kx. \qquad (1.87)$$

For the wave reflected from a free surface ($w_2 = 0$), acoustic pressure and oscillatory velocity are distributed in space in different manners (Figure 1.10,b) – the pressure node and velocity antinode are located at the interface.

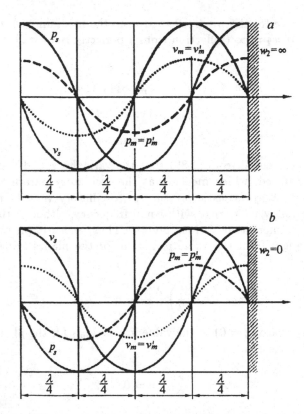

Figure 1.10.   Vibrational pressure and velocity distribution in the incident and standing wave upon the reflection of vibrations from (a) perfectly rigid and (b) free interfaces.

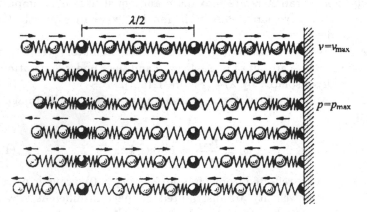

Figure 1.11.   Mechanical model for a standing wave.

As opposed to traveling wave, there is no energy transfer in the standing wave, but there is a local conversion of potential energy into kinetic, and vice versa. This is clearly seen from the comparison of models for traveling (Figure 1.6) and standing (Figure 1.11) waves. Each $\lambda/4$-thick section of space between the nearest nodes of pressure and velocity possesses constant energy and does not exchange it with the neighboring section. On the other hand, two times over the period, in each section there occurs the reversible transformation of kinetic energy related to the motion of particles and located near velocity antinodes (pressure nodes) into potential energy related to the medium elasticity and concentrated near pressure antinodes (velocity nodes).

1.3   Elastic Waves in Solids

Wave equation analogous to (1.59) can describe the propagation of elastic wave not only in fluids but also in solids. However, it should be noted that if the forces acting on the volume element in fluids are determined only by pressure, then in solids the relation between the forces and the strains they produce is more complex.

An external force can disturb the particles of a body from equilibrium, which gives rise to internal forces that tend to return these particles to equilibrium. These internal forces per unit area are known as stresses.

The stress arising in a volume element can be decomposed into the components that are normal or tangent to the surface (Figure 1.12,a). By their physical nature, the normal components (σ_{xx}, σ_{yy}, σ_{zz}*) are equivalent to pressure in fluids with a sole difference that their values depend on direction. The tangential components (σ_{xy}, σ_{yx}, σ_{xz}, σ_{zx}, σ_{yz}, σ_{zy}) tend to deform the volume element and are called the shear stresses. Based on the condition of equilibrium of the volume element, one can show that

$$\sigma_{xy} = \sigma_{yx}; \quad \sigma_{xz} = \sigma_{zx}; \quad \sigma_{yz} = \sigma_{zy}; \tag{1.88}$$

thus, the number of independent stress tensor components is reduced to six.

---

*The first subscript denotes the direction of the stress, while the second subscript indicates the direction of the normal to the surface area subject to a given stress component.

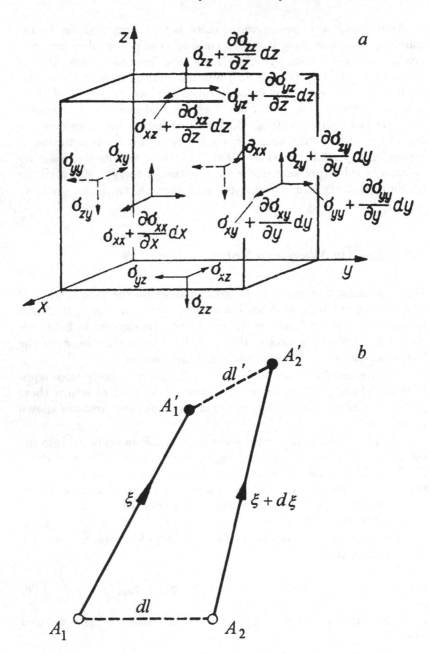

Figure 1.12.   Illustrations to determination of (a) stresses acting upon the surface of the elementary cell and (b) strains.

The stresses acting on the volume element can be represented as a tensor:

$$\|\Sigma\| = \left\| \begin{matrix} \sigma_{xx} & \sigma_{xy} & \sigma_{xz} \\ \sigma_{yx} & \sigma_{yy} & \sigma_{yz} \\ \sigma_{zx} & \sigma_{zy} & \sigma_{zz} \end{matrix} \right\| = \{\sigma_{ik}\}. \tag{1.89}$$

The stresses arising in a body tend to change its volume and shape, i.e., produce strains.

Under the action of a stress or external force, the point A_1 (Figure 1.12,b) with the coordinates x, y, z is displaced to the position A_1' with the coordinates x', y', z':

$$x' = x + \xi_{x1}; \quad y' = y + \xi_{y1}; \quad z' = z + \xi_{z1}, \tag{1.90}$$

where ξ_{x1}, ξ_{y1}, and ξ_{z1} are the components of the displacement vector at the point A_1. The point A_2, distant from A_1 by dl, initially has the coordinates $x + dx$, $y + dy$, and $z + dz$, the distance between A_1 and A_2 being

$$dl^2 = dx^2 + dy^2 + dz^2 = \sum_{i=1}^{3} dx_i^2. \tag{1.91}$$

When the point A_2 shifts to the position A_2', its coordinates change into $x + dx + \xi_{x2}$; $y + dy + \xi_{y2}$; $z + dz + \xi_{z2}$. Here ξ_{x2}, ξ_{y2}, and ξ_{z2} are the components of the displacement vector at the point A_2. After displacement, the distance between these points becomes equal to dl'

$$(dl')^2 = \sum_{i=1}^{3}(dx_i + d\xi_i)^2, \tag{1.92}$$

where $d\xi_i = \xi_{i2} - \xi_{i1}$.

The points A_1 and A_2 can move in any directions; therefore, $d\xi_i$ is a function of ξ

$$\left. \begin{aligned} d\xi_x &= \frac{\partial \xi_x}{\partial x}\, dx + \frac{\partial \xi_x}{\partial y}\, dy + \frac{\partial \xi_x}{\partial z}\, dz \\ d\xi_y &= \frac{\partial \xi_y}{\partial x}\, dx + \frac{\partial \xi_y}{\partial y}\, dy + \frac{\partial \xi_y}{\partial z}\, dz \\ d\xi_z &= \frac{\partial \xi_z}{\partial x}\, dx + \frac{\partial \xi_z}{\partial y}\, dy + \frac{\partial \xi_z}{\partial z}\, dz \end{aligned} \right\} \tag{1.93a}$$

or in an abridged representation

$$d\xi_i = \frac{\partial \xi_i}{\partial x_k}\, dx_k. \tag{1.93b}$$

By substituting (1.93) into (1.92), one can find that

$$
\begin{aligned}
(dl')^2 &= dx_i^2 + 2\frac{\partial \xi_i}{\partial x_k}\, dx_k\, dx_i + \frac{\partial \xi_i}{\partial x_k}\frac{\partial \xi_i}{\partial x_l}\, dx_k\, dx_l \\
&= dl^2 + \left(\frac{\partial \xi_i}{\partial x_k} + \frac{\partial \xi_k}{\partial x_i} + \frac{\partial \xi_l}{\partial x_k}\frac{\partial \xi_l}{\partial x_i}\right) dx_i\, dx_k. \quad (1.94)
\end{aligned}
$$

Thus, the increment of the square of the distance between two close points is

$$
(dl')^2 - (dl)^2 = 2\mathcal{E}_{ik}\, dx_k\, dx_i, \quad (1.95)
$$

where

$$
\mathcal{E}_{ik} = \frac{1}{2}\left(\frac{\partial \xi_i}{\partial x_k} + \frac{\partial \xi_k}{\partial x_i} + \frac{\partial \xi_l}{\partial x_k}\frac{\partial \xi_l}{\partial \xi_i}\right) \quad (1.96)
$$

is the strain tensor.

For small displacements and their derivatives, when the squares of these values can be neglected, one may use a simplified linearized representation for the strain tensor

$$
\mathcal{E}_{ik} = \frac{1}{2}\left(\frac{\partial \xi_i}{\partial x_k} + \frac{\partial \xi_k}{\partial x_i}\right), \quad (1.97)
$$

whose components have a simple physical meaning. The diagonal components possessing two identical subscripts (for example, $\mathcal{E}_{xx} = \partial \xi_x/\partial x$ or, introducing new designations $x \to 1$, $y \to 2$, $z \to 3$, $\mathcal{E}_{11} = \partial \xi_1/\partial x_1$) are equal to a relative change in dl in the direction of respective axis. Such strains are called compressive and tensile. Other components ($\mathcal{E}_{12}$, etc.) describe changes in the position of the element dl. Depending on the character of changes in the position of the neighboring elements of the medium, the strain can be either shear or rotational. If there is no rotation, the non-diagonal tensor components describe deformation of the right angle between respective coordinate axes.

Typically, vectors dl and $d\xi$ are directed differently. However, there are three mutually perpendicular directions corresponding to the principal axes of the tensor for which dl and $d\xi$ are parallel. In this coordinate system, only the diagonal tensor elements are non-zero, i.e. shear strain is absent. The mentioned directions are called principle, and strains in these directions are designated as $\mathcal{E}_1$, $\mathcal{E}_{11}$, $\mathcal{E}_{111}$. The following expression is valid for the compressive strain

$$
\mathcal{E}_1 + \mathcal{E}_{11} + \mathcal{E}_{111} = \eth. \quad (1.98)
$$

As mentioned above, for fluids there exist simple relations (1.53) between strain (change in density) and stress (change in pressure). For solids, one may use Hooke's law suggesting a linear relation between stresses and strains

$$\sigma_i = c_{ik}\mathcal{E}_k, \tag{1.99}$$

where c_{ik} are the moduli of elasticity.

Equations (1.99) involve 36 elastic moduli. For an elastic medium, the strain energy can be considered as a function of state, which leads to $c_{ik} = c_{ki}$ and, thus, reduces the number of independent moduli to 21. This number is characteristic of a triclinic system of a lower symmetry possessing only a first-order symmetry axis. The description of elastic crystals of a higher symmetry requires a smaller number of independent elastic moduli (in general, five moduli for hexagonal crystals and three for cubic ones).

In the case of an isotropic body, the moduli of elasticity do not depend on the direction of the coordinate axes, which leads to

$$\left. \begin{aligned} c_{11} &= c_{22} = c_{33} \\ c_{44} &= c_{55} = c_{66} = \mu \\ c_{12} &= c_{13} = c_{23} = \lambda \end{aligned} \right\}, \tag{1.100}$$

all other moduli being zero. The quantities μ and λ, known as the Lamé elastic constants, entirely characterize the elastic properties of isotropic bodies. The constant μ, known also as the shear modulus, represents the ratio of the shear stress to the shear strain and can be quite easily determined experimentally.

The constant λ characterizes the extent of the volume element changes in the plane perpendicular to the acting force. Experimental measurement of this constant presents a problem, therefore it is rarely used in calculations.

Taking into account that for isotropic media $c_{44} = \frac{1}{2}(c_{11} - c_{12})$ [20], one can write

$$c_{11} = \lambda + 2\mu. \tag{1.101}$$

The Lamé constants are related to the volume modulus of elasticity as follows

$$K = \lambda + \frac{2\mu}{3}. \tag{1.102}$$

Two other quantities, Young's modulus E and the Poisson coefficient ν, are widely used in calculations.

Both coefficients can be determined experimentally by means of a loading of a constant-section cylindrical rod along its axis. The normal stress σ_1 applied lengthwise produces the strain $\mathcal{E}_1$, all other stresses in

Table 1.4 Relationship between the elastic constants.

Parameter	Relation to other constants				
	λ, μ	K, μ	μ, ν	E, ν	E, μ
λ	λ	$K - \dfrac{2}{3}\mu$	$\dfrac{2\mu\nu}{1-2\nu}$	$\dfrac{E\nu}{(1+\nu)(1-2\nu)}$	$\dfrac{\mu(E-2\mu)}{3\mu-E}$
μ	μ	μ	μ	$\dfrac{E}{2(1+\nu)}$	μ
K	$\lambda + \dfrac{2}{3}\mu$	K	$\dfrac{2\mu(1+\nu)}{3(1-2\nu)}$	$\dfrac{E}{3(1-2\nu)}$	$\dfrac{E\mu}{3(3\mu-E)}$
E	$\dfrac{(3\lambda+3\mu)\mu}{\lambda+\mu}$	$\dfrac{9K\mu}{3K+\mu}$	$2(1+\nu)\mu$	E	E
ν	$\dfrac{\lambda}{2(\lambda+\mu)}$	$\dfrac{3K-2\mu}{6K+2\mu}$	ν	ν	$\dfrac{1}{2}\dfrac{E}{\mu}-1$
$\lambda + 2\mu$	$\lambda + 2\mu$	$K + \dfrac{4}{3}\mu$	$2\mu\dfrac{1-\nu}{1-2\nu}$	$\dfrac{E(1-\nu)}{(1+\nu)(1-2\nu)}$	$\dfrac{4\mu-E}{3-E/\mu}$

the rod being close to zero. Young's modulus is defined as the stress-to-strain ratio

$$E = \frac{\sigma_1}{\mathcal{E}_1}. \tag{1.103}$$

Additionally to the longitudinal strain in the rod, there is also the transverse (i.e. normal to the axis) strain, $-\mathcal{E}_2 = \mathcal{E}_3$. The ratio $\mathcal{E}_2/\mathcal{E}_1 = -\nu$ is known as the Poisson coefficient.

The relationship between elastic constants is presented in more detail in Table 1.4.

To deduce the wave equation describing the propagation of oscillations in solids, one has to calculate the forces acting on the volume element and relate them to accelerations imparted by these forces, as it has been done for fluids.

As a result, we can arrive at the expression, known as the Lamé equation, which describes the motion of an isotropic solid in the general form*

$$\rho\frac{\partial^2 \boldsymbol{\xi}}{\partial t^2} = (\lambda + \mu)\,\text{grad div}\,\boldsymbol{\xi} + \mu\Delta\boldsymbol{\xi}. \tag{1.104}$$

---

*The symbol Δ denotes the Laplacian of the vector $\boldsymbol{\xi}$

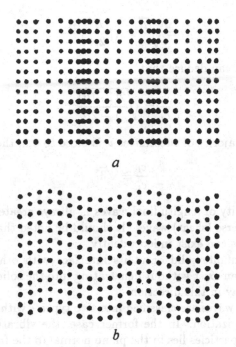

a

b

Figure 1.13. Longitudinal (a) and shear (b) waves in an infinite solid.

Based on this equation, one can deduce expressions describing the propagation of longitudinal compression-dilatation waves

$$\rho\frac{\partial^2\xi}{\partial t^2} = (\lambda + \mu)\Delta\xi \qquad (1.105)$$

and shear waves $(\eth = 0)$

$$\rho\frac{\partial^2\xi}{\partial t^2} = \mu\Delta\xi. \qquad (1.106)$$

When a shear (transverse) wave propagates through a medium, its particles oscillate in the direction perpendicular to the direction of the wave propagation (Figure 1.13).

The velocity of longitudinal and shear waves* are, respectively,

$$c_l = \sqrt{\frac{\lambda + 2\mu}{\rho}} = \sqrt{\frac{E}{\rho}\frac{1-\nu}{(1+\nu)(1-2\nu)}}, \qquad (1.107)$$

---

*Hereinafter the subscripts "t" and "l" are used to mark shear (transverse) and longitudinal waves, respectively.

$$c_t = \sqrt{\frac{\mu}{\rho}} = \sqrt{\frac{E}{2\rho} \frac{1}{1+\nu}} \qquad (1.108)$$

from which one can deduce that

$$\frac{c_l}{c_t} = \sqrt{\frac{2(1-\nu)}{1-2\nu}}.$$

Since for the majority of substances ν is close to 0.3, then

$$\frac{c_l}{c_t} \cong \sqrt{3}.$$

Thus, the velocity of longitudinal waves is always greater than the velocity of transverse (shear) waves. It should be noted that longitudinal and shear waves can exist independently.

In a semi-infinite solid or at its interface with other media, the surface waves can exist without penetrating into the solid bulk because of a strong decay in this direction.

The surface waves are classified into two types – with a vertical and horizontal polarization. In the former case, the vibrational displacement vector of particles lies in the plane normal to the free (boundary) surface, while in the latter case it is parallel to this surface and normal to the direction of wave propagation.

The most frequent surface waves are those with the vertical polarization (the Rayleigh waves), which propagate over the free surface. The energy of these waves is concentrated in the subsurface layer with a thickness of $(1-2)\,\lambda_R$ (λ_R is the wavelength). The wave particles move along elliptical trajectories, the major axis of which, ξ_z, is perpendicular to the surface, while the minor axis, ξ_x, is parallel to the direction of wave propagation. The Rayleigh waves can be obtained in plates whose thickness is much greater than the wavelength.

The velocity of the Rayleigh waves propagation can be calculated by formula

$$c_R = \frac{0.87 + 1.12\nu}{1+\nu} \sqrt{\frac{\mu}{\rho}} \qquad (1.109a)$$

or, taking into account that $\nu \sim 0.3$

$$c_R \cong 0.9 \sqrt{\frac{\mu}{\rho}} = 0.9 c_t. \qquad (1.109b)$$

Waves with the vertical polarization can also propagate over the boundary of a solid half-space with a liquid or solid medium.

Waves with the horizontal polarization (Love waves) travel over the interface of a solid half-space with a solid layer. These waves are completely transverse, as their elastic strains are shear.

In finite-size solid bodies (e.g., in plates and rods), elastic waves may exist as well. These waves known as normal waves are determined by shape elasticity. As opposed to elastic waves in infinite media, the normal waves satisfy both the equations of elasticity and boundary conditions at the body surfaces. These conditions mainly involve the absence of surface mechanical stresses, which leads to a considerably more sophisticated distribution of displacements and stresses over finite bodies than in the case of infinite media.

The normal waves in plates are classified into two groups – Lamb waves and transverse waves. Vibrational displacements of particles in the Lamb wave occur both in the direction of wave propagation and normally to the plate surface (Figure 1.14).

There are symmetrical (s) and antisymmetrical (a) Lamb waves. In symmetrical waves, the particles trajectories are symmetrical with respect to the central plane $z = 0$ (Figure 1.14,a). In the upper and lower parts of the plate, the displacements ξ_x are of the same sign, while the displacements ξ_z have opposite signs. In the case of antisymmetrical Lamb waves (Figure 1.14,b), the reverse is true.

Transverse normal waves possess only one displacement component ξ_y (absent in the Lamb waves) which is parallel to the plate surface and normal to the direction of wave propagation (Figure 1.14,c). The transverse normal waves can also be symmetrical and antisymmetrical.

One of the basic properties of the Lamb and transverse normal waves is that, at given values of frequency ω and thickness $2h$ of the plate, the symmetrical and antisymmetrical waves traveling in this plate are limited in their number and differ from each other in the velocity of propagation and in-depth distribution of displacements and stresses. Wave number is the greater, the higher is $\omega h/c_t$ (the Lamb waves) or $\omega h/c_t \pi$ (transverse waves).

Only two zero-order Lamb waves – longitudinal s_0 and bending a_0 – can exist in a thin $(\omega h/c_t \ll 1)$ plate.

The longitudinal wave in a plate is similar to the transverse wave in an infinite solid body: in both waves the longitudinal displacement component ξ_x prevails, and the displacement component ξ_z is $c_t/\omega h$ times smaller than ξ_x.

At $\omega h/c_t \pi < 1/2$, only single normal symmetrical transverse wave travels in the plate, the velocity of its propagation being c_t.

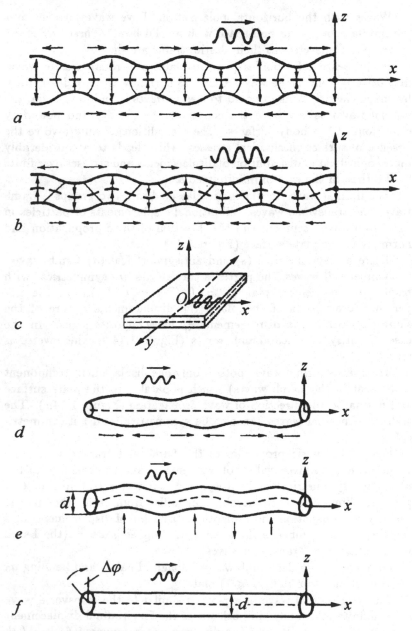

Figure 1.14. Schematic representation of vibrational motion in (*a*) symmetrical and (*b*) antisymmetrical Lamb waves, (*c*) transverse waves in plates, (*d*) longitudinal, (*e*) bending, and (*f*) torsional waves in rods. Arrows indicate the direction of displacement.

Table 1.5 Formulas for the speed sound in finite solids.

Waves	Formulas for sound velocity
Bending waves	A rod of radius r: $$c_b = \sqrt{\pi r f}\left(\frac{E}{\rho}\right)^{1/4}$$ A $2h$-thick plate: $$c_b = \sqrt{\pi h f}\left(\frac{E}{3(1-\nu^2)\rho}\right)^{1/4}$$
Longitudinal waves in rods	$c_l = \sqrt{\dfrac{E}{\rho}}$
Lamb waves in plates	$c_L = \sqrt{\dfrac{E}{\rho}\dfrac{1}{1-\nu^2}}$
Torsional waves in rods	$c_r = \sqrt{\dfrac{\mu}{\rho}}$

Normal waves in rods are analogous to the Lamb and transverse waves in plates. The normal waves in rods can be classified as longitudinal (compression), bending, and torsional.

In the longitudinal normal wave (analogous to the symmetrical Lamb wave), oscillations are symmetrical around the rod axis x with the maximum longitudinal displacement component ξ_x (Figure 1.14, d).

In the bending normal wave (analogous to the antisymmetrical Lamb wave), the axis x is curved, and the transverse displacement component ξ_z is prevalent (Figure 1.14,e).

In the torsional normal wave (analogous to transverse waves in a plate), there is only one azimuthal displacement component ξ_φ. Oscillations in this wave, representing the rotation of the rod particles around the rod axis x, are symmetrical relative to this axis (Figure 1.14,f). Expressions for the velocity of longitudinal, bending, and torsional normal waves are given in Table 1.5. It should be noted that the bending wave velocity is frequency-dependent.

Table 1.6 presents elastic constants and the velocities of propagation of longitudinal and shear waves in metals and some other materials.

The mechanism of energy transfer in infinite solids in an adiabatic thermodynamical approximation is analogous to the above considered mechanism for fluids.

Table 1.6   Elastic constants and the velocities of elastic waves in some solids at 20°C [15–17].

Material	Density ρ, g/cm³	Young's modulus $E \times 10^{-3}$, MPa	Shear modulus $\mu \times 10^{-3}$, MPa	Poisson coefficient ν	Longitudinal waves		Wave resistance $w_0 = \rho_0 c_l \times 10^{-5}$, g/cm s*	Velocity of shear wave $c_t \times 10^{-5}$, cm/s
					Velocity $c_l \times 10^{-5}$, cm/s			
					infinite medium	rod		
				Metals				
Aluminum	2.7	7.1	2.6	0.34	6.32	5.08	13.7	3.08
Beryllium	1.8	31.5	15.0	–	–	13.00	24.1	8.85
Vanadium	6.1	13.0	4.7	–	–	4.60	28.1	2.76
Bismuth	9.8	3.2	1.2	0.33	2.18	1.79	17.6	1.10
Tungsten	19.1	36.0	14.5	0.35	5.46	4.31	82.3	2.62
Iron	7.8	21.0	8.2	0.28	5.85	5.17	40.4	3.23
Gold	19.3	8.1	2.8	0.39	3.24	2.03	39.2	1.20
Cadmium	8.6	6.0	2.3	0.30	2.78	2.40	20.1	1.50
Cobalt	8.8	21.4	6.0	–	–	4.91	43.2	2.60
Magnesium	1.7	4.0	1.7	0.33	5.77	4.90	8.5	3.10
Manganese	8.4	21.4	–	–	4.66	–	–	2.35
Copper	8.9	12.5	4.5	0.35	4.70	3.71	33.0	2.26
Molybdenum	10.2	33.6	11.9	–	–	5.74	58.6	3.40
Nickel	8.9	20.5	7.7	0.31	5.63	4.78	42.5	2.96
Niobium	8.6	11.0	3.8	–	–	3.58	30.7	2.10
Tin	7.3	5.5	1.5	–	3.32	2.73	19.8	1.67
Platinum	21.3	17.0	6.2	0.27	3.96	2.80	59.6	1.67
Mercury	13.6	–	–	–	1.46	–	19.8	–
Lead	11.4	1.6	0.6	0.45	2.16	1.20	13.7	0.70
Silver	10.5	7.5	2.8	0.37	3.60	2.64	27.7	1.59
Antimony	6.8	7.9	2.0	–	–	3.40	23.2	1.73

Table 1.6 Continued.

Tantalum	16.6	19.0	6.3	–	3.35	55.5	2.00
Titanium	4.5	10.5	4.6	–	5.00	22.5	3.20
Chromium	7.2	25.9	–	–	6.00	43.2	–
Zinc	7.1	10.5	3.5	0.17	3.81	27.1	2.41
Nonmetallic materials							
Natural rubber	0.95	–	–	–	0.03	0.03	–
Soft rubber	0.90	–	–	1.48	–	–	–
Hard rubber	1.20	–	–	2.40	–	–	1.99
Ice	1.00	–	–	3.98	0.33	3.98	3.98
Paraffin wax	0.83	–	–	2.20	–	1.80	1.80
Organic glass	1.18	–	–	2.67	0.35	3.20	1.12
Polystyrene	1.06	–	–	2.35	0.32	2.49	1.12
Polyethylene	1.10	–	–	2.48	–	2.70	–
Porcelain	2.41	5.9	–	5.40	–	13.00	3.12
Quartz glass	2.60	–	–	5.57	0.17	14.50	3.51
Teflon	2.20	–	–	1.35	–	3.00	–
Graphite	2.22	1.1	–	–	–	5.10	–
Ebonite	1.20	–	–	2.40	–	1.90	–
Rocks							
Basalt	2.72	–	–	5.93	0.30	16.2	3.14
Shale	2.74	–	–	6.50	0.28	17.8	3.61
Gypsum	2.26	–	–	4.79	0.34	10.8	2.37
Marble	2.66	–	–	6.15	0.30	16.4	3.26
Mica	2.81	–	–	7.76	0.46	21.8	2.16
Granite	2.62	–	–	4.45	0.18	11.6	2.78

*The values for wave resistance refer to the case of longitudinal waves traveling in metal rods and infinite media (nonmetallic materials and rocks).

When an elastic wave is incident on the interface between two solids, the arising phenomenon is not a simple reflection and refraction of waves in liquids. Thus, a longitudinal (or transverse) wave incident obliquely on the interface between two solids I and II produces two (longitudinal and transverse) reflected waves in the solid I and two (longitudinal and transverse) refracted waves in the solid II. With account for boundary conditions, one can determine the amplitudes and phase angles for all four resultant waves.

1.4   Absorption of Ultrasonic Waves in Fluids and Solids

Propagation of ultrasonic waves in real media is accompanied with their attenuation dependent on geometrical factors (such as divergence of an acoustic beam), scattering, and energy dissipation.

Attenuation due to sound divergence is a drop in the intensity of vibrations proportional to the square of the distance from the sound source.

The character of sound dissipation depends on the nature and the extent of medium inhomogeneity [18]. Attenuation of low-amplitude harmonic oscillations due to absorption follows the exponential law: equal portions of acoustic energy are adsorbed over a given path interval

$$I_x = I_0 e^{-2\alpha x}, \tag{1.110a}$$

where I_0 is the intensity of oscillations at the source; I_x is their intensity at the distance x from the source; α is the coefficient of absorption; x is the distance covered by the wave.

The displacement amplitude, pressure, and other linear wave parameters obey the exponential law. For instance,

$$\xi_{mx} = \xi_{m0} e^{-\alpha x}, \tag{1.110b}$$

where ξ_{m0} and ξ_{mx} are the displacement amplitudes at the source and at the distance x from it.

Coefficient of absorption is determined by the matter properties and physically it is a measure of a drop in the amplitude of vibrations because of their absorption over a 1-cm path.

In some cases, the time coefficient α_τ is also used to characterize absorption. If, at a zero time, the traveling wave has the displacement amplitude $\xi_{m\tau}$, then it changes with time according to the law

$$\xi_{m\tau} = \xi_{m0} e^{-\alpha_\tau \tau}. \tag{1.111}$$

The absorption coefficients α and α_τ are related as follows

$$\alpha = \frac{\alpha_\tau}{c}. \tag{1.112}$$

The absorption of sound is determined by a diversity of physical phenomena. Below we will consider the absorption related to viscosity and finite heat conductivity of real gaseous, liquid, and solid media.

A fluid differs from a solid in that, when being applied with an arbitrarily small shear stress, some of its parts begin to slip with respect to others. There are no shear stresses in an equilibrium fluid, however they can arise in a non-equilibrium fluid subject to pressure. The shear forces in fluids, proportional to the rate of their deformation, are due to viscosity (internal friction) of fluids.

Based on the relationship between stress and strain (1.99) and bearing in mind that for fluids $\mu = 0$, one can write for the tangential and normal components of a stress

$$\sigma_{ik} = \eta \frac{\partial \mathcal{E}_{ij}}{\partial t}, \tag{1.113a}$$

$$\sigma_{ll} = \lambda \vartheta + \eta' \frac{\partial \vartheta}{\partial t} + \eta \frac{\partial \mathcal{E}_{ll}}{\partial t} = \lambda \vartheta + \left(\eta' + \frac{\eta}{3} \right) \frac{\partial \vartheta}{\partial t}, \tag{1.113b}$$

here η is the coefficient of proportionality known also as the shear-viscosity coefficient characterizing a dissipative stress when fluid layers glide over each other (sometimes the coefficient of kinematic viscosity $\nu = \eta/\rho_0$ can be conveniently used instead of η); η' is the coefficient of volume viscosity which measures a dissipative pressure resulting from the uniform compression of the medium.

Designating strain-dependent terms in expressions (1.113a) and (1.113b) through σ'_{ik} and accounting for the stress-pressure relationship, one can write the general expression

$$\sigma_{ik} = -p\delta_{ik} + \sigma'_{ik}. \tag{1.114}$$

where δ_{ik} is the Kronecker delta ($\delta_{ik} = 1$ if $i = k$ and $\delta_{ik} = 0$ if $i \neq k$).

The terms σ'_{ik} can be more conveniently represented by using the fluid velocity instead of the strain rate. Indeed, one can show that the shear stress $\sigma'_{xy} = \Delta F_x / \Delta S$ due to the fluid motion with the velocity v_x in the x direction is described by expression (Figure 1.15)

$$\sigma'_{xy} = \eta \frac{\partial v_x}{\partial y}. \tag{1.115a}$$

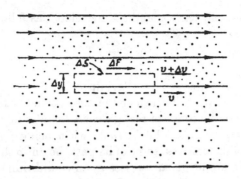

Figure 1.15. Relationship between shear stress and the motion of viscous liquid.

In the case of an incompressible fluid, the general expression for σ'_{ik} has the form

$$\sigma'_{ik} = \eta \left(\frac{\partial v_i}{\partial x_k} + \frac{\partial v_k}{\partial x_i} \right). \qquad (1.115b)$$

In the case of a compressible fluid, the stress tensor σ'_{ik} can be conveniently written as

$$\sigma'_{ik} = \eta \left(\frac{\partial v_i}{\partial x_k} + \frac{\partial v_k}{\partial x_i} - \frac{2}{3} \delta_{ik} \frac{\partial v_l}{\partial x_l} \right) + \eta' \delta_{ik} \frac{\partial v_l}{\partial x_l}. \qquad (1.115c)$$

The component of a viscous force acting on the unit volume in the x_i direction is given by

$$(F_v)_i = \frac{\partial \sigma'_{ik}}{\partial x_k} = \eta \frac{\partial^2 v_i}{\partial x_k^2} + \left(\eta' + \frac{\eta}{3} \right) \frac{\partial}{\partial x_i} \frac{\partial v_l}{\partial x_l}. \qquad (1.116)$$

Substituting F_v into (1.37), the equation of the low-amplitude vibrational motion of a viscous fluid may be written in the vector representation

$$\rho_0 \frac{\partial v}{\partial t} = -\operatorname{grad} p + \eta \Delta v + \left(\eta' + \frac{\eta}{3} \right) \operatorname{grad} \operatorname{div} v. \qquad (1.117a)$$

For an incompressible fluid, $\operatorname{div} v = 0$; therefore, the last term on the right-hand side of (1.117a) vanishes and the equation of motion of a viscous incompressible fluid (the Navier–Stokes equation) has the form

$$\frac{\partial v}{\partial t} = -\frac{1}{\rho_0} \operatorname{grad} p + \frac{\eta}{\rho_0} \Delta v. \qquad (1.117b)$$

After differentiating with respect to time, the equation (1.117a) for a plane wave takes on the form

$$\frac{\partial^2 v}{\partial t^2} = \rho_0' c_0^2 \frac{\partial^2 \eta}{\partial x^2} + \left(\frac{4}{3}\eta + \eta'\right) \frac{\partial^3 v}{\partial x^2 \partial t}. \qquad (1.118)$$

Equation (1.118) differs from the wave equation for a lossless medium (e.g., (1.59)) in having an additional term $(4/3\,\eta + \eta') \times \partial^3 v/(\partial x^2 \partial t)$.

Harmonic solution to the equation (1.118) is given by

$$v = v_m e^{-\alpha_f x} e^{j(\omega t - k_f x)}. \qquad (1.119)$$

If $\left(\dfrac{4}{3}\eta + \eta'\right) \dfrac{\omega^2}{\rho_0} \ll c_0^4$, then the coefficients α_f and k_f are equal to

$$\alpha_f = \frac{\omega^2}{2\rho_0 c_0^3} \left(\frac{4}{3}\eta + \eta'\right), \qquad (1.120)$$

$$k_f = \frac{\omega}{c_0} + j\alpha_f. \qquad (1.121)$$

Another reason for energy dissipation in a wave is levelling off the adiabatic temperature changes produced by compressions and rarefactions.

At a complete temperature levelling, when the sound propagates isothermally, this type of absorption is absent as in the case of adiabatic propagation. Actually, temperature is levelled only partially and hence absorption always takes place.

The mechanism of temperature levelling is related to heat emission and heat conduction. In real fluids, the absorption due to heat emission is small as compared with the absorption due to finite heat conduction.

Analysis of the mechanism of heat-conduction absorption with account for the adiabatic character of heating and isobaric heat transfer makes it possible to obtain an expression for the coefficient of absorption α_T:

$$\alpha_T = \frac{\omega^2}{2\rho c_0^3} \kappa_t \left(\frac{1}{C_V} - \frac{1}{C_P}\right), \qquad (1.122)$$

where κ_t is the thermal conductivity of a fluid. The dependence of α_T on frequency, density, and sound velocity is the same as for the viscous mechanism of energy loss (see (1.120)). An account for these two mechanisms of energy loss leads to the following expression for α_F

$$\alpha_F = \alpha_f + \alpha_T = \frac{\omega^2}{2\rho c_0^3} \left[\frac{4}{3}\eta + \eta' + \kappa_t \left(\frac{1}{C_V} - \frac{1}{C_P}\right)\right] = \frac{\omega^2}{2\rho c_0^3} b. \qquad (1.123)$$

Table 1.7   Absorption of ultrasound in air, water, and liquid metals [19].

Substance	Temperature, K	Coefficient of absorption $\times 10^{17}$, cm^{-1}s^2			Volume-to-shear viscosity ratio η'/η
		α_f/f^2	α_T/f^2	α_F/f^2	
Air	298	0.87×10^4	0.37×10^4	1.24×10^4	
Water	298	8.50	0.0064	8.51	
Sodium	373	1.24	8.31	11.50	2.0 ± 0.5
Potassium	348	2.41	26.90	29.90	0.4 ± 1
Zinc	723	0.75	3.48	3.70	-0.9 ± 2
Cadmium	633	0.48	10.80	14.50	9 ± 9
Mercury	298	0.98	4.29	5.70	0.6 ± 0.2
Gallium	303	0.37	1.04	1.58	0.6 ± 0.2
Tin	513	0.48	3.80	5.63	3.5 ± 1
Lead	613	1.13	6.44	9.40	2.1 ± 0.5
Bismuth	553	1.07	4.66	8.05	2.9 ± 0.5

Here, a quadratic frequency dependence of absorption coefficient is worth noting. Tables frequently give the values α_F/f^2 rather than α_F (Table 1.7). In gases, the viscous and heat-conduction mechanisms of absorption contribute almost equally to the attenuation of sound, while in water and organic solvents viscosity plays a major part. In liquid metals, possessing a high heat conductivity and relatively small viscosity, the heat-conduction absorption is 5 to 10 times greater than the viscous absorption. The value for η'/η estimated both theoretically and experimentally is about 0.5–3. Temperature dependence of ultrasonic absorption indicates that volume viscosity drops with increasing temperature, while the ratio η'/η is temperature-independent.

Absorption in a solid is due to physical processes occurring during the propagation of an elastic wave. These processes are of various nature and range from thermoelastic (or heat) losses to dislocational internal friction, magnetomechanical loss, and so forth [18]. Being different in their physical mechanism, all types of losses can be formally divided into three groups related to relaxation, hysteresis, and resonance. In each of these groups, losses show different dependence on frequency and amplitude, which will be discussed in detail below.

The propagation of an elastic wave through a solid is accompanied with all types of losses, although the contribution from each type is different.

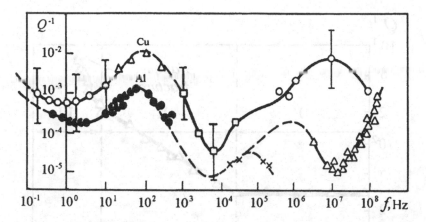

Figure 1.16. Frequency dependence of the elastic wave absorption (internal friction) in polycrystalline aluminum and copper at 20°C [20].

Provided that ambient conditions such as temperature and pressure are the same, each kind of losses is more pronounced within its own frequency range and at a certain strain amplitude.

Several peaks are observable on the $\alpha(\omega)$ curve as the frequency of vibrations varies from infrasonic to hypersonic values (Figure 1.16) Changes in the dislocational and granular structures as well as in the phase composition of a substance affect its absorption spectrum. Unfortunately, there are no experimental methods that would allow the absorption spectrum to be measured in a very broad frequency range, for instance, from 10^{-4} to 10^{10} Hz.

Depending on the amplitude of a vibrational strain, one can distinguish three main ranges of internal friction due to the motion of dislocations in metals (Figure 1.17). For small strain amplitudes ($\varepsilon_m < 10^{-6} - 10^{-5}$), absorption does not depend on ε_m (Figure 1.17, region I) – it is the so-called amplitude-independent absorption. Within the range $10^{-5} \leq e_m \leq 10^{-4}$, the coefficient of absorption increases linearly with the strain amplitude (region II), which corresponds to amplitude-dependent internal friction. When the strain amplitude is greater than 10^{-4}, the internal friction drastically increases (region III).

Till now there is no comprehensive theory of absorption; nevertheless, some aspects of this problem will be treated below.

In above consideration of elastic waves in solids it was assumed that there is a linear single-valued relationship between a stress and the strain it produced (see expression 1.99); in other words, that they are always in phase so that the application of a vibrational stress imme-

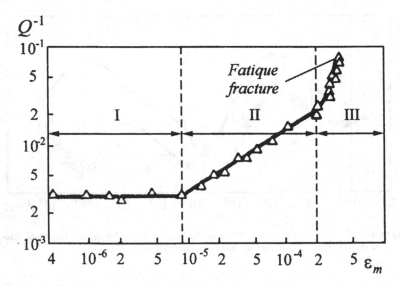

Figure 1.17. Amplitude dependence of absorption (internal friction) in polycrystalline lead, $f = 18$ kHz [21].

diately causes the respective vibrational strain, while the stress relief leads to an immediate removal of the strain. In reality, such a reversibility takes place only for infinitely slow processes, whereas the real wave motion in an elastic medium always occurs at a finite velocity and is not equilibrium, although there are processes bringing about the equilibrium. This is manifested in the irreversibility of wave motion and the dissipation of vibrational energy.

Formally, energy will dissipate if there is no linear and/or single-valued relationship between a stress and the strain it produced. This takes place when the equation relating stress and strain involves their time derivatives. As a result, the strain lags behind its stress (the effect of hysteresis).

A nonlinear relationship between stress and strain causes the dissipation of energy, as such relationship implies the interaction of an elastic wave with, for example, thermal vibrations of a lattice and hence the redistribution of vibrational energy. The extent of this interaction depends on the strain amplitude.

A nonlinear relationship between stress and strain varying in time leads to energy dissipation dependent on the strain amplitude.

In the case of an ideal elastic solid body, the stress and strain obey Hooke's law (see expression (1.99)). For simplicity (in a given case this does not virtually reduce a generality of the approach), one can consider relatively simple strains, such as pure shear, uniaxial stretching, uniform compression. Then, Hooke's law can be written in the form

$$\sigma = M\varepsilon \tag{1.124a}$$

or

$$\varepsilon = Y\sigma, \tag{1.124b}$$

where $M = 1/Y$ is the elasticity modulus known as the shear modulus μ in the case of shear strain, Young's modulus E in the case of uniaxial strain, and bulk modulus K in the case of hydrostatic strain; Y is the compliance modulus.

Hooke's law implies the fulfillment of three conditions defining an ideally elastic body:

(1) the strain produced by an applied stress (and vice versa) is single-valued and equilibrium;

(2) the equilibrium is established immediately;

(3) the response is linear.

In the case of a real solid, depending on its nature and ambient conditions (temperature and so on), only some of the above three conditions are fulfilled. If the third condition is violated, this results in a nonlinear elasticity. In the case of the first and third conditions violation, the material is instantaneously plastic, which is manifested as a flow of plastic crystals under a strong stress. The violation of the second condition is known as rigidity. The violation of the first and second conditions corresponds to a linear viscoelasticity. It should be noted that the term 'inelasticity' (rigidity) refers only to part of the entire response involving also an instantaneous elastic component.

Below we will consider the characteristics of absorption caused by thermal and viscoelastic losses and dislocational friction. In physics of metals, these different mechanisms of the elastic energy conversion into the thermal one are combined under a common name – internal friction.

As a measure of internal friction, the value Q^{-1} reciprocal of the quality factor (see (1.31)) is conventionally used.

As was mentioned above, during the propagation of elastic waves in a real solid, the strain lags behind the stress by some angle φ. Assuming that $\sigma = \sigma_0 \cos \omega t$ and $\varepsilon = \varepsilon_0 \cos(\omega t - \varphi)$ and taking into account that

$$\Delta W = \frac{2\pi}{\omega} \int_V \sigma \dot{\varepsilon} d\, dV = \pi \sin \varphi \int_V \sigma_0 \varepsilon_0 \, dV, \qquad (1.125)$$

$$W = \frac{1}{2} \int_V \sigma_0 \varepsilon_0 \, dV, \qquad (1.126)$$

one can show that

$$Q^{-1} = \sin \varphi. \qquad (1.127a)$$

This relation is valid for such angles φ that $\cos \varphi \sim 1$. Then (1.127a) can be rewritten as

$$Q^{-1} = \mathrm{tg}\varphi. \qquad (1.127b)$$

The angle φ is known as loss angle.

The strain can be written in the complex representation (with allowance for its lag behind the stress) as

$$\varepsilon = (\varepsilon_0' - j\varepsilon_0'')e^{j\omega t}, \qquad (1.128)$$

where ε_0' is the strain component that is in phase with the stress, while ε_0'' is the component lagging behind the stress by 90°.

Then

$$\mathrm{tg}\varphi = \frac{\varepsilon_0''}{\varepsilon_0'}. \qquad (1.129)$$

The value $M_\omega = \sigma_0/\varepsilon_0'$ is known as dynamic modulus, while the reciprocal value $Y_\omega = 1/M_\omega$ as dynamic compliance.

Two other quantities are used in relevant considerations – the complex modulus M^* and compliance Y^*

$$M^* = \frac{\sigma_0}{\varepsilon_0' - j\varepsilon_0''} = \frac{\sigma_0}{\varepsilon_0'(1 - j\mathrm{tg}\varphi)} \cong \frac{\sigma_0}{\varepsilon_0'}(1 + j\mathrm{tg}\varphi) \cong M_\omega(1 + j\varphi),$$

$$(1.130a)$$

$$Y^* \cong Y_\omega(1 - j\varphi). \qquad (1.130b)$$

The ratio of the imaginary part of M^* or Y^* to respective real parts is tangent of the phase angle φ. It is a measure of internal friction

$$\mathrm{tg}\varphi = \frac{\mathrm{Im}\, M^*}{\mathrm{Re}\, M^*} = -\frac{\mathrm{Im}\, Y^*}{\mathrm{Re}\, Y^*}. \qquad (1.131)$$

If a longitudinal wave propagates along the axis of a rod whose diameter is small in comparison with the wavelength, then, in terms of the complex Young's modulus $E^* = E' + jE''$, the wave equation can be written as

$$E^* \frac{\partial^2 \varepsilon}{\partial x^2} = \rho \frac{\partial^2 \varepsilon}{\partial t^2}. \tag{1.132}$$

The solution to this equation has the form

$$\varepsilon = \varepsilon_0 e^{-\alpha x} e^{j(\omega t - k_\alpha x)}. \tag{1.133}$$

After some transforms, one can get real and imaginary parts of the modulus

$$E' = \frac{\rho c^2 \left(1 - \dfrac{\alpha^2 c^2}{\omega^2}\right)}{\left(1 + \dfrac{\alpha^2 c^2}{\omega^2}\right)^2}, \tag{1.134a}$$

$$E'' = \frac{2\rho c^2 \alpha \dfrac{c}{\omega}}{\left(1 + \dfrac{\alpha^2 c^2}{\omega^2}\right)^2}. \tag{1.134b}$$

It follows from (1.134a) that if absorption is small then

$$E' \cong \rho c^2. \tag{1.134c}$$

From expressions (1.129) and (1.134) one can easily deduce the relationship between $\mathrm{tg}\varphi$, Q^{-1}, and α

$$Q^{-1} \approx \mathrm{tg}\varphi = \frac{E''}{E'} \sim \frac{2\alpha c}{\omega} = \frac{\lambda \alpha}{\pi}. \tag{1.135}$$

Sometimes the attenuation decrement $\Delta = \lambda \alpha$ is used for the estimation of absorption.

As for the non-linear relationship between stress and strain, in the most general form it can be given by

$$a_0 \sigma + a_1 \dot{\sigma} + a_2 \ddot{\sigma} + \ldots = b_0 \varepsilon + b_1 \dot{\varepsilon} + b_2 \ddot{\varepsilon} + \ldots, \tag{1.136}$$

where the upper dots conventionally denote the time derivatives of strain and stress.

The equation (1.136) is basic for the description of the acoustic wave propagation in a solid. One can show that the attainment of a steady-state value of some parameters (for instance, strain) in response to a change in other parameters (in a given case, applied stress) takes some time known as relaxation time. From the condition of normalization, the relaxation time is defined as the time during which the strain of a body applied with an instantaneous stress increases from zero to $(1 - 1/e) = 0.6321$ of its steady-state value.

To visualize the behaviour of a solid during the propagation of an elastic wave, some mechanical models described by its own differential equation of strain are used.

If in equation (1.136), only the coefficients a_0 and b_0 are non-zero, then it is reduced to the simplest form describing an ideally elastic body. The respective mechanical model is a spring obeying Hook's law. The energy of this spring is reversible: when the force (stress) is removed, the displacement (strain) returns to zero.

The internal energy loss is described by a Newtonian damper, representing a piston moving in an ideally viscous fluid.

The velocity of such piston is proportional to an applied force, and the work produced is dissipated completely as heat. The relationship between stress and strain, which are used instead of force and displacement for description of a solid body, is given by

$$\sigma = \eta\dot\varepsilon, \qquad (1.137)$$

where η is the viscosity of damper's fluid.

In mechanical models, their elements (spring and damper) can be either series or parallel.

Going on from models to a real solid, it is necessary, first of all, to define a physical meaning of the parameter η. If in the case of models this parameter means the viscosity of a damper, then in the case of a real solid it means the energy loss caused by viscosity and finite heat conductivity of a medium

$$\eta = \eta_f + \eta_T \qquad (1.138)$$

where η_f is the viscosity due to friction forces; and η_T is the viscosity due to finite heat conductivity.

In the simplest case of propagation of a longitudinal wave in a rod, η_f is given by

$$\eta_{fl} = \frac{4}{3}\mu' + \lambda' \qquad (1.139a)$$

and in the case of a transverse wave

$$\eta_{ft} = \mu', \qquad (1.139b)$$

here λ' and μ' are the imaginary parts of the first and second Lamé constants, whose dimensionality is the same as that of the coefficient of dynamical viscosity.

During the propagation of transverse waves the density of the medium and its temperature are constant, which results in the absence of energy loss due finite heat conductivity. At the same time, the propagation of longitudinal waves is associated with volume alterations and, consequently, temperature changes. The extent of these changes is determined by frequency and the nature of a solid, and in the case of a finite-size body also by its shape and dimensions. For a longitudinal wave propagating in a rod, η_T is given by [23]

$$\eta_T = \frac{\kappa_t T \alpha_T^2 c_l^2}{C_p^2} \left(1 - \frac{4c_t^2}{3c_l^2}\right)^2 \tag{1.140}$$

where κ_t is the heat conductivity; α_T is the coefficient of thermal expansion; C_p is the heat capacitance at constant pressure.

In metals, classical heat losses are considerably less than those caused by other factors. This is due to a high heat conductivity of metals, which leads to an isothermal propagation of waves at low and medium frequences.

The contribution from these losses becomes noticeable only at frequences of the order of several megahertz.

At much higher frequences, when the wavelength is very small, the regions of compression and rarefaction become closer, which facilitates heat exchange and approaches the process of wave propagation to be again isothermal.

Based on (1.139) and (1.140), one can write for the coefficient of absorption of longitudinal waves in a rod

$$\alpha_l = \frac{\omega^2}{2\rho c_l^3} \left[\left(\frac{4}{3}\mu' + \lambda'\right) + \frac{\kappa_t T \alpha_T^2 c_l^2}{C_p^2} \left(1 - \frac{4c_t^2}{3c_l^2}\right)^2 \right] \tag{1.141a}$$

and for transverse waves

$$\alpha_t = \frac{\omega^2 \mu'}{2\rho c_t^3}. \tag{1.141b}$$

A confirmation of the validity of these expressions is intricate, since it is difficult to distinguish the thermoelastic absorption from other kinds of losses – dislocational internal friction, scattering, and so on. Figure 1.18 gives representative curves for the coefficient of absorption of elastic waves in monocrystalline zinc. The calculated data agree

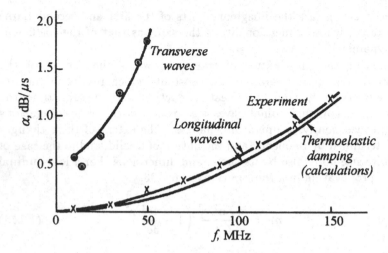

Figure 1.18. Frequency dependence of the absorption of ultrasonic waves propagating in zinc monocrystals along the hexagonal axis [22].

well with the experimental results obtained under the conditions of the absence of dislocational absorption and scattering.

The above considerations are valid for homogeneous isotropic materials. In the case of polycrystalline substances, there is a scattering of elastic waves from grains, which leads to an additional energy absorption. Depending on the grain size-to-wavelength ratio, one or another kind of losses would prevail.

If frequency is less that 1 MHz, then the wavelength is several times greater than the mean grain size, and the main contribution to absorption comes from thermal microprocesses. Because of the material inhomogeneity related to a random disposition of constituting anisotropic grains temperature distribution over grains during the propagation of a wave is not uniform. The situation worsens further because the intercrystallite boundary possesses less heat conductivity than crystallites themselves, which makes the temperature gradient near boundaries steeper.

Experiments indicated that absorption in a polycrystalline material is several times greater than in a homogeneous material possessing identical mechanical and thermal properties.

In ferromagnetic and ferroelectric materials, absorption is influenced not only by 'mechanical' inhomogeneities, but also by 'magnetic' and 'electric' ones. In particular, stresses in ferromagnetic materials depend on both the strain amplitude and magnetic induction.

References

1. J. Rayleigh, *Theory of Sound*, MacMillan & Co., London (1926).
2. F. Morse, *Vibrations and Sound* (Russian translation), Gostekhteoretizdat, Moscow (1949).
3. G. Lamb, *Dynamic Theory of Sound* (Russian translation), Fizmatgiz, Moscow (1960).
4. G. S. Gorelik, *Vibrations and Waves* (in Russian), Fizmatizdat, Moscow (1958).
5. M. A. Isakovich, *General Acoustics* (in Russian), Nauka, Moscow (1973).
6. F. Crawford, *Waves* (Russian translation), Nauka, Moscow (1974).
7. E. Skuchik, *Fundamentals of Acoustics* (Russian translation), 2 volumes, Moscow (1976).
8. L. Bergman, *Ultrasound and its Application in Science and Technics* (Russian translation), Inostr. Liter., Moscow (1955).
9. *Physical Processes of Ultrasonic Technology* (in Russian), ed. L. D. Rozenberg, 3 volumes, Nauka, Moscow (1967, 1968, 1970).
10. *Ultrasound Technology* (in Russian), ed. B. A. Agranat, Metallurgiya, Moscow (1974).
11. V. F. Nozdrev and N. V. Fedorishchenko, *Molecular Acoustics* (in Russian), Vysshaya Shkola, Moscow (1974).
12. O. V. Abramov, Sh. G. Khorbenko, and Sh. Shvegla, *Ultrasonic Processing of Materials* (in Russian), Mashinostroenie, Moscow (1984).
13. *Physical Acoustics* (Russian translation), ed. W. Mason, 4, Part B, Mir, Moscow (1970).
14. M. B. Gitis and I. G. Mikhailov, *Akustich. Zh.*, 12, 145 (1966).
15. *Physicochemical Properties of Chemical Elements* (in Russian), ed. G. V. Samsonov, Naukova Dumka, Kiev (1965).
16. *Handbook of Chemistry and Physics*, Chemical Rubber Publishing Co.
17. *Ultrasonics. A Small Encyclopedia* (in Russian), Sov. Entsiklop., Moscow (1979).
18. R. Truel, C. Elbaum, and B. Chick, *Ultrasonics in Solid-State Physics* (Russian translation), Mir, Moscow (1972).
19. J. Weber and R. Stefens, Absorption in Liquids, in *Ultrasonics in Quantum and Solid-State Physics* (Russian translation), Mir, Moscow, p. 75 (1970).
20. V. S. Postnikov, *Internal Friction in Metals* (in Russian), Metallurgiya, Moscow (1974).

21.   W. P. Mason, *J. Acoust. Soc. Am.*, **28**, 1207 (1956).
22.   K. Lücke, *J. Appl. Phys.*, **27**, 1433 (1956).
23.   L. D. Landau and E. M. Lifshitz, *Mechanics of Continuous Media* (in Russian), Gostekhizdat, Moscow (1953).

Chapter 2
The Propagation of High-Intensity Ultrasonics in Fluids

Chapter 1 was concerned with the propagation of vibrations in ideal media – gases, liquids, or solids. It was assumed that disturbances in an equilibrium medium caused by these vibrations were small, i.e., vibrations were considered to be of a low amplitude or intensity.

The propagation of finite-amplitude or high-intensity ultrasonic waves is accompanied with a variety of effects whose intensity depends on the amplitude of vibrations (the so-called nonlinear effects). Thus, strong ultrasonic fields in liquids give rise to nonlinear effects in their bulk such as cavitation, acoustic streaming, and radiation pressure. The extent to which effects are nonlinear in a given medium under given ambient conditions (pressure, temperature, etc.) depends on the acoustic Mach number

$$M_a = \frac{v_m}{c_0}, \tag{2.1}$$

where v_m is the vibrational velocity amplitude; c_0 is the sound velocity in an undisturbed medium.

The greater the Mach number, the stronger the nonlinear effects occurring during the propagation of elastic waves through the medium.

For an adequate description of these effects, the nonlinear terms of hydrodynamic equations (1.37–1.39) have to be taken into account. During analysis of elastic waves in Chapter 1 (sections 1.1–1.3), it was assumed that $M_a \ll 1$ and that vibrations are not damped, which allowed the equations of hydrodynamics and state to be linearized.

It should be noted that nonlinear effects can arise during the propagation of elastic waves even at $M_a < 1$, which is related to the fact that nonlinear corrections to the linear acoustic equations involve, along with the terms of the order M_a, the terms describing additive effects. Thus, in the case of a plane wave traveling along the x axis, the magnitude of these terms will be proportional to $M_a kx$ or $M_a \omega t$.

The Mach number is not more than $\sim 10^{-2}$ even for the greatest intensities of elastic vibrations attainable now and is not above 10^{-3} for high-power ultrasonic vibrations, which makes it possible to substantially simplify the initial set of hydrodynamic equations and to develop an approximated nonlinear theory of propagation of finite-amplitude elastic waves [1–6].

In this and in the following chapters, nonlinear effects arising in a liquid during the propagation of a high-power ultrasound will be mainly considered. Nonlinear effects in gases will be outlined in the sections describing ultrasonic applications.

2.1   Absorption of Finite-Amplitude Ultrasonic Waves in Liquids

The expression (1.123) of Chapter 1 describes the absorption of ultrasonic waves of quite a low intensity when the equations of hydrodynamics and state can be linearized. However, when the waves are rather intense, the characteristics of their propagation and absorption may change.

The propagation of finite-amplitude waves is described by a set of equations of hydrodynamics (1.37) and (1.38), state (1.39), and heat transfer

$$\rho T \left(\frac{\partial s}{\partial t} + v \nabla s \right) = \sigma_{ik} \frac{\partial v_i}{\partial x_k} + \kappa_t \Delta T, \qquad (2.2)$$

where s is the entropy of a unit mass.

With allowance for the quadratic terms, the equation of state can be written as

$$p = P - P_0 = c^2 \rho' + \frac{1}{2} \left(\frac{\partial c^2}{\partial \rho} \right)_s (\rho')^2 + \left(\frac{\partial \rho}{\partial s} \right)_p s'' \qquad (2.3a)$$

or

$$p = A \frac{\rho'}{\rho_0} + \frac{B}{2} \left(\frac{\rho'}{\rho_0} \right)^2, \qquad (2.3b)$$

where $A = \rho_0 c_0^2$, $B = \rho_0^2 (\partial c^2 / \partial \rho)_s$. Instead of B/A, the quantity γ is frequently used

$$\gamma = \frac{B}{A} + 1 = \left(\frac{\partial c^2}{\partial \rho} \right)_s \frac{\rho_0}{c_0^2} + 1. \qquad (2.4)$$

This quantity is equal to the ratio of heat capacities C_p/C_v in the case of gases or to the exponent n in the empirical equation of state in the case of condensed media (1.50).

The set of equations (1.37), (1.38), (2.2), and (2.3) describes the propagation of finite-amplitude waves.

It should be noted that the dimensionless parameter

$$N = \left[1 + \frac{\rho}{c}\left(\frac{\partial c}{\partial \rho}\right)_s\right]\frac{p_m}{b\omega} \tag{2.5}$$

enables one to assess the validity of application of the set of linearized equations (1.44), (1.48), and (1.51): $N \ll 1$ and $N \gg 1$ correspond to low-amplitude and finite-amplitude waves, respectively.

If the intensity of a sinusoidal wave near the source is rather high, its shape changes with distance because of the difference in the velocity of various wave regions. Regions that correspond to compression travel faster than the rarefaction regions. This leads to the sharpening of wave fronts and transformation of the initially sinusoidal wave into a sawtooth wave, and perhaps to discontinuities (pressure steps) in each wave period. On the other hand, the viscosity and finite heat conductivity of medium can bring about smoothing of the wave profile and reduce the velocity and temperature gradients.

A combined action of these opposite effects may eventually stabilize the waveform, which is illustrated in Figure 2.1.

The distortion of the waveform is associated with the change in its spectral composition: high-frequency harmonics appear and then diminish during the propagation of an initially monochromatic wave.

Mathematical analysis of the finite-amplitude wave propagation shows that the extent and character of the waveform distortion are largely determined by the acoustic Reynolds number Re_a which is a measure of the relative contributions from nonlinear and dissipative effects

$$\mathrm{Re}_a = 2\varepsilon\mathrm{Re} = \frac{\varepsilon}{\pi}\frac{v_m \lambda \rho_0}{b}, \tag{2.6}$$

where $\mathrm{Re} = v_m \lambda \rho_0 / 2\pi b$ is the hydrodynamic Reynolds number, and

$$\varepsilon = \frac{\gamma + 1}{2}. \tag{2.7}$$

The acoustic Reynolds number is proportional to the ratio of the nonlinear term in the equation of ultrasonic wave propagation to the term accounting for the medium viscosity and heat conductivity.

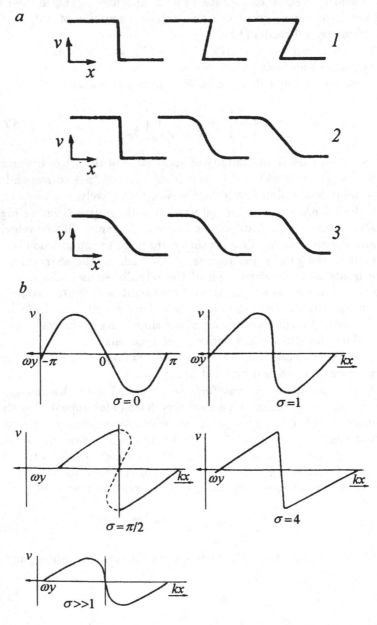

Figure 2.1.   Profiles of (a) a single compression step traveling along the x axis as a result of (1) nonlinear, (2) dissipative, (3) nonlinear plus dissipative effects, and (b) finite-amplitude wave at different σ [7].

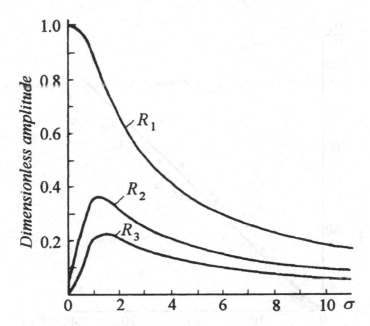

Figure 2.2. Amplitudes of the first, second, and third harmonics for Re$_a \gg 1$ versus the distance [7].

At Re$_a \gg 1^*$, the distortions of a sinusoidal waveform near the source ($\sigma = r/L \ll 1$, where $L = 1/k\varepsilon M_a = c_0^2/\varepsilon\omega_m$) are small, but increase with the distance so that at $\sigma = 1$, in the vicinity of $v = 0$, the wave profile becomes abrupt and, at $\sigma = \pi/2$, multi-valued, which implies the wave discontinuity and formation of a shock wave. The discontinuity will diminish with the distance and the wave will take a sawtooth form ($\sigma \sim 4$). The law, according to which the vibrational velocity amplitude of a sawtooth wave decreases, becomes especially simple at high σ

$$v_p = v_m \frac{\pi}{1 + \sigma}. \tag{2.8}$$

Analysis of this expression shows that, unlike a low-amplitude wave, a finite-amplitude sawtooth wave is attenuated nonexponentially, the extent of attenuation being increased with the wave amplitude ($\sigma \sim v_m$). The attenuation is associated with the conversion of part of the initial sinusoidal wave energy into the energy of arising high-frequency harmonics, rather than with the absorption of energy. Figure 2.2 shows the dimensionless amplitude A_n of the first three harmonics versus the

*The conditions $M_a \ll 1$ and Re$_a \gg 1$ are real and not mutually exclusive.

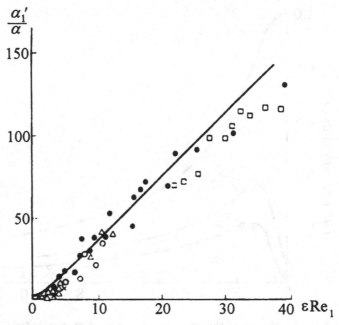

Figure 2.3. Dependence of the absorption coefficient α' of the first harmonic on the Reynolds number [7]. Points represent experimental data.

distance covered by the wave. Near a discontinuity the second and third harmonics increase and then gradually fall.

In the regions of the greatest wave distortions, the absorption of energy is maximum, which can be explained by the transformation of a sinusoidal wave into its high-frequency harmonics that are absorbed stronger than the sinusoidal wave itself.

The basic wave attenuation near $\sigma = \pi/2$ can be described by

$$\alpha_1' = \alpha\sqrt{1 + \tilde{R}e_{a1}} = \frac{1}{v_1}\frac{\partial v_1}{\partial r}, \qquad (2.9)$$

where $v_1 = v_m \frac{2}{Re_a}\left(\mathrm{sh}\frac{1+\sigma}{Re_a}\right)^{-1}$; and $\tilde{R}e_{a1} = 2pc_0 v_1 \varepsilon/b\omega$ is the instant value of the acoustic parameter that changes due to v_1. It is seen from expression (2.9) that absorption depends on small and great $\tilde{R}e_{a1}$ in different manners.

At $Re_{a1} \gg 1$, the attenuation coefficient of the basic harmonic grows quadratically and then linearly with Re_a and may significantly exceed the absorption coefficient of low-amplitude waves (Figure 2.3)*.

---

*Experiments with mercury showed that attenuation of finite-amplitude waves

High-frequency harmonics are attenuated as the first harmonic provided their number n is not too high and the condition $\mathrm{Re}_{a1}^2 \gg n^2$ is fulfilled.

But if the wave is low-intense, then $\mathrm{Re}_{a1} \ll 1$ and, based on (2.9), one can get

$$\alpha_1' = \alpha \left(1 + \frac{1}{2} \tilde{\mathrm{Re}}_{a1}^2 \right), \tag{2.10}$$

from which it follows that the attenuation coefficient of the first harmonic grows slower than the absorption coefficient of a low-amplitude wave and differs from those of higher harmonics.

The propagation of elastic vibrations in a liquid containing gas bubbles is accompanied with their additional attenuation related to the energy loss for the heating of gas in oscillating bubbles and the subsequent transfer of this heat to liquid, to the scattering of part of acoustic energy from the bubbles, and to the energy loss caused by acoustic streaming.

The extent of damping of ultrasonic oscillations can be characterized by the effective cross-sectional area σ_e that is the area of the cross section perpendicular to the direction of wave propagation, for which transmitted acoustic energy is equal to the sum of energies absorbed by and scattered from a bubble

$$\sigma_e = \sigma_a + \sigma_s, \tag{2.11}$$

where σ_a and σ_s are the effective cross-sectional areas of absorption and scattering, which can be determined from expressions [8]

$$\sigma_a = \frac{4\pi R^2 \left(\dfrac{\delta}{\lambda_R} - 1 \right)}{\left(\dfrac{f_r^2}{f^2} - 1 \right)^2 + \delta^2}, \tag{2.11a}$$

$$\sigma_s = \frac{4\pi R^2}{\left(\dfrac{f_r^2}{f^2} - 1 \right)^2 + \delta^2}, \tag{2.11b}$$

where R is the bubble radius; f_r is the resonant frequency of the bubble with the radius R_r; δ is the attenuation constant of the bubble; $\lambda_R = 2\pi R/\lambda$.

Hereinafter the subscript r indicates the resonant values of respective quantities.

at $p_m \sim 10$ MPa is by 5 orders of magnitude greater than the absorption of low-amplitude waves of the same frequency [6].

At $\xi_m \ll R$, the resonant frequency of a bubble can be determined from expression [9]

$$f_r \cong \frac{1}{2\pi R_r} \sqrt{\frac{3\gamma \left(P_0 + \dfrac{2\sigma_L}{R_r} \right)}{\rho_L}}, \tag{2.12}$$

where P_0 is the hydrostatic pressure in a liquid; $\gamma = C_p/C_v$ for a gas.

The attenuation constant is determined by absorption due to a finite heat conductivity of a medium, δ_κ, the energy loss due to radiation, δ_r, and medium viscosity, δ_η:

$$\delta = \delta_\kappa + \delta_r + \delta_\eta. \tag{2.13}$$

The mechanism of thermal absorption is associated with the heat flow from a gas-filled bubble to a liquid, which appears because the work of a gas for liquid displacement is greater during the bubble shrinkage than during its expansion.

In [10], the quantity δ_κ was estimated for a resonant bubble

$$\delta_{\kappa r} = 2 \left[\frac{\sqrt{\dfrac{16}{9(\gamma-1)^2} \dfrac{F\alpha}{f_r} - 3} - \dfrac{3\gamma-1}{3(\gamma-1)}}{\dfrac{16}{9(\gamma-1)^2} \dfrac{Fz}{f_r} - 4} \right], \tag{2.14}$$

where

$$F = \frac{\gamma P_0}{V_0 d},$$

$$\alpha = 1 + \frac{3(\gamma-1)}{2\Phi R} \left[1 + \frac{3(\gamma-1)}{2\Phi R} \right],$$

$$\Phi = \sqrt{\frac{\omega}{2D}},$$

$$z = 1 + \frac{2\sigma_L}{RP_0} - \frac{2\sigma_L}{3RP_0}\gamma,$$

V_0 is the equilibrium bubble volume; D is the diffusion coefficient; d is the coefficient of proportionality dependent on the material density.

Figure 2.4 presents the frequency dependence of $\delta_{\kappa r}$ for air bubbles in water calculated by expression (2.14).

Parameter δ_r characterizes energy loss due to radiation emission. At resonance, it is described by expression

$$\delta_{rr} = \frac{\omega_r R}{c_L}. \tag{2.15}$$

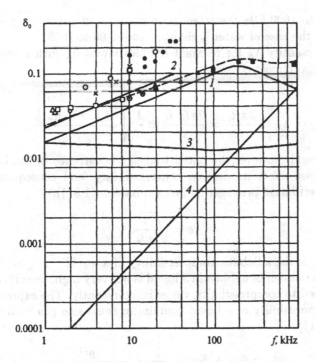

Figure 2.4. Calculated components ($\delta_{\kappa r}$, curve *1*; δ_{rr}, curve *2*; $\delta_{\eta r}$, curve *3*) and the attenuation coefficient δ_r (curve *4*) vesus frequency for a resonant air bubble in water. Points are experimental data [10].

For large bubbles, δ_{rr} does not depend on frequency.

The third type of energy loss is related to viscosity. During expansion of a bubble, the spherical interface with a liquid is deformed so that its thickness decreases, while the area increases. Upon bubble shrinkage, the reverse takes place. If the liquid is incompressible, then only shear viscous stresses appear, and the interface volume does not change. This leads to a situation that the energy of bubble compression exceeds the energy gained by the bubble during its expansion. The quantity $\delta_{\eta r}$ is described by expression

$$\delta_{\eta r} = \frac{8\pi\eta}{3\gamma P_0} f_r \left(\frac{\alpha}{z}\right)^{1/2}. \qquad (2.16)$$

The frequency dependence of $\delta_{\eta r}$, as well as the calculated and experimental values of the attenuation constant $\delta_0 = \delta_{\kappa r} + \delta_{rr} + \delta_{\eta r}$ are shown in Figure 2.4. Analysis of expressions (2.14), (2.15) and dependences in this Figure indicates that within the frequency range from 20 to 1000 kHz the attenuation constant changes insignificantly. Thus, in the case of water it changes from 0.08 to 0.13.

The quantity λ_R for resonant bubbles vibrating with a low amplitude is also almost constant

$$\lambda_R = \frac{2\pi R_r}{\lambda} = \frac{2\pi R_r f_r}{c} \cong \frac{1}{c}\sqrt{\frac{3\gamma\left[P + \dfrac{2\sigma_L}{R_r}\right]}{\rho}}. \qquad (2.17)$$

It should, however, be noted that the occurrence of bubbles in a liquid must affect its density, compressibility, and consequently the rate of oscillation propagation (see expression (1.57)):

$$c = \sqrt{\frac{1}{\rho\beta_s}}, \qquad (2.18)$$

where β_s is the adiabatic compressibility of a liquid.

If the number of bubbles in a liquid is not very high, then its density, as opposed to compressibility, varies insignificantly. The expression for the compressibility of a liquid containing equal-size gas bubbles is as follows [11]

$$\overline{\beta}_S = \beta_{SL} + \beta_{SG} = \frac{1}{\rho c_L^2} + \frac{gU}{1 - \dfrac{f^2}{f_r^2} + j\delta\dfrac{f}{f_r}}, \qquad (2.19)$$

where β_{SL} is the liquid compressibility; β_{SG} is the gas-filled bubble compressibility; c_L is the speed of sound in a liquid lacking bubbles; U is the volume concentration of a free gas in the liquid; g is the coefficient dependent on the liquid properties (for the air bubble in water it equals 1×10^{-7}).

In deducing this equation, the bubble vibrating with a low amplitude was considered as a zero-order radiator.

Bubbles with dimensions less than the resonant dimensions oscillate in phase with acoustic wave pressure [11]. But if bubbles are greater than the resonant ones, they vibrate in antiphase with acoustic pressure. In this case, acoustic frequencies lower than the resonant frequency of bubbles enhance the compressibility of the liquid and reduce sound velocity, while higher frequencies act to decrease the compressibility of liquid.

With allowance for (2.19), the complex speed of sound can be written as

$$\bar{c} = \sqrt{\frac{1}{\rho\beta_s}} \qquad (2.20)$$

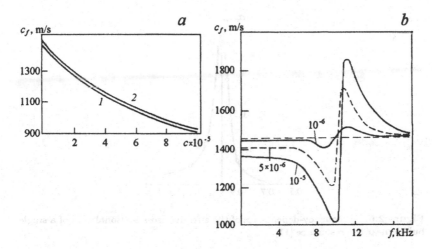

Figure 2.5. The sound phase velocity in water as a function of (*a*) the concentration of a free gas: (*1*), $t = 18°C$; (*2*) $t = 28°C$; and (*b*) frequency (water contains equal-size bubbles with $f_r = 10$ kHz). Numerals near curves indicate the concentration of a free gas in parts per one water part.

and the phase velocity as

$$c_f = \frac{1}{\mathrm{Re}\left(\dfrac{1}{\bar{c}}\right)}. \tag{2.21}$$

Figure 2.5 shows the phase velocity of sound in water versus the concentration of free gas and the frequency of oscillations [11].

Assuming that δ_r and λ_R weakly depend on frequency, one can show that at a given bubble radius effective cross-sectional areas $\sigma_e(\sigma_a, \sigma_s)$ are maximum when $f = f_r$. It should be emphasized that the resonant values $\sigma_{er}(\sigma_{ar}, \sigma_{sr})$ considerably exceed the geometrical cross section of a spherical bubble (πR^2). In the case of bubbles in water with resonant frequencies 40–100 kHz ($\delta_r \sim 0.1$), the ratio of respective cross-sectional areas is

$$\frac{\sigma_{er}}{\pi R_r^2} = 3 \times 10^3.$$

This is illustrated by Figure 2.6 showing the effective cross-sectional area of a single bubble oscillating near the resonance.

By estimating the contributions from absorption and scattering into attenuation and using expressions (2.11a) and (2.11b), one can show that

$$\frac{\sigma_s}{\sigma_a} = \frac{\lambda_R}{\delta - \lambda_R}. \tag{2.22}$$

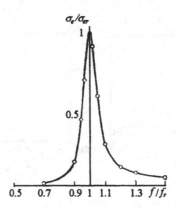

Figure 2.6. Frequency-dependence of the effective cross-sectional area of a single bubble near the resonance [11].

Taking into account the values for δ and λ_R, one can infer that bubbles absorb acoustic energy considerably stronger than scatter it.

If the absorption coefficient is known, one can estimate the attenuation of sound in the liquid containing gas bubbles [8]. In the case when bubbles have equal size and their number changes in the direction of wave propagation, the sound intensity difference over the distance dr will be

$$dI = -n(r)\sigma_e I(r)\, dr, \qquad (2.23)$$

where n is the number of bubbles in unit volume; r is the distance in the region containing gas bubbles covered by the acoustic wave.

Integrating, one can get

$$I(r_e) = I_0 e^{-\sigma_e N(r_e)}. \qquad (2.24)$$

Here I_0 is the initial intensity of sound in the region containing bubbles; $N(r_e) = \int_0^{r_e} n(r)\, dr$ is the total number of bubbles in a liquid column of the unit cross section and length r_e.

Then the attenuation coefficient can be written as

$$\alpha_e = \overline{n}\sigma_e r_e, \qquad (2.25)$$

where

$$\overline{n} = \frac{1}{r_e}\int\limits_0^{r_e} n(r)\, dr = \frac{1}{r_e}N(r_e).$$

If the bubbles in a liquid are of various size, it is necessary to determine their net cross section of damping over the whole liquid volume

$$\Sigma_e = \int_{R_{\min}}^{R_{\max}} \frac{4\pi R^2 n \frac{\delta}{\lambda_R}}{\left(\frac{f_r^2}{f^2} - 1\right) + \delta^2} \, dR, \qquad (2.26)$$

where $R_{\max}$ and $R_{\min}$ are the maximum and minimum radii of bubbles, respectively.

The integral (2.26) can be calculated by using some assumptions based on a prevailing contribution from the resonant bubbles into damping

$$\Sigma_e = 346.5 \, V(R), \qquad (2.27)$$

where $V(R) = \frac{4}{3}\pi R^3 n(R)$ is the distribution density of bubbles.

It follows from experiments with a gas-saturated water [12] that the distribution density of bubbles obeys the law

$$n(R) = \frac{A}{R^b}, \qquad (2.28)$$

where A and b are the numerical constants; $b \cong 3 - 3.5$. In the case of the distribution (2.28), the contribution from non-resonance bubbles into the value Σ_e could be neglected. Then

$$\alpha_e = C n(R_r) R_r^3, \qquad (2.29)$$

where C is the constant equal to 6.3×10^5 for water, if α is expressed in dB/m.

Experimental data on attenuation of acoustic waves in liquids containing gas bubbles agree quite well with respective theoretical estimations [13], which is illustrated by Figure 2.7 showing the damping of 20-kHz ultrasonic vibrations in water containing gas bubbles. The solid line shows the theoretical estimate, and dashed lines indicate the experimental data.

It should be emphasized that the above consideration is valid only when the number of bubbles in a liquid is relatively small and the total attenuation is a sum of separate 'attenuations'. But when the number of bubbles is quite great, their effective cross sections overlap.

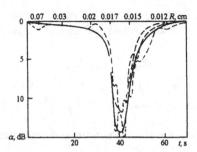

Figure 2.7. Damping of 20-kHz ultrasonic vibrations in a liquid layer containing bubbles of different size (0.016-cm bubbles are resonant). Solid and dashed lines show theoretical and experimental data, respectively.

The estimations performed in [14] show that this takes place when the distance between bubbles $l' < 100R$, which corresponds to the volume concentration of resonant bubbles equal to 5×10^{-6} parts of gas per one part of water. At a higher gas concentration, the energy absorbed by gas bubbles will be less than that predicted by expression (2.29).

2.2   Ultrasonic Cavitation

Analysis of the effect of high-power ultrasonic vibrations on liquids indicates an important part played by cavitation in ultrasonic effects in the processed media.

In spite of the absence of a general theory of cavitation, the phenomena arising in liquids under the action of ultrasound have been presently well studied. Thus, B. Noltingk and E. Neppiras, S. Herring, H. Flinn, J. Kirkwood, H. Bethe, F. Gilmore, W. Lauterborn, and others obtained differential equations describing the dynamics of single bubbles and their ensembles in an ultrasonic field. While L. D. Rozenberg, M. G. Sirotyuk, A. V. Akulichev, W. Lauterborn, L. Crum, and others investigated experimentally the nascent and developed ultrasonic cavitation.

Cavitation lies in the formation of tiny discontinuities or cavities in liquids followed by their growth, pulsation, and collapse. Cavities appear as a result of the tensile stress produced by an acoustic wave in the rarefaction phase.

If the tensile stress persists after the cavity has been formed (negative pressure of the ultrasonic field), the cavity will expand to reach the sizes severalfold in excess of the initial size. In this case, the cavity (bubble) retains its spherical form, as confirmed by a high-speed filming. Further behavior of the cavity is highly diverse – it can pulsate around its equilibrium radius either linearly or nonlinearly, or may, being enlarged to some maximum size, rapidly shrink until the growing internal pressure ceases the collapse. The rate of cavity collapse can be so high that it produces a hydraulic shock wave with pressures reaching tens of thousands of MPa. At this final stage, the sphericity of the cavity can be disturbed and it may even break down into smaller cavities.

The behavior and the motion of cavities are determined by the properties of both the liquid and ultrasonic field.

During cavitation in an ultrasonic field, many cavities appear simultaneously at distances less than the ultrasonic wavelength to form a cavitation region where relatively low-density energy of ultrasonic field transforms into high-density energy inside and near collapsing cavities. Owing to this effect, acoustic cavitation may cause erosion of solid surfaces occurring in the cavitation region, excite luminescence of the liquid, and initiate some chemical reactions.

There are a great many of works, including monographs and reviews, devoted to cavitation [12, 15–23]; however, most of them are concerned with the behavior of a single cavity. At the same time, the processes that take place in the cavitation region are yet imperfectly studied, especially theoretically.

To disrupt a liquid and form a cavity it is necessary that the amplitude of acoustic pressure be above some value specific for a given liquid and determined by its tensile strength.

According to kinetic theory, the tensile strength of an ideally pure liquid is determined by molecular forces and, according to theoretical estimates, must reach several thousands of MPa. Accounting for the probability of a spontaneous formation of cavities because of thermal fluctuations of vapor bubbles reduces the above estimate to 10^2 MPa [24].

However, this estimate is again in significant excess of the experimentally observed values. Thus, the cavitation threshold for water lies within the range 0.01–0.5 MPa at frequences up to 10^2 kHz.

Such a discrepancy between the theoretical and experimental tensile strengths is explained by the occurrence in a real liquid of various solid, vapor, and gaseous microinclusions, which are weak points of the liquid representing cavity nuclei.

The estimations of the cavitational strength of a liquid containing vapor or gas nuclei (tiny bubbles) showed that when the peak sound pressure p_m is below some critical value P_c^v, the bubbles are stable; but above this value they become unstable, begin to grow, and finally burst [15]. The expression for the tensile strength of a liquid containing a bubble with the initial radius R_0 is as follows

$$P_c^v = P_0 - P_v + \frac{2}{3\sqrt{3}} \sqrt{\frac{(2\sigma_L/R_0)^3}{P_0 - P_v + 2\sigma_L/R_0}}, \qquad (2.30)$$

where P_v is the saturation vapor pressure.

The amplitude of acoustic pressure, P_c^v, is known as the vapor cavitation threshold. The expression (2.30) was obtained by Blake from the theory of an equilibrium vapor- or gas-filled bubble in a static-pressure liquid.

In the absence of an acoustic field, the vapor- or gas-filled bubble disappears due to gas diffusion into the liquid, even if it is saturated with gas.

On the other hand, the pulsating bubble in an acoustic field can increase in size because of gas diffusion from the liquid, even if it is not saturated with gas [25, 26]. According to the Hsieh-Plesset theory, there is a peak pressure threshold P_c^g at which the bubble nucleus begins to grow through a rectified diffusion

$$P_c^g = \sqrt{\frac{2}{3}P_0} \left\{ 1 + \frac{2\sigma_L}{R_0 P_0} - \frac{c_\infty}{c_0} \right\}^{1/2}, \qquad (2.31)$$

where c_0 is the equilibrium concentration of a dissolved gas at pressure P_0; c_∞ is the apparent gas concentration in liquid far away from the nucleus.

Figure 2.8,*a* shows the cavitation thresholds P_c^v and P_c^g versus the nucleus radius, calculated for $P_0 = 0.1$ and 1.0 MPa and various c_∞/c_0 ratios.

It is seen that the cavitation thresholds P_c^v and P_c^g diminish with increasing nucleus radius. At R_0 ranging from 10^{-5} to 10^{-4} cm, the thresholds of vapor-induced cavitation and that induced by rectified diffusion are virtually coincide. In this case, the nucleus size will drastically increase once the peak sound pressure reaches the values 0.1–0.3 MPa.

At R_0 beyond the range 2×10^{-5}–10^{-4} cm, the nucleus will grow due to rectified diffusion until pressure reaches the threshold P_c^g.

It should be noted that expressions for the cavitation thresholds reported in [15, 25, 26] are frequency-independent because of their derivation in terms of the steady-state bubble theory. The consideration of cavities in dynamics makes it possible to account for the influence of

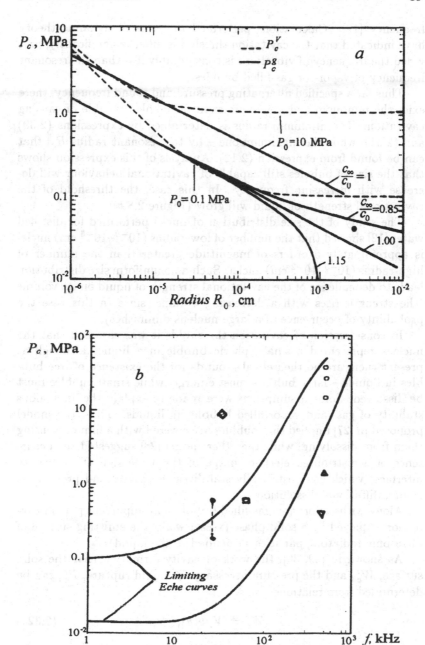

Figure 2.8. Cavitation threshold versus (*a*) nucleus radius, (*b*) frequency. Various symbols refer to experimental cavitation thresholds reported by various authors [18].

frequency [12]. Calculations performed in terms of dynamic theory
have indicated that the cavitation thresholds differ insignificantly, provided the frequency of vibrations is considerably less than the resonant
frequency of vapor- or gas-filled bubbles.

Thus, at a specified alternating pressure and sound frequency, there
exist the minimum and maximum radii of bubbles capable of causing
cavitation. The minimum radius is determined by expressions (2.30)
and (2.31), while the maximum one – by the resonant radius R_r, that
can be found from expression (2.12). Analysis of this expression shows
that the size of bubbles still capable of cavitational behaviour will decrease with increasing frequency. In this case, the threshold of the
cavitational strength of liquid will grow (Figure 2.8,*b*).

The study of the size distribution of nuclei performed for distilled
water [12] showed that the number of low-radius $(10^{-6}-10^{-5}$ cm) nuclei
is approximately 5 orders of magnitude greater than the number of
high-radius $(10^{-3}-10^{-2}$ cm) nuclei. Such a nonuniform size distribution
leads to dependence of the cavitational strength of liquid on its volume
(the strength rises with a decreasing volume, since in this case the
probability of occurrence of a large nucleus diminishes).

In consideration of cavitation thresholds it was assumed that the
nucleus represented a small spheric bubble in a liquid, although at
present there are no theoretical grounds for the existence of free bubbles in liquids – large bubbles must emerge, while small bubbles must
be dissolved. Some assumptions were made to explain the anomalous
stability of gas- and vapor-filled bubbles in liquids. Thus, the model
proposed in [27] implied that bubbles are covered with a film preventing
them from dissolving, while the other model [28] suggested the occurrence of a distributed electric charge of the same sign at the cavity
interface, which electrostatically stabilizes the cavity and prevents it
from a diffusional dissolution.

Along with vapor- and gas-filled bubbles, an important part in cavitation is played by a solid phase (vessel walls, the emitting surface of
ultrasonic radiators, particles suspended in the liquid).

As shown in [29, 30], the work of cavity formation near the solid
surface, W_c^s, and the pressure necessary for liquid rupture, P_c^s, can be
determined from relations

$$W_c^s = W_c \Phi(\theta), \qquad (2.32a)$$

$$P_c^s = P_c \left[\Phi(\theta)\right]^{1/2}, \qquad (2.32b)$$

where $W_c = 16\pi\sigma_L^2/3P_c^2$ is the work of cavity formation in a homogeneous liquid; $P_c = -[16\pi\sigma_L^2/(3kT \ln NkT/h)]^{1/2}$ is the pressure

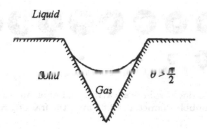

Figure 2.9.   Model of a gaseous nucleus on a solid unwettable particle.

necessary for liquid rupture; k is the Boltzmann constant; N is the Avogadro number; h is Plank's constant; $\Phi(\theta)$ is the trigonometric function of the wetting angle θ whose values for a plane solid surface, spherical particle, spherical particle with a conical or spherical hollow ($\Phi_p(\theta)$, $\Phi_s(\theta)$, $\Phi_{cs}(\theta)$, and $\Phi_{ss}(\theta)$, respectively) are different [30–33].

E. Harvey suggested a model that combined the ideas of consideration of solid and gaseous inclusions as cavitation nuclei [34]. According to this model, a solid nonwettable particle with a crack filled with an insoluble gas can serve as a cavitation nucleus. Due to nonwettability of the crack wall, the liquid in the crack has a convex meniscus relative to the gaseous phase (Figure 2.9), thus tending to reduce gas pressure in the crack because of surface tension.

This model could well be applied for the walls of the vessel in which the liquid is placed, provided that these walls are nonwettable and possess microscopic cracks.

This idea was experimentally supported by E. Harvey *et al.* [34] and later by R. Knapp [17]. One has to conclude that the cavitational strength of a liquid is determined by its physicochemical properties and primarily by "weak spots" it contains.

A correct experimental determination of the cavitation threshold presents a problem, which explains a significant disagreement between the relevant experimental values obtained by various authors (Figure 2.8). It should be noted that in some cases the measuring systems employed may modify the medium properties and thus influence the size of nuclei and their distribution in the liquid bulk.

At present, the most commonly used methods for determining cavitation thresholds are visual, acoustic, and erosive.

The first method consists in the visual observation of cavities.

In the acoustic method, the investigator determines the instant of appearance of a specific cavitation noise that originates from shock

Figure 2.10.   Laser-induced spherical cavitation bubble in water far from boundaries. The maximum bubble diameter is 2.65 mm, the framing rate is 75 000 frames per second [21].

waves generated by collapsing cavities. This instant usually corresponds to the appearance of the first subharmonic in the acoustic spectrum.

In the erosion test, the onset of cavitation is referred to the beginning of cavitational erosion of special samples placed in ultrasonically treated liquid.

The cavities formed in a liquid pulsate under the action of ultrasonic field (the pulsation of a single bubble is illustrated in Figure 2.10 [21]). After reaching the maximum diameter of 2.65 mm, the bubble undergoes a strong collapse and then oscillates with a deviation from a spherical shape. The mathematical description of cavitation could be conveniently done if consider cavities as spherical bubbles filled with a gas-vapor mixture and naving dimensions much smaller than the ultrasonic wavelength. Such cavities could be described by hydrodynamic equations.

The cavity in an incompressible liquid with a constant density and infinitely large sound velocity can be described by equation [35]:

$$R\frac{d^2 R}{dt^2} + \frac{3}{2}\left(\frac{dR}{dt}\right)^2 + \frac{1}{\rho_L}\left[P_\infty - P(R)\right] = 0, \qquad (2.33)$$

here R is the cavity radius; P_∞ is the pressure infinitely far from the cavity; $P(R)$ is the pressure on the cavity surface.

In the simplest case, when $P_\infty = P_0 = \text{const}$ and $P(R) = 0$ (empty bubble), the solution to equation (2.33) has the form

$$U^2 = \frac{dR}{dt} = \frac{2}{3}\frac{P_0}{\rho_L}\left(\frac{R_{\max}^3}{R^3} - 1\right), \qquad (2.34)$$

where U is the velocity of the bubble boundary motion; $R_{\max}$ is the maximum cavity radius at the collapse onset.

Based on expression (2.34), one can obtain formula for the duration of collapse in a pressure field:

$$\tau_c = 0.915 R_{max} \sqrt{\frac{\rho_L}{P_0}}. \qquad (2.35)$$

In a more general case of a gas-vapor-filled cavity in an acoustic field, the pressures at the bubble boundary and at infinity are functions of time. The pressure inside of such a bubble will be described by formula

$$P(R) = \left(P_0 + \frac{2\sigma_L}{R_0} \right) \left(\frac{R_0}{R} \right)^{3\gamma} + P_v - \frac{2\sigma_L}{R}, \qquad (2.36)$$

where $2\sigma_L/R$ is the pressure determined by surface tension; $\gamma = 1$ and $4/3$ for isothermal and adiabatic pulsations, respectively; P_v is the vapor pressure which is constant at a given temperature.

When the cavity is subjected to an ultrasonic field with the pressure amplitude p_m and frequency ω, pressure at infinity is given by

$$P_\infty = P_0 - p_m \sin \omega t. \qquad (2.37)$$

By substituting expressions (2.36) and (2.37) into equation (2.33), one can obtain the so-called Noltingk–Neppiras equation [36]

$$R\frac{d^2 R}{dt^2} + \frac{3}{2}\left(\frac{dR}{dt}\right)^2 + \frac{1}{\rho_L}\left[P_0 - P_v - p_m \sin \omega t + \frac{2\sigma_L}{R} \right.$$
$$\left. - \left(P_0 + \frac{2\sigma_L}{R_0} \right) \left(\frac{R_0}{R} \right)^{3\gamma} \right] = 0. \qquad (2.38)$$

This equation is unsolvable in the general form, but can be solved numerically.

The Noltingk–Neppiras equation describes quite well the pulsation of cavities in an acoustic field (stable cavitation). However, at the final stage of the cavity collapse, when the velocity of the cavity boundary becomes comparable with the sound velocity in liquid, the assumption as to its incompressibility seems to be invalid. Therefore, this equation does not allow an adequate determination of collapse velocities and related peak pressures in cavities.

Besides, the Noltingk–Neppiras equation does not account for energy loss and consequently can describe only undamped motion of a cavity.

The allowance for liquid compressibility and viscosity (Herring–Flinn approximation) enables a satisfactory description of stable cavities whose radii increase not more than severalfold as compared with the initial ones [18]

$$R\left(1 - 2\frac{U}{c_0}\right)\frac{d^2R}{dt^2} + \frac{3}{2}\left(1 - \frac{4}{3}\frac{U}{c_0}\right)\left(\frac{dR}{dt}\right)^2 + \frac{1}{\rho_L}\left[P_0 - P_v\right.$$

$$- p_m \sin\omega t + \frac{2\sigma_L}{R} + \frac{4\eta U}{R} - \left(P_0 + \frac{2\sigma_L}{R_0}\right)\left(\frac{R_0}{R}\right)^{3\gamma}\right]$$

$$+ \frac{R}{\rho_L}\frac{U}{c_0}\left(1 - \frac{U}{c_0}\right)\frac{dP(R)}{dR} = 0. \tag{2.39}$$

The Herring–Flinn equation (2.39) more adequately describes the collapse velocity and minimum radius of the cavity than the Noltingk–Neppiras equation does; however, at $U/c_0 > 1$, it describes the collapsing cavity only qualitatively.

The motion of a cavity with an arbitrary velocity is described by the Kirkwood–Bethe–Gilmore equation [37]:

$$R\left(1 - \frac{U}{c}\right)\frac{d^2R}{dt^2} + \frac{3}{2}\left(1 - \frac{U}{3c}\right)\left(\frac{dR}{dt}\right)^2$$

$$- \left(1 + \frac{U}{c}\right)H - \frac{U}{c}\left(1 - \frac{U}{c}\right)R\frac{dH}{dR} = 0, \tag{2.40}$$

where

$$H = \int\limits_{P_\infty}^{P(R)} \frac{dP}{\rho_L}$$

is the free enthalpy at the surface of a spherical cavity.

By substituting expressions for $P(R)$ and P_∞ from (2.36) and (2.37) into the expression for free enthalpy and deriving ρ_L from the equation of state in the form (1.50), one can obtain

$$H = \frac{n}{n-1}\frac{A^{1/n}}{\rho_L}\left\{\left[\left(P_0 + \frac{2\sigma_L}{R}\right)\left(\frac{R_0}{R}\right)^{3\gamma} - \frac{2\sigma_L}{R} + B\right]^{(n-1)/n}\right.$$

$$\left. - [P_0 - p_m\sin\omega t + B]^{(n-1)/n}\right\}. \tag{2.41}$$

The sound velocity c is related to H as

$$c = \left[c_0^2 + (n-1)H\right]^{1/2}. \tag{2.42}$$

The Kirkwood–Bethe–Gilmore equation should be solved as a set of equations (2.40)–(2.42).

The description of the cavity collapse by the Noltingk–Neppiras (2.38), Herring–Flinn (2.39), and Kirkwood–Bethe–Gilmore (2.40–2.42) equations can be understood from Figure 2.11,*a* which plots analytically derived velocities of the Rayleigh cavity collapse in terms of the three above approximations and numerical solutions to equations (2.40–2.42). It is seen that the cavity collapse is most adequately described by the Kirkwood–Bethe–Gilmore approximation [18].

At the same time, the experimentally observed pulsation of a cavity agrees well with the results of calculations performed in terms of the Herring–Flinn approximation (2.39) for a bubble with the initial radius $R_0 = 10^{-3}$ cm, acoustic pressure $p_m = 0.12$ MPa, and frequency $\omega = 28$ kHz (Figure 2.11,*b*) [12].

It should be noted that the lack of general analytical solutions to the Noltingk–Neppiras, Herring–Flinn, and Kirkwood–Bethe–Gilmore equations does not allow a complete description of the cavity motion as a function of ultrasonic field parameters (in particular, its frequency and vibrational pressure amplitude) and those of the liquid (cavity size, viscosity, surface tension, etc.). Nevertheless, some important conclusions as to the influence of frequency, vibrational pressure amplitude, and the size of cavities on their motion can be drawn based on analysis of numerical solutions.

Thus, analysis of the cavity behavior as a function of acoustic pressure allowed Apfel [38] to reveal the regions of development of stable and transient cavitation (Figure 2.12), in which cavitation depends on frequency and peak sound pressure in different fashions.

In the region of stable cavitation, the bubble motion at different frequencies can be illustrated by the graphs in Figure 2.13 derived from the Gilmore equations [39]. When the vibrational frequency approaches the resonant frequency of a cavity, the pulsation amplitude $(R_{\max} - R_0)/R_0$ rises (Figure 2.14) [40].

When the acoustic pressure amplitude is below a certain value, the character of bubble pulsation changes insignificantly (Figure 2.15); but above this value there is a structural instability in the solution of equations describing the bubble motion [39, 40].

The transition to a transient cavitation depends not only on the acoustic pressure but also on frequency and the initial radii of cavities.

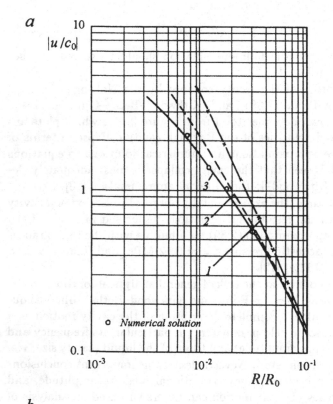

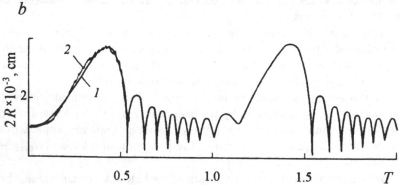

Figure 2.11. (a) The rate of bubble collapse approximated according to (1) Noltingk–Neppiras, (2) Herring–Flinn, (3) Kirkwood–Bethe–Gilmore and (b) behaviour of the bubble estimated in terms of (1) the Herring–Flinn model [18] and (2) from the experimental results [12].

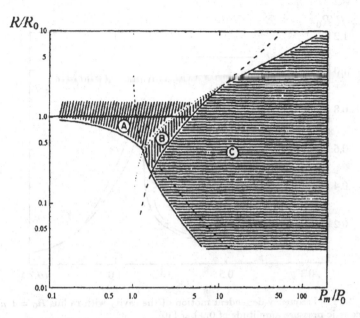

Figure 2.12. Regions corresponding to stable and transient cavitation upon the motion of a bubble with $R_r = 3.25$ μm and $f = 1$ MHz [38]. A is the region of stable cavitation (the growth of bubbles is defined by rectified diffusion). B is the intermediate region. C is the region of transient cavitation.

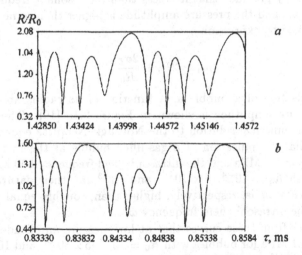

Figure 2.13. Radius–time curves of bubble oscillations for $R_0 = 10$ μm and $p_m = 80$ kPa: (a) $f = 70$ kHz; (b) $f = 120$ kHz. The solid dots correspond to times $t_k = k/f$ ($k = 1, 2, 3, \ldots$) when the period of the sinusoidal driving has elapsed [39].

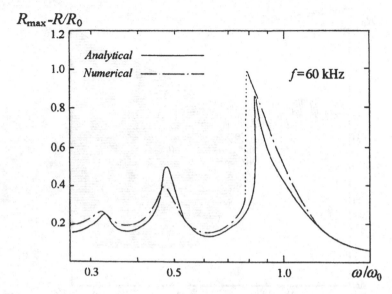

Figure 2.14. Frequency-dependent motion of the cavity with radius $R_0 = 1$ μm at an acoustic pressure amplitude of 0.9 bar [40].

It was found that the pulsation of a cavity depends little on ultrasonic frequency provided that it is less than the resonant frequency of the cavity, ω_r, and the pressure amplitude is higher than some critical value specified by expression

$$p_m \geq 2P_0 + \frac{2\sigma_L}{R_0}. \tag{2.43}$$

The pulsation of a bubble as a function of its radius and vibrational pressure amplitude is shown in Figure 2.16 [19]. The curves represent a numerical solution to the Noltingk–Neppiras equation describing adiabatic pulsation of a gas-filled bubble at the hydrostatic pressure $P_0 = 0.1$ MPa and the ultrasonic field frequency 500 kHz. For bubbles with $R_0 = 10^{-4}$, 5×10^{-4}, and 10^{-3} cm, their natural resonant frequency ω_r is, respectively, higher than, roughly equal to, and less than the ultrasonic field frequency ω.

It follows from these curves that at low pressure amplitudes (0.02, 0.05, and 0.1 MPa for bubbles with $R_0 = 10^{-4}$, 5×10^{-4}, and 10^{-3} cm, respectively), when $p_m < P_c$, the bubbles pulsate nonlinearly without collapse. Bubbles with the radius less than or equal to its resonant value (10^{-4} and 5×10^{-4} cm) pulsate with a period close to that of the ultrasonic wave (T). Bubbles with the radius greater than its resonant

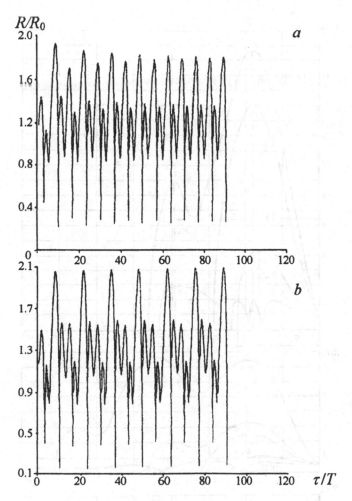

Figure 2.15. Motion of cavity with the radius $R_0 = 50$ μm at a frequency of 60.5 kHz and pressure: (a) 0.65; (b) 1.15 bar [40].

value pulsate with a period longer than T and nearly equal to the period of natural vibrations of bubbles.

For $p_m > P_c$, bubbles with $\omega_r > \omega$ grow throughout the negative pressure half-period and during part of the positive half-period. Bubble grows up to the maximum radius R_m over time τ_l and then collapses over time τ_c (designations are shown in the Figure). Both R_m and the collapse time τ_c rise with p_m, the onset of collapse being possible in the phase of negative pressures. Above a certain value of

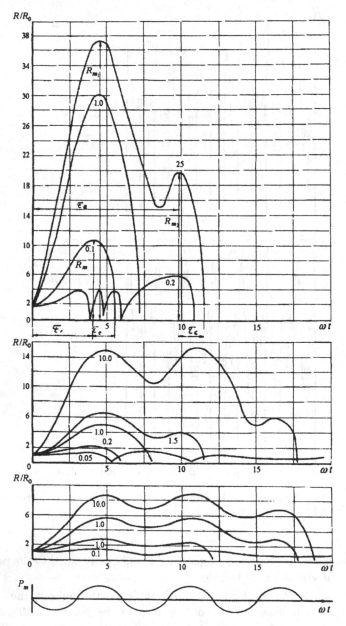

Figure 2.16.   Pulsation of cavitation bubbles with radius R_0 equal to (a) 10^{-4}, (b) 5×10^{-4}, and (c) 10^{-3} cm at different acoustic pressures p_m indicated by numerals near the curves (in MPa). Changes in p_m in the course of time are shown by the diagram d.

p_m (2.5 and 1.5 MPa for bubbles with $R_0 = 10^{-4}$ and 5×10^{-4} cm, respectively), a bubble, after having been enlarged to some extremum radius R_{m1}, begins to collapse, but then resume its growth in the rarefaction phase. This growth continues until the bubble radius reaches the second extremum value R_{m2}, after which the bubble will collapse completely. With increasing p_m, the bubble can attain three, four, and more extremum dimensions. In this case, the time of its growth is given by

$$\tau_c = 0.75T + (i-1)T.$$

When bubbles are large so that $\omega_r < \omega$, they can possess several extremum dimensions even at small pressure amplitudes (exceeding, however, P_c). In this case, one can speak about structural (or phase) instability of the solution to equation (2.38).

Such a structural instability arising during the description of the motion of a cavity was studied in detail by Lauterborn [39, 41]. The instability of solutions is determined by both the amplitude of acoustic pressure and the frequency of vibrations (Figure 2.17) [39].

The collapse time can be quite correctly estimated by the formula of Akulichev [19] that differs from formula (2.35) in having the sum $P_0 + p_m$ instead of P_0. Analogously, the collapse velocity is described by the formula similar to (2.34), but with $P_0 + p_m$ instead of P_0. The account for liquid compressibility in the Herring–Flinn and Kirkwood–Bethe–Gilmore approximations somewhat sophisticates the expressions for collapse velocity, but still does not allow one to determine the minimum radius $R_{\min}$ of collapse cessation as a result of compression of vapor and gas inside the cavity. When the cavity radius is close to $R_{\min}$, the values for collapse velocity determined by the above-described formulas will significantly differ from those obtained by solving the respective differential equations.

The time of collapse, estimated in terms of the Gilmore model [42], showed a dependence on the initial radius of the cavity and acoustic pressure (Figure 2.18).

As mentioned above, pressure in a collapsing cavity can reach large values. As this takes place, the cavity begins to radiate finite-amplitude spherical waves which, propagating through a liquid, transform into shock waves capable of producing various cavitation-related effects.

A. Vogel and W. Lauterborn [43] were able to experimentally determine pressures arising from the laser-induced cavity collapse (Figure 2.19). They observed a rapid (∼40 ns) increase in pressure followed by its slow (∼100 ns) exponential fall. The peak pressures in such shock

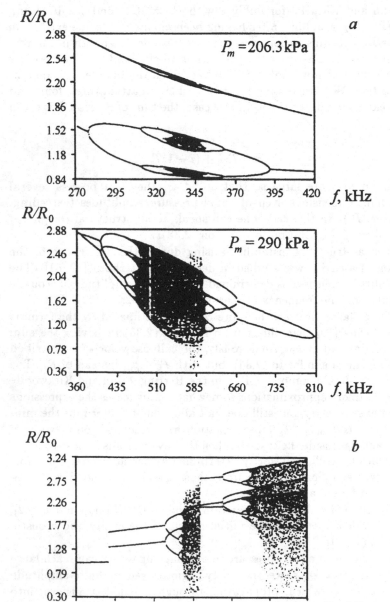

Figure 2.17. Bifurcation diagrams showing the motion of a cavity with $R_0 = 10 \ \mu$m as a function of (a) frequency at different acoustic pressure amplitudes (206.3 and 290 kPa) and (b) acoustic pressure amplitude at a frequency of 600 kHz [39].

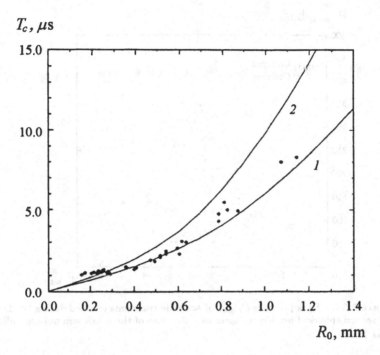

Figure 2.18. Collapse time T_c as a function of the initial radius R_0 of an air bubble calculated in terms of the Gilmore model. Dots are experimental data [42]. (*1*) $P_m = 65$ Mpa; (*2*) $P_m = 102$ MPa.

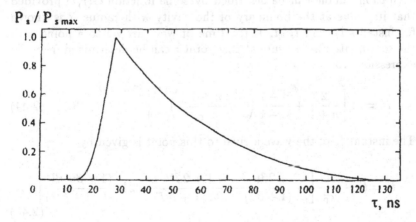

Figure 2.19. Pressure profile of the acoustic transient resulting from cavitation bubble collapse [43].

P $_{s\,max}$, bar

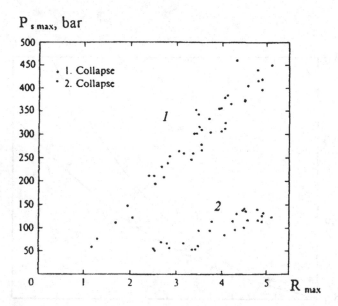

Figure 2.20. Peak pressure P_s $_{max}$ of acoustic transients emitted during the first and second spherical bubble collapses as a function of the maximum bubble radius R_{max} [43].

waves depend on the maximum radius of the collapsing bubble (Figure 2.20).

The propagation of finite-amplitude spherical waves emitted by a cavitation bubble can be described by some function $G(r,t)$ provided that its value at the boundary of the cavity with radius R is known for instants t_R and t_s (t_s is the time of wave arrival to a point with the coordinate r). Pressure at this point r can be determined from the expression [19]

$$P_s = A\left[\frac{2}{n+1} + \frac{n-1}{n+1}\left(1 + \frac{n+1}{rc_0}G\right)^{1/2}\right]^{2n/(n-1)} - B. \quad (2.44)$$

The instant t_s of the wave arrival to this point is given by

$$t_s = t_R + \frac{2G}{c_0^3}\left[\frac{1+2\beta u}{\beta u(1+\beta u)} - \frac{1+2\beta U}{\beta U(1+\beta U)} - 2\ln\frac{(1+\beta u)\beta U}{\beta u(1+\beta U)}\right], \quad (2.45)$$

where

$$G(R,t_R) = G(r,t) = R\left(H + \frac{U^2}{2}\right),$$

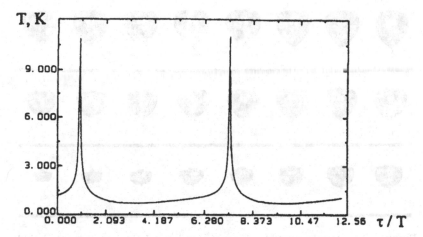

Figure 2.21. Temperature curve calculated for a collapsing cavitation bubble with the initial radius 50 μm at acoustic pressure 1.15 bar and frequency 60.5 kHz [40].

$$\left. \begin{array}{l} \beta u = \dfrac{1}{2}\left[\left(1 + \dfrac{n+1}{rc_0^2}G\right)^{1/2} - 1\right] \\[3mm] \beta U = \dfrac{1}{2}\left[\left(1 + \dfrac{n+1}{Rc_0^2}G\right)^{1/2} - 1\right] \end{array} \right\},$$

H and U can be found from the solution to the Kirkwood–Bethe–Gilmore equations (2.40–2.42) for a pulsating cavity.

Analysis of expression (2.44) performed by Akulichev [19] indicates that, at high r, the shape of a propagating wave is spoiled: if near the cavity the function $P_s(t)$ represents a sharp pulse, then some distance apart there is a discontinuity in the function corresponding to the shock wave front. The amplitude of shock waves is large only near the collapsing cavity, being decreased with distance by the law $1/r$. This explains why cavitation-related processes (for instance, erosion) are most pronounced near collapsing cavities.

Experiments and theory indicate that instantaneous values of pressure in a shock wave can be as high as tens of MPa. That is why shock waves may cause microdestruction of solid surfaces and impart a considerable acceleration to particles suspended in the liquid.

In addition to the formation of shock waves, the last stage of bubble collapse is characterized by a drastic increase in temperature. The calculations yielded the value of about 10 000 K (Figure 2.21) for water [40], and 5000 K for silicone oil [44].

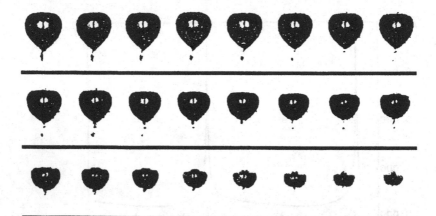

Figure 2.22. Development of a ring vortex and cumulative jets during the bubble collapse near the solid boundary. The frame size 4.5 mm × 3 mm, framing rate 300 000 s^{-1}, $\gamma = S/R_{\max} > 2$ (S is the distance from the solid boundary visible as a dark stripe at the lower edge of each frame) [45].

If a cavitation bubble occurs near a solid wall, the last stage of its collapse involves the distortion of its spherical form, development of its instability, and the formation of cumulative jets (Figure 2.22) with the velocity governed by the initial radius of the bubble (Figure 2.23) [42], while pressure depends on the distance from the wall S (Figure 2.24) [45].

A pulsating bubble would finally collapse, giving rise to small bubbles. The process of cavity "reproduction" is schematically represented in Figure 2.25. It is the avalanche-like reproduction of cavities by the chain mechanism [46] that can be considered as a developed cavitation with a specific cavitation zone (Figure 2.26,a) in the form of strands (Figure 2.26,b) moving in an ultrasonic field.

The formation of cavitation zone is accompanied with the appearance of a cavitation noise, whose intensity rises with that of ultrasound (Figure 2.27).

The specific feature of liquid in an ultrasonic field is its luminescence (the so-called sonoluminescence) [48] that appears when the intensity of ultrasound surpasses a certain level (Figure 2.28). The spectral characteristics of sonoluminescence depend on the properties of the liquid and the gas dissolved in it, among which of particular importance are the heat conductivity of gas and its polytropic constant ($\gamma = C_p/C_v$). The intensity of sonoluminescence is a function of the acoustic pressure amplitude (Figure 2.29).

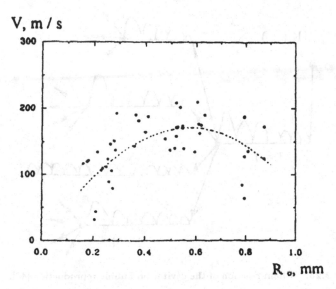

Figure 2.23. Average jet velocity versus the initial radius of the bubble 5 μs after its collapse [42].

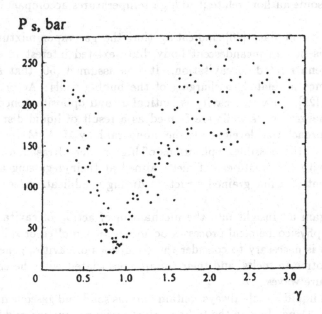

Figure 2.24. Pressure amplitude P_s after the first collapse as a function of $\gamma = S/R_{max}$ ($R_{max} = 3.5 \pm 0.6$ mm) [45].

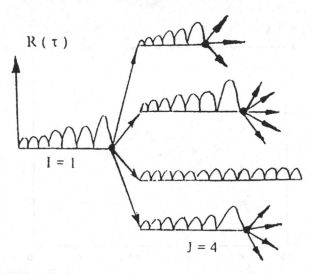

Figure 2.25.   Chain reaction of the cavitation bubble reproduction [46].

The origin of sonoluminescence is yet imperfectly understood, although some authors relate it to high temperatures accompanying collapse.

The recent data [46], indicating that the gas-vapor mixture in a bubble is not an incandescent body, have excited interest in electric phenomena related to cavitation. It was assumed [50] that sonoluminescence is related to charging of the bubble walls. According to Frenkel [29], a nascent cavity is lenticular, and opposite noncompensated charges on its walls are formed as a result of liquid disruption. This approach was developed in the monograph by M. A. Margulis [46].

One of the possible applications of high-power ultrasound is associated with the treatment of melts aimed at their degassing and development of a fine-grained structure during solidification (see Chapter 10).

To gain an insight into the mechanism of action of cavitation on various physicochemical processes occurring in a molten or solidifying metal it is necessary to consider the conditions of cavities generation, their motion in melts, and their collapse associated with the emission of pressure waves.

Real liquid metals always contain various solid and gaseous microinclusions. Depending on the technology of metal production and the relative content of such inclusions, the local disruption of a liquid metal (with the formation of cavities) under the action of tensile stresses can

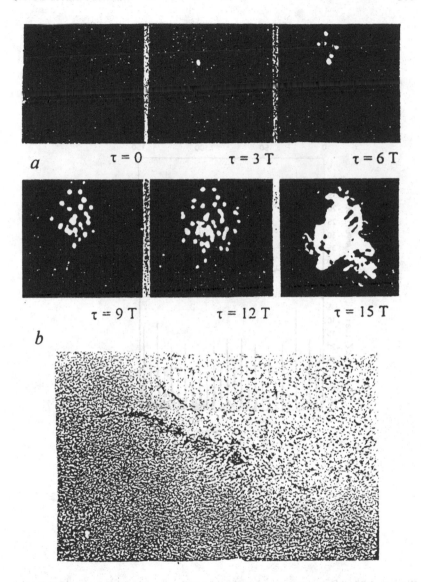

Figure 2.26. (a) Development of a cavitation zone in water applied with 20-kHz vibrations at the pressure amplitude 0.5 MPa [20] and (b) formation of cavitation strands ($f = 23.1$ kHz) [41].

occur at the melt interface either with a gaseous phase (bubbles) or with a solid phase (solid inclusions and vessel walls). Generally, the solid defects differing in nature and dimensions, surface has a diversity of

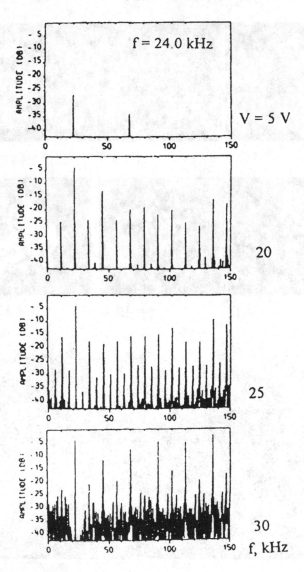

Figure 2.27. Power spectra of cavitation noise at various ultrasound intensities dependent on the voltage V applied to the piezoelement [47].

which depends on material smoothness and method of its processing. Thus, the typical surface defect of electropolished molibdenum alloy TsM-2A is a spherical cavity with the mouth radius $r_d \sim 2 \times 10^{-5}$ cm [51].

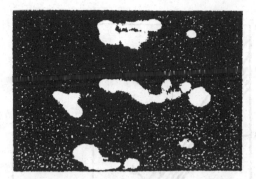

Figure 2.28. Sonoluminescence in water excited with 20-kHz ultrasonic vibrations with the pressure amplitude 0.5 MPa [40].

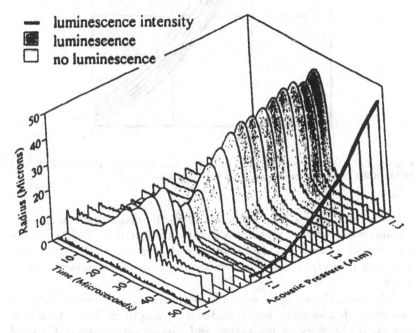

Figure 2.29. The relationship between bubble pulsation, acoustic pressure amplitude, and sonoluminescence intensity in the water containing argon [49].

As in water, the presence of bubbles may considerably affect the cavitation threshold in melts. In [52], the cavitational strength of melts of some low-melting-point metals containing bubbles with $R_0 = 10^{-6}$–10^{-3} cm was estimated by formula (2.30). The estimated strengths are

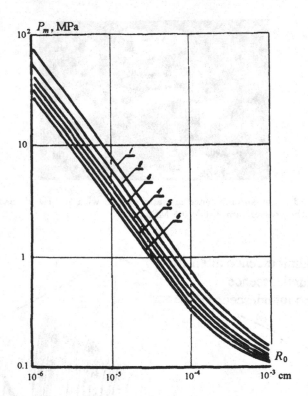

Figure 2.30.   Cavitation threshold in Zn (*1*), Cd (*2*), Sn (*3*), Pb (*4*), Bi (*5*), and In (*6*) melts as a function of the initial bubble radius.

presented in Figure 2.30 as a function of R_0. It is seen that cavitation is possible when acoustic pressure is about some fractions of MPa and the melt contains vapor-gas-filled bubbles with a radius of 10^{-4}–10^{-3} cm. Such conditions take place when elastic vibrations are supplied to metal melts.

The part played by a solid phase (particles and vessel walls) in cavitation development was considered at length in [53]. Astashkin estimated the work of nucleation and disruption pressures for solid particles and plane surfaces with allowance for the defects (spherical and conical depressions) that contribute to nucleation.

Estimations of the time the various particles (Al_2O_3 with $\rho_s = 3.97$ g/cm^3 and M_0 with $\rho_s = 10.2$ g/cm^3) are suspended in low-temperature melts with $\rho_L = 6 \div 8$ g/cm^3, performed with allowance for Stokes' forces, gravity, and forces responsible for Brownian motion, have shown that this time can be quite long (several seconds), if the

Table 2.1 Threshold pressure of cavitation in a tin melt containing solid phase [51]*.

Contact angle of wetting, θ, grad,	Threshold pressure, MPa			
	Plane surface, P_{cp}^s	Spherical surface, P_{cs}^s	Plane surface containing a conical depression of radius r_s	
			r_s, cm	P_{ccp}^s
60	6.03×10^3	6.52×10^3	2×10^{-5}	17.0
130	1.56×10^3	3.20×10^3	1×10^{-3}	4.7×10^{-2}
174	1.09×10^2	3.97×10^2	1×10^{-3}	1×10^{-3}

* In a pure tin melt, $P_c \cong 5.37 \times 10^3$ MPa.

particle radius R_s is not above 5×10^{-5}–10^{-4} cm. That is why the influence of a solid phase on the cavitational strength of melts is considered below for particles whose radii are not in excess of the stated values. The work of nucleation and the disruption pressure near particles and various defects were calculated for the solid phase either wetted ($\theta = 60°$) or unwetted ($\theta = 150°$) with a melt. Both cases are possible in practice. Thus, in the systems consisting of oxides and transition metals (Ni, Co, Fe), the contact angle θ is within 120–150°; in metal-carbide systems it is within 0–60° [54].

Calculations were performed for metals possessing various surface tensions (tin, aluminum, and iron) and water as a reference.

It was found that in the presence of a wetted solid phase (plane surface, spherical particles), the work of nucleation and disruption pressure are higher than those in respective homogeneous melts, and that in the case of spherical particles these parameters increase as the radius of particles, R_s, approaching some critical value R_t (Table 2.1). Conversely, in the presence of an unwetted solid phase, the work of nucleation and disruption pressure are lower than those in respective homogeneous melts and drop with the increasing R_s/R_t ratio.

Interestingly, the calculated disruption pressure of melts containing an unwetted perfect solid phase is quite high. Therefore, considerably smaller experimental disruption pressures cannot be explained by the occurrence of perfect particles in melts.

Activation of a depression is usually suggested to take place when a new nucleus can spontaneously grow on the surface around the depression [55].

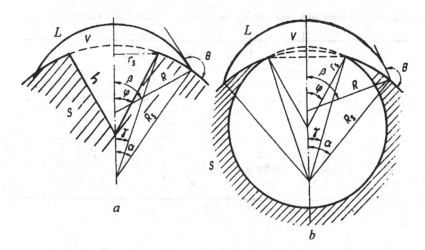

Figure 2.31. Illustrations to formulas (2.46) describing the formation of a cavitation nucleus in (*a*) conical or (*b*) spherical hollows.

The pressure necessary for activation of a wetted conical depression containing a gas-vapor-filled bubble can be estimated using expression [33]

$$P_a \cong 0.4 \left(\frac{2\sigma_L \cos \theta}{h \sin \varphi} \right)^{3/2} P_{vgo}^{-1/2}. \qquad (2.46a)$$

For unwetted depressions, this expression has the form

$$P_a \cong \frac{2\sigma_L \sin \delta}{r_s}, \qquad (2.46b)$$

where P_{vgo} is the initial pressure of the gas-vapor mixture inside the bubble, and

$$\delta = \pi - \theta.$$

Other geometric designations are given in Figure 2.31.

Estimations indicate that, at small r_s or $h \sin \varphi$, activation pressure is very high. However, it falls with increasing θ and r_s to be as low as 10^{-2} MPa at $r_s = 10^{-3}$ cm.

These estimates, valid for static tensile pressures, give grounds to consider that the presence of microdefects with characteristic dimension 10^{-3} cm on the radiator surface is equivalent to contamination of a melt. Kikuchi [56] has experimentally confirmed that cavitation threshold in an ultrasonic field strongly depends on the degree of finishing of the emitting surface of the radiator.

Table 2.2 Effect of a rectified diffusion ($t = 3T$) on the hydrogen content of a bubble in aluminum melt [57].

Amplitude of sound pressure p_m, MPa	Hydrogen content of a cavity $\times 10^{-12}$, kg		
	$R_0 = 10^{-4}$ cm	$R_0 = 10^{-3}$ cm	$R_0 = 10^{-2}$ cm
0.2	4×10^{-3}	4×10^{-4}	4×10^{-7}
1.0	5×10^{-3}	5×10^{-3}	7×10^{-7}
5.0	0.3	0.3	0.2
10.0	2.0	1.0	0.65

Such a 'contamination' can, to a certain extent, mask the contribution from real cavitation nuclei in the liquid bulk (in particular, bubbles) to a reduction in the cavitational strength of liquids.

It should be noted that the disruption pressure for water is smaller than that for any metal melts.

In considering cavitation in molten and solidifying metals, it seems expedient to discuss the motion of cavities and to estimate pressures resulting from their collapse.

The dynamics of cavities in an aluminum melt was treated in [57] by invoking the Rayleigh equation (2.35) describing the motion of a bubble in a incompressible constant-pressure liquid and the equation relating hydrogen diffusion into the bubble with its dimensions and internal pressure. Calculations were performed for cavities with the initial radius $R_0 = 1 - 100$ μm and sound pressures $p_m = 0.12 - 10$ MPa (the ultrasonic frequency 18 kHz).

In the absence of cavitation ($p_m = 0.2$ MPa), bubbles of all sizes perform nonlinear pulsations and do not collapse over the time of observation. In this case, the internal pressure in bubbles changes insignificantly. At $p_m = 1.0$ MPa, bubbles behave as typical cavities, collapsing by the end of the second period of the acoustic wave. At still higher sound pressures ($p_m = 5 - 10$ MPa), the dynamics of bubbles exhibits several extrema and their collapse is completed by the end of the third period. It should be noted that when bubbles expand tens or hundreds times, the internal pressure drops by several orders of magnitude to become smaller than 10^{-2} MPa.

The internal pressure will be significantly higher, if gas diffusion into bubbles is taken into account. Estimates for the amount of hydrogen diffused from the melt into a cavity are listed in Table 2.2. The estimates indicate that cavitation can affect the kinetics of gas liberation from melts.

Because of experimental difficulties in determining shock pressures in melts, their theoretical estimation is of great importance. The most valuable information is that about the final stages of the cavity collapse. For general reasons, one can assume that mathematical description of the bubble moving in a liquid [19] is valid also for metal melts.

As shown above, the rate of the cavity collapse and the number of pressure pulses exhibited by a collapsing cavity are closely related to the values of static pressure and acoustic field parameters, as well as to the properties of the melt. In the final analysis the identity of emitted waves is determined by the cavity collapse rate.

Therefore, it seemed reasonable to estimate the rate of cavity collapse and the value of pressures arising in the melts of readily fusible metals in comparison with water and to elucidate the parts played by the acoustic field and melts in pulsation of cavities [52].

This was done based on a numerical solution to the Kirkwood–Bethe–Gilmore equations (2.40–2.42, 2.44, 2.45) for molten gallium, tin, bismuth, zinc, cadmium, indium, and lead.

The intensities of cavitation were compared with respect to the peak pressures $P_{s,\max}$ in a finite-amplitude wave arising during the collapse of a cavity with $r = R_{\min}$.

The quantities R_0, p_m, η, σ_L, ρ_L, and c_0 were varied, as parameters of equations (2.40)–(2.42) and (2.44)–(2.45), within the ranges real for the chosen low-melting-point metals. The influence of vapor pressure was neglected because of its insignificancy for the chosen metals around their melting points.

In calculations, all the quantities characterizing metals were referred to $t = 1.1T_0$.

For an appropriate choice of the initial radius R_0 of the cavity for which cavitation might be expected, its resonant radius $R_{r,\max}$ was estimated by the Minnaert formula (2.12). Calculations were carried out for a 20-kHz ultrasound. The radius of a resonant cavity was found to lie within the range $(5.4–6.8) \times 10^{-2}$ cm for all metals studied. As shown by V. A. Akulichev [19], stronger cavitational effects are to be expected for cavities with $R_0 < R_{r,\max}$. The experimentally observed threshold cavitation pressures in melts ranged from 0.1 to 0.7 MPa.

Bearing in mind these two circumstances, calculations were performed mainly for cavities with $R_0 < 0.1R_{r,\max}$ and acoustic pressure amplitudes $p_m = 0.1–0.2$ MPa.

It should be noted that the peak pressure thresholds p_{mc} at which bubbles become unstable and begin to emit finite-amplitude waves appear to be of particular interest.

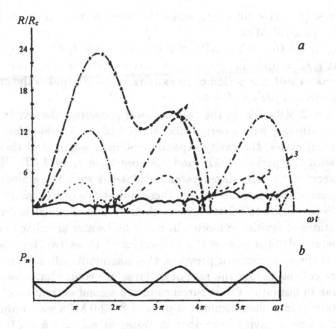

Figure 2.32. (a) Pulsation of a cavitation bubble with $R_0 = 10^{-3}$ cm in bismuth melt at the sound pressure amplitude 0.1 (curve *1*), 0.2 (curve *2*), 0.7 (curve *3*), 2.0 (curve *4*) MPa and (b) time dependence of pressure amplitude p_m.

At $p_m \sim 0.1$ MPa, the cavity in molten bismuth pulsates nonlinearly and the pressure exerted on its boundary is within tenths of MPa (Figure 2.32).

At $p_m \sim 0.2$ MPa, the cavity pulsation becomes unstable, the collapse is more rapid, and the internal pressure $P_{s,\max}$ is three orders of magnitude higher that actually corresponds to the onset of cavitation. The threshold of cavitation thus estimated agrees quite well with experimental data. Analogous results were obtained for other metal melts.

At $p_m = 0.2$ MPa, the cavity behaves in such a manner that after the initial collapse it pulsates nonstationary with a period $\sim \pi/2$ (in ωt units). In this case, the internal pressure may reach a value of several tens of MPa; therefore, these pulsations cannot be neglected during the estimation of the intensity of cavitation.

At p_m above 0.2 MPa, cavities oscillate almost harmonically with a period $\sim 2\pi$.

At the final stage of the cavity collapse, the velocity of its walls increases more than twofold over a period of about 10^{-9} s and reaches the value $\sim (3-4) \times 10^5$ cm/s, while the internal pressure rises from 3×10^3 to 15×10^3 MPa.

At p_m equal to 1.8–2.0 MPa, the collapse time τ_c attains a value such that $\tau_c/T = 0.75$. In this case, the cavity collapse is not complete and alternates with the period of its expansion, which makes the cavity to pulsate with a period of $\sim 4\pi$.

At $p_m = 2$ MPa, like in the case of lower pressures, the cavity pulsates nonstationary with a period close to that of its natural pulsations. After several cycles, the cavity expands, collapses again, and then oscillates with the period $\sim 3T$, and collapse time $\tau_c \sim 1.5T$. Thus, cavity executes virtually degenerated oscillations with the emission of low-pressure waves. In other words, these pulsations of the cavity contribute to cavitation to a lesser extent than do pulsations at lower p_m.

Pulsations of cavities in molten bismuth and water are alike in spite of a substantial difference in the properties of these two liquids, although, at the same acoustic pressure, the maximum radius of the cavity before collapsing and the rate of collapse are somewhat greater in water than in bismuth. The occurrence of the second extremum of cavity oscillations in molten bismuth at $p_m = 1.8-2.0$ MPa was confirmed by calculations of cavity behaviour in water at $p_m = 1.5-2.0$ MPa [19].

The above consideration pertains to cavities with the initial radius $R_0 = 10^{-3}$ cm. Since in real liquids bubbles are of various sizes, it is of interest to estimate how the initial radius may affect cavitation. Analysis indicates that smaller initial radii of a cavity at a given pressure $p_m = 0.3$ MPa not only considerably enhance peak pressure in the emitted waves, but can also greatly modify the character of cavity pulsation. When the initial radius decreases from 3×10^{-3} to 5×10^{-4} cm, peak pressure rises almost threefold. In this case, the collapse time τ_c falls and, after the first collapse, the cavity with $R_0 = 5 \times 10^{-4}$ cm executes two oscillation cycles followed by a second expansion (the period of these oscillations is two times shorter than that for the cavity with $R_0 = 10^{-3}$ cm). After the first collapse, nonlinear oscillations of the cavity with $R_0 = 3 \times 10^{-3}$ cm are symmetrical relative to the instant $\tau = \tau_{R,max}$, the expansion time τ_e and collapse time t_c being equal.

When the initial radius of the bubble increases, R_{max} falls and R_{min} rises, which implies that the amplitude of bubble pulsation diminishes. This effect is related to an enhanced content of gases in the bubble defined by formula [15]

$$\delta = \frac{P_0 + 2\sigma_L/R_0}{P_0} \left(\frac{R_0}{R_{max}} \right)^3. \tag{2.47}$$

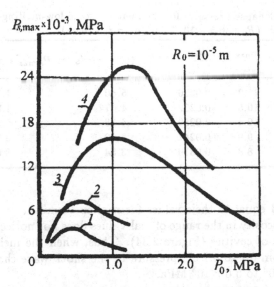

Figure 2.33. Peak pressure during the cavitation bubble collapse vs static pressure P_0 at different values of sound pressure p_m: (*1*) 0.1; (*2*) 0.2; (*3*) 0.3; (*4*) 0.5 MPa.

The gas content δ increases with the initial radius of the bubble and surface tension, which must reduce $P_{s,\max}$ in the shock wave.

Along with acoustic pressure and the initial radius of the cavity, its behavior can be significantly altered by static pressure P_0. Let us quantitatively estimate the effect exerted by static pressure on the pressure in shock wave. The curve $P_{s,\max} = f(P_0)$ at $p_m = 0.3$ MPa has a maximum at pressures close to atmospheric. The maximum is shifted to greater P_0 at higher acoustic pressures (Figure 2.33).

It is of practical interest to compare pulsations of cavities with the same initial radius R_0 at constant p_m and P_0 in melts of various metals. For correctness, calculations were carried out for $p_m = 0.3$ MPa, when cavities would not perform unstable pulsations after the first collapse. The results are summarized in Table 2.3.

It follows from the tabulated data that the extent of pulsations, $R_{\max}/R_{\min}$, collapse rate U, and the pressure $P_{s,\max}$ are of the same order of magnitude for all investigated melts. Therefore, it would be impossible to elucidate from these data the relationship between the properties of melts, the behavior of cavities, and pressure $P_{s,\max}$.

In this connection, calculations were performed with parameters η, σ_L, ρ_L, c_L, n varied in the ranges characteristic of low-melting-point metals. When one of these parameters was varied, the others were taken

Table 2.3   Collapse pressures for the melts of some low-melting-point metals ($R_0 = 6.5 \times 10^{-4}$ cm, $p_m = 0.3$ MPa).

Metal	$\dfrac{R_{max}}{R_0}$	$\dfrac{R_{min}}{R_0}$	$U \times 10^{-5}$, cm/s	$P_{s\ max} \times 10^{-4}$, MPa
Tin	11.9	0.0226	5.80	1.16
Bismuth	10.5	0.0232	4.07	1.07
Cadmium	10.4	0.0250	4.38	0.98
Lead	9.9	0.0247	3.62	0.95
Indium	8.8	0.0310	3.54	0.65

constant and equal to their values for molten bismuth. It was found that melt viscosity in the range of real values does not noticeably affect the behavior of cavities (Figure 2.34). Thus, when the melt viscosity increased from 0.2 to 0.7 cP, pressure in the shock wave changed only from 5.5×10^3 to 5.0×10^3 MPa.

The effect of surface tension on cavity pulsation and peak pressures is quite sophisticated and questionable. According to Flinn [18], surface tension promotes the collapse of bubbles, while Agranat [58] showed that varying of σ_L in the range 0.02–0.1 N/m^2 enhances the erosion activity of cavities in water.

A detailed analysis of surface tension σ_L as a parameter entering into equations (2.40)–(2.42) has shown that its influence varies in the course of collapse, and the overall effect can hardly be predicted unambiguously. The growth of σ_L within the range characteristic of readily fusible metals (0.1–1.0 N/m^2) leads to a decrease in the peak pressure after the first collapse (Figure 2.34, curve *5*). At the same time, the increased instability of the system gives rise to subsidiary pressure pulses close in time to the first collapse, which must obviously enhance the erosive effect of cavitation.

Similar analysis of the peak pressure $P_{s,max}$ and behavior of cavities as functions of density ρ_L, nonlinear parameter n, and sound velocity c_L (in these calculations, the equality $A.n = \rho_L c_L^2$ was fulfilled) indicates that peak pressure in the emitted wave grows with decreasing ρ_L and increasing c_L and n (Figure 2.34, curves *1*, *2*, and *3*). It was established that the phase instability of the system is determined primarily by the liquid inertia. In other words, the instability grows with decreasing liquid density, due to which the number of nonstationary subsidiary pulsations of the collapsing cavity rises. It was also found that the maximum radius of cavity upon its expansion, as well as the rate of its collapse, increase with decreasing density of melts.

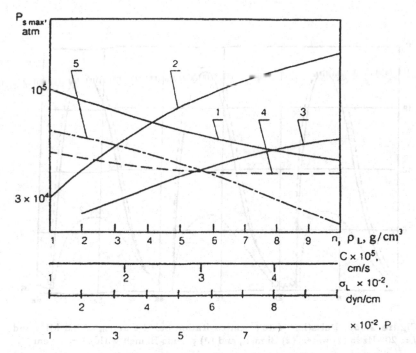

Figure 2.34. Peak pressure during the cavitation bubble collapse versus the melt properties: (1) density ρ_L; (2) sound velocity c; (3) nonlinear parameter n; (4) viscosity η; (5) surface tension σ_L.

A determining role of density in pulsations also follows from the comparison of functions $R/R_0 = f(t)$ for water (Figure 2.35, curve 1), bismuth (curve 2), and a pseudo-metal with density $\rho_L = 1$ g/cm^3 (curve 3). It is seen that curves for water and pseudo-metal are virtually coincident.

The collapsing cavity emits finite-amplitude spherical waves that transform into shock waves during the propagation through the liquid. Figure 2.36,a, illustrating the generation of a shock wave in bismuth melt, represents calculated profiles of this wave at various distances r from the cavity boundary as functions of the delay time $t_s = (t - t_R)$. It is seen that pressure is a multiple-valued function of the delay time, which is because of neglecting energy dissipation during the shock wave propagation.

Peak pressure in the shock wave propagating in a molten metal is inversely proportional to the distance covered, which agrees with the data from [19]. Figure 2.36,b shows the dependence of $P_{s,\max}$ on the

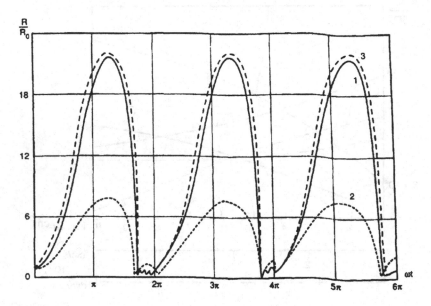

Figure 2.35.   Pulsation of 10^{-3} cm cavitation bubbles at $p_m = 0.3$ MPa and $f = 20$ kHz in (*1*) water, (*2*) Bi melt, and (*3*) pseudo-Bi melt with $\rho_L = 1\ \mathrm{g\,cm^{-3}}$.

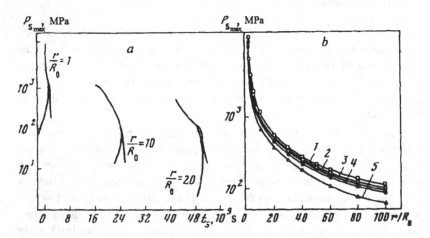

Figure 2.36.   (*a*) Calculated shock wave profile in bismuth and (*b*) changes in the shock wave pressure $P_{s\ max}$ as a function of the distance from a collapse bubble in metal melts: (*1*) Sn; (*2*) Bi; (*3*) Cd; (*4*) Pb; (*5*) In.

distance covered by the shock wave in liquid metals. It is seen that already at $r \sim 5 \times 10^{-4}$ cm from the collapsing cavity the pressure $P_{s,\max}$ decreases by a factor of several hundreds and is about 100 MPa. In other words, shock waves are very local. It should be also noted that the allowance for the function discontinuity in the shock wave will lead to a still greater fall of $P_{s,\max}$ with the distance [19].

Thus, in the melts of low-melting-point metals, cavitation appears at acoustic pressures ~ 0.2–0.5 MPa. Collapsing cavities of a certain initial radius (10^{-3}–10^{-4} cm) give rise to shock waves with peak pressures of the order of thousands of MPa. Shock waves exert their action mainly near collapsing cavities.

Cavitation in molten metals is difficult for experimental studying, which is related to their high temperatures, opaqueness, chemical activity, as well as to the necessity of accounting for contact phenomena (wettability) at the liquid-solid interface (on the surface of a radiator). As shown above, cavitation near the interface is largely determined by the wettability of the solid-state surface. In turn, wettability depends on interfacial physicochemical interactions that can be affected by cavitation due to a local increase in thermodynamic parameters, such as pressure and temperature. Therefore, a direct measurement of cavitation thresholds in an ultrasonic field necessary for determination of cavitational strength of liquids would be, to a certain extent, incorrect. The cavitational strength of melts could be more correctly determined under the conditions providing minimal interfacial physicochemical interactions, for example, by the method of focusing cylindric radiators.

To avoid restrictions associated with high temperatures of melts and their chemical activity, the authors [59] used in experiments gallium (the melting point $T_0 = 303$ K), Wood's alloy ($T_L = 89°C$), and the radiator made of a titanium alloy VTZ-1*.

It is known [12, 18] that the instants of cavitation onset can be determined by the appearance of the first subharmonic in acoustic spectrum and by the appearance of the continuous spectrum of cavitation noice. In this connection, it was expedient to measure two pressure thresholds in melts – the pressure p_{md}, at which the first discrete components p_{md} appear in the acoustic spectrum, and p_{mc} corresponding to the appearance of the continuous spectrum of cavitation noise.

In the case of low driving voltage of a transducer ($U \sim 10 - 20$ V), the form of the sound pressure signal was close to sinusoidal, but at higher voltages the sinusoidal form was distorted (sometimes quite

---

*Weight losses due to titanium corrosion in molten gallium at 300°C comprised some fractions of mg per year.

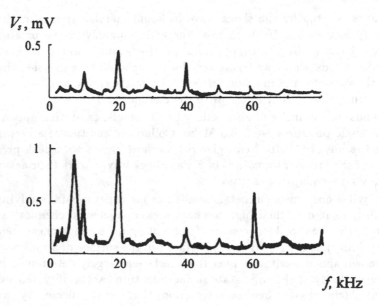

Figure 2.37. Cavitation noise spectrum in gallium melt at $p_m \sim P_c$ (a) and $p_m > P_c$ (b).

strongly) even in the absence of cavitation. In both cases, the signal spectrum exhibited discrete components close to the subharmonic $f/2$, harmonics $2f$ and $3f$, and ultraharmonics $3f/2$ and $5f/2$. The appearance of these components could be explained by multiple reflections in the melt.

A further increase in driving voltage led to piercing clicks in a hydrophone (Figure 2.37), which could be related to a nonstationary character of cavitation when acoustic pressure is close to its threshold value. It can be suggested that each sound click corresponds to a disruption event in the melt or collapse of a cavity ensemble.

Analysis of the hydrophone signal spectrum showed that acoustic clicks are associated with the appearance of discrete spectral components with frequences f_c whose ratios to f make up a random series.

At higher acoustic pressures in the melt, the spectrum becomes continuous with a pronounced discrete subharmonic $f/2$ and ultraharmonics $3f/2$ and $5f/2$ (Figure 2.37,b), which probably corresponds to the formation in the melt of a great number of cavities whose pulsations are unstable in phase and amplitude; in other words, a cavitation region is formed. The corresponding value of acoustic pressure can be taken as the threshold p_{mc}. In this case, the sound field pattern is complex, with local areas of an enhanced cavitation noise, which can be ascribed to the formation of cavitation strands in the melt.

The threshold cavitation is almost frequency-independent below 25 kHz; however, above this value the threshold rises with frequency.

Cavitation in melts was studied on readily fusible metals, such as Sn, Bi, Pb, In, and Cd (99.999%), using a setup that made it possible to maintain the melt at a constant temperature ($t = 1.2T_0$) and to supply vibrations of a desired intensity [60]. In this case, the measured values were acoustic power of vibrations supplied to a load, load resistance, and cavitation noise. The cavitation thresholds were determined by the appearance of discrete components in the cavitation noise spectrum. This method yielded two cavitation thresholds – P_{cb}, corresponding to the onset of cavitation near the radiator-melt interface, and P_{cz}, corresponding to the onset of cavitation in the melt bulk. To measure the first threshold, the metering probe tip was placed 1–2 mm from the radiator, while in the second case it was 10–20 mm apart.

An alternative method for estimating the cavitation threshold in melts is based on a cavitation-dependent drop in the load resistance.

Similar to water, the melts of readily fusible metals exhibited a nonlinear behavior of the load resistance at an enhanced intensity of vibrations supplied to melts. Figure 2.38 gives the values of the delivered acoustic power and the load resistance of molten tin[*].

In spite of a significant thickness of the melt ($\sim 10\lambda_L$), the ultrasound absorption was not great. It was assumed that the emission of elastic vibrations into the melt is governed by the quality of acoustic contact between the melt and the radiator rather than by the ratio of their wave resistances. Indeed, in the first experiment with a new radiator, the power of vibrations supplied to the melt was about zero up to $\dot{\xi}_m^2 = \dot{\xi}_{m1}^2$. The resistance to radiation in this case was also low (dashed sections of the curves). When the vibrational velocity surpassed a certain value, the intensity of vibrations increased sharply. In the following experiments, the dependences I, $(R_l/S)_p = f(\dot{\xi}_m^2)$ behaved differently as compared with the first experiment – the ultrasonic power supplied to the melt grew with the amplitude of vibrational velocity, while the resistance to radiation changed insignificantly.

---

[*]Parameter $\dot{\xi}_{max}^2$ was used to exclude from a consideration the problems associated with the conversion of electric energy into mechanical one.

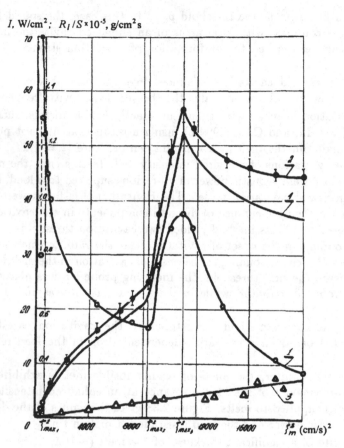

Figure 2.38. Load resistance (1) and the intensity of ultrasound absorbed by a rod radiator and tin melt (2), rod radiator (3), and the melt (4), as functions of the square of the vibrational velocity amplitude ξ_m^2.

These results enable the supposition that when vibrations supplied to the load are of low intensity ($\xi_m < 1\ \mu m$), the acoustic contact between the "nontinned" radiator and the melt is either absent or poor, which can be accounted for by a low wettability of the radiator material (large contact angle θ)* or by the occurrence of an oxide layer on the emitting surface of the radiator. However, when the intensity

---

*It was established in separate experiments that the angle of wetting of titanium, molibdenum, niobium, and tantalum probes with a tin melt is, indeed, very great (~110–$160°$), which testifies to the absence of steady-state physicochemical interactions at the interface.

of vibrations supplied to the load is in excess of the value characteristic of a given melt-radiator pair, the acoustic contact between them
becomes more perfect. In this case, the melt is as though connected
to the vibrational system, which results in a drastic increase in the
resistance to radiation. Nevertheless, its value remains much less than
the wave resistance, obviously because of still imperfect acoustic contact.

A further increase in the vibrational velocity caused an elevation of
the acoustic power supplied to the load, a drop in the load resistance
(Figure 2.38, region II), and some increase in the values R_l/S and I,
which was due to the onset of cavitation, primarily, at the radiator-
melt interface (Figure 2.38, region III). In view of this, the quantity
P_{cl}, corresponding to such value of R_{ll}/S above which the radiation
resistance grew with ξ_{max}, was taken as a cavitation threshold. The
growth of radiation resistance was induced by cavitation both near the
radiator and in the melt bulk. The cavitation noise pattern, followed
with a special probe, confirmed the formation of a cavitation zone in
the melt.

In the first set of experiments, tin and bismuth melts ultrasonically
irradiated with the aid of titanium, molibdenum, niobium, tantalum,
and steel probes were used.

The smallest values were obtained for the threshold P_{cb} determined
spectrally and assigned to cavitation at the radiator-melt interface,
and the largest values were obtained for the threshold P_{cz} referred to
the cavitation zone. The values for the threshold P_{cl} estimated by
the method of load resistance lie between P_{cb} and P_{cz}, which is likely
due to the fact that P_{cl} referred to the developed cavitation with the
expanded cavitation zone, increased number of pulsating cavities, and
their altered compressibility.

Comparison of the cavitation noise oscillograms obtained in the
range of 1–20 kHz near the radiator-melt interface and in the cavitation
zone has shown that in the latter case the relative content of the basic
excitation frequency drops, and that of a continuous spectrum and
discrete components increases. This means that at $p_m > p_{cz}$, cavitation
is more developed from the viewpoint of both the cavitation zone size
and the character of cavitation processes.

The thresholds P_{cb}, P_{cz}, and P_{cl} for the metals and probes
studied lie in the range 0.1–0.4 MPa (Table 2.4), the threshold
P_{cz} being independent of the probe material (as opposed to P_{cb} and
P_{cl}).

To elucidate the reason for the influence of probe material on cavitation thresholds, they were compared in their contact angles.

Table 2.4 Cavitation thresholds in fusible metal melts.

Metal	Radiator material	Contact angle θ, °	Cavitation threshold, MPa			
			P_{cb}	P_{cl}	P_{cz}	P_c (calculated for $R_0 = 10^{-4}$ cm)
Tin	Titanium	130	0.28	0.35	0.48	0.40
	Niobium	140	0.20	0.30	"	"
	Tantalum	150	0.18	0.30	"	"
	Molibdenum	155	0.18	0.30	"	"
Bismuth	Titanium	115	0.30	0.40	0.48	0.25
	Niobium	125	0.25	0.90	"	"
	Tantalum	155	0.20	0.25	"	"
	Molibdenum	160	0.18	0.25	"	"
Cadmium	Titanium	115	0.15	0.55	0.72	0.60
Indium	"-"	120	0.22	0.40	0.50	0.20
Lead	"-"	130	0.30	0.35	0.54	0.30
Zinc	"-"	130	0.40	0.60	0.80	0.70

These angles were measured by the method of a lying drop [61] in a vacuum (the residual pressure 10^{-5} mm Hg) at temperatures equal to $1.2T_0$ of respective metals. It was found that the cavitation thresholds P_{cb} and P_{cl} were the higher, the stronger were physicochemical interactions at the radiator-liquid interface (that is, the lower was contact angle). At the same time, P_{cz} and P_{cl} were weakly dependent on the properties of the melt itself.

The experimentally determined values for P_{cz} showed a satisfactory agreement with the cavitation thresholds P_c calculated for cavities with $R_0 = 5 \times 10^{-4}$–10^{-3} cm in the melts of readily fusible metals. In this case, P_{cz} depends, to a certain extent, on the properties of the melt, in particular, on its surface tension.

Real liquid metals are known to contain various solid and gaseous microinclusions, whose content is governed largely by the technology of metal production and its purity. It could be reasonable to suggest that cavitation threshold should depend on the amount of dissolved gases and nonmetal impurities in melts. Indeed, experiments with pure aluminum have shown that the threshold of developed cavitation decreases with increasing hydrogen content in the melt (Figure 2.39) [62].

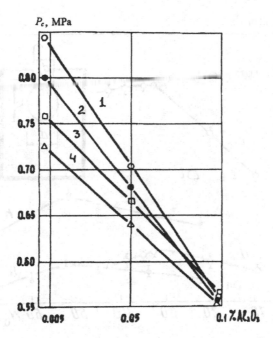

Figure 2.39. Cavitational strength of aluminum melt versus the content of alumina at various concentrations of hydrogen: (*1*) 0.1, (*2*) 0.2, (*3*) 0.3, and (*4*) 0.4 cm^3 per 100 g of melt [62].

The threshold showed a still stronger dependence on the occurrence of Al$_2$O$_3$ impurity particles in the melt.

Of practical interest is the size of cavitation sone produced by ultrasound. This size was estimated for bismuth and tin melts as a function of acoustic power supplied to the melt. The radiator diameter was d_r, the temperature of melts being 20°C above their melting points. To estimate the dimensions of cavitation zone, a frame with a 10-μm titanium foil was introduced into the melt in such a manner that the frame plane was parallel to the probe axis. Ultrasonic vibrations of a certain intensity were supplied to the melt for 15 min, the frame was then withdrawn from the melt to measure the dimensions d and h of the cavitation region with traces of cavitational erosion.

It was found that the dimensions of cavitation region in both melts initially grew with a supplied ultrasonic power to reach their maximum values $d = 0.9 \div 1.1 d_r$ and $h = 2.6 \div 4.3 d_r$ and then somewhat decreased (Figure 2.40), which was presumably due to the increasing number of cavities per unit volume.

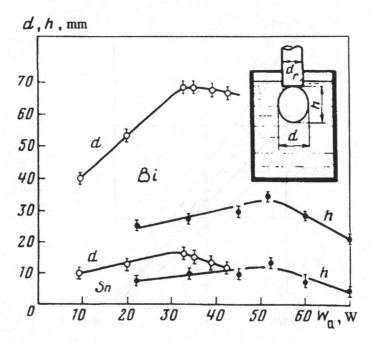

Figure 2.40. Cavitation region dimensions as functions of the acoustic power of vibrations supplied to bismuth and tin melts.

It should be noted that the above foil method for measuring the dimensions of cavitation region is not precise, since the introduced foil can somewhat modify cavitation.

2.3 Acoustic Streaming

The propagation of finite-amplitude ultrasonic waves in a liquid is accompanied, along with other nonlinear effects, by stable acoustic streaming that develops in a free nonuniform acoustic field or near various obstacles (interfaces) due to energy loss in the sound waves. In this case, viscous forces act to stabilize streamings that are always turbulent. The literature devoted to acoustic streaming is vast and originates from the first experimental work of M. Faraday [63] and the first theory of stationary flows by Rayleigh [2]. Relevant investigations embraced the analysis of turbulent motion in a free acoustic field or near vibrating cylinders and spheres, in cylindrical tubes, near a hole through which an acoustic beam passes, near pulsating bubbles, in a gap between a plane solid surface and vibrating rod, and so on.

The theoretical and experimental works on acoustic streaming were first reviewed in the late 1960s [64, 65].

The scale and velocity of streaming in a medium depend on its properties as well as on the shape and structure of its boundaries. Three types of acoustic streaming have been most closely investigated:

(1) large-scale streaming arising in a free inhomogeneous acoustic field whose inhomogeneities are significantly greater than the sound wavelength;

(2) streaming in a medium confined by rigid walls;

(3) low-scale vortices arising in a viscous boundary layer near obstacles.

This section deals with slow acoustic streamings whose velocities increase with the intensity of vibrations, being, however, considerably less than the vibrational velocity.

All kinematically diverse acoustic streamings arise as a result of irreversible losses of the wave energy and momentum. The structure of acoustic streaming is determined by the distribution of ultrasonic energy and absorption coefficient over the medium, and boundary conditions for the hydrodynamic velocity field [64, 65], which, in turn, defines the distribution of volume density of adsorbed energy and the magnitude and direction of the energy flux vector αJ, (α is the coefficient of absorption).

The streaming outflows from an ultrasonic source with the maximum $|\alpha J|$ (forward streaming) and inflows to the source in directions with the minimum $|\alpha J|$ (backward streaming).

The theory of acoustic streaming is based on the hydrodynamic equations of viscous compressible liquid (1.37), (1.38), (1.39).

To describe its motion, it is necessary to determine the internal viscous forces F_v in equation (1.37). Based on expression (1.116) for viscous forces, one can write the equation of motion for a real liquid in the absence of external forces*

$$\rho \left\{ \frac{\partial v}{\partial t} + v\,\mathrm{grad}v \right\} = -\mathrm{grad}P + \eta \Delta v + \left(\eta' + \frac{\eta}{3} \right) \mathrm{grad\,div}\,v. \quad (2.48)$$

_____

*The Hamiltonian ∇ and Laplacian Δ will be used in the following calculations (see footnote on page 15).

In the case of stable acoustic streaming, the basic nonlinear hydrodynamic equations (2.48), (1.38), (1.39) can be solved by successive approximations. The recent literature is mainly concerned with a second-order approximation.

Parameters defining the state of a medium – density, pressure, and velocity – can be represented as expansions into series

$$P = P_0 + p' + p'' + \ldots$$
$$\rho = \rho_0 + \rho' + \rho'' + \ldots \qquad (2.49)$$
$$v = v' + v'' + \ldots,$$

where P_0 and ρ_0 are, respectively, the pressure and density of an undisturbed liquid; ρ', p', v' are the first-order approximations; ρ'', p'', v'' are the second-order approximation.

The quantities p', ρ', and v' change harmonically with the frequency ω and characterize the acoustic field in the first-order approximation.

The second-order approximation involves time-independent correcting terms and those varying with a single and double sound frequency. In particular, streaming velocity is characterized by the value $v'' = V$.

In the method of successive approximations, the first-order approximation v' is first determined to be then used for calculation of the second-order approximation.

By substituting expansions (2.49) into expressions (2.48) and (1.38) and grouping the terms of the same order of magnitude, one can obtain hydrodynamic equations in the first-order approximation

$$\rho_0 \frac{\partial v'}{\partial t} = -\operatorname{grad} p' + \left(\frac{4}{3}\eta + \eta'\right) \operatorname{grad} \operatorname{div} v' - \eta \Delta v', \qquad (2.50)$$

$$\frac{\partial p'}{\partial t} + \rho_0 \operatorname{div} v' = 0. \qquad (2.51)$$

Taking into account the equation of state for a liquid, written as

$$p' = c^2 \rho' \qquad (2.52)$$

and eliminating the quantities p' and ρ', one can obtain the equation for v'

$$\left(2k^{-2} + \frac{jb}{\eta}\beta^{-2}\right) \operatorname{grad} \operatorname{div} v' + 2v' = -j\beta^2 \nabla \times \nabla \times v', \qquad (2.53)$$

where

$$\beta^2 = \frac{\omega\rho}{2\eta}.$$

The hydrodynamic equation in the second-order approximation can be obtained similarly, by substituting (2.49) into (2.48) and (1.38):

$$\rho_0 \frac{\partial v''}{\partial t} + \eta \Delta v'' - \left(\frac{4}{3}\eta + \eta'\right) \operatorname{grad} \operatorname{div} v'' = -\operatorname{grad} p'' - F'', \quad (2.54)$$

$$\frac{\partial \rho''}{\partial t} + \operatorname{div}(\rho' v') + \rho_0 \operatorname{div} v'' = 0, \quad (2.55)$$

where

$$F'' = \frac{\partial(\rho' v')}{\partial t} + \rho_0 v' \operatorname{div} v' + \rho_0(v' \operatorname{div})v'.$$

Expressions (2.54) and (2.55) averaged with respect to time over several periods are given by

$$\eta \Delta v'' - \left(\frac{4}{3}\eta + \eta'\right) \operatorname{grad} \operatorname{div} v'' = -\operatorname{grad} p'' - \overline{F}'', \quad (2.56)$$

$$\operatorname{div} v'' = -\frac{1}{\rho_0} \overline{\operatorname{div}(\rho' v')}, \quad (2.57)$$

where

$$\overline{F}'' = \rho_0 \overline{[v'\nabla v' + (v'\nabla)v']}.$$

It is seen from (2.56) and (2.57) that the time-independent quantity $\overline{F}''$ is obtained by averaging (with respect to time) the expression made up from v' and its derivatives; therefore, $\overline{F}''$ is basically determined if the first-order velocity v' is known.

Equations (2.56) and (2.57) are basic in the theory of slow stationary streaming. It should be noted that these equations are similar to the basic equations (1.37) and (1.38) for streaming in a low-viscous medium subject to external forces, and the quantity F'' is an exact analog of the driving force field.

The streaming velocity v'' can be conveniently found from the equation (2.58) obtained from (2.54) by the rotational transform of its both sides

$$\Delta \Omega = \frac{1}{\nu} \overline{\nabla f''}, \quad (2.58)$$

where

$$\Omega = \nabla \times v'', \quad f'' = \frac{F''}{\rho_0}, \quad \nu = \frac{\eta}{\rho_0}.$$

The convenience of equation (2.58) in comparison with equation (2.54) lies in the absence of the second-order static pressure p''. Therefore, in equation (2.58), only the components of v'' are dependent variables.

Equation (2.58) is the flow equation in the Poisson form, while the function $1/\nu \cdot \overline{\nabla f''}$ is the rotational source density.

In some cases, equation (2.58) can be somewhat simplified. Thus, in the case of a rotationless free isotropic acoustic field, $\nabla \times v' = 0$, and

$$\overline{\Delta\Omega} = \frac{1}{\nu}\overline{[\nabla\nabla v' \times v']}. \qquad (2.59)$$

In the case of a solenoidal field, $\nabla v = 0$, and

$$\overline{\Delta\Omega} = -\frac{1}{\nu}\overline{\nabla \times [v' \times (\nabla \times v')]}. \qquad (2.60)$$

The condition $\nabla v' = 0$ can be also fulfilled in an anisotropic field, if the obstacles are much less than the wavelength of a standing sound wave.

The method of successive approximations, when applied to analysis of acoustic streaming, imposes some restrictions on the amplitude of sound vibrations.

Expansions (2.49) imply that each of the next terms is much smaller than the previous one. The estimation of terms in (2.54) and (2.56) indicates that the quantities $\eta v''/(L'')^2$ (L'' is the stable flow scale) and $\rho v_m^2/L'$ (v_m is the amplitude of v'; L' is the sound field scale in the first-order approximation) are of the same order of magnitude. Taking into account the condition $v''/v_m \ll 1$ that is valid for slow streaming, one can show that the theory of stable large-scale ($L'' \gg 1$) streaming is applicable for small acoustic Reynolds numbers

$$\mathrm{Re}_a = \frac{\varepsilon\rho_0 v_m \lambda}{\pi b} \ll \frac{\lambda L'}{(L'')^2} \sim 1. \qquad (2.61)$$

For low-scale vortices ($L' \sim L'' < \lambda$), the acoustic Reynolds numbers can be great.

In terms of the theory considered, the ratio of the streaming velocity to the vibrational velocity amplitude is of the order of $M_a\Phi$, where $M_a = v_m/c_0$ is the acoustic Mach number, Φ is the geometric factor proportional to the dimensionless streaming scale.

By their spatial scale (the ratio of the field dimension to the length of a longitudinal wave in an infinite medium), the wave processes can be distinguished as occurring in a free medium, finite-thickness layer, and in a thin layer.

In a free space, all the considered distances from the wave source are small as compared with distances to the space boundaries. In this case, the wave is traveling.

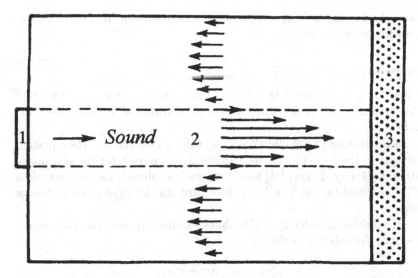

Figure 2.41. The Eckart streaming: (*1*) sound source, (*2*) cylindrical tube, (*3*) absorber.

In a finite-thickness layer, all of its dimensions L_n satisfy the condition

$$L_n < 30\lambda.$$

In a thin layer, its dimension in the direction of wave propagation satisfies the condition

$$L_1 \ll \lambda,$$

while all other dimensions satisfy the previous condition.

In order to find the velocity of stable streaming in a rotationless (potential) field of the first-order velocities, one has to solve the equation (2.59). This problem has been first solved by Eckart [66] for the case of the traveling wave propagation through a free isotropic acoustic field. For a well-collimated sonic beam, the product $\nabla \times v'$ is infinitely large at its boundary and equals zero at all other points. In the case of megahertz waves, when their source size is much greater than the wavelength, the volume of a rotational region at the sonic beam boundary is small as compared with the volume of the whole acoustic field. In this case, the contribution from the surface vortices can be neglected.

The wave traveling over a cylindrical tube, one end of which accomodates the wave source, and the other has an absorber (Figure 2.41), possesses only one component of vibrational velocity directed along the cylinder axis z. This component can be written in the cylindrical coordinates r, φ, z as

$$v_z = v(r)e^{-\alpha z}\sin(\omega t - kz), \tag{2.62}$$

where $\alpha = (\omega^2/2p_0c_0^3)b$ is the sound absorption coefficient, and $v(r)$ characterizes the radial distribution of vibrational velocity over the tube.

Substituting (2.62) into (2.59) and taking into account the equation of state in the form (2.52), one can obtain a solution for the stationary stream velocity. Eckart [65] has found such a solution for the case when $e^{-\alpha z} \sim 1$ and the integral liquid flux through the tube cross-section is zero.

If vibrational velocity is distributed uniformly over the tube cross-section, the solution is simple

$$v(r) = v \quad \text{at} \quad r < r_1,$$
$$v(r) = 0 \quad \text{at} \quad r_1 < r < r_0, \tag{2.63}$$

where r_1 and r_0 are the radii of the sound beam and tube, respectively.

The transverse distribution of the streaming velocity is given by

$$v_z'' = v_z = \begin{cases} V\left\{\dfrac{1}{2}\left(1-\dfrac{\psi^2}{\chi^2}\right)\left(1-\dfrac{1}{2}\chi^2\right)(1-\psi^2)-\ln\chi\right\}; & 0 \leq \psi \leq \chi \\ -V\left\{\left(1-\dfrac{1}{2}\chi^2\right)(1-\psi^2)+\ln\psi\right\}; & \chi \leq \psi \leq 1 \end{cases}, \tag{2.64}$$

where $\psi = \dfrac{r}{r_0}, \quad \chi = \dfrac{r_1}{r_0},$

$$V = \frac{b}{4\eta c_0}\, v_m^2\,(kr_1)^2. \tag{2.65}$$

The typical pattern of streaming velocities is shown in Figure 2.41. Within the sound beam ($r \leq r_1$), streaming is directed from the source, whereas in the opposite direction outside the beam ($r_1 < r < r_0$). The velocity is maximum on the sound beam axis.

Analysis of expression (2.65) indicates that the streaming velocity is proportional to the square of the vibrational velocity times the square of the frequency times the volume-to-shear viscosities ratio. Figure 2.42 illustrates the streaming velocity in water 40 cm from a 1.2-MHz ultrasonic source versus the sound pressure amplitude near the source [67]. When the sound pressure $p_m \sim 0.7$ MPa, there occurs the generation

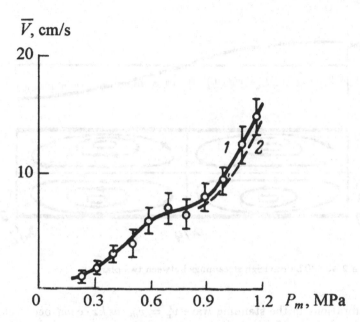

Figure 2.42. The velocity of the Eckart streaming in water versus the sound pressure amplitude: (*1*) experimental results [67], (*2*) calculations [68].

of a sawtooth wave at some distance from the source, which alters the dependence of the streaming velocity on the intensity of vibrations.

In a low-viscous medium, the ratio of the streaming velocity to the vibrational velocity peak is about $M_a(kr_1)^2$. In this case, the streaming scale is significantly greater than the wavelength; therefore, according to (2.61), the above solution is valid only for the acoustic Reynolds numbers much less than 1. For kilohertz and megahertz ranges and relatively low-viscous liquids (~ 0.01 P), this imposes a rather strong restriction on the sound pressure amplitude. When $\mathrm{Re}_a \geq 1$, at some distance from the source the sinusoidal wave is distorted and possesses of a sawtooth form, which enhances the wave absorption. The streaming velocity will no longer satisfy the condition of slow streaming.

At present, the theory of rapid acoustic streaming is still in its infancy. The streaming of this type can be regarded as a streaming caused by dissipative losses dependent on vibrational amplitude.

In other particular case of streaming confined by rigid walls, the velocity field can be considered, in the first approximation, as solenoidal. An example of such streaming is a two-dimensional Rayleigh streaming [2] produced by a standing wave between two planes (Figure 2.43).

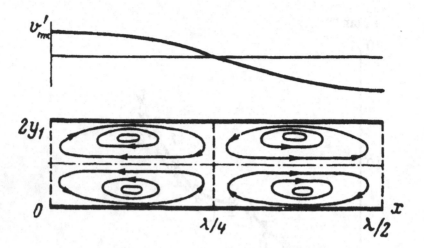

Figure 2.43.   The Rayleigh streamings between two planes.

Vibrations in the standing wave $v'_x = v_m \cos kx \cos \omega t$ occur along the ox axis. One of the planes has the ordinate $y = 0$, and the other – $y = 2y_1$. For $0 \le y \le y_1$, the vibrational velocity components have the form

$$\left.\begin{aligned}
v'_x &= v_m \cos kx \left[-\cos \omega t + e^{-\mu} \cos(\omega t - \mu) \right] \\
v'_y &= v_m \frac{k\delta}{\sqrt{2}} \sin kx \left[\left(1 - \frac{\mu}{\mu_1}\right) \cos\left(\omega t - \frac{\pi}{4}\right) - e^{-\mu} \cos\left(\omega t - \frac{\pi}{4} - \mu\right) \right]
\end{aligned}\right\},$$
$$(2.66)$$

where $\delta = (2\nu/\omega)^{1/2}$ is the acoustic boundary layer thickness; $\mu = y/\delta$, $\mu_1 = y_1/\delta$.

In deriving the relations (2.66) it was taken that $k\delta \ll 1$ and $y_1 \gg \lambda$. Away from the boundaries, where the terms with $e^{-\mu}$ can be neglected, the expressions for the streaming velocity are given by

$$\left.\begin{aligned}
v''_x &= -\frac{3v_m^2}{16c_0} \sin 2kx \left[1 - 3\left(1 - \frac{\mu}{\mu_1}\right)^2 \right] \\
v''_y &= -\frac{3v_m^2}{16c_0} ky_1 \cos 2kx \left[\left(1 - \frac{\mu}{\mu_1}\right) - \left(1 - \frac{\mu}{\mu_1}\right)^2 \right]
\end{aligned}\right\}.$$
$$(2.67)$$

Schematically, the streaming of this type is shown in Figure 2.43. It consists of vortices along the ox axis $\lambda/4$ apart each other. Near the

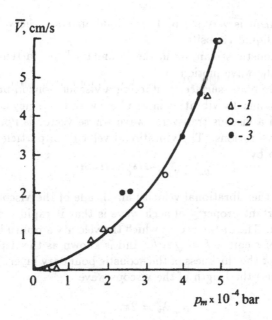

Figure 2.44. The Rayleigh streaming velocity in air as a function of the sound pressure amplitude [69]. The velocity is measured by means of: (*1*) visualization of streamings with suspended powder particles, (*2*) thermoanemometer, (*3*) Pitot tube.

boundary planes, the medium flows from the antinodes to nodes of the vibrational velocity, while in the central region – in the opposite direction. The streaming velocity does not depend on viscosity and is proportional to v_m^2/c_0.

Figure 2.44 illustrates the dependence of the maximum stream velocity in air on the sound pressure, from which it follows that the velocity is proportional to the square of the sound pressure [69].

As in the case of the Eckart streaming, the Rayleigh streaming is also characterized by the ratio of rotational and vibrational velocities in the acoustic wave, which is proportional to $M_a\Phi$. Since the streaming scale $L'' \sim \lambda$, the solutions obtained are valid for the acoustic Reynolds numbers smaller that 1. For larger Reynolds numbers the streaming pattern is similar, although the rotational velocity is greater than it follows from (2.67) [69].

It should be noted that the Rayleigh solution fails to give a correct value for the streaming velocity in the boundary layer ($\mu \to 0$).

The streaming velocity of a real liquid near solid walls must be zero. The velocity vanishes in a thin boundary layer, across which the velocity gradient is very steep. In the final analysis, the velocity drop is related to liquid viscosity.

As for acoustic streaming in the boundary layer, attention has to be given to the wave motion in it.

An infinite plane surface, contacting a viscous noncompressible liquid and harmonically vibrating in its plane with frequency ω, generates in this liquid a viscous transverse wave whose vector is normal to the direction of vibrations. The vibrational velocity of particles in such a wave is given by

$$v_v = v_{v0} e^{-x/\delta} e^{j(x/\delta - \omega t)}, \qquad (2.68)$$

where v_{v0} is the vibrational velocity amplitude of the viscous wave.

An important property of such wave is that it rapidly dies out inward a liquid. The distance over which the velocity amplitude decreases by a factor of e equals $\delta = \sqrt{2\nu/\omega}$ and is known as the depth of wave penetration or the thickness of the acoustic boundary layer. This quantity determines the length of the viscous wave

$$\lambda_v = 2\pi\delta. \qquad (2.69)$$

Parameter $1/\delta$, known as the absorption coefficient of the viscous wave, is 15–17 orders of magnitude greater than the absorption coefficient of longitudinal waves in Newtonian liquids [70].

In view of the steep velocity gradients in the boundary layers, the "viscous" terms in the equation of motion, containing the velocity derivatives with respect to coordinates, are great even if the kinematic viscosity is low. Hence, the acoustic wave momentum drastically changes in the boundary layer, giving rise to considerable forces and thereby streaming.

Analysis of acoustic streaming in the boundary layer can be performed by invoking the hydrodynamic Prandtl equations (see, for instance, [3]):

$$\left.\begin{array}{l} \dfrac{\partial v_x}{\partial t} + v_x \dfrac{\partial v_x}{\partial x} + v_y \dfrac{\partial v_y}{\partial y} + \nu \dfrac{\partial^2 v_x}{\partial y^2} = \dfrac{\partial V}{\partial t} + V\dfrac{\partial V}{\partial x} \\[2mm] \dfrac{\partial v_x}{\partial x} + \dfrac{\partial v_y}{\partial y} = 0 \end{array}\right\}, \qquad (2.70)$$

where $v(x,t)$ is the streaming velocity far away from the boundary; v_x and v_y are the respective components of vibrational velocity.

These equations for the boundary layer near the plane $y = 0$ are derived from the Navier–Stockes equation (the equation of motion of a viscous incompressible liquid) (1.117a) and equation of continuity

(1.38) for a two-dimensional circumfluence of the plane xz (streaming along the x axis; z-component of the velocity equals zero).

Equations for the boundary layer are valid if $\lambda > \lambda_v$. The wavelength ratio of viscous to acoustic waves can be represented as

$$\frac{\lambda_v}{\lambda} = k\delta = \left(\frac{3}{2}\frac{M_a}{\mathrm{Re}_a}\right)^{1/2}. \qquad (2.71)$$

The condition $\lambda \gg \lambda_v$ is fulfilled at $\mathrm{Re}_a \gg M_a$ and does not limit the amplitude of vibrations at which the theory of stable streaming near the boundary is valid.

In deriving the Prandtl equations (2.70) it was assumed that (i) the velocity gradient is considerably steeper normally to the boundary than in the parallel direction, (ii) the boundary layer lacks the transverse pressure gradient ($\partial P/\partial y = 0$), and (iii) pressure in the boundary layer, $p(x)$, is equal to that in the streaming bulk.

Using expansions (2.49) for the velocity components, one can write equation for the boundary layer in the second-order approximation

$$\left.\begin{array}{l} \dfrac{\partial v_x''}{\partial t} - \nu\dfrac{\partial^2 v_x''}{\partial y^2} = V\dfrac{\partial v}{\partial x} - v_x'\dfrac{\partial v_x'}{\partial x} - v_y'\dfrac{\partial v_x'}{\partial y} \\[2mm] \dfrac{\partial v_x''}{\partial x} + \dfrac{\partial v_y''}{\partial y} = 0 \end{array}\right\}. \qquad (2.72)$$

The boundary conditions for the half space $y > 0$ are

$$v_x|_{y=0} = v_y|_{y=0} = 0 \quad \text{and} \quad v_x \to V(x,t) \quad \text{at} \quad y \to \infty.$$

These equations were solved by Schlichting [71, 72] for a standing wave: $V(x,t) = -v_m \cos kx \cos \omega t$. Near the boundary ($\mu \ll 1$), the velocity components has the form (within the accuracy to the terms $\sim \mu^2$)

$$\left.\begin{array}{l} v_x'' = v_x = \dfrac{v_m^2}{4c_0}(\mu - \mu^2)\sin 2kx \\[3mm] v_y'' = v_y = \dfrac{v_m^2}{4c_0}k\delta\mu^2\cos 2kx \end{array}\right\}. \qquad (2.73)$$

Based on analysis of these expressions one should highlight some characteristics of Schlichting's streaming. At the points where $V_m = -v_m\cos kx = 0$ or $\partial V_m/\partial x = 0$ the streaming velocity has only a normal component, i.e., at the nodes and antinodes of vibrational velocity the liquid streamings either toward or from the boundary. The components v_x and v_y change their signs at $\mu \cong 1$ and 1.9, respectively [71, 72], which determines the dimensions of the boundary vortices, $\lambda/4 \times 1.9\delta$.

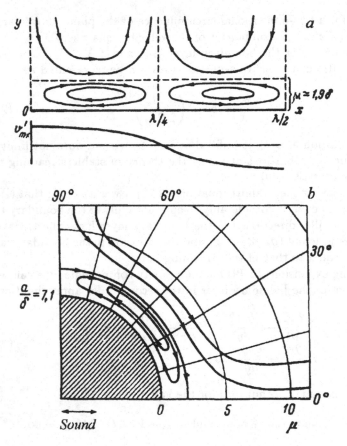

Figure 2.45. The Schlichting streaming near (*a*) a plane boundary and (*b*) a vibrating cylinder. Arrows indicate the direction of the source vibrations [69].

The streaming pattern is shown in Figure 2.45. One should note that vortices in the boundary layer and those in the Rayleigh streaming rotate in the opposite directions.

The solution to the problem of the acoustic boundary layer near the plane surface can be used to describe the streaming near the curved surfaces with curvatures significantly smaller than the length of the viscous wave [73].

According to the Westerfeld theorem, the boundary streaming generally represents two vortices (this is also the case when liquid vibrates near the stationary solid boundary). These vortices are depicted in Figure 2.46 where the streaming dimension L'' and dimensionless pa-

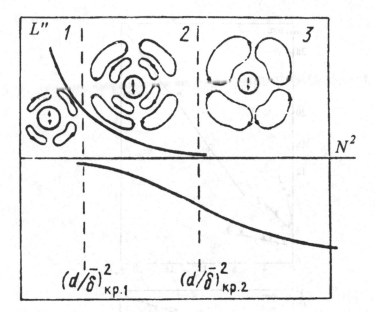

Figure 2.46. Structure of two-vortex acoustic streamings in the field of a sound dipole [70].

rameter $N^2 = d^2/\delta^2$ (d is the vibrating body dimension in the direction of vibrations) are plotted on the ordinate and abscissa, respectively. In the case of the two-vortex streaming, the distance from the vibrating surface to the separatrix between the two vortices (region 2) in the direction of vibrations was taken as a streaming dimension, whereas in the extreme cases of one-vortex streaming (regions 1 and 3) this was the distance to the farthest, but still detectable streamline. The quantities $(d/\delta)^2$ depend on the body shape.

Figure 2.47 shows changes in the dimensions of the inner and outer vortices as a function of the displacement amplitude [70]. It is seen that these dimensions grow virtually linearly with the displacement amplitude over the entire range investigated.

The acoustic streaming velocity in the boundary layers depends on the power of vibrations, absorption properties of the medium, as well as on the acoustic and hydrodynamic field patterns. The function $v(\xi_m/\delta)$ is linear in the range $\xi_m/\delta = 1 \div 22$ (Figure 2.48).

At small ratios ξ_m/δ, the function of the streaming velocity on the displacement amplitude is quadratic [70].

Of great applied and theoretical interest is the microstreaming around air bubbles in a vibrating liquid.

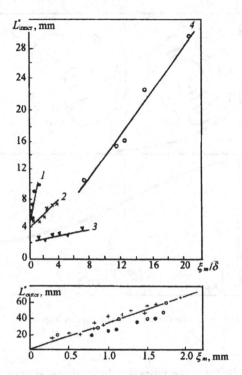

Figure 2.47. The size of (*a*) inner and (*b*) outer vortices as a function of the radiator displacement amplitude (the radiator represents a 36-mm diameter sphere, $f = 50$ Hz): (*1*) glycerol, (*2*) 90% glycerol, (*3*) 75% glycerol, (*4*) 50% glycerol, (*5*) water [70].

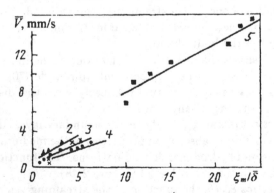

Figure 2.48. Acoustic streaming velocity as a function of the radiator displacement amplitude (the radiator represents a 36-mm diameter sphere, $f = 50$ Hz). Designations as in Figure 2.47 [70].

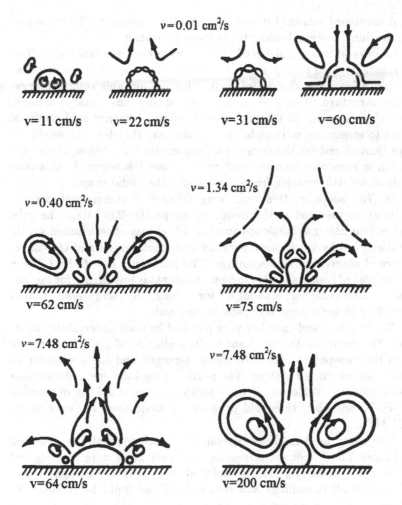

Figure 2.49. Microjets arising near vibrating air bubbles [74].

At a given frequency, this microstreaming depends on the liquid viscosity and vibrational velocity amplitude. Figure 2.49 illustrates the microstreaming in water-glycerol mixtures as a function of their viscosity and vibrational velocity amplitude.

The sound-induced streaming in the boundary layer reduces its thickness (as compared with common hydrodynamic streamings) and thus appreciably induces the heat and mass transfer. As a result, the streaming can promote cleaning of contaminated surfaces, atomization of fuels, the rate of heterogeneous reactions, and so on.

A combined action of streaming of all types enhances the mixing of the medium, thereby leading to its homogenization.

Of interest is the problem of acoustic streaming in metal melts. The literature devoted to this problem is scarce. It is, however, known that stirring affects the kinetics of melt solidification and resulting crystalline structure. Therefore, acoustic streaming must present interest from the viewpoint of stirring intensification. Experimental study of acoustic streaming is impeded by cavitation; therefore, it would be expedient to restrict the considered frequencies from below, since cavitation is known to decrease with rising sound frequency. In this connection, we will consider frequencies above the audio range.

In [76], acoustic streamings were studied in transparent organic melts at 45 kHz and in the frequency range 100–250 kHz. The substance (eutectic naphthalene-camphor alloy) was chosen based on its similarity in physicochemical parameters to metal melts and the convenience of observation of streamings. The parameter $\alpha = L/R$ (L is the molar heat of fusion and R is the gas constant) is less than 2 for metals, close to 2 for camphor, is about 6.5 for naphthalene, and usually ranges from 8 to 16 for nonmetal organic compounds.

Naphthalene and camphor were purified by multiple crystallization, since the purity of the resultant molten alloy is of great importance from the viewpoint of its cavitational strength and the possibility to obtain an overcooled system. The purity of naphthalene and camphor was considered sufficient, if their molten alloys could be overcooled by 4–5°C and their threshold pressures of cavitation, P_c, were above 0.01 MPa.

The setup used in these experiments is schematically represented in Figure 2.50. Acoustic streaming and cavitation were investigated in the alloy containing 35% of naphthalene and 65% of camphor ($\alpha = 3.5$), in which streamings and convective flows could be reliably observed.

In some cases, other alloys possessing various temperatures of crystallization and structure of crystallization front were used for investigating specific features of the interaction of crystallization fronts and streamings.

Since it was found that cavitation and streamings behave in a frequency-independent fashion in the range 100–250 kHz, the results obtained at 45 and 139 kHz will be considered.

Experiments showed that the ultrasonic field acted upon the melts gives rise to streamings of various kinds and scales, which depend on ultrasonic intensity and the temperature of melts actually determining their viscosity. Besides, streamings significantly depend on the place of

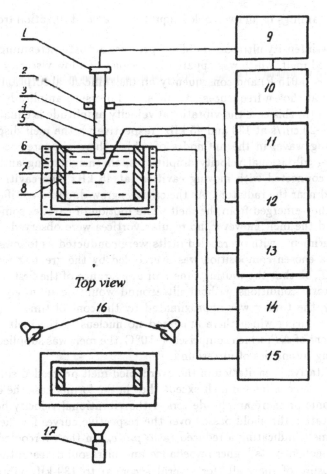

Figure 2.50. Schematic representation of a setup for investigating of acoustic streaming in overcooled melts: (*1*) transducer, (*2*) electrodynamic probe, (*3*) metering waveguide, (*4*) radiator, (*5*) cuvette with a melt, (*6*) water bath, (*7*) hydrophone, (*8*) refrigerator, (*9*) ultrasonic generator, (*10*) frequency meter, (*11, 12, 13*) electron voltmeters, (*14*) heating thermostat, (*15*) cooling thermostat, (*16*) lamps, (*17*) microscope.

their occurrence; in other words, whether they occur near the crystallization front or in a free (overcooled or overheated) liquid phase.

In this connection, it is expedient to divide the results into three groups:

(1) streamings in an overheated liquid ($T - T_L > 5°C$);

(2) streamings in an overcooled liquid ($T < T_L$, $\Delta T = T_L - T > 2°C$);

(3) streamings in an overcooled liquid near a crystallization front.

A low-intensity ultrasound did not produce acoustic streaming in an overheated melt, which was apparently because of a low viscosity of the melt ($\mu \sim 0.015$ P) and consequently an insignificant absorption of ultrasound at chosen frequences. At the sound pressure amplitude about 0.01 MPa (in this case the vibrational velocity amplitude is equal to 2–4 and 10–20 cm/s at 139 and 45 kHz, respectively), the melt displayed a standing wave, in the antinodes of which there appeared cavities. At higher vibrational velocity amplitudes, the neighboring antinodes became connected with moving cavities (at 45 kHz, the cavities also appeared near the radiator). As the cavities grew due to a rectified diffusion, they emerged from the melt. The motion of cavities somewhat disturbed the melt; however, no regular vortices were observed.

Experiments with overcooled melts were conducted as follows. The melt of a chosen composition was overcooled by the greatest possible value ΔT, so that the waiting time τ of appearance of the first nucleus under static conditions without ultrasound would be infinitely large. Actually, the time τ was approximated by the control time τ_0 (equal to ~ 1 h) during which there appeared no nucleus in the melt. After attainment of ΔT, which comprised 4–10°C, the melt was supplied with increasing intensities of ultrasound.

Qualitatively, cavitation in the overcooled melt proceeded similarly to that in the overheated melt except that in the former case the curves of the sonic pressure amplitude versus the vibrational velocity beyond the cavitation threshold passed over the respective curves for the overheated melt, indicating a reduced ratio p_m/v_m in the overcooled melt and consequently its higher impedance and ultrasound absorption.

Exposure of the melt (for several seconds) to 139-kHz vibrations with the vibrational velocity amplitude above the cavitation threshold gave rise to nucleation near cavities. It should be noted that the control time $\tau_0 \sim 1$ h $(\tau/\tau_0 = 10^{-3} \div 10^{-4})$ seems to be sufficient.

Further ultrasonic irradiation of the overcooled melt somewhat increased the number and size of new solid-phase inclusions. As this took place, the effective viscosity of the melt obviously increased too, and after about 10 s acoustic vortices appeared and developed in the zone of motion of cavities and crystallites (Figure 2.51).

As follows from the photographs taken in the 139-kHz experiments, the vortices change during ultrasonic treatment. When the solid phase content is small (Figure 2.51,a), the dominating vortices are those with $L \sim \lambda$, whereas larger vortices with $L \sim (2.5 \div 3)\lambda$ are rare. The development of solid phase (Figure 2.51,b) acts to reduce the number

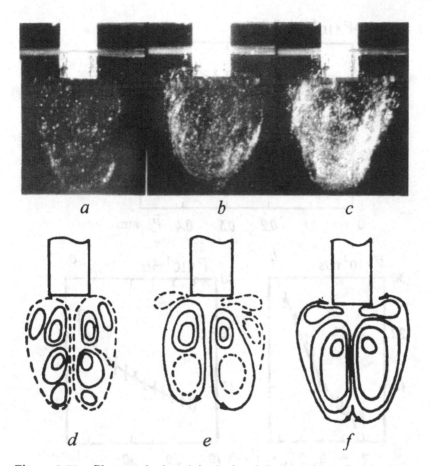

Figure 2.51. Photographs (×1.5) (a, b, c) and diagrams (d, e, f) of developing acoustic streamings in an overcooled melt ($\Delta T = 4°C$), 10 (a, d), 20 (b, e), and 30 (c, f) seconds after the onset of 139-kHz vibrations with the velocity amplitude $v_m = 5$ cm/s.

of small vortices with $L \sim \lambda$ and increase the number of larger vortices. At still higher solid phase content (Figure 2.51,c), large vortices with $L > \lambda$ dominate, and additional vortices appear near the radiator.

In could be suggested that the solid-phase particles appearing in the melt significantly increase its viscosity, thus reducing the sound wave momentum due to absorption and (eventually) acoustic streaming.

At 45 kHz, the mechanism of the streaming appearance and development was basically analogous to that in the case of 100- to 250-kHz vibrations.

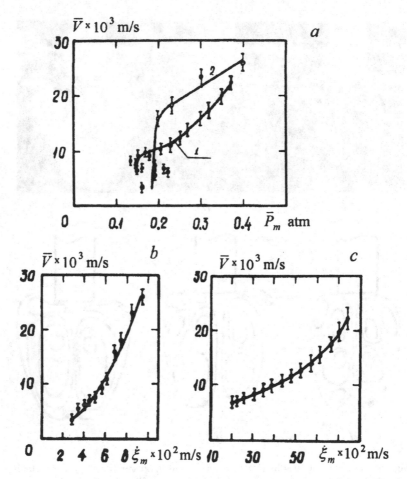

Figure 2.52. Acoustic streaming velocity in an overcooled melt ($\Delta T = 4^\circ$C) as a function of (a) sound pressure amplitude (1, 45 kHz; 2, 139 kHz) and vibrational velocity amplitude at (b) 139 and (c) 45 kHz.

Analysis of photographs and cinemagraphic frames showed that streamings in the vessel produced by 45-kHz ultrasonic irradiation were stable only when the fine-grained solid phase content was rather high, i.e., when the melt was sufficiently viscous. At low frequences, when the ratio of the vessel thickness to the wavelength was smaller, the acoustic streaming tended to be flat. In this case, the time τ of the streaming establishment was 20–30 s.

The acoustic streaming velocities versus the vibrational pressure peak and velocity are presented in Figure 2.52. The dependences were plotted based on analysis of photographs and frames that were taken after a 10–30-s ultrasonic treatment of the melt, i.e. when the acoustic vortices had sufficient time to develop.

Estimations of the maximum streaming velocity $(\overline{V}_{max})$, scale L, acoustic Mach number $M_a = v_m/c_L$, and the ratio of the streaming velocity to vibrational one (V_{max}/v_m) carried out for 45- and 139-kHz vibrations enabled the inference that in both cases slow large-scale streamings with $L > \lambda$ were predominant:

Frequency, kHz	$\overline{V}_{max}$	L/λ	M_a	V_{max}/v_m
45	2.3	1.2	3×10^{-4}	0.03
139	2.8	2.5–5	0.6×10^{-4}	0.3

These data, in particular the values M_a for 139-kHz vibrations, qualitatively agree with the literature data for other liquids. Analysis of the streaming velocity V_{max} versus the pressure amplitude indicates that streamings are formed at a certain pressure amplitude threshold p_{mt} that is lower for 45 kHz than for 139 kHz. The comparison of p_{mt} with the thresholds of cavitation and crystal nucleation shows their agreement within the accuracy of the experiment.

Thus, cavitation and acoustic streaming developed in melts induce the formation and transfer of solid-phase nuclei, favoring crystallization and solidification of overcooled melts.

Of great practical interest is the study of formation and development of streamings during melt crystallization, i.e. in the presence of the solid-liquid interface (the crystallization front). In this case, acoustic streaming must depend (additionally to the above-mentioned factors such as the extent of melt overcooling and intensity of supplied vibrations), on the distance l between the crystallization front and the radiator, as well as on the two-phase zone width Δl.

This width depends on the properties of the melt and conditions of its solidification. By varying the composition of the melt and the conditions of its cooling, one can change, to a certain degree, the two-phase zone width and thus make the crystallization front more or less even.

It was expedient to analyse two extreme cases: the crystallization front is far from $(l \gg \lambda)$ or close to $(l \leq \lambda)$ the radiator. In the latter case, the streaming must be different in melts with a narrow $(\Delta l < l)$ and wide $(\Delta l \sim \lambda)$ two-phase zone. In experiments, the extent of cooling was chosen from the condition of an insignificant velocity of the crystallization front $(< 0.005 \text{ cm/s})$.

In the absence of ultrasonic field, the melt moved convectively (including the region near the crystallization front).

Irradiation with the vibrational amplitude above the threshold value gave rise to streaming. When the crystallization front was sufficiently far from the radiator ($l \gg \lambda$), the streamings developed in the same fashion as in the case of free liquid.

It should be noted that in some cases the occurrence of a liquid-solid interface in the melt led to reduction of the cavitation threshold. Thus, 139-kHz vibrations supplied to the naphthalene(35%)-camphor(65%) melt caused cavitation already at the vibrational velocity amplitude equal to 1–1.3 cm/s. A reduction in the cavitation threshold can be related to a lower solubility of gases in a solid phase than in liquid one, as a result of which some amount of gas is liberated at the crystallization front giving rise to gaseous bubbles that can serve as cavitation nuclei. In the melts with a higher content of naphthalene, the effect of threshold reduction was less pronounced.

When the crystallization front is very close to the radiator, the initial stage of streaming occurs similarly: above the cavitation threshold bubbles appear at the pressure antinodes and near the crystallization front. But in the following the process proceeds depending on the two-phase zone width. If it is narrow ($\Delta l \sim 0.1$ cm $< \lambda$), cavitation is accompanied by the appearance of solid-phase particles. As they are formed in the overcooled region (thermal or concentrational) near the crystallization front, this favors heat transfer through the solid phase and promotes the growth of crystals from these particles. An increase in the viscosity of the system gives rise to streamings which, together with cavitation, disperse the crystals growing on crystallization front and carry them into the melt, thus increasing the effective viscosity of the melt still further. Eventually, the melt will possess both large-scale ($L \sim \lambda$) and relatively low-scale streamings near the crystallization front (Figure 2.53).

In the case of a wide two-phase zone ($\Delta l \sim 0.4$ cm), the developing cavitation disperse the growing crystals and nucleation near the pulsating cavities is weak (Figure 2.54). The crystallites splitted from the crystallization front either stick between dendrites of the two-phase zone, or grow in the overcooled region to join eventually to the crystallization front.

The large-scale streamings with $L \sim \lambda$ are established near the crystallization front 7–10 s after the onset of ultrasonic vibrations. In the course of time the streaming gradually occupy more of free liquid, the regular structure of streamings being disturbed near the crystallization front.

The behaviour of acoustic streamings and their interaction with convective flows are determined by the ultrasonic field parameters, compo-

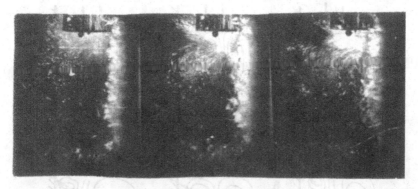

Figure 2.53. Acoustic streaming in the naphthalene(35%)-camphor(65%) melt at 139 kHz and vibrational velocity amplitude $v_m = 2$ cm/s. A narrow two-phase zone ($\Delta l \sim 0.1$ m). The interval between frames is 10 s.

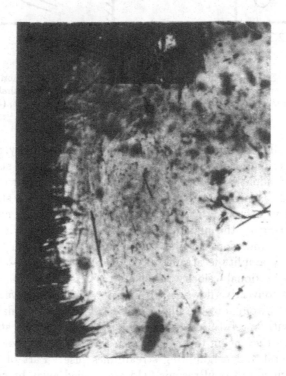

Figure 2.54. Dispersion of crystals from the crystallization front and their removal into the naphthalene(35%)-camphor(65%) melt at 139 kHz and vibrational velocity amplitude $v_m = 4$ cm/s. Magnification ×2.5.

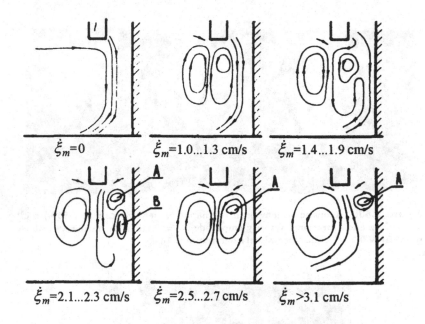

Figure 2.55. Effect of an increasing vibrational velocity amplitude on the development and interaction of acoustic and convective flows in the naphthalene(35%)-camphor(65%) melt at 139 kHz: (*1*) radiator, (*2*) crystallization front, (*3*) melt.

sition of the alloy, direction of convective flows, their velocity, as well as the two-phase zone width. Figure 2.55 represents acoustic streamings arising in the naphthalene(35%)-camphor(65%) melt under the action of 139-kHz ultrasonic vibrations and the interaction of these streamings with convective flows as functions of the vibrational velocity amplitude. In this experiment, the plane crystallization front was at a distance $l < \lambda$ from the radiator. Regular streamings appeare when the vibrational velocity amplitude attains its threshold value $v_m \sim 1.3$ cm/s. When the vibrational velocity amplitude is less than 1.4 cm/s, the acoustic and convective flows fail to interact, the convective flows being dominant throughout the liquid phase. At the vibrational velocity amplitude within 1.4–1.9 cm/s, the velocity and scale of streamings increased and they begin to interact with the convective flows so that their peripheral and neighboring streamlines fuse. Part of the crystal nuclei formed in the ultrasonic field are carried away by convective flows.

At still higher vibrational velocity amplitude ($1.9 < v_m < 2.3$ cm/s), the convective motion before the crystallization front become unstable.

The crystallites move along the convective streamlines and along the streamlines of vortex A and sporadical vortex B. At $v_m > 2.3$ cm/s, vortices A and B are stable in the narrow range of v_m beyond which vortex A disappeare, while vortex B is involved into the large-scale flow.

At $v_m > 3.1$ cm/s, the velocity of flows increase, the vortices close to crystallization front become unstable and disappeare. The acoustic streamlines are closed by the convective streamlines. The scale of the vortex faced to the free liquid somewhat increase. Acoustic streamings become dominant.

The mean velocities of streamings for various amplitude ranges were also determined from the analysis of photographs and cinemagraphic frames (Figure 2.56). With increasing v_m, the convective motion velocity (curve 1) gradually drops (this corresponds to diagrams I–IV in Figure 2.55) to reach zero at $v_m \sim 2.2$–2.5 cm/s. The velocity of low-scale vortices A and B is of the same order of magnitude as that of convective flows (curve 2), whereas the velocity of large-scale vortices at $v_m = 3$ cm/s somewhat exceed that of convective motion (the open-dot portion of curve 3). At $v_m > 3$ cm/s, the acoustic and convective flows actively interact, the velocity of convective flows being increased by an order of magnitude.

In spite of different experimental conditions (opposite directions of convective streamings, uneven front, different spacings between the crystallization front and radiator), if the intensity of vibrations is low, the acoustic and convective streamings did not interact. But if the vibrational velocity amplitude exceeds a certain value ($v_m > 3$–8 cm/s) dependent on the alloy composition, the streamings actively interact, and the velocity of streaming becomes 5–10 times higher than that of convective streamings.

The development of large-scale streamings act to level off temperature in the liquid bulk, to transfer the fine-grained solid particles which are nuclei of crystallization, and to disperse crystals. Thus, in the experiment with the naphthalene(35%)-camphor(65%) alloy (139-kHz ultrasound, vibrational velocity amplitude $v_m > 3.5$ cm/s, and velocity of the crystallization front motion 0.005 cm/s), the temperatures in the melt bulk and near the crystallization front (the distance between the measuring thermocouples was 20 mm) are virtually the same, while differed by $3.2°C$ before the onset of ultrasonic vibrations.

An increase in the amount of fine-grained solid particles and their transfer by acoustic streamings results in a more uniform fine-grained structure of the crystallizing ingot.

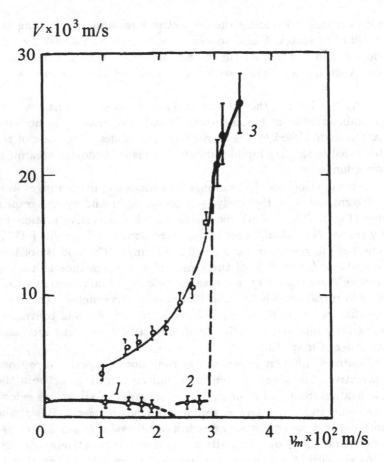

Figure 2.56. The velocity of convective (*1*) and acoustic (*2, 3*) flows in the naphthalene(35%)-camphor(65%) melt versus the vibrational velocity amplitude.

Besides being indicative of the relationship between such phenomena as cavitation, nucleation, and acoustic streaming, the data obtained elucidate the conditions for streamings development in the organic melts simulating crystallizing metals and the role of streamings in the formation of crystalline structure.

2.4 Radiation Pressure

Along with acoustic streaming, the propagation of a finite-amplitude ultrasound in liquids is associated with another nonlinear effect, namely the radiation or sound pressure, which lies in the occurrence of some nonzero net pressure exerted on any obstacle subject to an ultrasonic field.

The pressure on the obstacle in the ultrasonic field varies with the frequency of vibrations and, on the average, equals zero in a linear approximation. A nonzero pressure arises due to the effects of a second-order insignificancy and is small as compared with the sound pressure amplitude.

Thus, if the sound pressure amplitude in air is 10^3 dyne/cm^2, then the radiation pressure of the sound normally incident on the totally reflecting surface is about 1 dyne/cm^2 [75]. The radiation pressure is described by a set of equations (1.37), (1.38), (2.2), and (2.3a). With allowance for the terms quadratic with respect to amplitude and linear terms with respective dissipative coefficients, the equation (2.2) takes on the form

$$\rho_0 T_0 \frac{\partial S'}{\partial t} = \kappa_t \Delta T'. \tag{2.74}$$

Taking into account that

$$T' = \left(\frac{\partial T}{\partial p}\right)_S p', \tag{2.75}$$

one can obtain from (2.74)

$$S' = -\frac{\kappa_t}{T} \left(\frac{\partial T}{\partial p}\right)_S \operatorname{div} \boldsymbol{v}. \tag{2.76}$$

By substituting (2.76) into (2.3a) and in view of the known thermodynamic relations, one can derive

$$p' = c^2 \rho' + \frac{1}{2}\left(\frac{\partial c^2}{\partial \rho}\right)_S \rho'^2 - \kappa_t \left(\frac{1}{C_V} - \frac{1}{C_p}\right) \operatorname{div} \boldsymbol{v}. \tag{2.77}$$

Within the accuracy to the second-order infinitesimals, the equation of motion (1.37) can be rewritten as

$$\rho_0 \frac{\partial \boldsymbol{v}}{\partial t} + \rho' \frac{\partial \boldsymbol{v}}{\partial t} + \frac{\rho_0}{2}\nabla v^2 = -\nabla p' + \left(\frac{4}{3}\eta + \eta'\right) \nabla \operatorname{div} \boldsymbol{v}. \tag{2.78}$$

By averaging this equation with respect to time, integrating it over space, and taking into account that the term $\rho_0(\partial \boldsymbol{v}/\partial t)$ vanishes as a result of the averaging, equation for the time-averaged pressure $\overline{p}'$ can be derived

$$\overline{p}' = \frac{c^2 \overline{\rho'^2}}{2\rho_0} - \frac{p_0 \overline{v^2}}{2} + \left(\frac{4}{3}\eta + \eta'\right) \operatorname{div} v + \text{const.} \qquad (2.79)$$

This expression is basic for calculation of the radiation pressure exerted by an ultrasonic field.

The radiation pressure is a vector whose i-projection in a viscous heat-conducting medium is equal to*

$$P_{ri} = \overline{p}' n_i + \overline{\rho v_i (v_k n_k)} - \sigma'_{ik} n_k, \qquad (2.80)$$

where n is the unit vector of the internal normal to the surface.

In the case of an ideal liquid, the radiation pressure vector takes the form

$$\overline{p}_r = \overline{p}' n + \overline{\rho v (v n)}. \qquad (2.81)$$

On the surface of a solid-state obstacle, the normal component of velocity $(vn) = 0$ and

$$P_r = \overline{p}' n. \qquad (2.82)$$

In the general case, the radiation pressure vector has nonzero normal and tangential components. Thus, if a totally absorbing obstacle is positioned in such a manner that $n \| v$, then the radiation pressure components are as follows

$$P_{rn} = \overline{p}' + \overline{\rho v^2}, \quad P_{rt} = 0. \qquad (2.83a)$$

But if $n \perp v$ (the velocity is tangential to the surface), then

$$P_{rn} = \overline{p}', \quad P_{rt} = 0. \qquad (2.83b)$$

In the case of an obliquely incident ultrasonic wave, the normal and tangential components of radiation pressure are given by

$$P_{rn} = \overline{p}' + \rho v_n^2; \quad P_{rt} = p\overline{v_t v_n}. \qquad (2.83c)$$

Let us consider some particular cases of the radiation pressure estimation based on above equations. Thus, if vibrations propagate

---

*σ'_{ik} are the third-order infinitesimals, as well as the term $(4/3\eta + \eta')\operatorname{div} v$ in the expression (2.79).

through an infinite medium and decay at infinity, then the pressure $\overline{p}'$ can be calculated from expression (2.79). In this case, the term $b \operatorname{div} \boldsymbol{v}$ can be omitted as a third-order infinitesimal, and the constant has to be taken zero to satisfy the condition of the absence of a disturbance at infinity

$$\overline{p}' = -\frac{\rho v^2}{2} + \frac{c^2 \rho'^2}{2\rho}. \tag{2.84}$$

In the case of vibrations propagating in an infinite medium, it is of interest to estimate the radiation pressure exerted on obstacles, totally absorbing, or totally reflecting, or representing the interface between two liquid phases.

The radiation pressure exerted by a plane traveling acoustic wave on the totally absorbing obstacle is given by expression from [3]

$$\boldsymbol{P}_{ra} = \rho v(\boldsymbol{v}\boldsymbol{n}). \tag{2.85a}$$

In this case, $\rho' = \rho v/c$ and $\overline{p}' = 0$. At $\boldsymbol{n} \| \boldsymbol{v}$

$$\boldsymbol{P}_{ra} = \rho \overline{v^2} \boldsymbol{n} = 2\overline{E}_k \boldsymbol{n}, \tag{2.85b}$$

where E_k is the kinetic energy density in the wave.

If vibrations are totally reflected from a stationary solid obstacle, this gives rise to standing waves. In a linear approximation, the field of a plane standing wave near the obstacle is given by

$$\left. \begin{array}{l} v = -v_m \sin kx \, \cos \omega t \\[6pt] \rho' = \dfrac{\rho v_m}{c} \cos kx \, \sin \omega t \end{array} \right\}, \tag{2.86}$$

where the x axis is normal to and originates from the obstacle surface.

Substituting (2.86) into (2.84), one can get

$$\overline{p}' = \frac{\rho v_m^2}{4} \cos 2kx. \tag{2.87a}$$

At $x = 0$, this expression transforms into

$$\overline{p}' = \frac{\rho v_m^2}{4}. \tag{2.87b}$$

In the case of a reflecting obstacle, the radiation pressure is determined by the quantity $\bar{p}'$ [76]

$$\boldsymbol{P}_{rr} = \frac{c^2}{2\rho}\overline{p'^2}\boldsymbol{n}. \tag{2.88a}$$

By substituting expression (2.87b), one can derive

$$\boldsymbol{P}_{rr} = \frac{\rho v_m^2}{4}\boldsymbol{n}. \tag{2.88b}$$

When determining the radiation pressure exerted on the interface between two liquid media, one has to take into account that two waves – incident and reflected – propagate through the first medium, whereas the refracted wave travels through the second medium. In the case of the normally incident wave, the boundary conditions are as follows

$$\rho_1 c_1(v_1 - v_1') = \rho_2 c_2 v_2, \tag{2.89}$$

$$v_1 + v_1' = v_2,$$

where the subscripts 1 and 2 denote the first and second media, v_1 and v_1' are the velocities of the incident and reflected waves, respectively. In a linear approximation, the expressions (2.87) are valid for each of the waves.

Using expressions (2.87), (2.77a), and (2.84), one can get [75]

$$P_r = 2\rho_1 v_1^2 \left[\frac{\rho_1^2 c_1^2 + \rho_2^2 c_2^2 - 2\rho_1\rho_2 c_1^2}{(\rho_1 c_1 + \beta_2 c_2)^2} \right]. \tag{2.90}$$

Of interest is the case when plane standing waves are generated in the liquid confined between two solid walls.

Rayleigh [77] derived an expression for the sound pressure exerted on the stationary rigid wall

$$P_{rn} = \bar{p}' = \frac{\gamma + 1}{8}\rho_0 v_m^2, \tag{2.91}$$

where v_m is the standing wave velocity amplitude; $\gamma = 1+\rho/c^2(\partial c^2/\partial \rho)_S$ is the parameter describing nonlinear properties of the medium (for an ideal gas, $\gamma = C_p/C_V$).

The sound pressure dependent on the parameter γ is known as the Rayleigh pressure.

Of practical interest is the question concerning the forces acting upon the particle subject to the sound field, since they should be taken into account in consideration of such phenomena as coagulation in gases and liquids and degassing of liquids.

Some authors (see, for instance, [78, 79]) obtained expressions for the forces acting on the particle (spherical or discoidal) placed in the field of traveling or standing plane waves.

The simplest case is when a spherical particle with the radius much less that the wavelength ($R \ll \lambda$) is considered.

One can write for a compressible sphere subject to a traveling plane wave

$$F_{rx} = \frac{2}{9}\pi R^2 (kR)^4 \rho v_m^2 \left(a_1^2 + a_1 a_2 + \frac{3}{4} a_2^2 \right), \qquad (2.92)$$

where

$$a_1 = 1 - \frac{c^2 \rho}{c_s \rho_s}; \quad a_2 = 2\frac{\rho_s - \rho}{2\rho_s + \rho}.$$

In the case of a standing wave [78]

$$F_{rx} = \pi R^2 (kR) \rho v_m^2 f \left(\frac{\rho_s}{\rho} \frac{c_s}{c} \right), \qquad (2.93)$$

where

$$f\left(\frac{\rho_s}{\rho} \frac{c_s}{c} \right) = \frac{\rho_s + \frac{2}{3}(\rho_s - \rho)}{2\rho_s + \rho} - \frac{1}{3}\frac{c^2 \rho}{c_s^2 \rho_s}.$$

Here the subscript s indicates that parameters c and ρ are referred to the solid phase.

It can be inferred from these expressions that the radiation forces acting on a small particle are significantly stronger in the standing wave than in the traveling wave, since the small multiplier is to the first power in the former case and to the fourth power in the latter case. Moreover, the particles whose radii are greater than the resonant value are displaced to the velocity antinodes of the standing wave [80], and those with lesser radii – to the velocity nodes. The resonant particles (i.e., possessing the resonant radius) are at equilibrium.

The radiation forces acting on the particle are shown in Figure 2.57,a for arbitrary kR.

The relevant expressions for the force acting upon a solid free disk were obtained in [81, 82]. If the disk axis coincides with the direction of

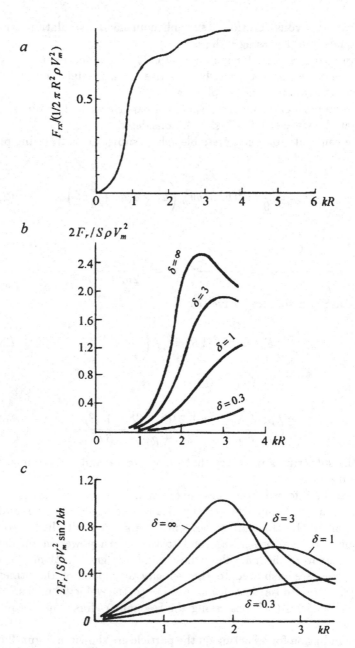

Figure 2.57. Radiation forces acting upon (a) a stationary spherical particle in the plane traveling wave field [82] and upon a discoidal particle in (b) plane traveling and (c) standing waves [83].

the vibrational velocity, then the radiation force in the traveling wave
is given by

$$F_r = \frac{8}{27\pi} S(kR)^4 \left(\frac{\delta}{\delta+1}\right)^2 \rho v_m^2 \qquad (2.94a)$$

and in the standing wave

$$F_r = \frac{2}{3\pi} S\, kR\, \rho v_m^2 \left(\frac{\delta}{\delta+1}\right)\left[1 + \frac{\delta}{\delta+1}\frac{(kR)^2}{5}\right]\sin 2kh, \qquad (2.94b)$$

where

$$S = \pi R^2,$$

$$\delta = \frac{3}{8}\frac{m}{\rho R^3},$$

m is the disk's mass; h is the distance from the disk surface to the
velocity node.

The radiation forces acting in traveling and standing plane waves
are shown in Figures 2.57,b,c [82] for $kR \sim 1$ and various values of
parameter δ.

References

1. L. K. Zarembo and V. A. Krasil'nikov, *Introduction into
    Nonlinear Acoustics* (in Russian), Nauka, Moscow (1966).

2. J. Rayleigh, *Theory of Sound*, MacMillan & Co., London (1926).

3. L. D. Landau and E. M. Lifshitz, *Mechanics of Continuous
    Media* (in Russian), Gostekhizdat, Moscow (1953).

4. O. V. Rudenko and S. I. Soluyan, *Fundamentals of Nonlinear
    Acoustics* (in Russian), Nauka, Moscow (1975).

5. J. Guckenheimer and P. Holmes, *Nonlinear Oscillations in
    Dynamical Systems and Bifurcations of Vector Fields* (1983).

6. J. Weber and R. Stefens, Absorption in Liquids, In: *Ultrasonics
    in Quantum and Solid-State Physics* (Russian translation), Mir,
    Moscow, p. 75 (1970).

7. K. A. Naugol'nykh, In: *Intense Ultrasonic Fields* (in Russian),
    Nauka, Moscow, p. 5 (1968).

8. *Physical Principles of Underwater Acoustics* (Russian
    translation), ed. V. I. Myasishchev, Sov. Radio, Moscow (1955).

9. H. Minnaert, *Phil. Mag.*, **16**, 235 (1933).

10. C. Devin, *JASA*, **31**, 1654 (1959).

11. E. Skuchik, *Fundamentals of Acoustics* (Russian translation), **2**, Inostr. Liter., Moscow (1959).

12. N. Sirotyuk, In: *Intense Ultrasonic Fields* (in Russian), Nauka, Moscow, p. 167 (1968).

13. E. L. Carstensen and L. L. Foldy, *JASA*, **19**, 481 (1947).

14. M. L. Eksner, *Problems of Contemporary Physics* (Russian translation), Inostr. Liter., Moscow, p. 118 (1953).

15. M. Kornfeld, *Elasticity and Strength of Liquids* (Russian translation), GINTL, Moscow, Leningrad (1951).

16. A. D. Pernik, *Problems of Cavitation* (in Russian), Sudostroenie, Leningrad (1966).

17. R. Knapp, J. Dayly, and F. Hammit, *Cavitation* (Russian translation), Mir, Moscow (1974).

18. G. Flinn, In: *Methods and Instruments for Study of Ultrasound* (Russian translation), **1**, part B, 7, Mir, Moscow.

19. V. A. Akulichev, In: *Intense Ultrasonic Fields* (in Russian), Nauka, Moscow, p. 129 (1968).

20. L. D. Rozenberg, In: *Intense Ultrasonic Fields* (in Russian), Nauka, Moscow, p. 221 (1968).

21. W. Lauterborn, *Proc. Symp. Finite-Amplitude Wave Effects in Fluids*, ed. L. Bjorno, Copenhagen (1973).

22. W. Lauterborn, *Cavitation and inhomogeneities in Underwater Acoustics* (1980).

23. F. R. Gilmore, Cal. Tech. Inst. Rep. 26–4 (1952).

24. Ya. B. Zel'dovich, *Zh. Eksp. Teor. Fiziki*, **12**, 525 (1942).

25. D.-Y. Hsieh and M. S. Plesset, *J. Acoust. Soc. Am.*, **33**, 206 (1961).

26. M. Strasberg, *J. Acoust. Soc. Am.*, **33**, 359L (1961).

27. F. F. Fox and K. F. Herzfeld, *J. Acoust. Soc. Am.*, **26**, 984 (1954).

28. W. E. Whybrew, G. D. Kinzer, and R. Gunn, *J. Geophys. Res.*, **57**, 459 (1952).

29. Ya. I. Frenkel', *Kinetic Theory of Liquids* (in Russian), AN SSSR, Moscow (1945).

30. J. C. Fisher, *J. Appl. Phys.*, **19**, 1062 (1948).

31. M. C. Plesset, *Cavitation State of Knowledge*, ASME Conference in Illinois, New York (1969).

32. D. A. Labuntsov, *Teploenergetika*, no. 7, 19 (1959).

33. S. G. Bankov, In: *Physics of Boiling* (in Russian), Mir, Moscow, p. 81 (1964).

34. E. N. Harvey, D. K. Barnes, W. D. McElroy, A. H. Whiteley, D. C. Pease, and K. W. Cooper, *J. Cell. Comp. Physiol.*, **24**, 17

(1944).

35. J. Rayleigh, *Phil. Mag.*, **34**, 94 (1917).
36. B. E. Noltingk and E. A. Neppiras, *Proc. Phys. Soc.*, **64B**, 1032 (1951).
37. R. Koul, *Underwater Bursts* (Russian translation), Inostr. Liter., Moscow (1950).
38. R. E. Apfel, *Br. J. Cancer*, **45**, 140 (1982).
39. U. Parlitz, V. Englisch, C. Scheffczyk, and W. Lauterborn, *J. Acoust. Soc. Am.*, **88**, 1061 (1990).
40. L. Crum, *Ultrasonics*, **22**, 215 (1984).
41. W. Lauterborn, *Encyclopedia of Physical Sciences and Technology, Acoustic Chaos*, 1, Academic Press, p. 104 (1992).
42. A. Philipp, M. Delins, C. Scheffczyk, A. Vogel, and W. Lauterborn, *J. Acoust. Soc. Am.*, **93**, 2496 (1993).
43. A. Vogel and W. Lauterborn, *J. Acoust. Soc. Am.*, **84**, 719 (1988).
44. B. Flint and K. Suslick, *J. Am. Chem. Soc.*, **111**, 6987 (1989).
45. A. Vogel, W. Lauterborn, and R. Timm, *J. Fluid Mech.*, **206**, 229 (1989).
46. E. A. Margulis, *Sonochemical Reactions and Sonoluminescence* (in Russian), Khimiya, Moscow (1986).
47. W. Lauterborn, E. Schmitz, and A. Judt, *Int. J. Bifurc. and Chaos*, **3**, 635 (1993).
48. H. Frenzel and H. Schultes, *Z. Phys. Chem.*, **27B**, 421 (1934).
49. S. Putterman, *Proc. ACS Colloid and Surface Science Symp.*, Stanford (1994).
50. Ya. I. Frenkel', *Zh. Fiz. Khim.*, **14**, 305 (1940).
51. Yu. S. Astashkin, In: *New Physical Methods in Technology* (in Russian), Metallurgiya, Moscow, p. 70 (1977).
52. O. V. Abramov, Yu. S. Astashkin, and V. A. Petrovskii, In: *New Physical Methods in Metallurgy* (in Russian), Metallurgiya, Moscow, p. 161 (1974).
53. Yu. S. Astashkin, In: *Acoustic Vibrations in Technology* (in Russian), Metallurgiya, Moscow, p. 26 (1981).
54. Yu. V. Naidich, In: *Surface Phenomena in Melts and Crystallizing Solid Phases* (in Russian), Nal'chik, p. 30 (1965).
55. V. I. Danilov, *Structure and Crystallization of Liquids* (in Russian) (1956).
56. E. Kikuchi, *VI-th Int. Congress on Acoustics*, Report NH-5-6, Tokyo, 4p (1968).
57. G. I. Eskin, A. I. Ioffe, and P. P. Shvetsov, In: *Technology of Light Alloys* (in Russian), VILS, Moscow, p. 8 (1974).

58. *Ultrasound Technology* (in Russian), ed. B. A. Agranat, Moscow, Metallurgiya (1974).

59. O. V. Abramov, Yu. S. Astashkin, and S. I. Pugachev, *Tekhnol. Sudostr.*, no. 1, 111 (1976).

60. O. V. Abramov and Yu. S. Astashkin, In: *New Physical Methods in Metallurgy* (in Russian), Metallurgiya, Moscow, p. 155 (1974).

61. Yu. N. Ivashchenko, V. B. Bogatyrenko, and V. I. Eremenko, In: *Surface Phenomena in Melts and Powder Metallurgy* (in Russian), AN USSR, Kiev, p. 391 (1963).

62. V. I. Dobatkin and G. I. Eskin, In: *Effect of High-Power Ultrasound on Metal Interface*, Nauka, Moscow, p. 6 (1986).

63. M. Faraday, *Phil. Trans. Roy. Soc. London*, **121**, 221 (1831).

64. W. L. Nybord, *Physical Acoustics*, **2** (1965).

65. L. K. Zarembo, In: *Intense Ultrasonic Fields* (in Russian), Nauka, Moscow, p. 89 (1968).

66. Eckart, *Phys. Rev.*, **73**, 68 (1948).

67. E. V. Romanenko, *Akust. Zh.*, **6**, 92 (1960).

68. Yu. G. Statnikov, *Akust. Zh.*, **13**, 146 (1967).

69. Yu. Ya. Borisov and Yu. G. Statnikov, *Akust. Zh.*, **11**, 35 (1965).

70. P. P. Prokhorenko, S. I. Pugachev, and N. G. Semenova, *Ultrasonic Metallization of Materials* (in Russian), Nauka and Tekhnika, Minsk (1987).

71. H. Schlichting, *Phys. Z.*, **33**, 327 (1932).

72. H. Schlichting, *Boundary Layer Theory*, New York (1955).

73. W. L. Nyborg, *JASA*, **30**, 3296 (1958).

74. O. V. Abramov, Yu. S. Astashkin, and V. S. Stepanov, *Akust. Zh.*, **25**, 180 (1979).

75. Z. A. Goldberg, In: *Intense Ultrasonic Fields*, Nauka, Moscow, p. 49 (1968).

76. A. A. Eihenwald, *Usp. Fiz. Nauk*, **14**, 552 (1934).

77. Lord Rayleigh, *Philos. Mag.*, **10**, 354 (1905).

78. L. P. Gor'kov, *Dokl. Akad. Nauk SSSR*, **140**, 88 (1961).

79. L. V. King, *Proc. Roy. Soc.*, **A147**, 212, 861 (1936).

80. K. Yosioka and Y. Kawasima, *Acustica*, **5**, 167 (1955).

81. H. Olsen, W. Romberg, and Wergeland, *JASA*, **30**, 69 (1959).

82. J. Awatani, *JASA*, **27**, 282 (1955).

Chapter 3

The Propagation of High-Intensity Ultrasonics in Solids

This chapter reviews nonlinear effects arising during the propagation of high-power ultrasound in solids such as amplitude-dependent internal friction and dislocational structure alteration, as well as related changes in the mechanical properties of ultrasonically treated solids and kinetics of diffusion processes.

3.1   Amplitude-Dependent Dislocational Internal Friction

In addition to the thermoelastic and other kinds of energy losses which were considered in section 1.4, the propagation of elastic waves in crystalline solids leads to losses due to the stress-induced movement of inherent structure defects. In this respect, absorption of vibrations due to the movement of dislocations is of prime interest.

Read [1] was the first to suggest that ultrasonic energy can be dissipated through the movement of dislocations. Koehler [2] refined this idea by proposing that a linear dislocation can vibrate under the action of a cyclic mechanical stress like a string imposed by forced vibrations. Later, Granato and Lucke [3] considerably improved the original model.

They assumed that dislocations are linear and pinned with interstitial atoms, point defects, and dislocation network nodes (Figure 3.1, scheme A). The pinning of dislocations with network nodes is superior

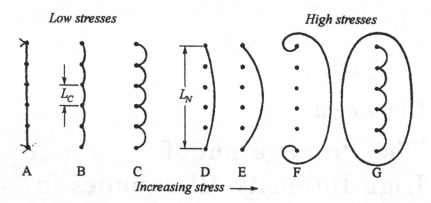

Figure 3.1. Sequential stages of bowing and breakaway of a pinned dislocation line under increasing applied stress [3].

to that with point defects and interstitials. Pinning with point defects controls the dislocation loop length L_p which is smaller than the loop length L_n between dislocation network sites.

If an external stress is applied to a crystal with dislocations, the resultant strain involves an elastic component ε_e and dislocation component ε_d, the latter being due to the dislocation motion

$$\varepsilon_d = bly, \tag{3.1}$$

where b is the Burgers vector, y the dislocation displacement, and l the dislocation segment length.

The relation between stress and dislocation strain is schematically illustrated in Figure 3.2.

In the range of very weak stresses (Figure 3.1, scheme B and respective point B in Figure 3.2), the dislocation loops L_p bow and break away when a particular stress level is achieved, which is accompanied by an abrupt increase in the dislocation loop length from L_p to L_n (Figure 3.1, schemes C and D, and respective points C and D in Figure 3.2). Ones a breakaway stress is surpassed, the dislocation strain dramatically increases, which results in a bowing of the loop L_n. The network nodes are assumed to be extremely strong locking points allowing no breakaway to occur at them.

A further increase in the applied stress leads to the excitation of the Frank–Read sources (Figure 3.1, schemes F and G) and, hence, to plastic strain.

In terms of the above model, one can distinguish two types of energy losses due to low-level stresses. The first type is due to the friction-

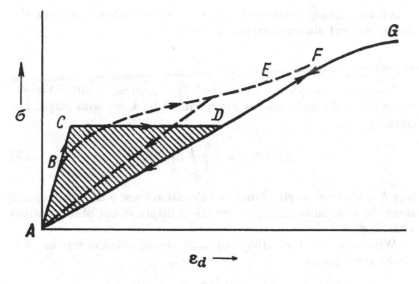

Figure 3.2. Stress-strain dependences for a model presented in Figure 3.1 [3].

related lag of strains behind stresses, which must lead to energy absorption. This type of frequency-dependent energy loss is maximum near the resonance determined by the loop length and vanishes at very high and low frequencies.

The second type of losses is conditioned by the elastic contraction of long dislocational loops during the relaxation phase (Figure 3.2, path D–A), which leads to hysteresis. The contracted loops are again pinned with point defects, and the process is repeated cyclically. This type of losses is proportional to the area embraced with the dislocational stress-strain curves. Since the relation between the dislocation stress and the strain it produced does not depend on frequency in a wide frequency range, this type of losses is frequency-independent.

All this is valid for zero temperature, but at non-zero temperatures dislocations can overcome a potential barrier at the expense of thermal excitation.

It is possible to estimate, in terms of the model considered, the absorption coefficient related to dislocation motion.

With allowance for (1.104) and strain-displacement relationship, one can write the following equation of motion

$$\frac{\partial^2 \sigma}{\partial x^2} - \rho \frac{\partial^2 \varepsilon}{\partial t^2} = 0. \tag{3.2}$$

As was already mentioned, the total deformation ε involves the elastic (ε_e) and dislocational (ε_d) components

$$\varepsilon = \varepsilon_e + \varepsilon_d.$$

The elastic component is described by expression (1.103). According to (3.1), the dislocational strain caused by loops with length l is given by

$$\varepsilon_d = \Lambda b \overline{S} = \Lambda \frac{b}{l} \int_0^l S(y)\, dy, \qquad (3.3)$$

here Λ is the total length of mobile dislocation lines; y is the coordinate along the dislocation line; $\overline{S}$ is the mean displacement of dislocations with length l.

With allowance for (1.103) and (3.3), the equation of motion (3.2) can be rewritten as

$$\frac{\partial^2 \sigma}{\partial x^2} - \frac{\rho}{\mu} \frac{\partial^2 \sigma}{\partial t^2} = \frac{\Lambda \rho b}{l} \frac{\partial^2}{\partial t^2} \int_0^l S(y)\, dy. \qquad (3.4)$$

The character of dislocation displacements under the action of cyclic stresses is defined by a chosen model. If the vibrating string model is used, then

$$A \frac{\partial^2 S}{\partial t^2} + B \frac{\partial S}{\partial t} - C \frac{\partial^2 S}{\partial y^2} = b\sigma, \qquad (3.5)$$

where $S = S(x, y, t)$ is the displacement of dislocation element from its equilibrium position, occurring in the slip plane; $A = \pi \rho b^2$ is the effective mass of dislocation unit length; B is the friction force per unit length; $C = 2\mu b^2 / \pi(1-\nu)$ is the effective tension of a bent dislocation; $b\sigma$ is the driving force per dislocation unit length, produced by an applied shear stress.

The friction constant B can be estimated by two ways: through the heat liberated near a vibrating dislocation [4] and through the scattering of thermal phonons from dislocations [5]. The constant B was estimated to be 5×10^{-5}–5×10^{-3}.

The boundary conditions are as follows

$$S(x, 0, t) = S(x, l, t) = 0. \qquad (3.6)$$

The equations (3.4) and (3.5) form a set of integral-differential equations. Among its possible solutions $\sigma = \sigma(x, y, t)$ and $S = S(x, y, t)$, of most interest are those for which σ is a periodical function of time, independent of the coordinate y.

The substitution of the solution

$$\sigma = \sigma_0 \exp(-\alpha_d x) \exp\left[j\omega\left(t - \frac{x}{c}\right)\right] \tag{3.7}$$

gives

$$S = \frac{4b\sigma}{\pi A} \sum_{n=0}^{\infty} \frac{1}{2n+1} \sin\frac{(2n+1)\pi y}{l} \frac{\exp[j(\omega t - \delta_n)]}{[(\omega_n^2 - \omega^2)^2 + (\omega d)^2]^{1/2}}, \tag{3.8}$$

where

$$d = \frac{B}{A}; \quad \omega_n = (2n+1)\frac{\pi}{l}\left(\frac{C}{A}\right)^{1/2}; \quad \delta_n = \text{arctg}\frac{\omega d}{\omega_n^2 - \omega^2}.$$

The above expressions for σ and S satisfy the set of equations (3.4) and (3.5), if

$$\alpha_d(\omega) = \frac{\omega}{2c}\frac{\Lambda D E^2}{\pi}\sum_{n=0}^{\infty}\frac{1}{(2n+1)^2}\frac{\omega d}{(\omega_n^2 - \omega^2)^2 + (\omega d)^2}, \tag{3.9a}$$

and

$$c(\omega) = c_0\left[1 - \frac{\Lambda D E^2}{2\pi}\sum_{n=0}^{\infty}\frac{1}{(2n+1)^2}\frac{\omega_n^2 - \omega^2}{(\omega_n^2 - \omega^2)^2 + (\omega d)^2}\right], \tag{3.9b}$$

where

$$c_0 = \left(\frac{\mu}{\rho}\right)^{1/2}; \quad D = \frac{8\mu b^2}{\pi^3 C}; \quad H = \frac{\pi^2 C}{A}.$$

As dislocations move only in slip planes, which generally are not parallel to the direction of wave propagation, then, instead of σ and ε, only their respective components must enter into equations (3.4) and (3.5). In this connection, the right-hand side of expression (3.9a) and the second term on the right-hand side of expression (3.9b) must be multiplied by the orientational factor Π representing the square of the ratio of the shear stress, normalized in the slip system, to the applied stress σ.

Analysis of the expressions obtained and comparison of calculated and experimental data indicate that already the first term ($n = 0$) of the series gives a good representation of the function for the majority

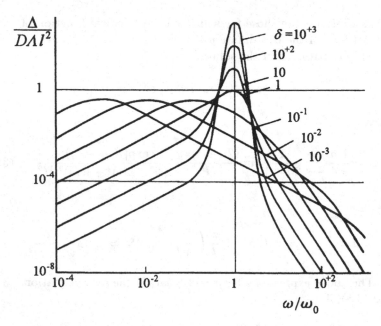

Figure 3.3. Frequency dependences of damping decrement for various damping constants and delta distribution of loops in length [3].

of interesting cases. The absorption coefficient α_d and attenuation decrement Δ are given by

$$\alpha_d = \frac{\Pi}{2\pi c} D\omega_0^2 \Lambda l^2 \frac{\omega^2 d}{(\omega_0^2 - \omega^2)^2 + (\omega d)^2}, \qquad (3.10\text{a})$$

$$\Delta = \alpha_d \lambda = \Pi D \Lambda l^2 \frac{d}{\omega_0} \frac{a^2}{(1 - a^2)^2 + a^2 (d/\omega_0)^2}, \qquad (3.10\text{b})$$

where $a = \omega/\omega_0$ is the normalized frequency.

The expression (3.10b) is written in the form convenient for analysis of the frequency dependence of attenuation decrements, which are shown in the reduced form $\Delta/D\Lambda l^2$ in Figure 3.3 for various values of the damping constant $\delta = d/\omega_0$.

When damping is small ($\delta \ll 1$), the decrement linearly increases with frequency almost to its resonant value, then drastically rises to reach its maximum, after which decreases proportionally to the third power of frequency. But if damping is great, the decrement grows linearly to its maximum value which is attained before the resonance. Near the resonance, the decrement falls as $1/\omega$ and then as $1/\omega^2$.

Thus, the maximum loss takes place at $\omega = \omega_0$ and $\omega = \omega_0^2/d$ in the cases of small and great damping, respectively. The resonant frequency depends on the material parameters ρ, μ, l.

In the case of great damping and $(\omega/\omega_0)^2 \ll 1$, the expression (3.10h) can be rewritten in the form

$$\Delta = \frac{\Omega D \Lambda l^2 \omega \tau_d}{1 + \omega^2 \tau_d^2}, \qquad (3.11)$$

where Ω is the orientation factor; and $\tau_d = Bl^2/\pi^2 C$.

The expression (3.11), as a function of frequency, is relaxational with a maximum

$$\Delta_m = \frac{\Omega D \Lambda l^2}{2} \qquad (3.12a)$$

at frequency

$$\omega_m = \tau_d^{-1} = \frac{\omega_0^2}{d} = \frac{\pi^2 C}{Bl^2}. \qquad (3.12b)$$

At frequencies much more than ω_m, so that $\tau_d \gg 1$, the decrement Δ decreases as ω^{-1}, and the absorption coefficient attains the value

$$\alpha_{d\infty} = \frac{4\Omega \mu b^2 \Lambda}{PBC} \qquad (3.13)$$

The expression (3.13) shows that the maximum absorption depends on the orientation factor, dislocation density, and damping constant, but does not depend on the dislocation loop length and the effective tension of dislocations.

In real solids, dislocational loops do not have an equal length l but are randomly distributed in length according to the distribution function

$$N(l)\,dl = \frac{\Lambda}{\bar{l}^2} \exp\left(-\frac{l}{\Lambda}\right) dl, \qquad (3.14)$$

where $N(l)\,dl$ is the number of loops whose lengths lie between l and $l+dl$, and $\bar{l}$ is the mean length of loops. Then the attenuation decrement is given by expression

$$\Delta = \int_0^\infty \delta(l) N(l)\,dl. \qquad (3.15)$$

The allowance for the length distribution of dislocation loops leads to an enhanced decrement maximum and its shift toward lower frequencies; in other words, the contribution into absorption from longer loops is greater than follows from their relative number.

Analysis of the schemes presented in Figures 3.1 and 3.2 indicates that losses due to hysteresis must be zero for the cyclic stress amplitudes less than some critical value corresponding to the point C. But once they are greater than this critical value σ_{mc}, the energy loss w over the period of vibrations is constant and equal to the hatched area in Figure 3.2. When the amplitude exceeds s_{mc}, the decrement Δ jumps from zero to its maximum value and then changes inversely proportional to the square of the stress amplitude ($\Delta = 2w\mu/\sigma_{mc}^2$).

It should be noted that the allowance for the length distribution of dislocation loops leads to a gradual, rather than abrupt, changes in the attenuation decrement with the stress amplitude. The breakaway of dislocations from pinning points occurs when the force applied to a dislocation segment overwhelms the force acting on it from the pinning point. For two neighboring dislocation segments of lengths l_1 and l_2, the former force equals $b(l_1 + l_2)\sigma/2$, and the latter is $\sim E_c/b$ (E_c is the binding energy of the pinning point). The pair of neighboring segments having the greatest total length breaks away first, which increases the total length of other two segments – adjacent to the first two and newly formed. This leads to a rapid development of the breakaway process.

To calculate the amplitude dependence of decrement, one should know or determine the distribution function for loop pairs. Assuming that all distances between the dislocation network nodes have the same length L_n and that the unstressed loops are distributed randomly in accordance with expression (3.14), Granato and Lucke [3] obtained the following expression for the amplitude-dependent decrement

$$\Delta_n = \frac{C_1}{\varepsilon_m} \exp\left(-\frac{C_2}{\varepsilon_m}\right), \qquad (3.16)$$

where ε_m is the strain amplitude; $C_1 = [\Omega D \Lambda L_n/(\pi L)]C_2$ and $C_2 = k\eta b/L$; k is the breakaway-related parameter accounting for the direction of propagation of vibrations relative to the slip plane, $k = \mu/(4RE)$; R is the coefficient of reduced shear stress; η is the Cottrell misfit parameter accounting for the relative difference in the sizes of pinning defects and material atoms.

In accordance with (3.16), the graph $\ln(\Delta_n \varepsilon_m)$ vs. $1/\varepsilon_m$ should represent a straight line with a slope proportional to the number of pinning sites and with the x-intercept proportional to dislocation density.

Analysis of the available literature data has revealed that a sharp increase in internal friction (Figure 1.17, region III) in metals occurs at the strain amplitudes greater than 10^{-4}. In this case, the acoustic energy loss is due to irreversible dislocation motion driven by the Frank–Read sources. For most metals, the acoustic stress amplitudes corresponding to $\varepsilon_m > 10^{-4}$ are above the critical stress necessary for dislocation reproduction.

Below the internal friction in metals and alloys is discussed for frequencies 20–40 kHz and strain amplitudes $\varepsilon_m > 10^{-4}$ to 10^{-3}, as reported by Kulemin [6]. According to his data, internal friction can be measured by the thermoacoustic method. The mechanism of internal friction at large strain amplitudes implies that energy losses depend on the time of vibrational treatment. In this case, the coefficient of internal friction of most metals except iron decreases (Figure 3.4). The plateauing of $Q^{-1}(\tau)$ plots is caused by a variation in the dislocation density during ultrasonic treatment (see section 3.2 in this chapter). In iron, high-intensity vibrations cause an initial increase in the internal friction and then its leveling off (Figure 3.4, curve *5*), which appears to be the result of a dynamic equilibrium between the processes of dislocation breakaway, reproduction, and pinning due to carbon diffusion.

At $\varepsilon_m > 10^{-4}$, the values of internal friction measured at increasing and decreasing strain amplitude differ, which can be explained by the fact that in this range of strain amplitudes material's properties change during the time of measurements.

Figure 3.5 shows the internal friction vs. strain amplitude plots for some metals, arrows indicating the direction of measurements. When the increasing ε_m surpasses some critical value (points A on the plots), the internal friction increases abruptly (curve *a*). During the phase of decreasing strain amplitude, the internal friction diminishes too, the respective curve *b* being situated below curve *a*. The critical amplitude ε_m is now slightly higher. If ε_m increases again, Q^{-1} values fit well curve *b*, but during a subsequent reduction in the strain amplitude, Q^{-1} plots fit curve *c* (for Cd) situated below curve *b*. In this case, the critical value for ε_m increases further, whereas the loop becomes slightly narrower.

Some metals exhibit special features in their $Q^{-1}(\varepsilon_m)$ curves. Thus, tin and zinc display peaks (curves *4* and *5*, respectively), which can be explained by the fact that, at some amplitude ε_m, the Frank–Read dislocation density reaches its maximum and would not grow with a further increase in ε_m, whereas the acoustic energy input continues to rise and can reduce Q^{-1}. It can be inferred that in metals lacking maxima on their $Q^{-1}(\varepsilon_m)$ plots, peak amplitudes were not attained.

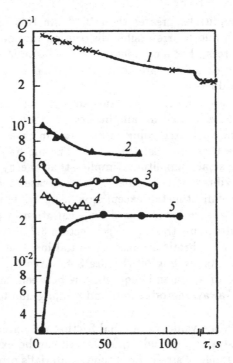

Figure 3.4. Internal friction versus the time of ultrasonic treatment ($f = 20$ kHz). (*1*) Cd, $\varepsilon_m = 0.92\times10^{-4}$; (*2*) Sn, $\varepsilon_m = 0.81\times10^{-4}$; (*3*) Zn, $\varepsilon_m = 2.62\times10^{-4}$; (*4*) Ni, $\varepsilon_m = 1.6\times10^{-4}$; (*5*) Fe, $\varepsilon_m = 2.84\times10^{-4}$. (*1*)–(*4*) $t = 20°$C; (*5*) $t = 525°$C; (*4*) and (*5*) $H = 280$ Oe [6].

In all metals except Fe, the values of internal friction obtained at rising ε_m were higher than those obtained at diminishing ε_m.

The behavior of internal friction in Fe is different (curve *6*), which can be explained by the fact that iron typically exhibits a stronger interaction between its dislocations and carbon atoms. When ε_m rises, the density of dislocations increases and they break away from the pinning carbon atoms, thus augmenting internal friction. Carbon diffusion is necessary for the pinning of free dislocations, although it takes some time to occur. Indeed, the effect of increasing Q^{-1} observed with diminishing ε_m disappears with time.

Various additives reduce Q^{-1} because of the increasing number of pinning points. The extent of a drop in Q^{-1} depends on the additive content and its interaction with the newly formed Frank–Read dislocations (Figure 3.6).

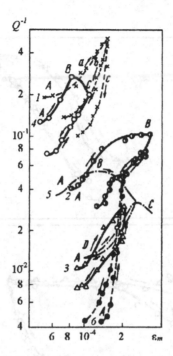

Figure 3.5. Internal friction of different metals as a function of the strain amplitude ($f = 20$ kHz). (*1*) Cd; (*2*) Cu; (*3*) Ni; (*4*) Sn; (*5*) Zn; (*6*) Fe. The curves for Cd, Sn, Zn, and Cu were obtained at 20°C, and those for Ni and Fe at 400 and 525°C, respectively [6].

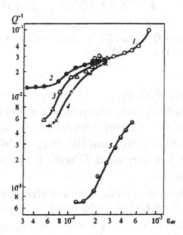

Figure 3.6. Effect of modifying impurities on internal friction of polycrystalline nickel at 20°C, $H = 280$ Oe, $f = 20$ kHz. (*1*) Ni (99.99%); (*2*) Ni + 0.1%Fe; (*3*) Ni + 1%Cu; (*4*) Ni + 1%Fe; (*5*) Ni + 10%Fe [6].

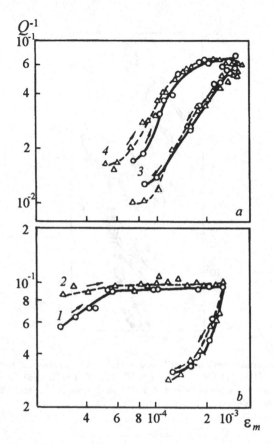

Figure 3.7.   Dependences $Q^{-1}(\varepsilon_m)$ for (*a*) copper and (*b*) aluminum ($t = 20°C$): (*1*), (*3*) $f = 20$ kHz; (*2*), (*4*) $f = 44$ kHz [6].

It was established that in the frequency range 20–45 kHz, the $Q^{-1}(\varepsilon_m)$ curves are considerably independent of frequency (Figure 3.7).

In experiments with different temperatures, Q^{-1} abruptly rises when a particular critical strain amplitude ε_m (which is the lower the higher is temperature) is surpassed (Figure 3.8). Similar results were obtained for other metals.

The internal friction in metals depends strongly on their state. A preliminary treatment of metals reduces Q^{-1} (Figure 3.9), which can result from the increased density of forest dislocations in the active slip system.

To conclude, sections 1.4 and 3.1 dealt with some types of losses that occur during the propagation of elastic waves in solids.

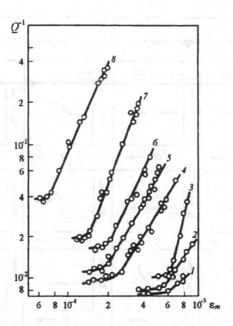

Figure 3.8. Internal friction in copper as a function of the strain amplitude at various temperatures [6]: (*1*) 93; (*2*) 113; (*3*) 143; (*4*) 293; (*5*) 373; (*6*) 473; (*7*) 623; (*8*) 823 K.

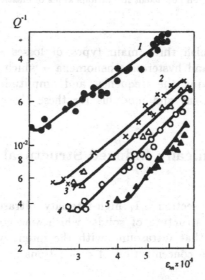

Figure 3.9. Internal friction of copper as a function of the strain amplitude: (*1*) after annealing; (*2*)–(*5*) after a relative deformation of 2.6, 3.6, 5.2, and 7.3%, respectively [6].

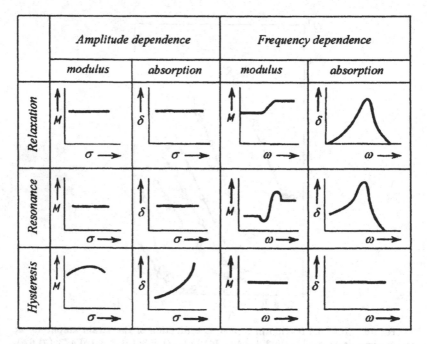

Figure 3.10. The behavior of frequency and amplitude dependences of elastic modulus and absorption coefficient for various kinds of elastic wave energy loss in solids.

One can distinguish three main types of losses – due to relaxation, resonance, and hysteresis phenomena – which are illustrated in Figure 3.10 showing the frequency and amplitude dependences of absorption coefficient and modulus for these three types of energy loss.

3.2   Ultrasonically Induced Structural Changes in Solids

As was noted in section 3.1, high-intensity ultrasound must affect the dislocational structure of solids, which was confirmed in experiments showing that ultrasound with the intensity exceeding some level increases the concentration of dislocations and point defects in solids.

Noticeable contributions to the investigation of these processes were made by N. A. Tyapunina [7–9], A. V. Kulemin [6], I. G. Polotskii [10–16], and others.

It was shown [7] that ultrasonically induced stresses in NaCl and LiF crystals with the amplitude exceeding their yield strength enhances dislocation density.

The frequency of ultrasound in these experiments was 100 kHz. The relative strain amplitudes comprised 2×10^{-6} and 2.7×10^{-4}, which corresponded to peak stresses of about 0.2 and 23 MPa. The first value was considerably less than the yield stresses of the investigated materials (5 and 2 MPa for LiF and NaCl, respectively), whereas the second value was above them. The ultrasonic treatment lasted for 1 h. The initial density of dislocations estimated by the etching pit method was 10^4 cm^{-2}.

Ultrasonic treatment with a strain amplitude of 2×10^{-6} did not affect the etching pattern, but the treatment with the strain amplitude 2.7×10^{-4} gave rise to new small pits.

Analogous results were obtained in the experiments with other alkaline halides [8, 17].

Experiments with LiF showed that the minimal ultrasonic stress able to produce new dislocations comprised 5.8 MPa, i.e. was about the yield stress of this material.

The density of newly generated dislocations at a given stress amplitude depended on the time of treatment in a particular range specific for this stress. Thus, at 8.5- and 27-MPa stresses, the saturation time was > 1 h and 5 min, respectively. The increase in saturated stress amplitude raised further the dislocation density.

A linear dependence of dislocation density on displacement amplitude was observed by Pines and Omel'yanenko [18] who established that the minimal ultrasonic stress necessary for generation of new dislocations in NaCl comprised 2.5 MPa, i.e. was again about the yield stress of the material.

These authors also investigated the polycrystalline metals with the fcc lattice (Al, Cu, and Ni) and found that dislocation density linearly increased beginning from a particular peak stress threshold ξ_{mt} close, at least for Cu and Ni, to their yield stresses [18].

The excitation of vibrations in the investigated metal specimens made of Al, Cu, Ni, Fe, and some steels was performed as shown in Figure 3.11 [19]. Electric oscillations of ultrasonic frequency were generated by a self-oscillator (*1–3*) enabling a continuous tuning of output power and frequency in the range 16–44 kHz with an accuracy of up to ±1 Hz. Changes in a load (induced, for instance, by its heating) did not affect the operation of the generator. Electric oscillations were converted by magnetostrictive transducer (*4*) into mechanical vibrations with the peak displacement up to ~ 15 μm. A set of horns (*5–7*) was

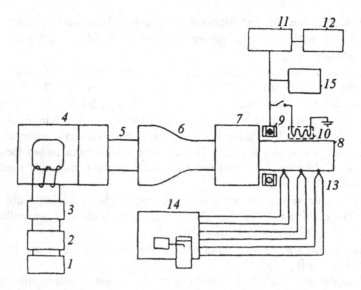

Figure 3.11.   Block diagram of installation for the excitation of various kinds of ultrasonic vibrations [19]. (*1*) self-oscillator; (*2*) intermediate amplifier; (*3*) output amplifier; (*4*) magnetostrictive transducer; (*5*) step-wise horn; (*6*) Fourier horn; (*7*) transducer of longitudinal vibrations into torsional and bending ones; (*8*) specimen under study; (*9*) displacement amplitude probe; (*10*) strain amplitude probe; (*11*), (*12*) units for amplification and recording of signals from probes (*9*) and (*10*); (*13*) thermocouples; (*14*) temperature recording unit; (*15*) frequency meter.

used to enhance the vibrational amplitude to 100–120 μm. The specimen (*8*) was made as an ultrasonic waveguide of the resonant length. The setup was equipped with gauges of displacement amplitude (*11*), strain (*12*), frequency (*15*), and temperature (*14*) measured with the aid of respective probes (*9, 10, 13*).

Experiments with polycrystalline copper (99.9%) and aluminum (99.99%) showed that, at $\sigma_m > \sigma_{mt}$, the dislocation density increased with the displacement amplitude; the dependence $N(\sigma_m)$ for Al could be approximated by a straight line, whereas that for Cu tended to some constant value. The quantity σ_{mt} comprised 0.4–0.5 σ_Y (Figure 3.12).

The dislocation density grew with the time of treatment first linearly, but then went into plateau (Figure 3.13,*a*) [18]. The saturated density of dislocations was linearly dependent on temperature (Figure 3.13,*b*), the temperature dependence of the threshold amplitude being inverse, which, in the authors' opinion, was due to a temperature-dependent reduction in the yield stress of the material.

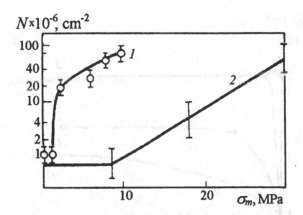

Figure 3.12. Dependence of dislocation density on the stress amplitude for (*1*) Al at 20°C and (*2*) Cu at 450°C [6].

A more detailed analysis of the effect of ultrasound on aluminum structure was carried out by other researchers.

B. Langenecker [20–24] showed that ultrasonic treatment acted to increase the density of dislocations and their wall-like alignment. He directly observed the reproduction of dislocations under the action of 40-kHz ultrasonic vibrations excited in the specimen placed in an electron microscope chamber [23]. This author observed also a distinct cellular structure in ultrasonically treated aluminum monocrystals (Figure 3.14) [24].

Of interest are the results of I. G. Polotskii *et al.* [10, 11] who found that the ultrasonic treatment of aluminum monocrystals augmented the density of dislocations and their accumulation in the slip planes oriented in the direction of elastic wave propagation. The ultrasonic treatment of aluminum monocrystals at elevated temperatures also caused the migration and splitting of subboundaries.

Anchev and Skakov [25] treated polycrystalline aluminum with 20-kHz vibrations at temperature ranging from −195 to 150°C and observed the formation of extended clusters of dislocations, composed mainly of prismatic dipoles and a great number of dislocation loops. There were many dislocation steps whose number, as well as the number of prismatic dipoles and dislocation loops, increased with temperature.

In our experiments [73], ultrasound induced the formation of prismatic dislocation loops in polycrystalline copper. Structural alterations were investigated by the etching pit method that preferably revealed

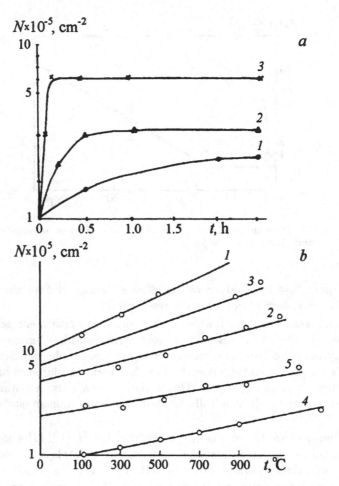

Figure 3.13.   Dependence of dislocation density in aluminum on the time of treatment at 20°C and displacement amplitudes (*1*) 13, (*2*) 15, and (*3*) 18 μm [18].

fresh dislocations emerged onto (111) planes and those differing from them within 2°.

Figure 3.15,*a* shows the microstructure of copper specimens subjected to ultrasonic treatment for 30 min at the strain amplitude $\varepsilon_m = 0.2 \times 10^{-3}$. In this case, the etching pits were very small and located in pairs 1–3 μm apart, either chaotically or regularly along some directions. The number of etching pits increased with the strain amplitude and treatment time, although their distribution pattern remained unchanged.

It was suggested that paired etching pits on the section surface if due to the emergence of dislocations of opposite signs constituting a dislocation loop. This suggestion was confirmed by layer-by-layer etching and crystallographic analysis.

Figures 3.15,*b,c* show the microstructure of the same spot of the section after its repeated layer-by-layer polishing and etching. It is seen that in this case the distance between the etching pits in pairs either increased or decreased in accordance with differing profiles of dislocation loops at different depths from the initial surface. This is schematically illustrated in Figure 3.15,*d*.

The results of crystallographic analysis of the central grain are presented in Figure 3.15,*e*. If one draws lines through the paired pits, then they will appear parallel, within some accuracy, to some directions indicated in the figure. Taking into account that etching reveals only the dislocations that emerge on the (111) planes or differ from them within $2°$, and estimating the mutual disposition of indicated directions, one can infer that these directions are parallel to (112) and (110) planes, i.e. to the intersection traces of (110) and (111) planes. With allowance for the fact that paired etching pits belong to dislocation loops, the lines through paired pits must coincide with the intersection traces of the section surface with dislocation loop planes. Analysis of the disposition of paired etching pits indicates that they pertain to $1–3$-μm long dislocation loops lying in (110) planes. One can believe that dislocation loops are formed as a result of the collapse of dislocation segments during their reciprocal motion and multiple crossing under cyclic loading, as well as a result of the condensation of vacancies [26].

X-ray diffraction analysis of ultrasonically treated copper monocrystals performed in [15, 16] has indicated an enhanced density of dislocations and the subgrain refinement.

Electron microscopic analysis [19] of the distribution and density of dislocations in the austenite 12Cr18Ni9Ti steel showed that the as-annealed dislocation density is relatively low, but increases up to $\sim 10^9$ cm^{-2} after ultrasonic treatment with the strain amplitude exceeding a particular threshold value. In this case, the structure becomes helicoidal. It should, however, be noted that increased dislocation density is observed only in the grains oriented most favorably with respect to the cyclic stress vector σ_m. Comparison of the dislocation structure of ultrasonically treated steel and that subjected to plastic deformation has indicated that in the former case dislocations are more curved with a greater number of steps and knees, which testifies to the ultrasonically induced processes of intersection, transverse slipping, and ascent of dislocations.

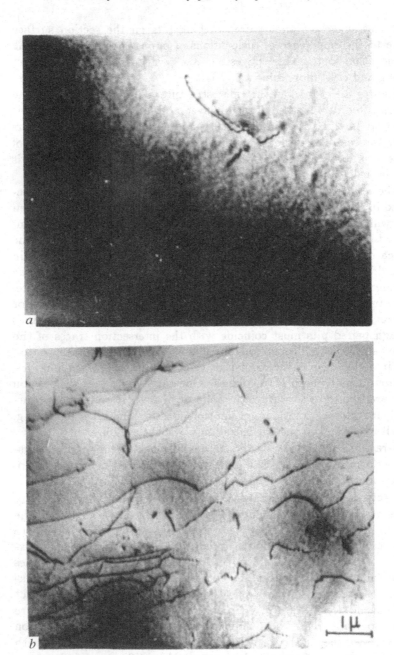

Figure 3.14.   Dislocational structure in monocrystalline aluminum (*a*) before exposure to vibrations and after exposure for (*b*) 1, (*c*) 4, and (*d*) 8 s [24].

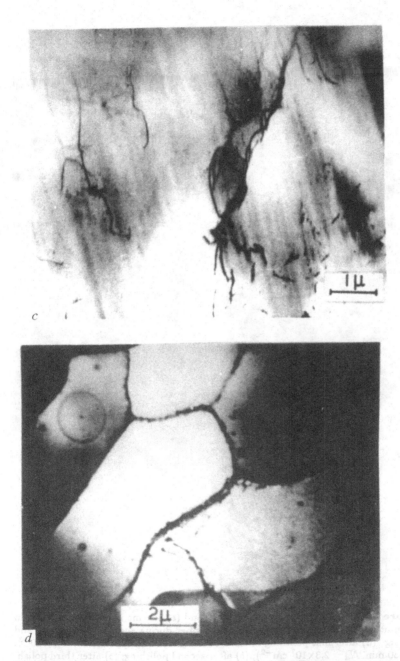

Figure 3.14. Continued.

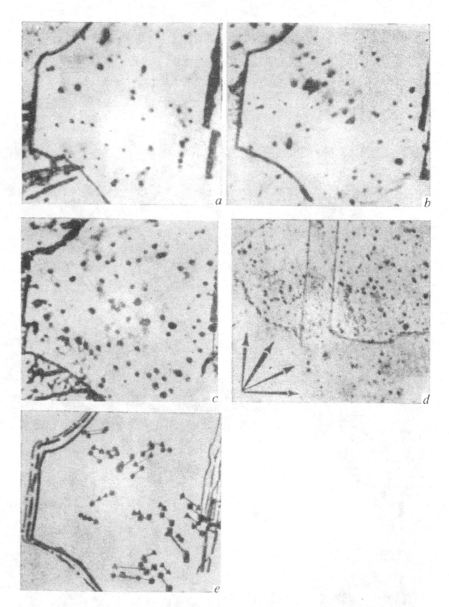

Figure 3.15. $(a–d)$ Micrographs (a, b, c) and (e) scheme illustrating changes in the mutual disposition of etching pits in copper during successive polishing and etching: (a) after first polishing of ultrasonically treated copper $(\varepsilon_m = 0.2\times10^{-3}$, $\tau = 30$ min, $N_d = 2.3\times10^7$ cm^{-2}); (b) after second polishing; (c) after third polishing; (d) an illustration to crystallographic analysis of the structure [6]. Magnification $\times1350$ [6].

The increasing temperature of ultrasonic treatment acts to augment the number of grains with an enhanced dislocation density. The number of dislocation steps and knees rises too, the helicoids straighten, and the structure tends to be cellular. The possible source of dislocations are grain boundaries and carbide segregates. The dislocation structure formed as a result of ultrasonic treatment at 920–960°C differs greatly from that observed after high-temperature rolling. The dislocations produced by ultrasonic treatment at 980°C can be efficiently annealed at 950°C.

Literature data on the influence of cyclic loading frequency on dislocation structure are scarce.

Polycrystalline copper specimens loaded with 160-Hz and 20-kHz vibrations displayed similar Wohler curves [27] and similar processes of the formation and evolution of dislocation structure (dislocation clusters and networks).

Similar experiments [28] were carried out with annealed nickel specimens loaded with low-frequency (50 Hz) and high-frequency (20 kHz) vibrations at 77 and 300 K. The relative strains ε_m comprised 6×10^{-4} and 8×10^{-4} at 300 and 77 K, respectively.

These frequencies produced similar effects on the dislocation structure which, in both cases, contained regions with low and high density of dislocations. In regions with a high dislocation density (up to 10^{11} cm^{-2}), the number of dislocation loops and dipoles was great.

Analogous results were obtained by other authors [27, 29] who revealed no noticeable structural differences in copper subjected to ultrasonic treatment and low-frequency fatigue tests.

W. Mason [30] investigated the effect of ultrasound on polycrystalline lead in a wide range of strain amplitudes and observed an enhanced density of structural defects in the region of plastic fatigue.

Of great practical interest is the investigation of iron behavior under cyclic loading. Some relevant works have been reviewed in the monograph [6].

Treatment of iron (carbon content 0.003%) with 20-kHz elastic vibrations was performed on a setup schematically represented in Figure 3.11. During the treatment, half-wave specimens were cooled with water. The level of applied cyclic stress was varied.

The dislocation density increased when the stress amplitude surpassed a particular level. The distribution of dislocations was not random: their density was considerably higher near the grain boundaries. As a rule, dislocations were concentrated in extended globular clusters

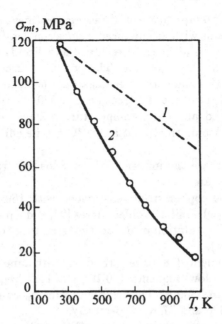

Figure 3.16. Temperature dependence of the critical stress necessary for dislocation motion in iron: (*1*) theoretical curve; (*2*) experimental curve [6].

displaying the occurrence of small dislocation loops and high density of steps.

It should be noted that such structure is typical for cyclic loading [31]. In iron, even at high stress amplitudes, one could observe the regions with either normal or considerably increased density of dislocations.

In these experiments, the threshold amplitude σ_{mt} drastically dropped with increasing temperature (Figure 3.16). The value for σ_{mt}, experimentally determined at 20°C, agreed well with the stress necessary for the Frank–Read source operation [32]. However, the experimental σ_{mt} decreased with temperature much more abruptly than the calculated value, which could be explained by the fact that calculations were performed without allowance for the interaction of dislocations with each other and with impurity atoms, which always takes place in real metals.

A number of works devoted to investigation of Armco iron (0.01% C, 0.006% N, 0.028% Mn, 0.005% P, 0.05% S, and 0.003% Si) and low-carbon steel (0.07% C, 0.006% N, 0.27% Mn, 0.013% P, 0.018% S, 0.03% Si, and 0.07% Cr) at high-frequency loading have been per-

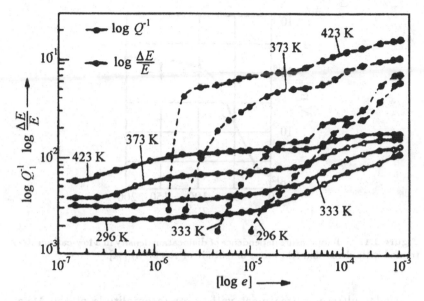

Figure 3.17. Dependence of the internal friction coefficient (solid lines) and modulus defect (dashed lines) on the strain amplitude at various temperatures [33].

formed by A. Puskar [33–36]. After annealing at various temperatures (870–970°C), the half-wave specimens were subjected to 23-kHz ultrasonic vibrations at the strain amplitudes from 1.5×10^{-7} to 4×10^{-4} and temperatures 23, 60, 100, 150, and 200°C. The number of applied vibration cycles was varied from 9×10^5 to 1×10^7. The specimens were tested for the coefficient of internal friction Q^{-1} and defect modulus $\Delta E/E$.

At 23°C and strain amplitudes from 1.5×10^{-7} to 2×10^{-6}, Q^{-1} did not change in ultrasonically treated Armco iron. Above $\varepsilon_{mt1} = 2 \times 10^{-6} Q^{-1}$ slowly grew with the strain amplitude, the growth being enhanced at $\varepsilon_{mt} = \varepsilon_{mt2} > 1 \times 10^{-5}$. As this took place, the elastic modulus of material also changed (Figure 3.17).

At 50, 100, and 150°C, the values for ε_{mt1} and ε_{mt2} were 1×10^{-6}, 4×10^{-7}, 2×10^{-7} and 4.5×10^{-6}, 2×10^{-6}, 1.4×10^{-6}, respectively.

When the number of applied vibration cycles exceeded a particular value (which was the lower, the higher was the temperature of treatment), Q^{-1} increased to an extent dependent on temperature.

The defect modulus was similarly dependent on the strain amplitude, the number of applied vibration cycles, and temperature.

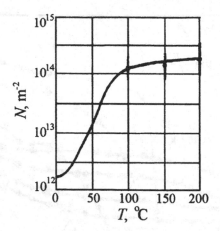

Figure 3.18. Temperature dependence of dislocation density in a low-carbon steel treated with ultrasound (the stress amplitude of 150 MPa) [34].

Under ultrasonic treatment with a stress amplitude of 150 MPa, the dislocation density in low-carbon steel with a mean grain size of 0.088 mm showed a nonlinear increase with temperature (Figure 3.18). At the same time, changes in the dislocation structure at 20°C were insignificant. At 100°C and above, globular entities and small dislocation loops were observed near the ferrite grain boundaries. The interior of grains had a relatively low density of dislocations. At 150°C, about 80% of grains possessed a well-defined cellular substructure. At 200°C, the number of such grains with a mean cell diameter of 2×10^{-7} mm increased to 95%.

I. G. Polotskii *et al.* [13, 37] investigated the influence of ultrasonic treatment on refractory metals with the bcc lattice (Mo, Cr, W).

The devices designed by the authors allowed them to excite vibrations with various frequency and strain amplitudes of order 10^{-4}–10^{-3} in rods and plates.

The electron microscopic studies of molybdenum monocrystals showed that dislocation density in as-annealed specimens was not more than 10^7–10^8 cm^{-2}, whereas a 15-min ultrasonic treatment caused the formation of dislocation groups with the density 2.7×10^9 cm^{-2}.

When the time of treatment was increased to 1 h, the dislocations were distributed more uniformly with a density of 7×10^9, comparable with that in molybdenum monocrystals upsetted by 10% at room temperature.

Ultrasound can modify the structure of metals with not only fcc or bcc lattices. Tyapunina and Shtrom [9] described the ultrasonically induced reproduction of dislocations in zinc monocrystals. Reproduction took place at the relative strain amplitudes above 10^{-5}. If the initial density of dislocations in zinc was 5×10^4 cm^{-2}, then after a 20-min ultrasonic treatment with the strain amplitude $\varepsilon_m = 0.8 \times 10^{-4}$ it increased up to 2×10^6 cm^{-2}. The amplitude-density dependence was nonlinear.

The newly formed dislocations were distributed quite uniformly over the section, nonplanar clusters being often observed. Such a pattern indicates that surface dislocations are of a screw type, since it is screw dislocations that do not form stable clusters in one plane but do participate in lateral slippage.

Ultrasound can also give rise to twinning dislocations whose width can increase 20 times after the ultrasonic treatment (our observations).

A. A. Durgaryan and E. S. Badayan [38] investigated the internal friction and Young's modulus in zinc and bismuth versus the ultrasonic displacement amplitude and found that when the internal friction is maximum, Young's modulus is somewhat reduced.

V. P. Grabchek and A. V. Kulemin [39, 40] employed optical microscopy for the investigation of ultrasonically induced dislocation nucleation in germanium monocrystals, measurement of the velocity and direction of motion of dislocations, as well as for the determination of the relationship between their density, stress amplitude, time, and temperature. It was found that ultrasonic treatment augmented the dislocation density (Figure 3.19). In this case, the selective etching revealed hexagonal rosettes resulted from dislocation half loops emerged to the surface. If vector σ_m is perpendicular to the {111} plane, the rosette rays are directed along $\langle 110 \rangle$. Ultrasonic treatment generated the Frank–Read sources on the crystal surface.

The ultrasonically induced motion of dislocations could be visualized by layer-by-layer etching of a chosen region of the specimen. Figure 3.19 presents the micrographs of the specimen structure after several cycles of ultrasonic treatment and subsequent etching. It is seen that some etching pits remain in their places (these are stationary dislocations), while others are displaced in a particular direction (mobile dislocations). It is noteworthy that the direction of dislocation motion is independent of the stress sign, which can be explained by the fact that moving dislocations leave behind them obstacles likely representing point defects. Analogous phenomenon was observed on ionic crystals [41].

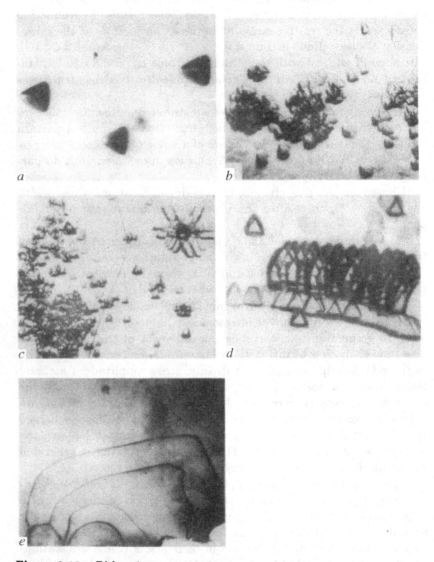

Figure 3.19. Dislocation structure of germanium (a) before ultrasonic treatment (magnification ×450) and after the treatment at 400°C and $\sigma_m = 3$ MPa for (b) 3 min and (c) 5 min; (magnification ×270); (d) treatment at 600°C and $\sigma_m = 3$ MPa for 10 min; (magnification ×60000); (e) layer-by-layer etching after each treatment cycle 0–5 min long [39].

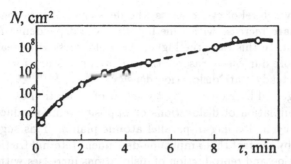

Figure 3.20. Dislocation density in germanium as a function of the time of ultrasonic treatment ($\sigma_m = 3$ MPa, $t = 700°C$) [39].

The estimation of the velocity of dislocations in germanium monocrystals versus temperature showed that this dependence can be described by the exponential function $e^{-U/kT}$, where U is the activation energy of dislocation motion. This agrees with the literature data concerning the effect of static loading on dislocation velocity [42]. It should be, however, noted that the velocity of dislocation motion induced by ultrasound is by an order of magnitude higher than that induced by the static loading producing an equal stress ($\sigma_0 = \sigma_m$).

The dislocation density increases monotonically with the time of treatment; and after the rise time τ_s goes into plateau (Figure 3.20).

Thus, the dislocational structure of materials subjected to ultrasonic treatment begins to change when the amplitude of stresses is above a particular value, specific for a given material and given ambient conditions.

The ultrasonically induced reproduction of dislocations in polycrystalline materials occurs at lower stresses than in the case of static loading. Based on the results of R. Keith and I. Gilman [43], A. Kulemin suggested that dislocations can be generated when the applied stress changes its direction [6]. The threshold amplitude, above which the reproduction of dislocations takes place, is related to the static yield stress as [41]

$$\sigma_{mt} = \sigma_{0.2} + \beta\omega^n, \tag{3.17}$$

where β and n are the constants specific for a given material. It should be, however, noted that Pines *et al.* [18] were unable to reveal any influence of frequency in the range 15–44 kHz on σ_{mt}.

At a given level of cyclic stress, the density of dislocations initially shows a sharp increase with time but then the dependence $N(\tau)$ goes into plateau. In this case, the higher the yield stress of annealed material, the more durable ultrasonic treatment is necessary for the attainment of a steady-state dislocation density.

As suggested by Kulemin [6], the effect of saturation can be related to the annihilation of dislocations of opposite signs produced in the process of cyclic loading in parallel atomic planes. This suggestion is confirmed by the fact that amplitude-dependent internal friction caused by the motion and reproduction of dislocations increases with the time of ultrasonic treatment if $\sigma_m > \sigma_{mi}$.

The specificities of ultrasonically induced evolution of dislocational structure in metals allows one to suppose that the formation of point defects will be enhanced.

Indeed, it was found [44] that ultrasonic treatment of NaCl crystals increased the concentration of both dislocations and point defects. Experiments were performed with a 100-kHz ultrasound and the strain amplitude $\varepsilon_m = 4 \times 10^{-4}$. The concentration of point defects was estimated from the photosensitivity of crystals.

The evolution of the dislocational structure of polycrystalline copper was also accompanied by changes in the concentration of vacancies [6], which was estimated from the data on lattice parameter and material density. The ultrasonic treatment was performed at 420°C for 30 and 60 min.

Exposure to ultrasound with $\sigma_m = 12$ MPa for 60 min reduced the material density and lattice parameter from their initial values of 8.9520 g/cm^3 and 3.6175 $\pm$ 0.0004 Å (as-annealed copper) to 8.9418 g/cm^3 and 3.6148 $\pm$ 0.0004 Å (ultrasonically treated copper). Changes in the density and lattice parameter took place when the stress amplitude surpassed some threshold value corresponding to an incipient reproduction of dislocations in copper induced by ultrasound. The dislocation density and concentration of vacancies, calculated for various amplitudes of ultrasonic stress from the density and lattice parameter data, are summarized in Table 3.1.

Analysis of these data indicates that ultrasonic treatment of copper at 420°C raises the concentration of vacancies, possibly arranged in the form of stable vacancy clusters.

Copper (99.999%) resistance measurements [45] showed that the low-frequency vibrational loading ($N \sim 10^4$, $\varepsilon_m = 2 \times 10^{-4}$) acted to increase the concentration of vacancies to $\sim 10^{-3}$, the attained concentration being retained up to the specimen failure. Such a concentration of vacancies is also typical of copper rolled at 77 K with a 700% de-

Table 3.1 Ultrasonically induced structural changes in copper [6].

Stress amplitude, MPa	Relative change in density $\Delta\rho/\rho$, %	Relative change in lattice parameter $\Delta a/a$, %	Dislocation density $N_d \times 10^8$, cm^{-2}	Concentration of dislocations Δn_v, at%
0	0	0	1	$<10^{-6}$
2.4	-0.01	0	20	$<10^{-6}$
72	-0.06	-0.02	30	3×10^{-4}
96	-0.09	-0.03	60	5×10^{-4}
120	-0.10	-0.06	100	10

formation. Analogous results were obtained in [46] for polycrystalline copper.

An increased concentration of vacancies was observed in ultrasonically treated monocrystalline [12] and polycrystalline [47] nickel. Belostotskii juxtaposed the vacancy concentration changes in response to the various kinds of treatment – plastic deformation, hardening, and ultrasonic irradiation:

$\Delta n_v \times 10^2$, at.%

Plastic deformation $\varepsilon_0 = 10\%$	0.7
Plastic deformation $\varepsilon_0 = 20\%$	0.7
Hardening	1.4
Ultrasonic treatment, $\sigma_m = 100$ MPa, $\tau = 8$ min	1.5
Ultrasonic treatment, $\sigma_m = 130$ MPa, $\tau = 1$ min	1.5

These data indicate that hardening and ultrasonic treatment increase the concentration of vacancies in nickel to a higher extent than conventional plastic deformation does.

As follows from works [32, 48], the increased concentration of vacancies can be related to an enhanced mobility of dislocations induced by cyclic loading. Excessive vacancies appear as a result of the recombination of moving dislocations and nonconservative dislocation jogs.

The possible mechanisms of formation of point defects caused by plastic deformation and vibrational loading were considered by Kulemin [6] who revealed the regularities relating the dislocational structure and concentration of vacancies to the specificities of elastic modulus and internal friction.

The arising point defects interact with dislocations to pin them. The dislocation pinning (the change in the mean distance L between pinning points) affects the elastic modulus E and internal friction Q^{-1} such as was shown in section 3.1

$$\frac{\Delta E}{E} = AL^2 N_d, \qquad (3.18a)$$

$$Q^{-1} = BL^4 N_d f^2, \qquad (3.18b)$$

where A and B are constants specific for a given material; $\Delta E = E_0 - E$; E_0 is the elastic modulus of an annealed material; E is the instantaneous elastic modulus.

If a substance contains nonequilibrium vacancies, then they must gradually diffuse to dislocations, pinning them and reducing L, which will increase E and diminish Q^{-1}. By increasing dislocation density and thus L, the low-level plastic deformation will enhance the internal friction and reduce the elastic modulus of crystalline materials. Some time after the deformation cessation, a partial restoration of the initial values of E and Q^{-1} will take place (the so-called Koster effect [49]).

Alternating strains that increase the concentration of dislocations and vacancies must alter the internal friction and elastic modulus. After a rest, there must be a partial restoration of E and Q^{-1}.

Relevant investigations were performed by A. W. Kulemin [6] with aluminum, copper, and nickel specimens. In all cases, cyclic strains led to a significant increase in the internal friction coefficient and a reduction in the elastic modulus, if the strain amplitude was above a threshold value (Figure 3.21). Thus, ultrasonic strains in aluminum with the amplitude $\varepsilon_m = 4.25 \times 10^{-4}$ increased Q^{-1} from 0.43×10^{-2} to 3.44×10^{-2} and decreased the elastic modulus by ~16%.

The exposure of samples to room and elevated temperatures led a gradual increase in elastic modulus (Figure 3.22) and decrease in internal friction. A sharp rise in E and Q^{-1} at room temperature persisted for 3 h, the extent of E restoration being the higher, the greater ε_m exceeded the threshold value ε_{mt}.

The restoration of E and Q^{-1} in the case of samples statically loaded at $\sigma_0 = \sigma_{m\ max}$ was significantly less pronounsed.

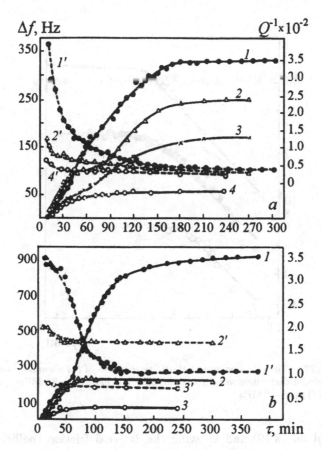

Figure 3.21. Dependences of resonant frequency (solid lines) and internal friction (dashed lines) of (a) nickel and (b) aluminum specimens on the time of ultrasonic treatment at 20°C with ε_m equal to: (a) 3.67×10^{-4}, 2.84×10^{-4}, and 2.08×10^{-4} (curves 1-3, respectively); (curve 4) static loading with $\sigma_0 m = 57$ MPa. (b) 4.25×10^{-4} and 0.98×10^{-4} (curves 1 and 2, respectively).

Kulemin [6] derived an expression relating the excessive concentration of vacancies Δn_v to the internal friction coefficient $Q^{-1} \times (\varepsilon_m, t)$

$$\Delta n_v = \frac{N \eta \varepsilon_m}{2\pi\beta} Q^{-1}(\varepsilon_m, t), \qquad (3.19)$$

where N is the number of applied vibrational cycles; η is the coefficient equal to $\sim 10^{-4}$ [50]; $\beta \leq 1$ is the factor accounting for the nonlinearity of a stress-strain hysteresis loop.

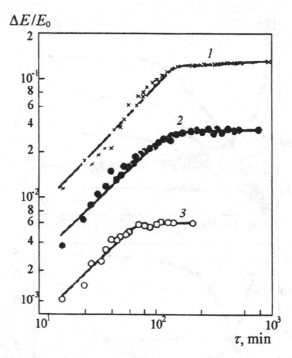

Figure 3.22. Time dependence of elasticity modulus of (*1*) aluminum and (*2, 3*) nickel loaded either ultrasonically or statically [6]: (*1*) $\sigma_m = 32$ MPa; (*2*) $\sigma_m = 75$ MPa; (*3*) $\sigma_0 = 57$ MPa.

Based on (3.19) and knowing the internal friction coefficient for a given metal, one can estimate the concentration of vacancies as a function of ε_m at various temperatures.

The estimates made agree quite well with the relevant experimental data reported in [6] and [51].

The estimation of ultrasonically induced vacancy concentration, performed based on the specificities of internal friction and elastic modulus behavior during the annealing of copper at 200–350°C, showed that it increased 7–8 times.

Thus, the ultrasonic treatment of metals induces the formation of excessive vacancies, if the cyclic stress amplitude is sufficient to cause the reproduction and motion of dislocations.

The mechanism of point defect formation under the action of ultrasound is still to be elucidated, although Kulemin [6] suggested that it should be analogous to that which is realized during plastic deformation.

Figure 3.23. Slip bands after the ultrasonic treatment of lithium fluoride at the stress amplitude higher that 8.5 MPa [7].

Ultrasonically induced changes in the dislocational structure of metals must cause their plastic deformation, which was a subject of research in some works.

As was mentioned above, a 100-kHz ultrasound with the stress amplitude equal to the yield strength of LiF crystals acted to enhance dislocation density [7].

Raising the ultrasonic stress amplitude to 8.5 MPa gives rise to slip bands which are shown in Figure 3.23. Analysis of the micrographs of the same spot of the crystal before and after ultrasonic treatment indicates that most of slip bands are formed by edge dislocations and only some of them by screw dislocations. The majority of newly formed bands are located near the existing single dislocations, dislocation loops, and slip bands.

Exposure of crystals to ultrasound with the stress amplitude above 17 MPa leads to the widening and diffusion of slip bands. In some cases, the diffuse bands of one system (for instance, screw) continue on the adjacent face of the crystal but as belonging to another system (for example, edge).

Ultrasound with the stress amplitude providing the formation of slip bands caused a plastic deformation of specimens. The slip bands were especially dense in the crystal middle, being absent at their edges. Such a nonuniform distribution of slip bands corresponds to the distribution of stresses in a standing wave.

In ultrasonically treated NaCl, single slip bands began to appear at stresses above 4.8 MPa, whereas diffuse slip bands took place above 7.2 MPa.

In the case of metals, effects are analogous. Thus, ultrasound with the stress amplitude exceeding a particular value causes the formation

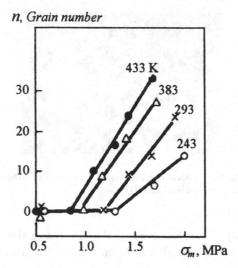

Figure 3.24.   The number of deformed grains in aluminum versus the amplitude of vibrational stress [6].

of slip bands on the aluminum surface. In the case of plastic deformation the arising slip traces look like separate lines, but after ultrasonic treatment they show high density to form slip bands. It should be emphasized that slip traces appear only on some grains, probably oriented most favorably relative to the applied stress. The surface density of grains (n) with slip traces versus the stress amplitude is shown in Figure 3.24 for various temperatures. The deformed grains appear only when the stress amplitude exceeds some threshold value σ_{mt}. The comparison of this value with that at which the dislocation density begins to increase showed that they coincide.

At low temperatures (293–300 K) and relatively small exposure time ($\tau \sim 60$ s), the slip traces represent straight lines going, in the limits of one grain, unidirectionally without intersections. At higher temperatures and exposure times, the slip lines become tortuous; one can observe secondary slip systems and extrusion of a metal.

With increasing stress amplitude and time of treatment, the slip band width increases, while the distance between bands decreases; that is, there occurs a microdeformation of metal grains leading finally to material fracture.

Point defects generated by dislocations migrate to sinks which, at enhanced temperatures, are mainly the boundaries of grains and blocks. The arising submicro- and microdefects are accumulated on grain boundaries and other interfaces to give rise to various micro- and macro-discontinuities (Figure 3.25,*a*, *1–2*). Micro- and macrodefects accumulated on grain boundaries lead to cracking and finally to failure of specimens. V. S. Biront [52] noted that in all cases the fracture of aluminum (99.99%) was preceded by a slippage over grain boundaries (Figure 3.25,*b*). Moreover, Biront observed the formation of a step on a polished section surface, along which the slippage took place (Figure 3.25,*c*, *1–3*).

A number of works devoted to analysis of the structural features of metals treated with high frequences and tested for fatigue were performed by B. Weiss and R. Stickler. They revealed the formation of slip bands on polycrystalline copper whose number increased with the number of applied ultrasonic cycles (N) and strain amplitude (Figure 3.26) [53]. Fractographic analysis has shown that the fracture was viscous, transcrystallitic, and independent of the grain size.

Slip traces on ultrasonically treated iron and low-carbon steels were investigated by Puskar [33–36]. The strain amplitudes ε_{mt}, at which Armco iron revealed slip traces after being loaded with 10^7 vibrational cycles, were 2.4×10^{-4}, 1.4×10^{-4}, 8×10^{-5}, and 4.6×10^{-5} at 23, 60, 100, and 150°C, respectively.

Puskar derived the temperature dependence of ε_{mt} as

$$\varepsilon_{mt} = A \exp\left[\frac{q}{RT}\right], \qquad (3.20)$$

where A is the constant specific for a given material and ultrasonic treatment parameters; q is the activation energy equal in the case of Armco iron to 3.25 kcal/mol.

Raising the strain amplitude above ε_{mt} leads to the growth of the number G_s of grains with slip bands (Figure 3.27), the number of slip bands per one grain, and to band lengthening. The dependences $G_s = f(\log \varepsilon_m)$ are parabolic, passing above each other with increasing temperature.

Slip traces in ultrasonically treated Armco iron are preferably oriented in the direction of wave propagation. They represent almost straight lines, sometimes branched, but become more tortuous with increasing temperature. Many slip lines are formed near the grain boundaries, but they do not cross these boundaries. A more detailed analysis of slip lines showed that they are not smooth but consist of a great number of small steps. Carbide inclusions occurring in metals are surrounded by thin, randomly oriented slip traces.

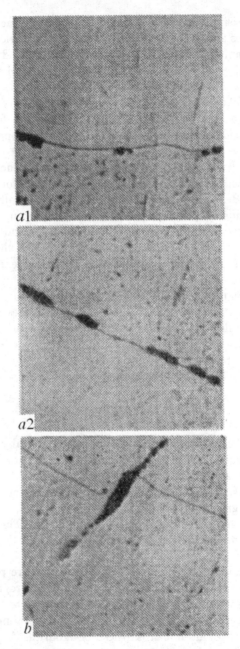

Figure 3.25. Sequential processes of the formation of (*a*) micropores, (*b*) slip traces, and (*c*) intergrain boundary steps in aluminum A99 during its ultrasonic treatment at 520°C (magnification ×260) [52].

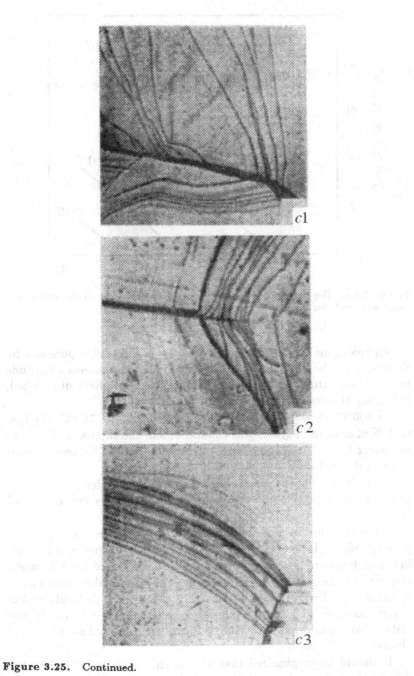

Figure 3.25. Continued.

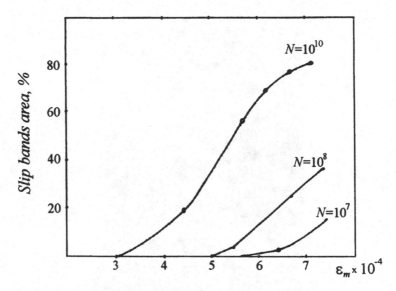

Figure 3.26. Dependence of the grain area with slip bands on the vibrational strain amplitude for low-carbon steel [53].

An analogous distribution of slip traces was observed in pure iron by Kulemin [6]. The slip band width grew with increasing stress amplitude and exposure time, whereas the distance between them diminished, indicating the development of deformation.

At 600°C, there occurred the metal extrusion. At 750°C, the mutual displacement of grains was enhanced, and the slip bands began to disappear to be almost nonobservable at 800–1000°C because of their considerable diffusion [54].

Some features of the plastic deformation of molybdenum ultrasonically treated at high temperatures were investigated by Polotskii *et al.* [13].

If the exposure to ultrasound was not ceased at the specimen temperature 850°C, it rapidly grew to 1200°C, leading to specimen fracture in the central part. Microstructure analysis has revealed the refinement of grains in the fracture region related to a considerable local plastic deformation. Farther from the fracture region, the dislocation density diminished, and no subboundaries were formed, so that 30 mm distant from this region no changes in the dislocational structure was observed.

It should be emphasized that a disoriented cellular structure was formed at vibrational amplitudes and exposure times such that micro-

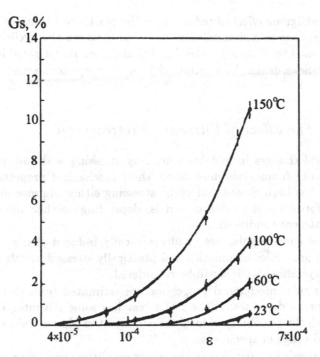

Figure 3.27. The number of grains with slip traces in Armco iron versus the strain amplitude at various temperatures of ultrasonic treatment [36].

cracks appeared in the surface layers of a specimen. The first cracks visible in an optical microscope appeared after a 10- to 15-s ultrasonic treatment with a strain amplitude of 5×10^{-3}.

3.3 Mechanical Properties of Ultrasonically Treated Metals

Mechanical properties of metals and alloys are mainly determined by structural imperfections occurring in crystals – atoms of dissolved chemical elements, vacancies, dislocations, grain boundaries, and particles of a second phase.

Ultrasonically induced changes in the dislocational structure of metals and alloys and the development of microplastic deformation must affect the mechanical properties of treated materials, which was the subject of research by A. V. Kulemin [6], I. G. Polotskii [55], and others.

An intriguing effect of reduction in the plastic resistance of materials in response to a simultaneous action of constant and cyclic stresses was revealed by B. Langenecker [56–58] and then investigated in detail by V. P. Severdenko, V. V. Klubovich, A. V. Stepanenko, and coworkers.

3.3.1   The Effect of Ultrasonic Pretreatment

Structural changes in metals induced by stressing with low and high (ultrasonic) frequencies must affect their mechanical properties. Indeed, it has been shown that cyclic stressing either improve or impair the performance of treated materials, depending on their initial state and treatment conditions.

Below some regularities of ultrasonically induced changes in mechanical properties of annealed and plastically stressed metals will different crystalline lattices will be considered.

As a rule, mechanical properties are estimated from the data of tensile or hardness tests. The latter test has some advantages associated with the possibility of local measurement of plastic resistance and repeated usage of specimens.

Ultrasonic vibrations with the stress amplitude exceeding a threshold value specific for a given material act to harden an annealed metal, the yield and breaking strengths being increased, whereas plasticity decreased.

Kulemin [6] investigated the effect of a 20-kHz ultrasound on the mechanical properties of metals with the fcc lattice. Elastic vibrations were supplied to a load with the aid of a waveguide system described in section 3.2 and schematically shown in Figure 3.11.

The exposure of annealed copper and aluminum specimens to a 20-kHz ultrasound at a fixed level of stress enhanced their yield strength (Figure 3.28). In this case, $\sigma_{0.2}$ monotonical increased for 10–30 s, depending on the metal treated, and then went to plateau. The plateauing was more rapid at higher stress amplitudes.

Ultrasonic loading with the amplitude above a threshold value enhanced the strength of metals and decreased their plasticity (Figure 3.29), the value σ_{mt} for the fcc-lattice metals being $\sim 0.5\sigma_{0.2}$.

Ultrasonic treatment also somewhat changed the stress–strain curves of metals (as compared with those for annealed metals) in a manner that after reaching the yield strength there was a stage of a slight slip, i.e. metal deformation without a noticeable increase in the load. This stage was more pronounced at higher vibrational amplitudes [59].

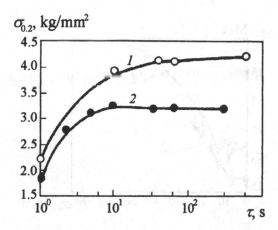

Figure 3.28. Dependence of the yield strength of copper (1, $\sigma_m = 67$ MPa) and aluminum (2, $\sigma_m = 16.4$ MPa) on the time of ultrasonic treatment [6].

The dependence of the resulting microhardness on the time of treatment (the number of applied vibrational cycles) and stress amplitude was analogous to those for the yield and tensile strengths.

Ultrasonically induced changes in the metal hardness take place if the stress amplitude is above a particular threshold value σ_{mt}, specific for each metal and equal to the stress at which dislocation density begins to grow (see section 3.2)

The amplitude threshold decreases with increasing temperature of ultrasonic treatment. Ultrasonic hardening is also temperature-dependent. At temperatures 0.35–$0.4\,T_m$, the strength and hardness rise proportionally to the stress amplitude, whereas at higher temperatures the linearity is disturbed (Figure 3.30).

Similar results were obtained for some other annealed metals with the fcc lattice [61–64].

Thus, Kralic and Weiss [62] described the hardening of metals (Al, Cd, Cu, Ni, Au, Ag) and alloys (Cu–Ni) under the action of ultrasound with the stress amplitude comprising 0.3–0.85 of the initial yield strength of a metal. The increase in the yield strength was quite great (6.5 times for gold and 2 times for silver), whereas the tensile strength changed insignificantly (by 5–10%). The extent of ultrasonically induced hardening of Cu–Ni alloys diminished with growing concentration of Ni in alloys, and the stress amplitude threshold increased. The temperature dependence of alloy hardening was the same as in the case of pure metals.

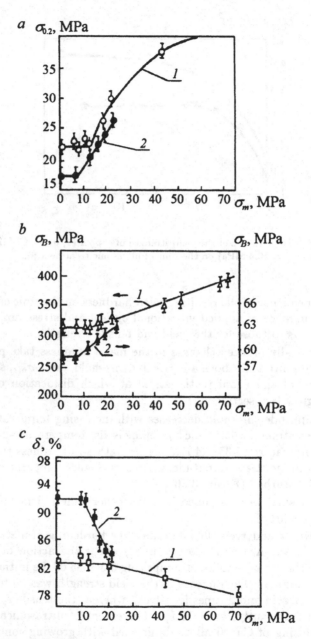

Figure 3.29. Dependences of (*a*) yield strength, (*b*) tensile strength, and (*c*) relative elongation of ultrasonically treated (*1*) copper and (*2*) aluminum on the stress amplitude. The time of treatment is 1 min [6].

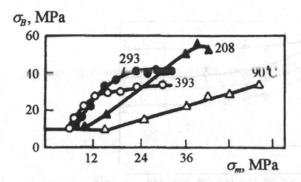

Figure 3.30. Strengthening of aluminum by its exposure to ultrasound at various temperatures [60].

Ultrasonic loading caused a reduction in the high-temperature creep rate of copper and aluminum [63, 64]. The stress-rupture strength of copper samples increased up to 3 times, but the steady-state creep rate decreased 7.5 times.

The above ultrasonically induced changes in the mechanical properties of metals concerned those with the fcc and hcp lattices.

A number of works [34, 65–72] were devoted to the investigation of the effect of ultrasound on the mechanical properties of iron and steels. When the vibrational stress amplitude did not exceed 180 MPa, the strength of these materials changed insignificantly. But at $\sigma_m =$ 220 MPa and temperatures above 200–250°C, the yield and tensile strengths of low-carbon steels (0.036% C) increased by 30 and 20%, respectively [68]. The stress amplitude threshold was between the yield and fatigue strengths of low-carbon steels.

The strength of high-carbon (~0.93% C) steels was insignificantly affected by ultrasound (within several percent).

The ultrasonic treatment of molybdenum was studied by Polotskii *et al.* [13]. The microhardness of polycrystalline molybdenum was influenced by ultrasound when its peak stress was above a particular value equal to 12.3 MPa.

To elucidate the mechanism of ultrasonic hardening of metals, it was necessary to study the temperature and time dependences of the properties of metals subjected to ultrasonic treatment and plastic deformation. Kulemin [6] has analysed the hardness data as concerned with isothermally annealed aluminum and copper samples subjected to ultrasound and plastic deformation. The degree of plastic deformation was chosen to be such that to equalize the hardness of the samples after

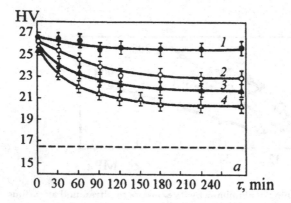

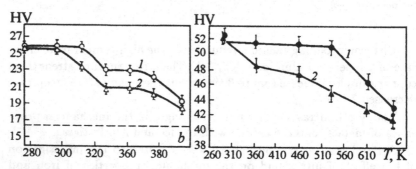

Figure 3.31. Dependence of hardness on (a) the time of Al annealing at $T = 308$ K (curves *1* and *2*) and $T = 353$ K (curves *3* and *4*) and the temperature of isochronal (3 h) annealing of (b) aluminum and (c) copper. The metals were hardened ultrasonically with $\sigma_m = 18$ MPa (curves *2* and *4*) or by plastic deformation with $\varepsilon_0 = 3.9\%$ (curves *1* and *3*). Dashed straight line indicates the boundary of annealed states [6].

ultrasonic treatment and plastic deformation. The hardness of samples the subjected to ultrasonic treatment was removed by their annealing at lower temperatures and more rapidly than in the case of tre samples subjected to plastic deformation (Figure 3.31). Analogous results were obtained in the experiments on the annealing of nickel specimens [47].

The temperature dependence of the hardness of ultrasonically treated metals allows the suggestion that hardening is due to the formation of both point defects (~60%) and dislocations (~40%).

The application of cyclic stress to a preliminary hardened metal leads to its softening and increased plasticity, of practical interest being the reduction of internal stress and alteration of its distribution.

Thus, a hardened material subjected to ultrasonic loading with the peak stress below its fatigue strength, did not fail in subsequent tests for longer time than that without such pretreatment [74]. Broom and Ham [75] showed that 106-MPa cyclic loading of stretch-hardened copper with the yield strength $\sigma_{0.2} = 210$ MPa reduced it to 160 MPa. This result was obtained with a temperature of 293 K, whereas at lower temperatures (90 K), the effect of softening was considerably less pronounced.

Softening effects were studied in most detail for low-frequency loading ($f \leq 100$ Hz) [13, 74–78], although ultrasonic softening was also investigated [79–84].

The ultrasonic treatment ($\sim$20 kHz) of fcc-lattice metals, preliminary hardened by plastic deformation, reduced their hardness and strength. In the case of a polycrystalline copper, its hardness diminished by 40% [79], yield strength 2.4 times, and tensile strength 2.8 times [80, 81]. Ultrasonic loading also shortened the recrystallization lag of copper 10–125 times [82, 83].

Kulemin [6] described experiments involving the ultrasonic loading of polycrystalline aluminum specimens that were 5% predeformed by rolling or stretching. Some time after exposure to ultrasound, the specimens were tested for hardness and internal friction coefficient. Figure 3.32 illustrates the microhardness curves of aluminum versus the time of treatment, as well as the amplitude dependences of ultrasonically induced hardness of annealed aluminum. Three regions of stress amplitudes could be distinguished, within which the kinetics of microhardness had its own peculiar features. At the stress amplitudes below a particular threshold value σ_{mt}, the rate of microhardness change was the same in ultrasonically treated and control (untreated) specimens (Figure 3.32, curves 1, 3). If the amplitude σ_m was more than σ_{mt1} but less than the second threshold value σ_{mt2}, the microhardness of ultrasonically treated specimen dropped monotonically (curve 4). At $\sigma_m > \sigma_{mt2}$, changes in microhardness were not monotonic: on the background of net reduction in microhardness, there were at least two successive phases of hardening and softening (curves 5 and 6).

Changes in microhardness were accompanied by concurrent changes in internal friction that increased upon softening and decreased upon hardening.

A gradual transfer from the softening phase to the phase of hardening under a prolonged treatment with cyclic stresses ($f = 100$ Hz) was observed also for polycrystalline copper [85] and some fusible metals [86].

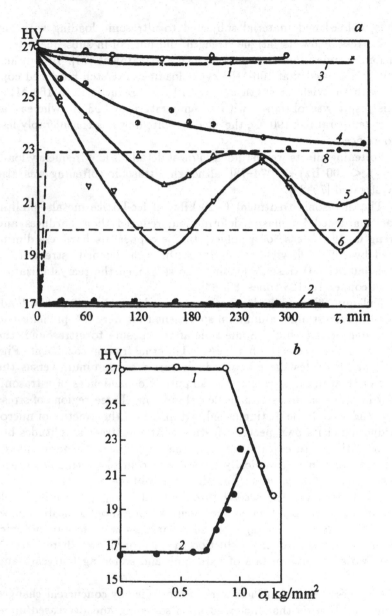

Figure 3.32. Dependence of the microhardness of aluminum at 33°C on (a) the time of ultrasonic treatment and (b) stress amplitude (exposure for 150 min): (1) control, deformed sample; (2) control, annealed sample; (3–6) ultrasonically treated samples predeformed at σ_m = 3–5.6, 4–8.4, 5–10, and 6–11.2 MPa, respectively; (7) and (8) indicate the levels attained in as-annealed samples pretreated with ultrasound for 1 min at σ_m = 7–9 and 8–10 MPa, respectively [6].

Similar results were obtained for the bcc-lattice metals such as iron and steels. The ultrasonic treatment ($\xi_m = 14$ μm) of a deformed iron reduced its $\sigma_{0.2}$ by 62% and σ_B by 51% [80, 81]. The ultrasonic treatment of a CrWMn steel accelerated approximately 4 times the relief of its residual stresses during tempering [87]. The ultrasonic treatment ($\xi_m \sim 40$ μm; 20°C; 5 min) of a 12Cr18Ni10Ti steel (30% predeformed) reduced its residual stresses to a value typical for steel heated to 730°C [88].

Somewhat different results were reported in [6] where ultrasonic treatment decelerated the internal friction restitution in iron (0.003% C) 11% predeformed by rolling.

It should be again emphasized that ultrasonic treatment changes the distribution of internal stresses in hardened materials and reduces their level, which can be illustrated by the data of Klassen–Neklyudova and Kapustin [84] who established that ultrasonic vibrations with a frequency of 7000 kHz and intensity 4 W/cm² reduced the level of residual stresses in a TlBr-TlI solid solution and made the distribution of tantalum salts more uniform.

As for the possible mechanisms of successive hardening and softening, one should note that amplitude σ_{mt1} corresponds to an cyclic stress level at which slip lines appear in a deformed metal, that is, to the stress necessary for the activation of dislocation motion (in the case of annealed materials, σ_{mt1} is somewhat less).

A. V. Kulemin [6] attributed ultrasonic softening to the processes of polygonization. Then, an initially stressed lattice is reconstructed into unstressed blocks separated by subboundaries consisting of dislocations of the same sign, which must emerge from their own slip planes and migrate to parallel slip planes. This process must be associated with the migration of vacancies too. As shown above (section 3.2), ultrasonic treatment promotes the formation of vacancies and, hence, should also promote the formation of polygonal structure and softening of material. If the stress amplitude and the time of treatment are above particular values, the level of internal stresses diminishes, so that deformed metal approaches, by its structural features, the annealed metal. In this state, an cyclic stress begins to harden the metal. This phenomenon can take place only for such σ_m which can harden the annealed material to the extent not less than the extent of the predeformed material softening by the same cyclic stress. That is, the inequality $\sigma_m > \sigma_{mt2}$ (σ_{mt2} corresponds to the point on the σ_m axis at which curves *1* and *2* in Figure 3.32,*b* intersect) must be fulfilled. As curves *4*, *7*, and *8* in this figure do not intersect, stresses with amplitude $\sigma_m = 1$ MPa will cause only a monotonic reduction in material hardness.

3.3.2   Combined Static and Cyclic Loading

In 1995 F. Blaha and B. Langenecker published the results of their study of ultrasonically loaded zinc monocrystals [56].

In their experiments, a 800-kHz ultrasound with the stress amplitude exceeding elastic strength brings about a 40% reduction in the static stress (Figure 3.33, curve *1*, point A). Once the irradiation is ceased (point B), the static stress returns to its initial value. The effect is reproducible. If ultrasonic irradiation is accompanied by stretching, the stress-strain curve *2* passes much below.

Later [20, 21, 57, 58, 88–90], analogous results were obtained with cadmium and zinc monocrystals loaded with frequencies ranging from 15 kHz to 1.5 MHz, for which the reduction in the static yield strength was proportional to the intensity of vibrations but was independent of frequency. By choosing an appropriate ultrasonic intensity, various degrees of softening of low-carbon and stainless steels, berillium, and tungsten could be obtained. At a particular intensity of vibrations, a plastic deformation of specimens might occur at room temperature virtually without an external static loading. In this case, monocrystals were deformed by the mechanism of twinning. Polycrystalline specimens were fractured along grain boundaries.

The deformation of specimens stressed by ultrasound and heat was similar; however, a much less amount of acoustic energy was necessary

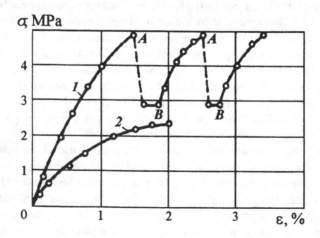

Figure 3.33. Effect of ultrasonic vibrations on the static tensile stress in zinc monocrystals irradiated (*1*) periodically or (*2*) continuously. Points *A* and *B* indicate the instants of the onset and cessation of vibrations, respectively.

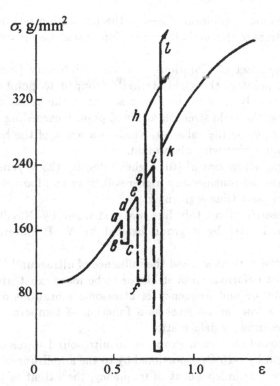

Figure 3.34. Changes in static shear stress induced by repeated stretching of zinc monocrystals with ultrasound of different intensity [90].

to reach the same extent of material softening. Thus, a zero static yield strength of aluminum was attained at the densities of thermal and acoustic energies equal to $\sim 10^{22}$ eV/cm^3 and only 10^{15} eV/cm^3, respectively. This phenomenon is explained by the focusing of acoustic energy to the crystalline lattice sites of a material that are the carriers of plastic deformation, i.e., dislocations and grain boundaries [89].

Instead of softening, high-level ultrasonic loading may cause the hardening of materials [90, 97], which is illustrated in Figure 3.34 by the curve showing the static stretching of a zinc monocrystal in combination with its repeated loading with ultrasonic vibrations of various intensity. The rate of stretching was constant, the frequency of vibrations was 25 kHz.

Vibrations with an intensity of $\sim$5-W/cm^2 reduced the static yield stress from point a to point b. In the interval bc, deformation occurred under ultrasonic irradiation. Ceasing the irradiation returned the deformation stress to the level observed before ultrasonic treatment (point *d*).

After the onset of vibrations of a higher intensity (more than 15 W/cm^2) at point *e*, the yield strength dropped to point *f*, that is, more considerably than in the first case. After the cessation of ultrasonic loading, the yield strength attained point *h* exceeding the normal stretch (point *q*) by the value *qh*, which is a result of the hardening of monocrystalline zinc with ultrasound.

Ultrasonic vibrations of still higher intensity (25 W/cm^2), applied at point *i*, caused considerably higher softening and hardening than in the first two cases (line segment *kl*).

The research along this line was extended by Nevill and Brotzen [91] and later by a group headed by V. P. Severdenko [92–100].

In [91], the authors studied the influence of ultrasound (15–50 kHz) on the plastic deformation of steels, or to be more exact, the influence of the amplitude and frequency of ultrasonic vibrations on the yield strength of a low-carbon steel as a function of temperature and the extent of preliminary deformation.

It was found that each exposure to ultrasound lowered the static yield strength by ~15% proportional by to the amplitude of vibrations. The increase was independent of frequency, the extent of preliminary deformation (within 16% of a residual elongation), and temperature (in the range 30–500°C).

Some other regularities of plastic deformation in an ultrasonic field were derived from the results of experiments with aluminum, copper, brass, technical-grade iron, and other materials [92–100]. Ultrasonic vibrations were found to lower both static yield stress and elongation, which could be related to the concentration of strains in a relatively small region of action of maximum cyclic loading. In the process of a combined (i.e., under ultrasonic irradiation) stretching of a specimen, it failed at the initial stage of neck formation in the section of maximum cyclic stress. The failure was fatigue.

In this case, the total stress in a stretched specimen is a sum of two components, static and dynamic (cyclic). The latter is distributed nonuniformly over resonant specimens (typically used in experiments). In the case of half-wave specimens, which are friendly-to-use and give the results that can be easily processed, the maximum cyclic stresses appear at the specimen midpoint, i.e., at the displacement amplitude node. In the case of aluminum half-wave specimens, ultrasound with a displacement amplitude of 12 μm and frequency 20 kHz produces at the displacement node (stress antinode) an cyclic stress of ~60 MPa. At the same time, a 5% stretching of samples without ultrasonic irradiation produces a stress of ~70 MPa. The stretching of specimens in an ultrasonic field enables one to obtain the same extent of elongation at a static stress of 5–7 MPa. Therefore, the total stress during a combined, cyclic plus static, loading comprises 65–67 MPa, which is close to the value developed upon entire static stretching.

The effect of ultrasound on the rheological properties of a diversity of materials was investigated by Polukhin *et al.* [101]. Experiments were performed on a cam plastometer at the rate of deformation varied from 0.005 to 25 s^{-1}. The ultrasonic frequency was 20 kHz, the amplitude of vibrations ranged within 1–10 μm.

The authors investigated alloys based on light (aluminum, magnesium, titanium) and heavy (zinc, copper, etc.) non-ferrous metals and some steels and nickel-based alloys.

It was found that ultrasound noticeably lowered the yield curves of all the investigated alloys (Figure 3.35), the extent of this lowering being different. Thus, at the vibrational amplitude 10 μm and relative deformation 0.2 varying at a rate of 0.5 s^{-1}, the relative reduction in the yield stress comprised 74, 43, and 28% for zirconium alloy TsZhKhV, copper alloy MH-40, and refractory nickel superalloy VZh98, respectively.

As a rule, the effect of ultrasound diminishes with increasing rate of deformation (Figure 3.36); the extent of this reduction is high for titanium alloys and relatively low for aluminum alloys, bronzes (BrNB8-0.5, Br013), and high-strength, low-alloy steels (for instance, 10Mn2NbW).

The application of ultrasound lowers the ductility of the majority of investigated alloys, the effect being more pronounced at higher amplitudes of vibrations. At the same time, the ductility of some bronzes increases with the amplitude of vibrations.

For elucidation of the mechanism of ultrasonic effect on plastic deformation, of great importance are the results of the works [6, 102, 104] devoted to the investigation of the creep of fcc-lattice (aluminum, copper) and bcc-lattice (iron) metals.

Experiments were performed with polycrystalline copper and aluminum specimens, annealed or ultrasonically treated. The static stress σ_0 and cyclic stress σ_m were varied and the relative deformation ε equal to the difference between the total deformation ε_t and instantaneous deformation ε_i was measured.

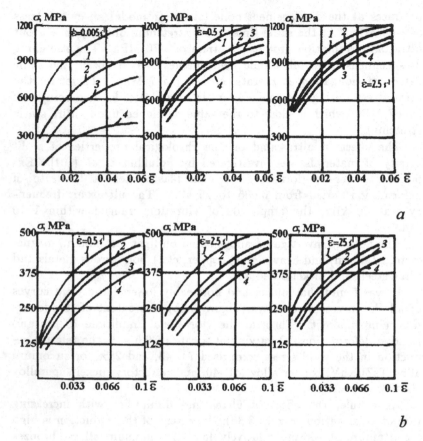

Figure 3.35. Yield curves for (*a*) titanium alloy VT3-1 and (*b*) aluminum alloy D16AT tested without ultrasonic irradiation (curve *1*) and under ultrasonic irradiation with the displacement amplitude equal to 3 (curve *2*), 5 (curve *3*), and 10 μm (curve *4*) at different deformation rates [101].

Figure 3.37 presents the creep curves of aluminum and copper obtained under various experimental conditions.

If the amplitude σ_m exceeds a particular threshold value σ_{mt}, and $\sigma_0 < \sigma_{0.2}$, ultrasound will sharply increase the rate of transient creep (the initial stage of deformation). The value σ_{mt} decreases with growing static stress and vanishes at $\sigma_0 = 60$ MPa and $\sigma_{0.2} = 50$ MPa ($\sigma_0 > \sigma_{0.2}$).

The rate of equilibrium creep under a continuous irradiation with ultrasound, $\dot{\varepsilon}_e$, is close to that of control specimens (Figure 3.37, curves *1* and *2*). In the case of cyclic irradiation (curve *3*), the rate of non-

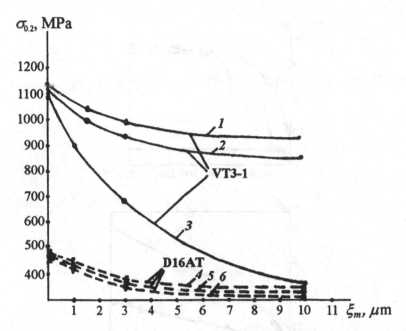

Figure 3.36. Dependence of the yield strength of ultrasonically treated VT3-1 ($\varepsilon = 0.04$) and D16AT ($\varepsilon = 0.06$) alloys on the amplitude of vibrations at the deformation rates equal to 0.005 (curve *1*), 0.5 (curves *2* and *4*), 2.5 (curves *3* and *5*), and 25 s^{-1} (curve *6*) [101].

equilibrium creep, $\dot{\varepsilon}_{ne}$, sharply increases each time after the onset of vibrations and considerably exceeds the steady-state creep rate attained over the same time as in the case of continuous irradiation. With increasing number of applied irradiation cycles, the effect of ultrasound drops. Nevertheless, the total deformation of materials loaded both statically and vibrationally is much greater than that attainable under static stress alone.

Analogous results were obtained at elevated temperatures (Figure 3.37). The activation energy of creep, estimated from the dependence $\ln \dot{\varepsilon} = f(1/\tau)$, appeared to be somewhat less in the case of a combined loading than in the case of static stress alone.

The rate of creep increases with static stress (Figure 3.38), being increased still further under the action of ultrasound.

One of the questions that is essential for the understanding of the mechanism of ultrasonic effect under discussion, is whether the creep acceleration caused by ultrasound is a consequence of the trivial summing of static and cyclic stresses (this mechanism

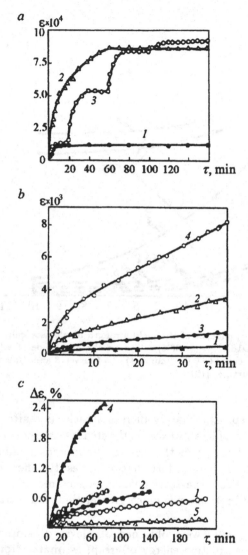

Figure 3.37. (a) Creep curves for copper at 20°C: (*1*) control sample, $\sigma_0 = 30$ MPa; (*2*) continuous exposure to ultrasound with $\sigma_m = 2.6$ MPa at $\sigma_0 = 30$ MPa; (*3*) repeated loading with ultrasound (20-min exposure, 20-min rest) with the same values of σ_0 and σ_m as in the case of curve *2*). (b) Creep curves for copper tested at 110°C (curves *1* and *2*) and 350°C (curves *3* and *4*) either statically with $\sigma_0 = 30$ MPa (curves *1* and *3*) or in a combination with ultrasound with $\sigma_m = 4$ MPa (curves *2* and *4*). (c) Creep curves for aluminum at 40°C: (*1*) control sample statically loaded with $\sigma_0 = 10$ MPa; (*2*)–(*4*) combined static ($\sigma_0 = 10$ MPa) and ultrasonic loading with $\sigma_m = 0.6$, 1.3, and 2 MPa, respectively; (*5*) creep test after a preliminary 60-s exposure to ultrasound with $\sigma_m = 5.3$ MPa [6].

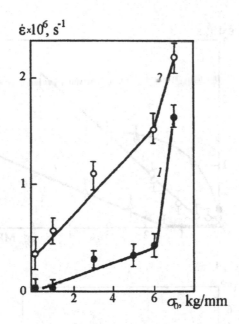

Figure 3.38. The rate of the transient creep of copper at 20°C as a function of static stress: (*1*) control sample; (*2*) combined static and ultrasonic loading with $\sigma_m = 4$ MPa [6].

was suggested in [92, 105]) or it depends on the stresses in a more complex manner. This question can be elucidated based on the results presented in Figure 3.39. It is seen that the creep rate $\dot{\varepsilon}_e$ rises with $\Delta\sigma$, if the static stress $\sigma = \sigma_0 + \Delta\sigma$ ($\Delta\sigma = \sigma_m$) is high; however, the rise is not so considerable as in the case cyclic loading (Figure 3.39,*b*).

Preliminary ultrasonic treatment with the amplitude exceeding the threshold value σ_{mt} (equal to 4–5 MPa for aluminum at room temperature) diminishes the rate of equilibrium creep (Figure 3.37, curve *5* and Figure 3.39,*b*, curve *1*).

The behavior of steady-state creep versus the time of ultrasonic treatment is reminiscent of that of the deformed sample hardness. Ultrasonic irradiation leads to a monotonic decrease in $\dot{\varepsilon}_e$, if it lasts less than 1 min. Upon longer exposure, the creep rate increases and then falls again. Analogous results were obtained in experiments with nickel-based alloys [106–108].

Smirnov and Baranova [104] investigated how a cyclic loading may modify the properties of a vacuum-annealed (800°C) iron (C + N =

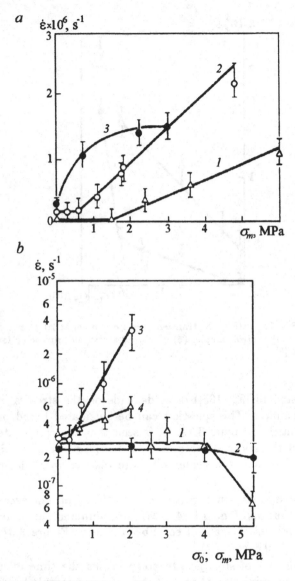

Figure 3.39. The rate of (a) the unsteady creep of copper ($\tau = 10$ min) at 20°C and (b) the steady-state creep of aluminum at 40°C versus the cyclic stress amplitude σ and additional static load $\Delta\sigma$ [6]. a: (1)–(3) $\sigma_0 = 2$, 30, and 60 MPa, respectively; b: (1) a preliminary 60-s ultrasonic loading with $\sigma_{m,\mathrm{pr}} = \sigma_m$ followed by a static loading with $\sigma_0 = 10$ MPa; (2) a preliminary 60-s static tensile loading with $\sigma_{0,\mathrm{pr}} = \sigma_{m,\mathrm{pr}}$ followed by creep test with $\sigma_0 = 10$ MPa; (3) simultaneous ultrasonic and static loading with $\sigma_0 = 10$ MPa; (4) creep test with $\sigma_\Sigma = \sigma_0 + \Delta\sigma = \sigma_0 + \sigma_m$ without ultrasound.

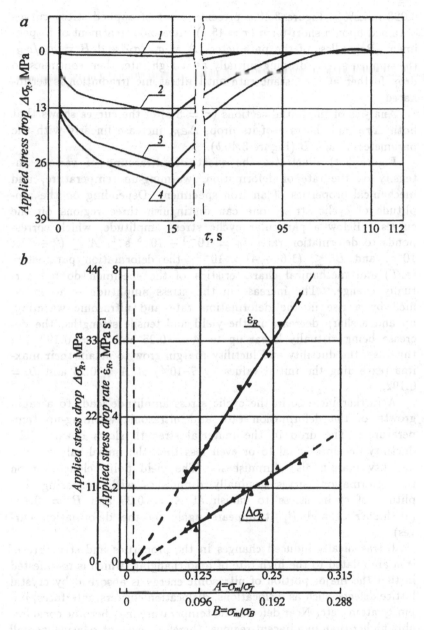

Figure 3.40. Effect of cyclic stress on (a) the time course of quantity $\Delta\sigma_R$ in iron at the deformation rate $\dot{\varepsilon} = 6 \times 10^{-4}\ \mathrm{s}^{-1}$: (1) $A = B = 0$; (2) $A = 0.1$, $B = 0.08$; (3) $A = 0.26$, $B = 0.20$; (4) $A = 0.29$, $B = 0.22$; and (b) on quantities $\Delta\sigma_R$ and $\Delta\dot{\varepsilon}_R$ at the initial stage of the process [104].

0.023 mass%). Figure 3.40,*a* presents a set of applied stress curves obtained upon a short-term ($\tau = 15$ s) ultrasonic treatment of a specimen. Regardless of the parameters $A = \sigma_m/\sigma_Y$ and $B = \sigma_m/\sigma_B$, the applied stress, droping initially at a high rate, then continues to drop further at a constant rate until ultrasonic irradiation is terminated.

Analysis of the initial sections ($\tau = 5$ s) of the curves shows that both $\Delta\sigma_R$ and the rate of its drop, $\Delta\dot{\varepsilon}_R$, increase linearly with the parameters A and B (Figure 3.40,*b*).

Figure 3.41 illustrates the effect of ultrasound of various intensity on the rate of deformation, warming-up temperature, and mechanical properties of an iron specimen. Depending on the amplitude of cyclic stress, one can distinguish three regions on the curves. Below a particular cyclic stress amplitude, which corresponds to deformation rates $\dot{\varepsilon}_0 = 10^{-3} \div 10^{-5}$ s^{-1}, $A \leq (2 \div 5) \times 10^{-2}$, and $B \leq (1.6 \div 4) \times 10^{-2}$, the deformation parameters ($\dot{\varepsilon}, T$) and mechanical characteristics of the specimen do not virtually change. The increase in the stress amplitude is accompanied by a rise in the deformation rate and ultrasonic warming-up and a sharp decrease in the yield and tensile strengths, the decrease being virtually linear up to $A = 0.25$ and $B = 0.192$. In this case, the ductility and luctility margin grow to attain their maxima (exceeding the initial values by 7–10%) at $A = 0.25$ and $B = 0.192$.

A further increase in the cyclic stress amplitude leads to a rapid growth of the deformation rate and ultrasonic warming-up temperature. The drop in the material strength slows down. The ductility becomes equal to or even less than the initial values. The ductility margin also diminishes. The yield point elongation on the deformation curves gradually diminishes with increasing amplitude of cyclic stress to vanish at $A = 0.25$ and $B = 0.192$ (at higher A and B, it appears again on the deformation curves).

Ultrasonically induced changes in the properties and structure of iron are related to the high rate of load changing, which is manifested in that the major portion of ultrasonic energy is absorbed by crystal lattice defects, such as dislocations, dislocation clusters, interfaces, impurity atoms, etc. Near defects, the temperature may become considerably higher than in adjacent regions; therefore, even at relatively small stress amplitudes (at least much less than the static elastic strength), the ultrasonic loading can lead to an intense local redistribution of defects.

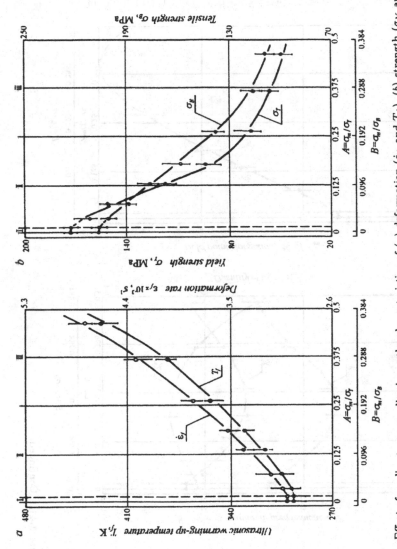

Figure 3.41. Effect of cyclic stress amplitude on the characteristics of (a) deformation ($\dot{\varepsilon}_f$ and T_f), (b) strength (σ_Y and σ_B), (c) plasticity (δ, Ψ, Ψ_*), and (d) yield point elongation (ε_{tot}) [104].

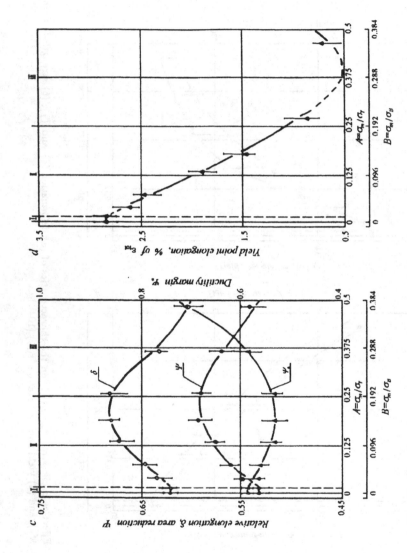

Figure 3.41.   Continued.

3.4 Diffusion in Ultrasonic Field

The works that appeared in 1930–1937s reported on an enhanced penetration of nitrogen, chromium, nickel, and carbon into steels as a result of excitation of high-frequency vibrations during thermochemical treatment of steel samples in induction furnaces.

The discovery of this effect marked the beginning of extensive research into the influence of mechanical vibrations on diffusion processes in solid metals.

The numerous relevant works published to date can be arbitrarily classified into two groups: (1) those dealing with direct measurements of diffusion coefficients and (2) with estimation of ultrasonic effect in the processes of heat and chemical treatment.

This section will review in detail the studies of the first group that are concerned with the regularities of carbon diffusion in nickel and iron [6], chromium diffusion in iron and steels [109, 110], diffusion of various impurities (Sb, Ga, In, Al, and Li) in germanium [111, 112], diffusion of zinc and aluminum in copper [113, 114], and zinc diffusion in iron [114].

In [6, 110, 112, 115, 116], the coefficient of carbon diffusion in iron and steels was measured over a wide range of temperatures and ultrasonic strain amplitudes.

To this end, half-wave specimens of a rectangular cross section were prepared from metals and alloys under study. One face of the samples was carburized (950°C) from a gas phase with carbon containing ^{14}C. After annealing, the depth distribution of radioactive carbon was profiled by means of layer-by-layer radiometry.

Figure 3.42 gives experimental curves for the coefficient of diffusion of radioactive carbon in nickel, D_C^{Ni}, as a function of the strain amplitude ε_m at different temperatures. For each temperature, there is a particular threshold value ε_{m1} below which the diffusion coefficient does not depend on ε_m and coincides with the diffusion coefficient in control samples. When the strain amplitude is more than ε_{m1}, the diffusion coefficient rises with ε_m to be maximum at a particular strain amplitude ε_{m2}. The exposure of the metal to vibrations with the strain amplitude greater than ε_{m2} acts to reduce D_C^{Ni}.

It should be noted that at 470°C and $\varepsilon_m > \varepsilon_{m1}$, there is only a monotonic increase in the coefficient of carbon diffusion in nickel, which is likely because the peak strains used in the experiments do not exceed ε_{m2}. The absence of ultrasonic effect on diffusion parameters at 700°C is most likely due to small ε_m, as compared to ε_{m1}.

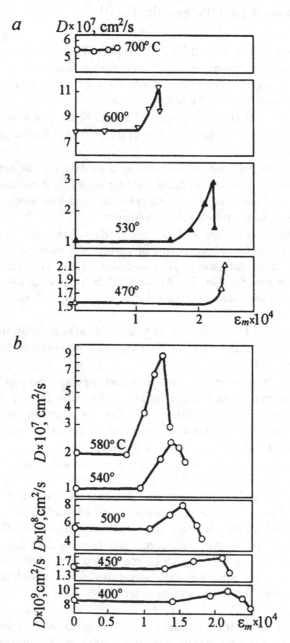

Figure 3.42. Coefficient of carbon diffusion in (a) nickel (99.9%) and (b) iron (0.002 mass% of C) as a function of strain amplitude at various temperatures [6].

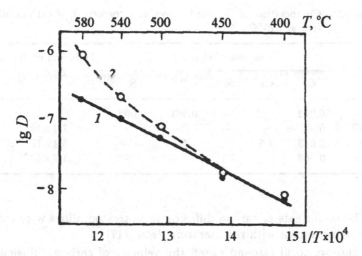

Figure 3.43. Temperature dependence of diffusion coefficient of carbon in iron at the strain amplitude ε_m: (1) 0; (2) 1.3×10^{-4} [6].

Similar behavior of the coefficient of carbon diffusion in α-Fe versus the strain amplitude was observed by Kulemin [6] (Figure 3.42,b).

At the same temperature, the thresholds ε_{m1} and ε_{m2} for α-Fe and Ni are different, the difference being decreased with increasing temperature.

Figure 3.43 presents the temperature dependences of the coefficient of carbon diffusion in control and ultrasonically treated samples of α-Fe, measured at a fixed amplitude $\varepsilon_{m1} < \varepsilon_m < \varepsilon_{m2}$.

For control samples, the dependence $\lg D$ vs $1/T$ is a straight line, and the dispersion coefficient is described by the Arrhenius equation

$$D_{\mathrm{C}}^{\alpha-\mathrm{Fe}} = 5.5 \times 10^{-2} \exp\left(-\frac{92}{RT}\right), \quad \mathrm{cm}^2/\mathrm{s}. \qquad (3.21)$$

The derived value of activation energy is in a satisfactory agreement with the literature data [117]. But for ultrasonically treated samples, the dependence $\lg D = f(1/T)$ does not obey the Arrhenius law.

The ultrasonic treatment enhances activation energy and pre-exponential factor, the effect of ultrasound being increased with temperature.

To clarify how impurities may affect the ultrasonic diffusion of carbon, Kulemin studied Fe-based alloys with different contents of carbon, chromium (carbide forming element), and silicon (forming no carbide) (Table 3.2).

Table 3.2   Composition of Fe-based alloys and coefficient of carbon diffusion at 525°C [6].

Alloy	Content of elements, %					Diffusion coefficient D, cm^2/s
	C	Cr	Si	P	S	
Fe	0.002	–	–	0.001	0.002	
Fe–C	0.02	–	–	_"_	_"_	7.4×10^{-8}
Fe–Cr	0.003	1.5	–	_"_	_"_	5.4×10^{-8}
Fe–Si	0.003	–	3.5	_"_	_"_	1.3×10^{-7}

The coefficients of carbon diffusion in untreated alloys were found to be in agreement with the literature data [117].

Exposure to ultrasound raised the velocity of carbon diffusion, if the strain amplitude surpassed a particular value specific for each alloy.

Increasing concentrations of carbon in alloys raised 3–4-fold the threshold amplitude ε_{m1} and thus reduced the effect of ultrasound (Figure 3.44). In the Fe–Si alloy, the diffusion coefficient began to rise at still higher ε_{m1}, the rate of the increase being less than in the case of iron with 0.02% C. The stage of a slow diffusion was not observed in this alloy, probably, because the experimental strain amplitudes were less than ε_{m2}. Conversely, the alloy that contained a carbide forming element (Cr) displayed a lower threshold amplitude ε_{m1} and thus substantially higher diffusibility of carbon atoms.

It is of particular interest to elucidate how the time of ultrasonic treatment affects the coefficient of carbon diffusion in iron and Fe–Cr alloys.

Figure 3.45 gives the time dependences of carbon diffusion coefficients in α-Fe and Fe–Cr. For both alloys, there exists a particular time interval, τ_0, over which diffusion coefficients grow linearly with time (at $\tau = 0$, the diffusion coefficients correspond to those in the samples annealed without ultrasound). When the exposure time is greater than τ_0, specific for each material, the diffusion coefficient will no longer change. The quantity τ_0 for Fe–Cr is approximately 2 times smaller than that for α-Fe.

The decrease in the strain amplitude of ultrasound somewhat reduced τ_0. It should be noted that the exposure time τ in all other experiments was taken to be more than τ_0, which corresponded to ordinary diffusion coefficients.

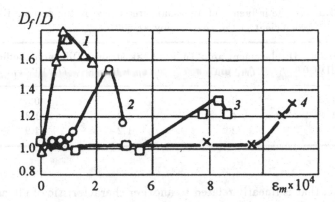

Figure 3.44. Dependence of the coefficient of carbon diffusion in iron and Fe-based alloys on the strain amplitude ε_m ($t = 525°C$, $\tau = 30$ min): (*1*) Fe + 1.5% Cr + 0.003% C; (*2*) Fe + 0.003% C; (*3*) Fe + 0.02% C; (*4*) Fe + 3.5% Si + 0.003% C [6].

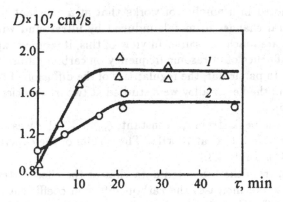

Figure 3.45. Dependence of the coefficient of carbon diffusion in (*2*) iron and (*1*) Fe + 1.5% Cr alloy at 560°C on the time of ultrasonic treatment: (*1*) $\varepsilon_m = 1.2 \times 10^{-4}$; (*2*) $\varepsilon_m = 1.9 \times 10^{-4}$ [6].

Qualitatively similar results were obtained in studies of the regularities of lithium diffusion in germanium monocrystals. Initially ($\tau \leq$ 10 min), ultrasound with the peak stress $\sigma_m = 3$ MPa at 350°C enhanced lithium diffusion by a factor of 30, but then this factor dropped to 2 [111].

Investigation of carbon diffusion in iron and nickel showed that the effect of ultrasound on diffusion coefficient depends on the strain ampli-

Table 3.3   To the influence of ultrasonic frequency on the carbon diffusion
coefficient in the Fe–Cr alloy at 500°C.

Frequency f, kHz	Displacement amplitude ξ_m, μm	Peak strain $\varepsilon_m \times 10^4$	Diffusion coefficient $D \times 10^7$, cm^2/s
0	0	0	0.84
20	4	1.2	1.8
40	2	1.2	1.9

tude ε_m, that is linearly related to another characteristic of ultrasound
field, the cyclic stress frequency f:

$$\varepsilon_m = k\xi_m f, \tag{3.22}$$

where k is the coefficient, constant for a given material under given
conditions.

It was noted in a number of works that at a constant peak strain
ε_m, structural changes in metals induced by ultrasound with various
frequency f are much the same. In view of this, it seemed expedient to
study the influence of ultrasonic frequency on carbon diffusion parameters in iron. In particular, the regularities of the diffusion of radioactive
carbon ^{14}C in the Fe–Cr alloy were studied at two frequencies ($\sim$20 and
$\sim$40 kHz) [6].

To have the peak strain ε_m constant, ξ_m at 40 kHz was taken to be
two times less than that at 20 kHz. The results of the experiments are
summarized in Table 3.3.

It is seen from the tabulated data that at the chosen strain amplitudes, ultrasound increases the carbon diffusion coefficient more than
twofold, the effect being virtually the same at 20 and 40 kHz.

A. V. Kulemin and V. P. Manaenkov investigated diffusion not only
in interstitial but also in substitution solutions [6, 112]. The effect of
ultrasound on the diffusion of substitutional atoms was studied during
diffusion of chromium in α-Fe and Fe–Cr alloy, as well as during self-
diffusion of iron. In the latter case, the radioactive isotope ^{59}Fe was
used.

Figure 3.46 and Table 3.4 present the dependences of the relative coefficient of chromium diffusion in α-Fe and Fe–Cr and the self-diffusion
of iron, respectively. As in the case of carbon diffusion, ultrasound
induces no changes in the diffusibility of chromium and iron atoms at
ε_m lesser than a particular value specific for a given alloy, ε_{m1}. But

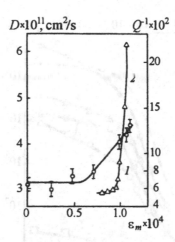

Figure 3.46. (*1*) Chromium diffusion coefficient and (*2*) internal friction in iron (0.003% C) as functions of the strain amplitude at 830°C [112].

at $\varepsilon_m > \varepsilon_{m1}$, the diffusion coefficient rises with increasing strain amplitude. The presence of chromium in the Fe–Cr alloy exerts the same effect on chromium diffusion as in the case of carbon diffusion, that is, chromium reduces the threshold amplitude ε_{m1} and increases the growth rate of diffusion coefficient.

An accelerated diffusion of iron and chromium in iron is observed only in the range of peak strains corresponding to enhanced internal friction. This implies that the effects mentioned are related to the ultrasonically induced motion and reproduction of structural defects.

The study of the effect of ultrasound on self-diffusion of iron in steels with different types of crystal lattice (bcc-type steel Fe + 0.003% C + 1.0% Mo and fcc-type steel Fe + 0.003% C + 7% Cr + 2.0% Ni) performed at 850°C has shown that irrespective of the lattice type, elastic vibrations accelerate self-diffusion if the peak strain exceeds a particular value.

Table 3.4 Self-diffusion of iron in a Fe-based alloy containing 1% Mo and 0.003% C (850°C).

Peak strain $\varepsilon_m \times 10^4$	0	1.0	1.7	2.0
Diffusion coefficient $D \times 10^{11}$, cm^2/s	1.34	1.37	2.9	5.4

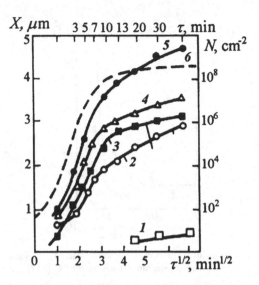

Figure 3.47.   Time dependences of the *p-n* junction depth in germanium mono-crystals at 600°C: (*1*) untreated sample containing Sb; (*2*)–(*5*) ultrasonically treated (stress amplitude 2 MPa) samples containing Sb, Ga, In, and Al, respectively. (*6*) Time dependence of dislocation density under the action of ultrasound with the stress amplitude 3 MPa [111].

The frequency of vibrations (20 and 40 kHz), just as in the case of carbon diffusion, produces no substantial difference in the diffusion of chromium in the Fe–Cr alloy.

In [118], the authors studied the effect of ultrasound on the diffusion of various impurities (Sb, Ga, In, and Al) in germanium monocrystals and found that the depth of occurrence of the *p-n* junction in ultra-sonically treated samples first drastically rises ($\tau \leq$ 10–13 min) and then decreases, the diffusion coefficient in ultrasonically treated samples being, however, higher than that in control samples (Figure 3.47, Table 3.5). It is seen that changes in the diffusion coefficient correlate with changes in the density of dislocations.

Ultrasonically accelerated diffusion in metals and semiconductors takes place when the peak strains (stresses) exceed a particular threshold value. The comparison of this value with those that provide an increase in dislocation density and vacancy concentration shows that these values coincide within the accuracy of experiments (see experimental data in the first sections of this chapter). Thus, it seems that changes in the kinetics of diffusion processes induced by ultrasound, similar to plastic deformation induced by irradiation with high-energy

Table 3.5   Diffusion coefficient of substitution impurities in germanium
monocrystals at 600°C [40, 111]. The vibration velocity vector coincides with the $\langle 111 \rangle$ direction.

Impurity	Diffusion coefficient D, cm^2/s		
	Control sample	Ultrasonically treated sample	
		Initial stage of treatment	Subsequent stage of treatment
Antimony	3.9×10^{-14}	8.0×10^{-11}	7.6×10^{-12}
Gallium	1.7×10^{-16}	2.2×10^{-10}	1.2×10^{-12}
Indium	1.3×10^{-15}	3.2×10^{-10}	9.0×10^{-11}
Aluminum	2.1×10^{-14}	4.2×10^{-10}	9.2×10^{-11}

particles, are due to the structural modification of materials. It would
be expedient to consider peculiarities of ultrasonically induced diffusion
processes from this viewpoint.

References

1.   T. A. Read, *Phys. Rev.*, **58**, 371 (1940).
2.   J. S. Koehler, *Imperfections in Nearly Perfect Crystals*, New York, (1952).
3.   A. Granato and K. Lucke, *J. Appl. Phys.*, **27**, 583 (1956).
4.   W. P. Mason, *J. Acoust. Soc. Am.*, **32**, 458 (1960).
5.   G. Leibfried, *J. Appl. Phys.*, **127**, 344 (1950).
6.   A. V. Kulemin, *Ultrasound and Diffusion in Metals* (in Russian), Metallurgiya, Moscow (1978).
7.   E. G. Shvidkovskii, N. A. Tyapunina, and E. P. Belozerova, *Kristallografiya*, **7**, 473 (1962).
8.   E. P. Belozerova, N. A. Tyapunina, and E. G. Shvidkovskii, *Kristallografiya*, **8**, 232 (1963).
9.   N. A. Tyapunina and V. E. Shtrom, *Fiz. Met. Metalloved.*, **23**, 744 (1967).
10. I. G. Polotskii, D. E. Ovsienko, E. L. Khodov, E. I. Sosnina, G. Ya. Bazelyuk, and V. Kushnir, *Fiz. Met. Metalloved.*, **21**, 727 (1966).
11. I. G. Polotskii, G. Ya. Bazelyuk, and S. V. Kovsh, In: *Defects and Properties of Crystalline Lattice* (in Russian), Akad. Nauk Ukr. SSR, p. 158 (1966).

12. V. F. Belostotskii, I. Ya. Dekhtyar, V. S. Mikhalenkov,
     I. G. Polotskii, and S. G. Sakharova, *Metallofizika* (in Russian),
     Naukova Dumka, Kiev, p. 31 (1971).

13. I. G. Polotskii, V. I. Trefilov, S. A. Firstov, G. I. Prokopenko,
     S. V. Kovsh, and V. A. Kotko, *The Effect of Ultrasonic and
     Low-Frequency Vibrations on the Dislocation Structure and
     Mechanical Properties of Molybdenum, Chromium, and Tungsten*
     (in Russian), Preprint IMF no. 4 (1973).

14. I. G. Polotskii, G. I. Prokopenko, V. I. Trefilov, and
     S. A. Firstov, *Fiz. Tverd. Tela*, **11**, 155 (1969).

15. G. Ya. Bazelyuk and I. G. Polotskii, *Ukr. Fiz. Zh.*, **19**, 208
     (1974).

16. I. G. Polotskii, N. S. Mordyuk, and G. Ya. Bazelyuk,
     Metallofizika (in Russian), Naukova Dumka, Kiev, p. 99 (1970).

17. R. W. Whithworth, *Phil. Mag.*, **5**, 425 (1960).

18. B. Ya. Pines and I. F. Omel'yanenko, *Dynamics of Dislocations*
     (in Russian), Proceedings of FTINT, Akad. Nauk Ukr. SSR,
     Khar'kov, p. 242 (1968).

19. A. V. Kulemin and S. Z. Nekrasova, In: *Effect of High-Power
     Ultrasound on Metal Interfaces* (in Russian), Nauka, Moscow,
     p. 139 (1986).

20. B. Langenecker, *Soc. Automat. Engin. Trans.*, **74**, 449 (1966).

21. B. Langenecker, *Proc. Amer. Soc. Test. Mater.*, **62**, 602 (1962).

22. K. H. Westmacott and B. Langenecker, *Phys. Rev. Let.*, **15**, 761
     (1965).

23. B. Langenecker, *Rev. Sci. Instr.*, **37**, 103 (1966).

24. B. Langenecker, *Proc. High Power Ultrasonics*, p. 32 (1971).

25. V. Kh. Anchev and Yu. A. Skakov, *Izv. Vuzov, Chernaya
     Metallurgiya*, **11**, 132 (1974).

26. T. Broom and R. R. Ham, In: *Vacancies and Point Defects*
     (Russian translation), Metallurgizdat, Moscow, p. 54 (1961).

27. B. Weiss and R. Maurer, *Z. Metallkunde*, **22**, 915 (1969).

28. I. A. Kindin, I. M. Neklyudov, M. P. Starolat, G. N. Malik, and
     O. I. Volchok, *Fiz. Tverd. Tela*, **12**, 2456 (1970).

29. A. Jobu, K. Z.Kuzumune, and K. Atsano, *Met. Inst. Sci. Ind.
     Res.*, Osaka Univ., **27**, 193 (1970).

30. W. P. Mason, *Bell Syst. Tech. J.*, **34**, 903 (1955).

31. V. M. Goritskii, V. S. Ivanova, L. G. Orlov, and V. F. Terent'ev,
     Dokl. Akad. Nauk SSSR, **205**, 812 (1972).

32. J. Fridel, *Dislocations* (Russian translation), Mir, Moscow
     (1967).

33. A. Puskar, *Ultrasonic*, **3**, 124 (1977).

34. A. Puskar, *Acta Metallurgica*, **24**, 861 (1976).

35. A. Puskar, *Eng. Fracture Mechanics*, **10**, 187 (1978).

36. A. Puskar, *Materials Sci. and Eng.*, **24**, 75–83 (1976).

37. S. V. Kovsh, V. A. Kotko, I. G. Polotskii, *et al.*, *Fiz. Met. Metalloved.*, **35**, 1199 (1973).

38. A. V. Durgaryan and E. S. Badayan, In: *Relaxation Phenomena in Metals* (in Russian), Metallurgiya, Moscow, p. 352 (1968).

39. V. P. Grabchak and A. V. Kulemin, *Akust. Zh.*, **22**, 838 (1976).

40. V. P. Grabchak and A. V. Kulemin, *Proc. IX All-Union Acoustic Conf.* (in Russian), Nauka, Moscow, p. 17 (1977).

41. N. A. Tyapunina, G. M. Zinenkova, and Ngu Ei, *Dynamics of Dislocations* (in Russian) (Proc. FTINT, Akad. Nauk Ukr. SSR), Khar'kov, p. 233 (1968).

42. B. F. Walker, V. A. Johnson, W. C. Hahu, and I. P. Wood, *Trans. Met. Soc. AiME*, **242**, 1233 (1968).

43. R. E. Keith and I. I. Gilman, *Amer. Soc. Test. Mater. Spec. Techn. Publ.*, **3**, 237 (1959).

44. O. P. Golovko, N. A. Tyapunina, and M. P. Maskal'skaya, *Fiz. Tverd. Tela*, **8**, 2416 (1967).

45. H. J. Dawson, *J. Appl. Phys.*, **39**, 3022 (1968).

46. E. G. Aizentson and N. K. Utrobina, *Izv. Vuzov, Tsvetnaya Metallurgiya*, **1**, 122 (1972).

47. V. F. Belostotskii, *Fiz. Tverd. Tela*, **33**, 651 (1972).

48. Van Buren, *Defects in Crystals* (Russian translation), Metallurgiya, Moscow, p. 584 (1962).

49. W. Koster, *Z. Metallkunde*, **32**, 282 (1940).

50. H. R. Peiffer, *J. Appl. Phys.*, **34**, 298 (1963).

51. V. F. Belostotskii and I. G. Polotskii, *Fiz. Met. Metalloved.*, **35**, 660 (1973).

52. V. S. Biront, *Employment of Ultrasound in Heat Treatment of Metals* (in Russian), Metallurgiya, Moscow (1977).

53. W. Hessler, H. Miillner, B. Weiss, and R. Stickler, *Metal. Sci.*, **5**, 225 (1981).

54. Yu. F. Balalaev, M. A. Gaponov, and V. S. Postnikov, *Fiz. Khim. Obrab. Metallov*, **1**, 108 (1970).

55. I. G. Polotskii, In: *Recent Developments in Ultrasonic Technology* (in Russian), LDNTP, Leningrad, p. 88 (1972).

56. F. Blaha and B. Langenecker, *Z. Naturwiss.*, **20**, 556 (1955).

57. F. Blaha and B. Langenecker, *Acta Metallurgica*, **7**, 93 (1959).

58. F. Blaha and B. Langenecker, *Null. Nat. Inst. of Sciences of India*, **14**, 16 (1959).

59. I. A. Gindin, G. N. Malik, I. M. Nekmodov, O. T. Razumny, *Izv. Vuzov, Fizika*, **2**, 51 (1972).

60. D. Oelschlagel, *Acta Phys. Austr.*, **8**, 175 (1964).

61. R. I. Garber and G. N. Malik, *Fiz. Met. Metalloved.*, **22**, 310 (1966).

62. G. Kralic and B. Weiss, *Z. Metallkunde*, **51**, 471 (1968).

63. G. Ya. Bazelyuk, G. Ya. Kozyrskii, I. Ya. Polotskii, and G. A. Petrunin, *Fiz. Met. Metalloved.*, **29**, 508 (1970).

64. G. Ya. Bazelyuk, G. Ya. Kozyrskii, I. Ya. Polotskii, and G. A. Petrunin, *Fiz. Met. Metalloved.*, **32**, 145 (1971).

65. Yu. F. Pon'kin, B. I. Volkov, L. P. Kudryakova, *et al.*, *Metalloved. Term. Obrab. Mater.*, **8**, 67 (1971).

66. V. N. Semirog-Orlik, A. S. Kuprina, and N. L. Pozen, *Problemy Prochnosti*, **6**, 63 (1971).

67. E. G. Konovalov and M. D. Tyavlovskii, *Dokl. Akad. Nauk Belorus. SSR*, **7**, 701 (1973).

68. S. Stauze, R. Mitsche, and B. Weiss, *Arch. Eisenhuttenwessen*, **41**, 867 (1970).

69. F. Bullen, A. Head, and W. Wood, *Proc. Roy. Soc.*, **216**, 1126 (1953).

70. V. A. Parfenov, *Metalloved. Term. Obrab. Mater.*, **4**, 40 (1961).

71. R. L. Kogan, In: *Cyclic Strength of Metals* (in Russian), Akad. Nauk. SSSR, Moscow, p. 54 (1962).

72. N. L. Pozen, V. N. Semirog-Orlik, and I. A. Troyan, *Fiz. Khim. Obrab. Materialov*, **5**, 112 (1969).

73. O. V. Abramov, A. I. Kovalev, O. M. Smirnov, *Fiz. Khim. Obrab. Mater.*, **4**, 142 (1974).

74. G. H. Sinclair, *Proc. Am. Soc. Test. Mat.*, 74 (1952).

75. T. Broom and R. R. Ham, *Proc. Roy. Soc.*, **242A**, 166 (1957).

76. M. Uempel, *Arch. Eisenhuttenwessen*, **22**, 425 (1951).

77. A. D. Kennedy, *J. Inst. Metals*, **87**, 145 (1958).

78. N. H. Polakowski and A. Palcoudhuri, *Proc. Am. Soc. Test. Mat.*, **54**, 701 (1954).

79. E. G. Konovalov and N. G. Dovgyalo, In: *Employment of Ultrasound in Machine Building* (in Russian), Mashprom, Minsk, p. 22 (1964).

80. E. G. Konovalov, V. A. Sinyaev, and S. I. Kovalev, *Izv. Akad. Nauk Belorus. SSR*, Ser. Fiz.-Tekh. Nauk, **4**, 9 (1970).

81. E. G. Konovalov, V. M. Drozdov and M. D. Tyavlovskii, *Dynamic Strength of Metals* (in Russian), Nauka i Tekhnika, Minsk, p. 304 (1969).

82. G. A. Hayes and I. C. Shyne, *Phil. Mag.*, **17**, 859 (1968).

83. G. A. Hayes and I. C. Shyne, *Metal. Sci. J.*, **5**, 19 (1971).

84. M. V. Klassen-Neklyudova and A. P. Kapustin, *Dokl. Akad. Nauk SSSR*, **17**, 1019 (1955).

85. D. S. Kemsley, *J. Inst. Metals*, **87**, 10 (1958).
86. A. I. Kennedy, *Nature*, **178**, 810 (1956).
87. K. M. Pogodina-Alekseeva and E. M. Kremlev, *Metalloved. Term. Obrab. Mater.*, **9**, 7 (1966).
88. B. Langenecker, *AIAA Journal*, **1**, 80 (1963).
89. F. Blaha, B. Langenecker, and B. Odschlagel, *Z. Metallkunde*, **51**, 636 (1960).
90. B. Langenecker, *IEEE Trans. of Sonics and Ultrasonics*, **SU-13**, 27 (1966).
91. G. F. Nevill and F. R. Brotzen, *Proc. Am. Soc. Testing Materials*, **57**, 751 (1957).
92. V. P. Severdenko and V. V. Klubovich, In: *Employment of Ultrasound in Machine Building* (in Russian), Nauka i Tekhnika, Minsk, p. 31 (1964).
93. V. P. Severdenko and V. V. Klubovich, *Tsvetnye Metally*, **11**, 111 (1965).
94. E. G. Konovalov and I. G. Dovgyalo, In: *Plasticity and Shaping of Metals* (in Russian), Nauka i Tekhnika, Minsk, p. 41 (1964).
95. V. P. Severdenko, V. V. Klubovich, and V. I. Elin, In: *Plasticity and Shaping of Metals* (in Russian), Nauka i Tekhnika, p. 103 (1966).
96. E. G. Konovalov and A. L. Skripchenko, In: *Plasticity and Shaping of Metals*, Nauka i Tekhnika, Minsk, p. 113 (1964).
97. V. P. Severdenko and V. V. Klubovich, *Dokl. Akad. Nauk Belorus. SSR*, **6**, 563 (1962).
98. V. P. Severdenko, V. V. Klubovich, and E. G. Konovalov, *Ultrasonic Treatment of Metals* (in Russian), Nauka i Tekhnika, Minsk, p. 235 (1966).
99. E. G. Konovalov and A. L. Skripchenko, In: *Plasticity and Shaping of Metals* (in Russian), Nauka i Tekhnika, Minsk, p. 23 (1964).
100. V. I. Elin and V. I. Severdenko, In: *Machine Building and Metal Processing* (in Russian), Belorussian Polytechnical Institute, Minsk, p. 37 (1968).
101. P. I. Polukhin, S. I. Perevalov, and A. M. Galkin, In: *Intensification of Technological Processes in an Ultrasonic Field* (in Russian), Metallurgiya, Moscow, p. 67 (1986).
102. O. M. Smirnov and A. V. Kulemin, In: *Employment of Novel Physical Methods for Intensification of Metallurgic Processes* (in Russian), Metallurgizdat, Moscow, p. 227 (1974).
103. O. M. Smirnov, V. A. Lazarev, and A. V. Kulemin, In: *Experience of Ultrasound Employment in Ferrous Metallurgy* (in Russian), Metallurgiya, Moscow, p. 32 (1977).
104. O. M. Smirnov and Z. V. Barantseva, *Fiz. Khim. Obrab.*

Mater., no. 1 (1988).

105. R. Pohlman and B. Lehfeldt, *Ultrasonic*, **5**, 56 (1966).

106. V. A. Kononenko and N. S. Mordyuk, *Fiz. Met. Metalloved.*, **26**, 1081 (1968).

107. N. V. Demchenko, G. Ya. Kozyrskii, V. A. Kononenko, and N. S. Mordyuk, *Fiz. Met. Metalloved.*, **29**, 657 (1970).

108. G. Ya. Kozyrskii, V. A. Kononenko, and N. S. Mordyuk, *Fiz. Khim. Obrab. Mater.*, **2**, 156 (1970).

109. O. M. Smirnov, V. A. Lazarev, and A. V. Kulemin, In: *Experience of Ultrasound Employment in Ferrous Metallurgy* (in Russian), Metallurgiya, Moscow, p. 30 (1977).

110. V. M. Golikov, A. V. Kulemin, and V. G. Borisov, In: *Diffusion in Metals* (in Russian), Izdat. Tula Polytechnical Institute, Tula, p. 63 (1976).

111. A. V. Kulemin, *Akust. Zh.*, **18**, 613 (1971).

112. V. M. Golikov, A. V. Kulemin, and V. P. Manaenkov, *Akust. Zh.*, **21**, 850 (1975).

113. A. V. Kulemin and A. M. Mitskevich, *Dokl. Akad. Nauk SSSR*, **189**, 518 (1969).

114. A. V. Kulemin and A. M. Mitskevich, *7-th Intern. Congress on Acoustics*, Budapest, p. 245 (1971).

115. V. M. Golikov, A. V. Kulemin, and V. A. Lazarev, *Fiz. Met. Metalloved.*, **35**, 785 (1973).

116. V. M. Golikov, A. V. Kulemin, and V. G. Borisov, In: *Diffusion Processes in Metals* (in Russian), Izdat. Tula Polytechnical Institute, Tula, p. 37 (1976).

117. J. J. Lander, M. E. Kern, and H. L. Beach, *J. Appl. Phys.*, **23**, 1305 (1952).

Chapter 4

Nonlinear Effects at Interfaces in Fluids

As shown in chapters 2 and 3, exposure to ultrasound may give rise to nonlinear effects in fluids and solids.

In addition to, or sometimes as a result of these nonlinear effects, there also arise other nonlinear interfacial phenomena whose influence on the kinetics of heat and mass transfer in fluids may be great.

This chapter gives a more or less detailed account of nonlinear phenomena developing on the free surface of a liquid and causing its atomization. Some aspects of the formation of gas bubbles and degassing of a homogeneous liquid as well as the preparation of emulsions from immiscible liquids will also be considered.

4.1   Free Surface in Liquids – Atomization Process

Ultrasonic atomization of liquids was discovered in 1927 by Wood and Loomis [1] who employed a glass horn representing a middle-narrowed tube. A thin liquid layer, flowing down from the upper end of the tube, was atomized in its narrowed part by 300-kHz ultrasound.

Later [2–8], it was established that atomization can occur not only in a thin layer, but also in a jet or in a fountain, vibrations being fed through liquid or gas (Figure 4.1).

In the case of atomization in a layer, droplets are formed on the surface of the liquid spread over a radiator.

A fountain arising on the surface of the liquid, to which ultrasonic megahertz vibrations are supplied, is atomized as a result of turbulent flows.

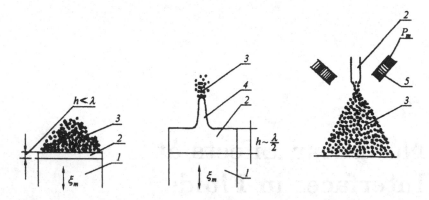

Figure 4.1. Schematic representation of the ultrasonic atomization of liquids in (a) layer, (b) fountain, and (c) jet: (1) radiator with the displacement amplitude ξ_m, (2) liquid, (3) liquid drops, (4) fountain, (5) vibrating gas jet with the pressure amplitude p_m.

A vibrating gas jet directed to a liquid stream can cause its atomization.

There exist cavitational [2] and capillary-wave [5–7] hypotheses of liquid atomization by acoustic vibrations supplied through a liquid.

Analysis, in terms of the capillary-wave hypothesis, of the behavior of a liquid on the solid surface vibrating with a frequency of 30 Hz led G. D. Malyuzhenets and V. I. Sorokin [9]* to establish that above a threshold value of the displacement amplitude, a system of capillary waves is formed on the surface of the liquid. At higher amplitudes, liquid droplets are separated from the wave crest to be thrown up to a sufficient height. Analogous results were obtained by W. Eisenmenger [10] in experiments upon atomization of liquids in a frequency range of 10–1500 kHz. Later, Stamm and Pohlman [11–12] derived an expression relating the size of ultrasonically atomized droplets and the capillary wavelength, which strongly supported the capillary-wave hypothesis.

In the 1960s, O. K. Eknadiosyants *et al.* [9, 13–20] published a number of works upon the ultrasonic atomization of liquids where proposed, in parallel with Pohlman *et al.* [21], a cavitation-wave hypothesis representing a compromise between the cavitational and capillary-wave hypotheses. It should be emphasized that the low-frequency ultrasonic atomization of liquid layers into drops may be due to [22]:

---

*See also V. I. Sorokin, Investigation of acoustic water-air resonators, Cand. Sc. (Phys.) Dissertation, Moscow, 1957.

(1) action of shock waves generated as a result of the collapse of cavities near the liquid-gas interface (the cavitational hypothesis);

(2) destruction of resonant gas- and vapor-filled bubbles pulsating near the liquid surface;

(3) splash of the liquid caused by pulsating subsurface gas- and vapor-filled bubbles;

(4) separation of liquid drops from the crests of the finite-amplitude standing capillary waves occurring on the surface of the hemispheric ledges formed by pulsating subsurface gas-vapor bubbles;

(5) separation of liquid drops from the crests of the finite-amplitude standing capillary waves occurring on the liquid surface in the absence of gas-vapor bubbles (the capillary-wave hypothesis).

Depending on experimental conditions and properties of atomized liquid, the contributions from these mechanisms to atomization can vary. Thus, in the case of low-viscous liquids, drops are mainly separated by the fifth mechanism, whereas in the case of fairly viscous liquids – by the fourth mechanism.

It should be noted that the cavitational hypothesis of acoustic atomization has not been proved until the work of Yu. Ya. Boguslavskii and O. K. Eknadiosyants [19].

To substantiate the cavitational hypothesis, they estimated the front pressure in the shock wave produced by collapsing bubbles and compared it with the threshold amplitudes for the formation of capillary waves and the separation of drops.

Theoretical analysis of capillary-gravitational waves and the accompanying processes of drops formation was first performed by G. D. Malyuzhenets and V. I. Sorokin [23] who considered the formation of the waves on the surface of a viscous liquid semispace $z < 0$ executing harmonical normal oscillations. Analogous analysis for ultrasonic frequences was carried out by W. Eisenmenger [10].

As was shown in chapter 1, the equation of continuity for an incompressible liquid is given by expression (1.48b). After substituting $v = -\text{grad}\Phi$ into this expression, it takes the form

$$\Delta\Phi = 0. \tag{4.1}$$

The motion of a liquid performing low-amplitude oscillations is, in a first approximation, potential; hence, velocity and pressure in a gravitational field are related as [24]

$$\frac{\partial \Phi}{\partial t} + \frac{v^2}{2} + \frac{P}{\rho_L} + gz = 0, \qquad (4.2a)$$

where g is the acceleration due to gravity.

In this case, the z-axis is defined to be directed upright, and the x, y-plane corresponds to the surface of a steady-state liquid.

The velocity of liquid particles was suggested to be so small that the term $v\,\mathrm{grad}\,v$ in the Euler equation (see expressions (1.37) and (1.43) of chapter 1) could be neglected in comparison with the term $\partial v / \partial t$. This condition is equivalent to the inequality $\zeta_m \ll \lambda_c$, where ζ_m is the amplitude of the gravitational-capillary wave and λ_c is its wavelength. In other words, the term $v^2/2$ in expression (4.2a) can also be neglected, and one can write

$$P = -\rho_L g \zeta - \rho_L \frac{\partial \Phi}{\partial t}, \qquad (4.2b)$$

here ζ is the z-coordinate of the liquid surface.

The liquid surface tends to equilibrium under the action of both gravitational forces and surface tension. The relationship between pressure and surface tension is given by the Laplace formula [24]

$$P = -\sigma_L \left(\frac{\partial^2 \zeta}{\partial x^2} + \frac{\partial^2 \zeta}{\partial y^2} \right). \qquad (4.3)$$

Juxtaposing (4.2b) and (4.3), one can write the condition of equilibrium for the liquid surface

$$\rho_L g \zeta + \rho_L \frac{\partial \Phi}{\partial t} - \sigma_L \left(\frac{\partial^2 \zeta}{\partial x^2} + \frac{\partial^2 \zeta}{\partial y^2} \right) = 0. \qquad (4.4)$$

Assume that the entire liquid oscillates in the z-direction

$$\xi(t) = \xi_m \cos \omega t. \qquad (4.5)$$

Then, the left-hand side of equation (4.4) should be supplemented with the term (4.6), that describes the accelerated motion of liquid

$$a(t) = -\xi_m \omega^2 \cos \omega t. \qquad (4.6)$$

Now expression (4.4) transforms into

$$\rho_L g \zeta + \rho_L \frac{\partial \Phi}{\partial t} - \sigma_L \left(\frac{\partial^2 \zeta}{\partial x^2} + \frac{\partial^2 \zeta}{\partial y^2} \right) = -\xi \rho_L \xi_m \omega^2 \cos \omega t. \qquad (4.7)$$

The insignificancy of the wave amplitude implies that the displacement ζ is small, and the vertical component of the surface velocity is

$$v_z = \frac{\partial \zeta}{\partial t}. \tag{4.8a}$$

On the other hand,

$$v_z = -\frac{\partial \Phi}{\partial z}; \tag{4.8b}$$

therefore,

$$\left. \frac{\partial \Phi}{\partial z} \right|_{z=\zeta} = \frac{\partial \zeta}{\partial t}. \tag{4.9}$$

In virtue of the displacement amplitude insignificancy, the condition (4.9) can be taken at $z = 0$ rather than an $z = \zeta$.

Thus, the boundary conditions for the potential Φ and the displacement ζ are defined by expressions (4.7) and (4.9).

Let us elucidate the probability of the standing sinusoidal wave formation on the liquid surface. For simplicity, one can restrict oneself to the case of the vibrational motion along the x-axis [10].

The solution to equation (4.1) can be sought in the form

$$\Phi = \Phi(t)e^{k_0 z}\cos k_c x. \tag{4.10}$$

Then, expression for the liquid surface displacement can be written as

$$\zeta = \zeta(t)\cos k_c x. \tag{4.11}$$

By substituting (4.10) and (4.11) into (4.8) and (4.9), one can obtain

$$\frac{\partial^2 \zeta(t)}{\partial t^2} + \left(\frac{k_c^2 \sigma_L}{\rho_L} + k_c g + \xi_m k_c \omega^2 \cos \omega t \right) \zeta(t) = 0. \tag{4.12}$$

As shown by Eisenmenger [10], the term that involves acceleration due to gravity can be omitted, if the frequency of vibrations is above 10 kHz. In this case and $\sigma_L \sim 100$ dynes/cm (0.1 N/m), the relative error is within some fractions of a percent. The frequency of capillary waves is given by

$$\omega_c^2 = \frac{k_c^3 \sigma_L}{\rho_L}. \tag{4.13}$$

The influence of viscosity can be accounted for by introducing the term $2\beta_e[\partial\zeta(t)/\partial t]$ into equation (4.12). In the case of the wave described by expression (4.10) and a sufficiently small viscosity coefficient η, the logarithmic damping decrement can be expressed as [24]

$$\beta_e = \frac{2\eta k_c^2}{\rho_L}. \tag{4.14}$$

By introducing a damping decrement, one can derive

$$\frac{\partial^2\zeta(t)}{\partial t^2} + 2\beta_e \frac{\partial\zeta(t)}{\partial t} + (\omega_c^2 + \xi_m k_c \omega^2 \cos\omega t)\zeta(t) = 0. \tag{4.15}$$

After some transforms and simplifications based on the estimation of the contributions from the terms of (4.15) to displacement, Eisenmenger got

$$\zeta = \zeta_m e^{[(k_c\omega/2)\xi_m - \beta_e]t} \sin\left(\frac{\omega}{2}t - \frac{\pi}{4}\right) \cos kx. \tag{4.16}$$

Analysis of this expression indicates that the amplitude of vibrations of the frequency $\omega_c = \omega/2$ exponentially increases with time, if

$$k_c \omega \xi_m > 2\beta_e. \tag{4.17}$$

Then, the threshold amplitude for the generation of capillary waves is given by

$$\xi_{m0} = \frac{4\eta k_c}{\rho_L \omega} = 2\eta \left(\frac{1}{\pi\sigma_L \rho_L f^2}\right)^{1/3}. \tag{4.18}$$

Taking into account the relationship between the frequencies of driving oscillations and capillary waves, one can get expression for the capillary wavelength

$$\lambda_c = \left(\frac{8\pi\sigma_L}{\rho_L f^2}\right)^{1/3}. \tag{4.19}$$

When the increasing amplitude of standing waves reaches the region of instability ($\xi_m \gg \xi_{m0}$), the process can no longer be described in terms of linear theory. Phenomena occurring in the range $\xi_{m0} < \xi_m < \xi_{md}$ were qualitatively described by G. D. Malyuzhents and V. I. Sorokin [9, 23]. Some time after the onset of driving vibrations, the increasing amplitude of capillary waves reaches some ultimate value, above which the motion would be periodical and stable. In this case, the standing wave crests are no longer sinusoidal, but have the form of slender tongues.

The upper limit ξ_{md} corresponds to the threshold amplitude, at which drops begin to separate from the crest.

At $\xi_m > \xi_{md}$, the amplitude of vibrations increases with time until drops separate from the crest at some initial velocity proportional to the amplitude ξ_m.

Further approximation to an adequate description of the atomization of liquid in a layer was done by P. Peskin and R. Raco [25] who considered capillary waves on the surface of a finite-thickness liquid layer. The thickness h of the layer was introduced into a consideration through the velocity potential Φ

$$\Phi = -\frac{1}{k_c} \frac{d\zeta(t)}{dt} \operatorname{ch} \left[k_c(z+h)\right]^{j k_c x}. \tag{4.20}$$

Analysis of the equations of motion enabled these authors to elucidate a relationship between the most frequent diameter of drops, D_d, frequency ω, displacement amplitude of the radiator surface, ξ_m, and the layer thickness h

$$\frac{D_d}{\pi \xi_m} = \left[\left(\frac{2\sigma_L}{\rho_L \omega^2 \xi_m^3} \right) 2 \operatorname{th} \left(\frac{\pi \xi_m}{D_d} \right) \left(\frac{h}{\xi_m} \right) \right]^{1/3}. \tag{4.21}$$

Functions $D_d/\pi\xi_m$ of the arguments h/ξ_m and $\rho_L\omega^2\xi_m^3/2\sigma_L$, calculated by formula (4.21) for the case of a liquid atomized by low-frequency, finite-amplitude ultrasonic vibrations, are shown in Figure 4.2,*a* and *b*, respectively. In the authors' opinion, the derived theoretical dependences agree well with experimental data. It should be noted that analysis of the literature data implies that a comprehensive nonlinear theory of the ultrasonic atomization of liquids that would relate ultrasonic parameters, technological factors, and the properties of the liquids, is still to be elaborated.

The capillary-wave theory allows the rate of the ultrasonic atomization of liquids to be estimated [11, 12]. In the consideration below, it is assumed that at least one drop of a constant diameter D_d would certainly separate from the crest of a finite-amplitude standing capillary wave.* Over a time period equal to $1/2T_c$ (T_c is the capillary wave period), on the area λ_c^2 there will appear two crests which give rise to

---

*Actually, the system must contain not only high-amplitude capillary waves from which drops are separated, but also waves with amplitudes less than the threshold value. Besides, some crests can give rise to several drops.

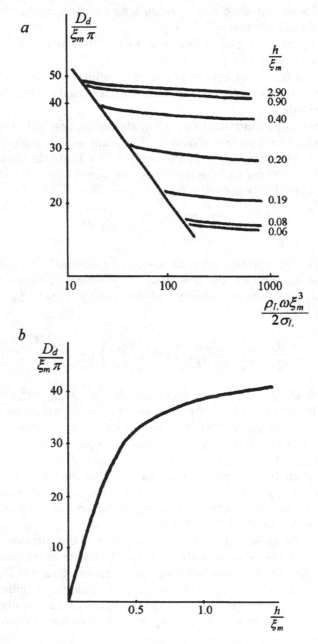

Figure 4.2. Relationship between the drop dimensions, amplitude of vibrations, and the thickness of the layer of an atomized liquid [25].

two drops. The overall number N of drops separated from the unit working area (1 cm^2) of an atomizer per unit time (1 s) will be

$$N = \frac{4}{T_c \lambda_c^2} = 2\frac{f_l}{\lambda_c}, \qquad (4.22)$$

and the estimated rate of atomization

$$\Pi_a' = 2\frac{f_c}{\lambda_c}\,\pi\frac{D_d^3}{6}, \quad \text{cm/s} \qquad (4.23a)$$

Taking into account (4.19) and assuming that $D_d = \alpha\lambda_c$, one can write

$$\Pi_a' = \frac{2}{3}\pi\alpha^3 \left(\frac{\pi\sigma_L}{\rho_L}\right)^{1/3} f^{1/3}, \qquad (4.23b)$$

where α is the coefficient of proportionality.

Analysis of expression (4.23b) indicates that the estimated rate of atomization weakly depends on the frequency of vibrations and such parameters of liquids as surface tension and density. As for the amplitude of vibrations and liquid viscosity, they do not enter into expression (4.23b) at all.

From the rate of atomization, one can estimate its efficiency, or the amount of a substance (in volume units) atomized by the total surface S_r' of an atomizer per unit time

$$\Pi_a = \Pi_a' \cdot S_r'. \qquad (4.24)$$

The least understood type of acoustic atomization is such a process in which acoustic energy is fed into the working zone via gas.

Since a rigorous theoretical description of such process is sophisticated, relevant works consider a linearized problem of a gas flowing around a liquid jet and its drops.

Some authors were able to get semiempirical formulas describing this type of atomization [26, 27]; however, its mechanism remains to be elucidated.

Some questions concerning the ultrasonically induced destruction of a cylindrical liquid jet and the deformation of its drops were considered in [8].

A liquid, issuing from a vessel under the action of a low excessive pressure, has the form of an almost cylindrical slightly converging jet. The jet breaks down because of increasing oscillations of the 'neck' type, which are due to surface tension, and of the bent type, which are due to the so-called Konstantinov's instability [29]. The jet instability may increase under the action of a cyclic external pressure. The jet of an inviscid, incompressible liquid moving at a constant longitudinal velocity rapidly undergoes a disturbance under the action of a cyclic external pressure provided that its frequency is above a particular value

$$\omega > \frac{4}{27} \frac{\rho_L U^3}{\sigma_L}, \tag{4.25}$$

where U is the jet velocity.

In the case of a liquid steel stream flowing at a velocity of 2 m/s, this value corresponds to $f_{min} \sim 1$ kHz. Higher external frequencies has no appreciable effect on the process of stream disintegration.

Drops formed as a result of jet disintegration are large (of the order of millimeters across), their size being weakly dependent on the disturbing frequency.

Due to liquid viscosity, there is a gradient of the longitudinal velocities of liquid particles in the cylindrical portion of a nozzle. Outside the nozzle, the velocity profile undergoes changes, which affects the jet and produces complex transients in it. Changes in the velocity profile take place within a distance $z = 0.416a$ (a is the jet radius) from the nozzle edge. On the axis of the jet, near the nozzle edge, there is a rarefaction (negative pressure), which may be responsible for an extreme sensitivity of this region of the jet to external influences. According to the laws of conservation of liquid discharge, momentum, and kinetic energy, the jet outside the nozzle must be in the form of a cone with an apex angle of $\sim 50°$.

These features of the issuing jet cannot, however, explain the formation of small (of the order of tens or hundreds of microns) liquid drops during a pneumoacoustic atomization.

One of the possible explanations can be the deformation and disintegration of primary fairly large drops into smaller ones [28]. In a streamline flow around a drop, the highest velocity is on its equator, and there appears a wake region behind the drop (Figure 4.3,a), which results in a pressure profile similar to that shown in Figure 4.3,b.

In the case of a pulsating sign-constant flow, the pressure difference between the 'pole' and 'equator' of the streamlined drop is

$$\Delta p = 0.45 \rho_G V^2, \tag{4.26}$$

which causes its deformation – the drop acquires the form of an ellipsoid of revolution. The drop will be changing until some critical deformation is achieved, after which the drop will be divided.

The necessary condition for the drop to be divided is that its surface energy should be greater than the sum of the surface energies of

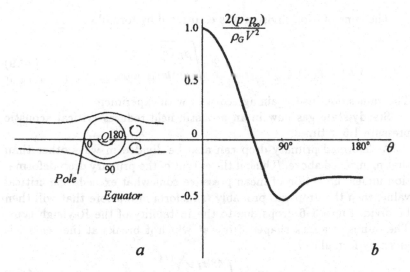

Figure 4.3. (a) Diagram of a gas flow around a spherical drop and (b) pressure distribution near the drop [8].

secondary drops. By equalizing the surface energy of an oblate ellipsoid of revolution and the net surface energy of n spherical secondary drops of the same total volume, one can estimate the acoustic pressure p_m sufficient to attain the critical deformation ε_n necessary for the breakup of the primary drop into n secondary drops

$$n = \frac{1}{8\varepsilon_n^3} \left[\varepsilon_n^3 + \frac{1}{\sqrt{\varepsilon_n^6 - 1}} \ln \left(\varepsilon_n^3 + \sqrt{\varepsilon_n^6 - 1} \right) \right]^3 , \qquad (4.27)$$

$$p_m = 2c_G \sqrt{\frac{2X_m \sigma_L \rho_g}{D_d}}, \qquad (4.28)$$

where

$$X_n = \frac{J(\varepsilon)}{\ln \varepsilon_n}; \quad J(\varepsilon) = \frac{\varepsilon^5}{5} - \frac{\varepsilon^2}{2} - \frac{1}{c} + \frac{1}{4\varepsilon^4} + 1.05.$$

In the case of the breakup of primary drops ($r = 3$ mm) of glycerol and bismuth and tin melts into three secondary drops (the oblate ellipsoid is reminiscent of three rather than two fused drops), calculations by formula (4.28) gave acoustic pressures close to experimental values (see Table 4.2).

The time of drop division was estimated by formula

$$\tau_3 = 1.26 \sqrt{\frac{\rho_L D_d^3}{8\sigma_L}} \qquad (4.29)$$

The values obtained again agreed well with experiment.

Steady-state gas flow in an acoustic field reduces critical acoustic pressure 1.5–2 times.

A deformed primary drop can also be divided in ways other than that mentioned above. Thus, if the extent of the primary drop deformation under the action of mean pressure somewhat exceeds the critical value, then the drop will probably transform into a tore that will then be divided into 3–6 drops due to the instability of the Rayleigh type. The radius of a disk-shaped drop, at which it breaks at the center, is given by formula

$$r_0 = \left(\frac{48\sigma_L \nu}{\pi \Delta p} \right)^{1/4} . \qquad (4.30)$$

The greater is the pressure difference, the sooner the disk-shaped drop breaks out and transforms into the tore.

The capillary waves generated in a drop under the action of cyclic pressure can also promote its breakup. The arising disturbances in thickness disrupt the layer with the thickness h into drops with the radius

$$r_z = \left(\frac{3h}{2\pi k_c^2} \right)^{1/3} . \qquad (4.31)$$

The radii of molten metal drops were calculated to be 50–100 μm, which is in a good agreement with experimental estimates.

Atomization of liquids was experimentally investigated with water and some organic liquids [13–20] to establish a relationship between the size of drops, properties of atomized liquid, and acoustic parameters.

High-speed filming (5000 frames/s) of the water layer atomized by 3.2-kHz vibrations showed that at a sufficient amplitude, they can cause the separation of single, approximately equal-size drops from the wave crests. Vibrations with higher amplitudes give rise to filaments on liquid's surface near the crests, which then break up into several drops of different sizes. The filming also showed that the high-amplitude vibrations excite capillary waves with rounded or acute crests, which will then transform into filaments.

Eisenmenger [10] experimentally estimated capillary wavelengths in the water layer subjected to vibrations with different frequencies, derived the frequency dependence of the amplitude threshold for the generation of capillary waves, and established the relationship between the amplitudes of exciting vibrations and capillary waves.

The capillary wavelength decreases with increasing exciting frequency (Figure 4.4, a) and is independent of the amplitude of capillary waves and surface tension in the range studied (0.065–0.085 N/m).

The amplitude threshold of exciting vibrations diminishes with increasing frequency (Figure 4.4, b), whereas the amplitude of capillary waves drastically increases when the amplitude of exciting vibrations exceeds a certain value (Figure 4.4, c).

Atomization of liquid in a fountain is characterized by the following features [16, 17]: low-amplitude vibrations of a focusing radiator cause the formation of a stream at the apex of a conical fountain, as it were composed of spherical beads; atomization occurs in the lower portion of the stream; the bead diameter decreases with increasing frequency of exciting vibrations.

Fog droplets are ejected from the fountain stream as rare momentary ($\sim$0.4 ms) bursts whose duration is considerably less than the recurrent time. Much larger drops may be ejected either simultaneously with the fog droplets or independently.

More intense vibrations make the stream cylindrical and uneven. The stream diameter increases with ultrasonic power, but weakly depends on frequency. In this case, the atomized region enlarges and shifts upward the stream.

Fog is ejected from the distinct zones of atomization that continuously migrate over the stream surface.

Atomization in fountain occurs above some threshold value of the displacement amplitude. The existence of the threshold and its dependence on temperature and properties of the liquid atomized were revealed by K. Sollner [2]. Later [17], it was established that the displacement amplitude threshold is proportional to the dynamic viscosity of liquid, but no unambiguous relationship between the threshold, liquid density, and surface tension was derived (regardless of whether atomization was in fountain or in layer).

Some authors [12, 30] observed a correlation between the droplet sizes and capillary wavelengths, which was governed by vibrational frequency and properties of an atomized liquid. In Figure 4.5, experimental plots fit well a straight line parallel to the frequency f-axis. The data were obtained by various investigators and refer to atomization in fountain or layer; therefore, the data scatter is largely due to different experimental conditions and methods of data processing.

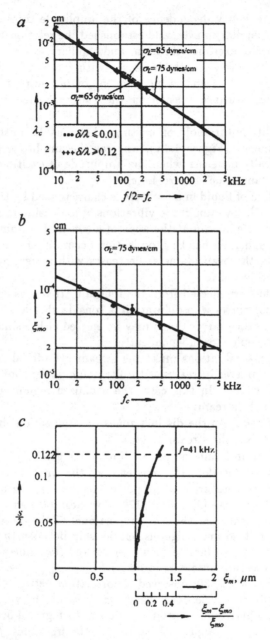

Figure 4.4. The relationship between (a) the capillary wavelength and frequency, (b) the threshold amplitude of capillary wave excitation and frequency, and (c) the amplitudes of capillary waves and exciting vibrations [10].

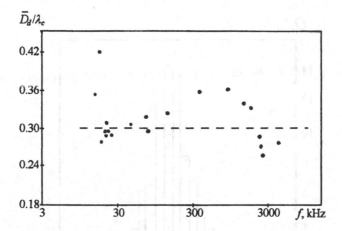

Figure 4.5. Relationship between the mean size of drops and the capillary wavelength during the atomization of liquids [9].

The mean diameter of drops, D_d, and the capillary wavelength λ_c are related as follows

$$D_d = \alpha \lambda_c, \qquad (4.32)$$

where the coefficient $\alpha = 0.3$ depends neither on the frequency of vibrations nor the properties of liquid and the method of its atomization.

Analysis of the drop size as a function of liquid density and surface tension suggests that the quantity $(\sigma_L/\rho_L)^{1/3}$ for real liquids changes little (from 2 to 5), as well as the size of drops of such dissimilar substances as organic liquids and molten metals atomized with the same frequency.

The value D_d can be widely varied only by changing the frequency of acoustic vibrations. The size distribution of drops is shown in Figure 4.6,a,b,c for the water atomized by vibrations with the amplitude 4 μm. It can be seen that the distribution peak, as well as the whole histogram, shift toward lower sizes with increasing frequency. Moreover, the histogram symmetry is somewhat disturbed due to the appearance of a subsidiary maximum at ~13 μm.

The drop spectrum noticeably widens and shifts toward larger sizes with the increasing amplitude of vibrations (Figure 4.6,d).

Comparison of the distribution patterns of the relative diameters D/D_d for liquids atomized in layer by low-frequency vibrations and in fountain by high-frequency vibrations indicates that D/D_d varies from 0.3 to 3.0 in both cases.

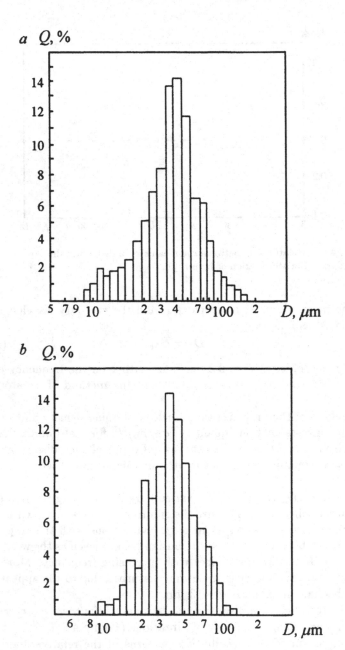

Figure 4.6. Size distribution of water drops atomized at a displacement amplitude of 4 μm and frequency (kHz): (a) 21, (b) 40, (c) 53. (d) Atomization at the amplitude 25 μm and frequency 21 kHz [12].

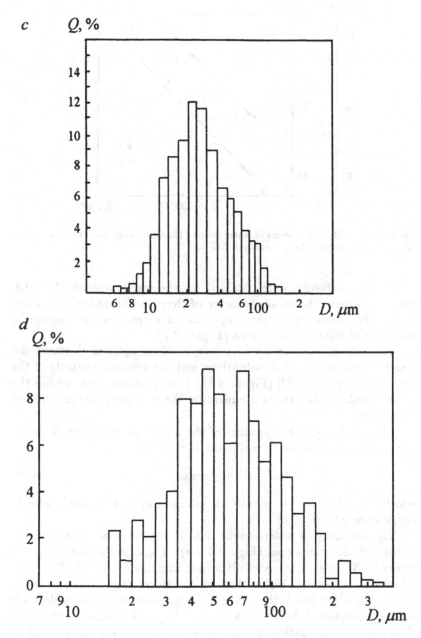

Figure 4.6. Continued.

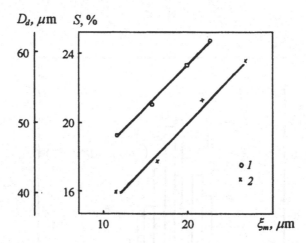

Figure 4.7. Mean diameters of water drops (*1*) and root-mean-square deviations (*2*) versus the amplitude of vibrations [9].

Analogous results were obtained for the water atomized by 20.8-kHz vibrations: the mean diameter of drops and standard deviation S linearly increase with the amplitude of vibrations, indicating the widening of distribution spectra (Figure 4.7).

The rate of atomization in the layer of an optimal thickness depends on the amplitude of vibrations and the dynamic viscosity of the liquid atomized [12, 22] (Figure 4.8). It can be seen that within the ranges studied, the rate of atomization shows a linear dependence on amplitude.

As shown in [12], the product of the maximum rate of atomization and the dynamic viscosity coefficient is constant

$$\Pi'_a \, \eta = \text{const}, \qquad (4.33)$$

which is also true for silicone oils whose coefficients of dynamic viscosity range from 0.65 to 20 cP.

The efficiency of water atomization in layer diminishes as the frequency of vibrations rises (Figure 4.9,*a*). Thus, at the same acoustic power 20 W, the efficiency of ultrasonic atomization by 20-kHz vibrations is about 8 times higher than by 50-kHz vibrations.

On the other hand, the efficiency of atomization strongly depends on the properties of the liquids atomized (Figure 4.9,*b*, the experimental data refer to atomization in fountain at 2.7 MHz). The behavior of all curves (corresponding to various liquids) with respect to the power of exciting vibrations is similar: there is a drastic increase in the efficiency

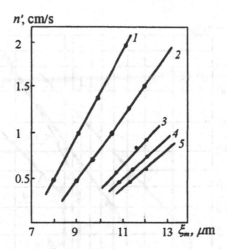

Figure 4.8. Amplitude dependences of the rate of atomization of glycerol-water mixtures with various viscosities (cP): (*1*) 1; (*2*) 2; (*3*) 3; (*4*) 4; (*5*) 5 [9].

of atomization at low powers followed by a stage of relatively slow increase with a tendency to saturation at higher powers.

At constant frequency (2 MHz) and power of vibrations, the efficiency of ultrasonic atomization depends only on such physicochemical parameters of atomized liquid as the saturation vapor pressure P_{vo}, coefficient of dynamic viscosity η, and surface tension σ_L [16]

$$\Pi_a = \frac{P_{vo}}{\eta \sigma_L}. \qquad (4.34)$$

Comparison of the calculated and experimental efficiencies of atomization shows that the ratio Π'_c/Π' ranges from 30 to 2 and directly correlates with the coefficient of dynamic viscosity (Table 4.1).

Of technological importance is the question as to the role of various parameters, first of all the layer thickness of the liquid atomized. As shown in [12, 22], atomization is most efficient at $h \ll \lambda/2$. However, there is a more delicate, albeit still obscure, dependence of the efficiency of atomization on the layer thickness. Thus, water is atomized by 20-kHz ($\lambda = 75$ mm) vibrations at the 30-μm displacement amplitude of the radiator, only if the following condition is fulfilled

$$0.4 \text{ mm} < h < 3.2 \text{ mm},$$

the maximum rate of atomization being observed at $h \sim 0.8$ mm [22]. There exists some critical thickness of the layer to be atomized [19].

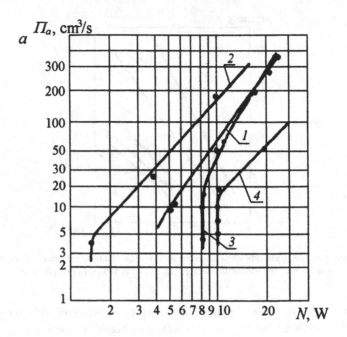

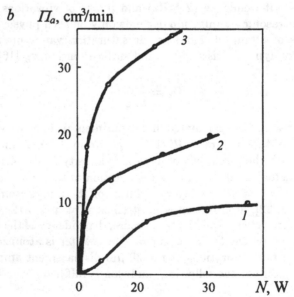

Figure 4.9. Dependences of the efficiency of atomization on ultrasonic power (*a*) for various frequencies (kHz): (*1*) 18.5, (*2*) 20, (*3*) 22.5, (*4*) 50 [9]; and (*b*) various liquids: (*1*) water, (*2*) BV-arol, (*3*) gasoline [5].

Table 4.1 Calculated and experimental rates of the atomization of liquids in layers [9]

Liquid	Frequency f, kHz	Amplitude ξ_m, μm	Density ρ_L, g/cm^3	Surface tension σ_L, dynes/cm	Viscosity η, cP	Atomization rate, cm/s	
						Calculated, Π'_c	Experimental, Π'
Water	20–21	10	1	74	1	5.5	1.3
Silicon oil	20–21	25	0.78	15.5	0.65	3.5	1.5
–”–	–”–	–”–	0.91	18	3	3.5	0.6
–”–	–”–	–”–	0.94	20	7	3.6	0.4
–”–	–”–	–”–	0.95	20	11.5	3.6	0.3
–”–	–”–	–”–	0.96	20	20	3.6	0.1
Fuel	20–21	25	0.80	30	2.0	4.3	0.8
–”–	–”–	–”–	0.85	27.9	2.5	4.2	0.8
–”–	–”–	–”–	0.85	27.3	8.1	4.1	0.3
Lead (T=350°C)	20–21	25	10.6	449	2.5	4.5	1.1

The most efficient atomization in fountain with the use of focusing systems takes place if the layer thickness is equal to the focal distance; i.e, when the radiator is focused on the surface of the liquid atomized.

Relatively few works dealing with pneumoacoustic atomization were aimed at studying its kinetics and the relationship between the fractional composition of drops, properties of the liquid atomized, acoustic parameters, and various technological factors. This problem is intricate because of a combined action of acoustic vibrations and gaseous streams.

M. A. Poger and O. K. Eknadiosyants [31] studied the pneumoacoustic atomization of water and found that the total time of drop transformation was 2.5 ms. After entering into atomization zone, the drop begins to drift to the radiator and to change its shape. Light spots on drop's surface indicate the occurrence of stresses. The drop gradually flattens into a film, normal to the gas flow, to be then disintegrated into particles carried by the gas flow away from the radiator. Based on these observation, the authors suggested that the flattening and fragmentation of drops occur under the action of ultrasonically generated shock waves [31].

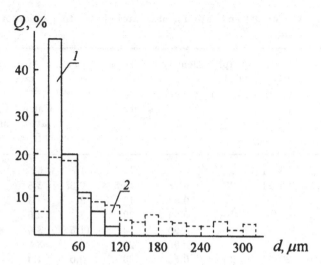

Figure 4.10.   Size distribution of drops during (*1*) pneumoacoustic and (*2*) pneumatic atomization of water [31].

During pneumatic atomization, the gas jet, acting upon a drop, gives rise to a bundle of slender liquid streams that disintegrate later into droplets (the disintegration begins from the side of the nozzle). The degrading drops are displaced by the gas flow away from the nozzle and initially retain their spherical form. In high-speed cinemagraphic frames, these drop show up white, which is probably due to the generation of surface waves.

Figure 4.10 shows the size distribution of water drops atomized pneumoacoustically and pneumatically (the pneumatic atomization was brought about by removing resonator from the radiator). As follows from the comparison of the mean diameters and standard deviations of droplets, pneumoacoustic atomization produces significantly smaller droplets than pneumatic atomization does (the mean diameters are 41.8 and 109.4 μm and standard deviations are 22.4 and 83.0 for droplets atomized pneumoacoustically and pneumatically, respectively).

The fractional composition of drops was estimated as a function of liquid discharge (Figure 4.11). The mean diameter of drops increases with the liquid discharge, and after reaching 300 l/h, the distribution spectrum significantly widens and shows an increase in the number of larger drops ($\sim$160 μm in diameter).

The authors of works [32–35] attempted to relate the level of acoustic pressure necessary for the pneumoacoustic atomization of drops and jets of various liquids to their physicochemical properties.

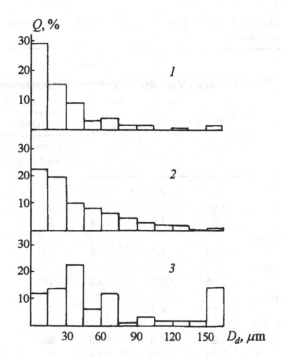

Figure 4.11. Size distribution of drops during the pneumoacoustic atomization of water at the rates (l/h): (*1*) 50, (*2*) 150, (*3*) 300 [31].

The quantity P_{mt} was measured for liquids atomized either without air flows (zone A of pure acoustic atomization), or under the combined action of air flows and acoustic vibrations (zone B of pneumoacoustic atomization).

The liquids investigated dramatically differed in their physicochemical properties, such as density, viscosity, and surface tension (Table 4.2). In particular, their density varied by more than an order of magnitude, dynamic viscosity – about 1500 times, and surface tension – 25 times.

Experiments were performed with drops and jets 2–3 and 0.8–1 mm in diameter, respectively.

Deformation of a drop before its breakup proceeds similarly for all liquids investigated (Figure 4.12,*a*). Initially, the drop transforms into a lens with the principal axis normal to the direction of the vibrational velocity vector. Then the lens acquires a bell-shaped form, there appear breaks, and finally the drop breaks down into small fragments. Prior to the breakup of a bell-shaped drop acted upon by a high-speed gaseous

Table 4.2 Physicochemical properties of liquids and acoustic pressure thresholds.

Substance	Parameter					
	Density $\rho_L \times 10^{-3}$, kg/m^3	Viscosity η, cP	Surface tension $\sigma_L \times 10^{-3}$, N/m	Atomization pressure $P_{mt} \times 10^{-2}$, Pa		
				Drops		Jets
				Experiment	Calculations	
Zone A						
Ethanol	0.80	1.2	22.0	30		
Water	1.0	1.0	72.8	52		54
Water + glycerol (9:1)	1.02	1.3	71.8	47		70
Water + glycerol (3:1)	1.06	2.1	70.2	46		72
Water + glycerol (1:1)	1.13	6.1	67.6	50		74
Water + glycerol (1:3)	1.20	36.5	65.0	60		
Glycerol	1.26	1489	62.4	62	51	100
Bismuth	9.8	1.6	378.0	122	127	120
Tin	7.0	1.9	530.0	150	150	130
Zone B						
Bismuth	9.8	1.6	378.0	63		
Wood's alloy	9.4	1.5	450.0	71		
Tin	7.0	1.9	530.0	76		
Indium	7.36	1.9	556.0	81		

flow, there occurs an intermediate stage involving the boundary layer separation similar to that observed in [31].

High-speed filming showed that the time of drop disintegration is 5–15 ms, or 100-300 vibrational periods.

The behavior of atomized liquid jets depends on the density and surface tension of the liquid (Figure 4.12,*b*).

Jets of heavy liquids with high surface tension (for instance, metal melts) disintegrate in a way similar to the disintegration of their drops. The breakup of jets occurs at a certain point, where acoustic pressure reaches a value specific for a given liquid. In this case, the jet will only slightly deviate from the vertical direction.

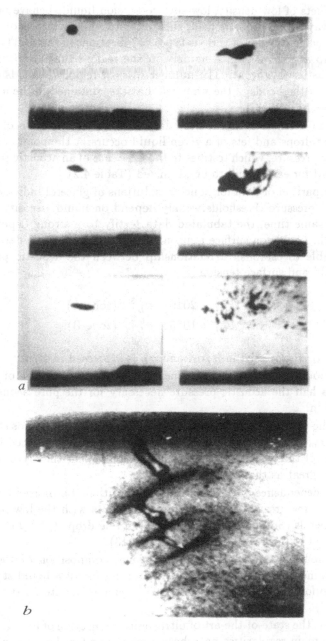

Figure 4.12. Disintegration of (a) bismuth drop (the interval between frames is 0.01 s) and (b) acetone jet.

The jets of low-density, low-surface-tension liquids behave in a different way. In an acoustic field, such jets are wavy, which is obviously due to the generation of various types of vibrations in them. These jets break up at several points, namely, in the regions that are parallel to the acoustic wave front. The number of such regions depends on the number of the bends of the wavy jet that are contained in the acoustic wavelength.

It can be inferred from these data that the atomization of equal-diameter drops and jets of a given liquid occurs at the points of equal acoustic pressure, which testifies to the existence of an acoustic pressure threshold for each liquid to be atomized (Table 4.2).

Comparison of data for aqueous solutions of glycerol indicates that acoustic pressure thresholds weakly depend on liquid viscosity $\eta^{0.035}$. At the same time, the tabulated data testify to a strong dependence of the thresholds on surface tension. Computation of these data made it possible to establish a relationship between the acoustic pressure threshold and surface tension

$$P_{mt} = 20161 \cdot \sigma_L^{0.5} \quad \text{(zone A)}$$
$$P_{mt} = 10554 \cdot \sigma_L^{0.5} \quad \text{(zone B)}$$

here, P_{mt} is expressed in N/m^2 and σ_L is expressed in N/m.

It should be noted that the pneumoacoustic atomization of liquids requires half the acoustic pressure necessary for the pure acoustic atomization.

In the case of jet atomization, the surface tension of liquids strongly affects P_{mt}, the effect of viscosity being also more pronounced than in the case of the atomization of drops. Thus, P_{mt} for the glycerol jet is twice as great as that for the water jet.

The dependence of the fractional composition of atomized liquid on acoustic pressure was evaluated in experiments with the low-melting-point metals (Sn and Bi). The yield Q of fine drops (¡ 50 μm in size) was found to increase with P_m (Figure 4.13).

It should be noted that the fractional composition of the powder obtained by the pneumoacoustic atomization of a liquid steel was more uniform than that in the case of the pneumatic atomization (Figure 4.14).

Thus, the state-of-the-art of ultrasonic atomization of liquids is such that its main regularities have been understood for the cases of ultrasound feeding via a gas and a liquid occurring in layer and in fountain.

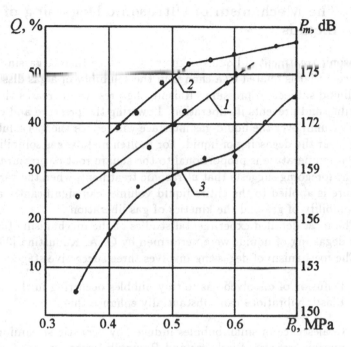

Figure 4.13. Correlation between the granulometric composition of (*1*) bismuth and (*2*) tin powders and (*3*) acoustic pressure [33].

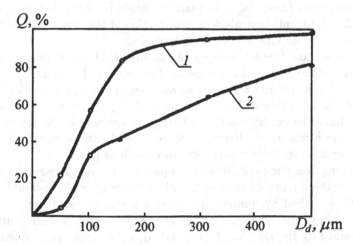

Figure 4.14. The integral size distribution of powders produced by (*1*) pneumatic and (*2*) pneumoacoustic atomization of molten steel R6M5 [34].

4.2 The Mechanism of Ultrasonic Degassing of Liquids

Ultrasonic treatment of liquids is known to induce their degassing [36], a process that is related to a change in the solubility of a gas dissolved in a liquid subject to pressure. Raising the pressure increases the gas solubility and prevents its liberation. Lowering the pressure and creating a vacuum over the liquid-gas interface will reduce the gas solubility and favour the degassing of liquid. For molten metals, gas solubility at a given temperature is proportional to the square root of pressure [37].

The foregoing suggests that ultrasonic treatment, when the variable pressure is applied to the entire liquid volume, can significantly affect the solubility of gas and the kinetics of gas liberation.

The most detailed experimental studies of the mechanism of ultrasonic degassing of liquids were performed by O. A. Kapustina [38–44].

The mechanism of degassing involves three successive stages:

(1) Diffusion of dissolved gas to tiny bubbles occurring in the liquid. Elastic vibrations can substantially enhance this process.

(2) Coalescence of small bubbles induced by acoustic streaming, radiation pressure, Bjerkness and Bernoulli forces.

(3) Emergence of enlarged bubbles.

In practice, these stages virtually coincide in time; therefore, only special experiments may allow the estimation of the contributions from each stage to the kinetics of liquid degassing.

To elucidate specific features of the dissolved gas diffusion to vibrating bubbles, one should consider the motion of a single spherical bubble. Let us restrict ourselves to analysis of the linear harmonic vibrations of a bubble with dimensions $R \ll \lambda_L$ (λ_L is the wavelength in the liquid to be degassed). In this case, the motion of the bubble subject to a continuous harmonic force $F_m e^{j\omega t}$ can be described by a linear second-order differential equation, such as (1.24).

Raising acoustic pressure above a particular level gives rise to cavitation in the liquid (see chapter 2), when the motion of the bubble can only be described by complex nonlinear differential equations.

At a zero time, when the bubble is at rest, its steady-state radius is determined by the amount of gas inside it, vapor pressure, and surface tension

$$P_g + P_V = P_0 + \frac{2\sigma_L}{R_0}, \qquad (4.35)$$

whereas gas concentration in the liquid near the bubble is given by [44]

$$C_s = C_p \left(1 + \frac{2\sigma_L}{R_0 P_0}\right), \tag{4.36}$$

where C_p is the steady-state gas concentration in the liquid.

If the gas concentration C_p differs from C_s, there appears a diffusional flow, whose value can be determined from the equation of diffusion [45]

$$\frac{\partial C}{\partial t} + (\overline{u}\nabla)C = D\nabla^2 C, \tag{4.37}$$

where $\overline{u}$ is the velocity of the medium motion.

In the absence of vibrations, the second term on the left-hand side of equation (4.37) can be omitted, and the expression for the gas flow into the bubble due to molecular diffusion takes the form

$$I_0 = 4\pi D_L R_0^2 (C_0 - C_s) \left\{ \frac{1}{R_0} + \frac{1}{\sqrt{\pi Dt}} \right\} \tag{4.38}$$

with the following initial and boundary conditions

$$\left. \begin{array}{ll} C(r, 0) = C_0 & r = R_0 \\ \lim_{r \to \infty} C(r, t) = C_0 & r > R_0 \\ C(R_0, t) = C_s & t \neq 0 \end{array} \right\} \tag{4.39}$$

Depending on the ratio C_0/C_s, the behavior of the bubble may vary: at $0 < C_0/C_s < 1$, when liquid is not saturated with gas, the bubble is gradually dissolved; but at $C_0/C_s > 1$, when liquid is supersaturated with gas, the bubble grows.

However, under the action of ultrasound, gas will diffuse from the liquid to the bubble even at $C_0/C_s \leq 1$, which can be explained as follows [44]. The concentration of gas during the phase of bubble shrinkage increases, and the gas diffuses from the bubble to the liquid. During the phase of bubble expansion, the gas duffuses in the opposite direction. Since the surface of the bubble is greater in the phase of its expansion than during its shrinkage, than the amount of the gas influent into the bubble during its expansion is greater than the amount of the gas effluent from the bubble during its shrinkage. Microstreaming, arising near vibrating bubbles, reduces the thickness of the acoustic boundary layer, thereby contributing to the kinetics of diffusion [46, 47]. Microstreaming is obviously involved in the delivery of gas-saturated liquid to the interface. Mathematical description of gas diffusion to a vibrating bubble should take into account that the velocity of medium motion $\overline{u}$ in a convective term of equation (4.37) is the sum of the vibrational velocity v and stationary velocity V.

During the estimation of the contribution from the convective term to the total diffusion flow, this term can be taken to be roughly equal to the sum of diffusion flows resulting from vibrations of the bubble itself, I_1, and from vibrations of the liquid near the bubble, I_2.

By assuming that $\langle (v\nabla)C \rangle \gg \langle (V\nabla)C \rangle$, the equation (4.37) can be reduced to the form

$$\frac{\partial C}{\partial t} + (v \sin \omega t \nabla)C = D_L \nabla^2 C. \tag{4.40}$$

The solution to this equation determines the diffusion gas flow with allowance for only vibrational motion of the bubble. If the bubble vibrates harmonically with a small amplitude, then

$$R(t) = R_0 \left(1 + \frac{\xi_m}{R} \sin \omega t \right), \tag{4.41}$$

where ξ_m is the amplitude of bubble vibrations. Then, the boundary conditions can be written as

$$\left. \begin{array}{l} C = C_0 \quad \text{at} \quad r \to \infty, \quad r > R_0 \\ C(t) = C_0 \left(I + \dfrac{P_m \sin \omega t}{P} \right) \quad \text{at} \quad r = R, \quad t \neq 0 \end{array} \right\}. \tag{4.42}$$

The solution to equation (4.41) has the form [45]

$$I_1 = \frac{24\pi D_L C_p}{\rho_L^2} \frac{P_m^2}{\omega^4 R^3 \left[\left(\dfrac{\omega_0^2}{\omega^2} - 1 \right)^2 + \delta^2 \right]}, \tag{4.43}$$

where ω_0 is the natural frequency and δ is the damping constant of bubble vibrations.

To estimate the diffusion flow I_2 with allowance for microstreaming, it should be assumed that $\langle (\overline{V}\nabla)C \rangle \gg \langle (v\nabla)C \rangle$. By averaging the diffusion equation (4.37) and boundary conditions (4.42) with respect to time, one can obtain

$$(\overline{V}\nabla)C = D_L \nabla^2 C \tag{4.44}$$

$$\left. \begin{array}{l} \lim C(r) = C_0, \quad r > R_0 \\ C(R) = C_s, \quad\quad t \neq 0 \end{array} \right\}. \tag{4.45}$$

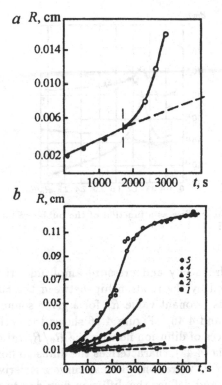

Figure 4.15. Time course of the mean radius of a bubble in water (a) without ultrasound (dark circles) and subject to a 26.5-kHz acoustic field with $p_m = 0.036$ MPa (open circles); (b) $p_m = 0$ (curve 1), 0.014 (curve 2), 0.04 (curve 3), 0.069 (curve 4), and 0.146 (curve 5) MPa [44].

The solution to equation (4.44) with the boundary conditions (4.45) makes it possible to find I_2 [48]:

$$I_2 = \frac{\sqrt{2\pi D_L}(C_0 - C_s + C_s\xi_m^2)P_m}{\rho_L\omega^{3/2}\sqrt{\left(\frac{\omega_0^2}{\omega^2} - 1\right)^2 + \delta^2}}. \tag{4.46}$$

The above expressions are valid for small Reynolds numbers (Re = $R_0 v/\nu_L \ll 1$). For Re > 1, the theory of mass transfer has not yet been developed.

Experimental observations [44] of the bubble behavior showed that, in an ultrasonic field, the bubble grows more rapidly, the mass transfer being considerably dependent on the acoustic pressure amplitude (Figure 4.15).

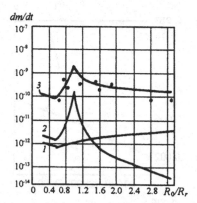

Figure 4.16. Diffusion flow as a function of the bubble radius at $f = 26.5$ kHz and $p_m = 0.0144$ MPa [44].

At constant frequency and pressure amplitude, the diffusion flow is largely governed by the relationship between the current radius of the bubble and its resonant value R_r for a given sound frequency (see expressions 4.43 and 4.46). Figure 4.16 shows theoretical and experimental dependences of diffusion flow on the R_0/R_r ratio. Curve *1*, calculated by formula (4.43), characterizes the diffusion flow to the bubble with allowance for its vibration, whereas curve *2* is derived with the aid of formula (4.46) and defines the diffusion flow due to microstreaming near the bubble surface. Dashed curve *3*, which is a sum of curves *1* and *2*, characterizes the total diffusion flow. The experimental points fit this curve quite well. Comparison of curves *1* and *2* indicates that under particular conditions the contribution from microstreaming to diffusion flow can be decisive.

A more detailed analysis shows that, at small acoustic pressures, the diffusion flow is mainly determined by microstreaming, while at acoustic pressures exceeding a particular value, the contribution from bubble vibrations becomes crucial. Thus, at $P_m = 0.1$ MPa, $f = 26.5$ kHz, and $R_0 = 7 \times 10^{-3}$ cm, the ratio I_1/I_2 is about 10^2.

At $C_0 < C_s$, the liquid is not saturated with gas and the bubble growth is possible only at a particular, critical amplitude of vibrations. The critical pressure P_m^* is calculated from the condition that the average diffusion gas flow into the bubble, which represents the sum of convective (acoustic) and molecular flows, is equal to zero

$$I_0 + I_1 + I_2 = 0. \tag{4.47}$$

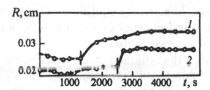

Figure 4.17. Time course of the mean radius of bubbles in glycerol [44]. Dashed portions of the curves correspond to static conditions (without ultrasound), and solid portions – to the effect of ultrasound with the frequency 26.5 kHz and peak pressure 0.05 MPa: (*1*) $R_0 = 0.0216$ cm, (*2*) $R_0 = 0.0266$ cm.

By deriving I_0, I_1, and I_2 from expressions (4.38), (4.43) and (4.46), one can obtain the expression for P_m^* [44]

$$P_m^3 + 3\chi P_m + 2q = 0, \tag{4.48}$$

where

$$\chi = \frac{(\rho_L \omega R)^2 k}{2\beta} \left\{ 2\alpha - \frac{384\pi D_L \omega}{\beta} \right\},$$

$$q = \frac{(\omega R)^3 \omega \rho_L \sqrt{\pi D_L \omega} k}{\sqrt{2}\beta} \left\{ \frac{5.5296 \times 10^4 \pi D_L \rho_L^2}{\beta^2} - \frac{32\omega R^2 \rho_L^2 \alpha}{\beta} - 8\alpha \omega R^2 \right\},$$

$$\alpha = \frac{C_0}{C_p} - \left(1 + \frac{2\sigma_L}{R_0 P_0} \right),$$

$$\beta = \frac{C_0}{C_p} + \left(1 + \frac{\sigma_L}{R_0 P_0} \right),$$

$$k = \left(\frac{\omega_0^2}{\omega^2} - 1 \right)^2 + \delta^2.$$

Analysis of this expression shows that at a given sound frequency, bubbles with dimensions close to resonant values possess the lowest critical pressure P_m^*. Smaller bubbles will obviously be dissolved. Bubbles with radii greater than R_r will shrink until approach the resonant dimensions. The radius of stable bubbles, corresponding to a diffusional equilibrium in the acoustic field, increases with the acoustic pressure amplitude.

Experiments support theoretical predictions about the existence of critical sound pressure and bubbles of a "stable" radius. Figure 4.17 presents curves illustrating changes in the mean radii of two bubbles with different initial radii $R_{01} = 0.0216$ and $R_{02} = 0.0266$ cm in an acoustic field. After the onset of vibrations, the rates of bubble growth are initially maximum, but then gradually diminish to become zero, which corresponds to diffusionally equilibrium bubbles with a constant radius. With increasing frequency of vibrations, this radius decreases, and the critical sound pressure rises.

It should be noted that mathematical description of diffusion processes in the case of developed cavitation presents a problem.

In addition to enhancing the formation of gas-filled bubbles, ultrasound also stimulates their effervescence, which is due to a combination of such effects as bubble enlargement, Bjerkness forces, acoustic streaming, and radiation pressure.

Let us consider two bubbles of radii R_1 and R_2, vibrating with the same frequency but at different velocities. Such bubbles interact due to the Bjerkness forces, the mean value of which is given by [49]

$$F_B = \frac{2\pi R_1 R_2}{\rho_L \omega^2} \frac{\cos\theta}{\sqrt{k_1}\sqrt{k_2}} \frac{P_m^2}{l^2},\qquad(4.49)$$

where

$$\theta = \operatorname{arctg}\frac{A-B}{1+AB};\quad A = \frac{\delta_1\omega_{01}}{\omega_{01}^2 - \omega^2};\quad B = \frac{\delta_2\omega_{02}^2}{\omega_{02}^2 - \omega^2};$$

$$k_1 = \left(\frac{\omega_{01}^2}{\omega^2} - 1\right) + \delta_1^2;\quad k_2 = \left(\frac{\omega_{02}^2}{\omega^2} - 1\right)^2 + \delta_2^2,$$

l is the distance between the bubbles.

The indices 1 and 2 refer to the first and second bubble, respectively. It follows from (4.49) that if the bubbles oscillate synchronously ($\theta = 0$), which is the case when the bubbles are of approximately the same size, the Bjerkness forces are attractive. But if vibrations are in antiphase ($\theta = \pi$), the bubbles are repelled.

Kazantsev derived an expression describing the distance between two oscillating bubbles [50]

$$\frac{\sqrt{l^3}}{C-t} = \frac{3\sqrt{3}}{2\sqrt{2}}\frac{\sqrt{R_1^3 + R_2^3}}{\pi\rho_L R_1 R_2}\sqrt{\frac{P_m^2}{f^2}}\frac{1}{\sqrt{k_1}\sqrt{k_2}},\qquad(4.50)$$

where C is the constant that depends on the frame of reference.

Experiments performed by Kazantsev [50] confirm the validity of expression (4.50) describing the distance $l(t)$ between two bubbles in an ultrasonic field (Figure 4.18).

Apart from to the Bjerkness forces, the liberation of gas-filled bubbles is also affected by acoustic streaming, which can change the velocity of translational motion of bubbles and, owing to velocity gradients

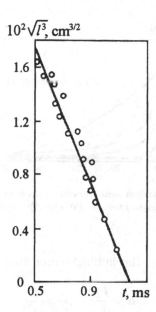

Figure 4.18. Time dependence of the distance between two gas-filled bubbles in the water subject to elastic vibrations with $f = 10.8$ kHz [50].

in the streaming field, increase the probability of bubble coalescence. Comparison of the velocities of the Ekkart and Rayleigh flows, capable of carrying away bubbles, and the velocity of bubble emergence under the action of buoyant forces shows that the effect of Rayleigh flows on the bubble motion is negligible [44]. At the same time, the Ekkart flow at high frequencies and intensities of ultrasound may substantially affect the motion of bubbles. Likewise, the probability of bubble coalescence due to velocity gradients in the streaming field may be significantly raised by the Ekkart flows but is virtually independent of the Rayleigh flows.

When a sound wave propagates in a liquid, the bubbles it contains are subject to a radiation pressure. In a standing sound wave, the radiation pressure exerted on a bubble with the radius less than the wavelength $(kR \ll I)$ is [51]

$$F_r = -\frac{4\pi}{k^2}\,\overline{E}\,\sin\,2kh\Phi(\alpha,\beta,k_r,R), \qquad (4.51)$$

where

$$\Phi = \frac{\alpha(k_r R)[3\beta - (k_r R)^2]}{\alpha^2(k_r R)^6 + [3\beta - (k_r R)^2]^2},$$

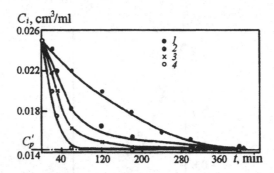

Figure 4.19. Time-dependent reduction in the concentration of air in water induced by ultrasound of various intensity (W s/cm³): (*1*) 3×10^{-6}, (*2*) 4×10^{-6}, (*3*) 7×10^{-6}, (*4*) 9×10^{-6} [38].

h is the distance from the bubble center to a velocity antinode or node;

$$\alpha = \frac{k_L}{k_g}; \quad \beta = \frac{\rho_g}{\rho_L}.$$

It follows from (4.51) that the behavior of a gas-filled bubble in a standing wave under the action of radition pressure depends on its radius: bubbles with radii less than its resonant value move to a velocity antinode, while those with radii greater than the resonant value move to a velocity node. This inference was experimentally underpinned in a number of investigations [52, 53]. The selectivity of radiation pressure with respect to various-size bubbles in a standing wave serves to a size distribution of bubbles and their accumulation at certain sites (nodes and antinodes of the standing wave), thereby increasing the probability of their coalescence under the action of the Bjerkness forces.

All these factors affect the kinetics of mass transfer in ultrasonic field. The investigation of water degassing in the ultrasonic field of a precavitation intensity showed that the extent of degassing increases with the time of treatment in certain limits (Figure 4.19) [40, 44].

Experiments were performed with 1.13-MHz ultrasound at acoustic powers from 2.5 to 45 W. The degree of degassing was defined as the amount of air liberated under the action of ultrasound from a unit volume. The experiments showed that the maximum degree of liquid degassing, $\mathcal{L}$, does not depend on the acoustic power applied. Time needed for the attainment of the maximum degassing, τ_0, falls with increasing acoustic power. The occurrence of the maximum degree of degassing is probably related to the attainment of an equilibrium state in the two-phase water-air system, when the liberation of air from the water is compensated for by its diffusion through the water-air interface.

The extent of degassing can be described by an empirical formula

$$\mathcal{L}(t) = \mathcal{L}_0(1 - e^{\alpha t}), \tag{4.52}$$

where $\mathcal{L}$ is the instantaneous value of degassing; t is the time of ultrasonic treatment; α is the coefficient of degassing.

The dependence of the degassing coefficient on applied acoustic power can satisfactorily be described by expression

$$\alpha = a(N - N_0)^n, \tag{4.53}$$

where N_0 is the experimental threshold acoustic power, at which degassing begins (in experiments with water it comprised 2 W, which corresponded to a power density of 3.3×10^{-3} W/cm^3 [40, 44]); a and n are constants for a given liquid and conditions of ultrasonic irradiation.

Based on formulas (4.52) and (4.53), one can calculate the time of the attainment of maximum degassing as a function of acoustic power

$$\tau_0 = -\frac{\ln(\mathcal{L}(t)/\mathcal{L}_0)}{a(N - N_0)n}, \tag{4.54}$$

where τ_0 and N are expressed in minutes and watts, respectively.

It should be noted that formulas (4.52) and (4.53) describe the kinetics of degassing for a particular case, i.e., for certain external pressure P_0, temperature, and the initial content of gas in the liquid.

At small treatment times, the extent of liquid degassing is linearly dependent on the applied vibrational power. A deviation from the linearity is related to the approaching of the maximum degree of degassing.

Analogous results were obtained in experiments with a 9.6-kHz ultrasound. At the vibrational velocity amplitude $v_m = 4.4$ cm/s, the rate of degassing decreases with the increasing time of treatment to attain, after 60 min, the maximum extent of degassing with the residual concentration of dissolved gas in the liquid comprising 31% of its initial value.

The occurrence of the maximum degree of ultrasonically induced degassing was also observed by Lindstrome [53], who revealed that the residual content of dissolved gas in liquid treated with a 700-kHz ultrasound comprised 50% of its initial content.

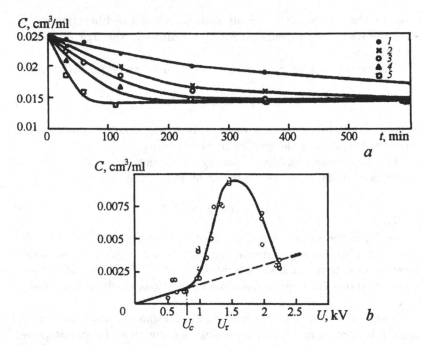

Figure 4.20.   Air concentration in water as a function of (*a*) the time of ultrasonic treatment at 22 kHz and various vibrational velocity amplitudes (cm/s): (*1*) 3.5, (*2*) 13, (*3*) 15, (*4*) 23, (*5*) 36 and (*b*) the focusing radiator voltage at $f = 500$ kHz [38].

Of particular interest is the part played by cavitation in the degassing of liquids. It was found that cavitation does not affect the kinetics of degassing. Figure 4.20 presents curves illustrating changes in the concentration of air in water during its degassing without (curve *1*) and with (curves *2–5*) cavitation. The vibrational velocity, corresponding to a cavitation threshold, amounted to 13 cm/s.

To compare between the mass transfer rates under the precavitation and cavitation conditions, a sample of a specified volume was ultrasonically treated for 10 min at various voltages across a focusing radiator. With the appearance of cavitation (at U_c), the rate of degassing increased approximately threefold as compared to that before cavitation (Figure 4.22,*b*). Increasing of the radiator voltage above a particular value (U_τ) leads to a decrease in the rate of degassing to values typical of precavitational degassing at higher voltages across the radiator (the dashed line shows the virtual course of degassing as a function of the radiator voltage provided that cavitation is suppressed).

Intensification of degassing by cavitation is related to an increased number of bubbles absorbing the dissolved gas. Increasing acoustic pressure promotes the growth of avities until, at some acoustic intensity, the time of bubble collapse becomes equal to a half-period of the acoustic pressure cycle. As this takes place, pressure at the final stage of collapse falls, causing a decrease in the number of secondary bubbles produced by a collapsing primary bubble [54].

Moreover, nonlinear vibrations of a cavitating bubble can be responsible for the elevated ratio of its maximal and minimal radii during the phases of bubble expansion and collapse, so that the interface through which the gas diffuse into the bubble drastically increases. In addition, the increased ratio acts to lengthen the phase of bubble expansion and to reduce the time of its collapse, which also contributes to the diffusion of gas into the bubble.

Thus, within a certain range of acoustic intensities, cavitation stimulates degassing of liquids due to the intensification of bubble formation and gas diffusion through the bubble-liquid interface. At the same time, cavitation does not affect the quasi-equilibrium concentration of dissolved gas.

Here, it would be of interest to consider how the rate of degassing depends on frequency. The relevant literature data are quite contradictory. According to Esche and Wenk [55], degassing is most intense at 35 kHz, which corresponds to a resonant frequency of most numerous bubbles ($R_0 = 0.01$ cm). On the other hand, the rates of degassing of some viscous liquids (oil, glycerol, etc.) at 40 and 50 kHz do not substantially differ [56, 57]. Similarly, degassing of a molten optical glass with 20- and 200-kHz ultrasonic vibrations occurs with the same rate [58, 59]. Sorensen showed that power necessary for the liberation of a certain amount of gas increases with frequency [60].

Experiments with the water degassed at 10, 22, 80, 500, and 1000 kHz showed that the rate of degassing can be approximated by formula [44]

$$\dot{C} = 2.3 \times 10^{-13} f^{1.43} e^{-6.65 \times 10^{-6} f}, \qquad (4.55)$$

with the maximum at $f \sim 200$ kHz (Figure 4.21).

Kapustina studied the effect of static pressure and temperature on liquid degassing in an ultrasonic field with a frequency of 1 MHz and acoustic power of about 3 W [38]. The rate of degassing was found to increase with decreasing static pressure; the residual concentration of dissolved gas was also affected. By comparing the residual concentration C' with the steady-state concentration of dissolved gas without ultrasound, C_p, one can estimate the effect of ultrasound on air

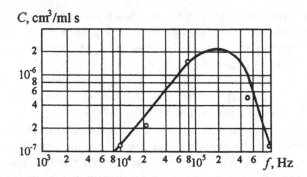

Figure 4.21. Frequency dependence of the rate of degassing at an acoustic pressure amplitude of 0.01 MPa [38].

solubility in water. The experimental dependence of the parameter $\gamma = (C' - C_p)/C_p$ (which characterizes a relative change in the gas solubility induced by ultrasound) on static pressure indicates that changes in gas solubility increase with decreasing static pressure. Therefore, ultrasonic degassing of liquids is favorable at a reduced static pressure (e.g., in a vacuum).

As for the effect of temperature on degassing, its rate rises with temperature. At the same time, in a temperature range of 20–60°C, the residual concentration of air in water is independent of temperature and amounts to about 30% of C_p observed at a static pressure of 0.1 MPa.

4.3 Ultrasonic Emulsification

Interaction of immiscible liquid phases in an ultrasonic field gives rise to an effect known as emulsification, when one of the immiscible liquids transforms into a phase dispersed over the other liquid. Ultrasound makes it possible to obtain fine, homogeneous, and chemically pure emulsions.

No comprehensive theory of ultrasonic emulsification has yet emerged, although some facets of the process have been well studied. In particular, it has been established that ultrasonic emulsification necessitates cavitation. The regularities of origination and development of cavitation determine the course of emulsification and its dependence on the intensity and frequency of oscillations, the presence of dissolved gases, as well as on temperature, pressure, density, viscosity, and surface tension of interacting liquids.

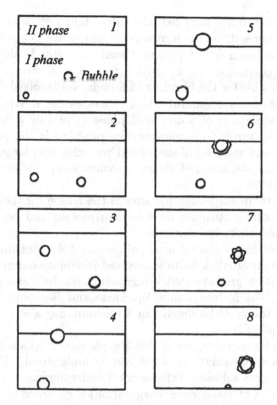

Figure 4.22. Schematic representation of emulsification.

The specific features of the initial stage of emulsification can be well understood from the diagram in Figure 4.22, representing processed film fragments taken from work [61]. The cavities arising in one (call it the first) of the contacting liquid phases can migrate to the interface and penetrate the second phase. The generated microstreamings act to 'scrape off' small droplets of the second phase from the interface with the first liquid phase. Upon further migration of the bubble to the first liquid phase, its surface will hold the droplets of the second liquid phase as a result of flotation.

Acoustic streamings carry bubbles over the body of the first liquid. As cavitation develops, these bubbles collapse to disperse the fine droplets of the second liquid in the first one with the formation of emulsion.

If acoustic pressure exceeds the level, at which cavitation may also develop in the second liquid phase, then the emulsion is 'inverse'.

Introduction of fine solid particles into a liquid phase subject to an ultrasonic field results in the formation of suspensions. The mechanism of this phenomenon is not presently well understood, although it is known that cavitation and acoustic streamings of various kinds and scales, responsible for the blending of liquids, are involved.

As the processes of emulsification or suspending develop, the coalescence (coagulation) of a dispersed phase (liquid or solid particles) begins to play a role. Coalescence (coagulation) is the processes of approaching and merging of suspended particles into larger ones (for instance, under the action of ultrasonic vibrations), due to which they may precipitate.

A small size of suspended particles is the reason for their high mobility: they can participate in Brownian motion and be carried by convective and hydrodynamic flows.

Additional forces arising in an ultrasonic field contribute to coalescence (coagulation): a particle involved in vibrational motion experiences radiation pressure that brings about its drift and transfer by streamings. The hydrodynamic Bjerkness and Bernoulli forces arising between the particles moving in a medium may also contribute to coalescence [62, 63].

It should be, however, noted that the physical mechanism of acoustic coalescence (coagulation) is not clearly understood. The existing hypotheses lack a sufficient experimental underpinning.

The kinetics of coalescence (coagulation) is governed by the dimensions and concentration of particles as well as by the ultrasonic field parameters and physicochemical properties of medium, in which the process takes place.

N. L. Shirokova [62] studied the influence of various geometric factors and ultrasonic field parameters on the coalescence of particles. Experiments were performed with both poly- and monoaerosols. As polyaerosols were used aluminum, bronze, and zinc oxide powders, polyvinyl chloride dust, and dimethylphthalate fog with particles from 1 to 40 μm in size. As monoaerosols were used lycopodium powder (particle size 20 μm) and Lycoperdon spores (particle size 3.5 μm). It was found that particles move in a strictly defined fashion. Namely, two particles approach each other, if the line through the particle centers is inside a cone whose vertex (the vertex angle $2\theta = 90°$) is in the center of one of the particles and axis is directed along the vibrational velocity vector. Conversely, they repel, if the other particle appears in the same cone but with the axis perpendicular to the vibrational velocity vector (Figure 4.23). The author inferred that the direction of particles interaction is governed solely by

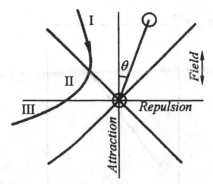

Figure 4.23. Interaction of particles in an ultrasonic field: (*I*) the region in which particles approach each other; (*II*) the region of the closest approach of particles and change of the direction of their motion; *III*, the region of particles repulsion.

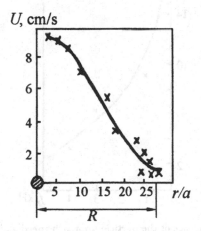

Figure 4.24. The speed of particles approach as a function of the distance between them [52].

the position of the particles relative to the vibrational velocity vector.

It should be noted that these results are inconsistent with the theoretical data of Konig [64].

The relative velocity of particles drastically rises, as the distance between them diminishes to a certain value known as the radius of interaction, R (Figure 4.24). This radius is proportional to acoustic pressure (Figure 4.25,*a*), whereas the frequency of vibrations and the relative velocity of particles are related reciprocally (Figure 4.25,*b*).

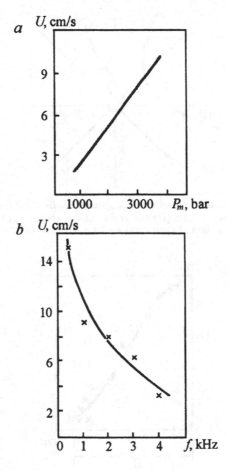

Figure 4.25. Dependence of the particle approach speed and the radius of inter-action on (a) peak sound pressure and (b) frequency of vibrations [65].

Similar results were obtained in studies of the coagulation of hy-drosols [61]. Along with hydrodynamic interactions, certain contribu-tions to this process are believed to be made by the flotation effect of cavitation bubbles, shock pressures resulting from their collapse, and acoustic streamings.

Acoustic flotation effect manifests itself in the concentrating of suspended particles around a pulsating cavity. The estimations made in [65] have indicate that in the case of 20-kHz vibrations, this effect is significant only for particles larger than 0.5 μm in diame-ter.

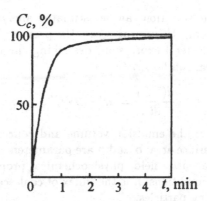

Figure 4.26. The extent of coagulation C_c of polystyrene latex as a function of the time of ultrasonic treatment [66].

The mechanism and the kinetics of coagulation are largely governed by the sign and magnitude of the charge of hydrosol particles, as well as by the degree of their solvation.

The study of the coagulation of polystyrene latex subject to an acoustic field [66] showed that hydrosol particles coagulate only under the action of cavitation, coagulation being initiated at the sites of maximum cavitation. The rate of acoustic coagulation varies with time (Figure 4.26).

Unlike the ultrasonic coagulation of 0.01-μm polystyrene latex particles, the coagulation of the 1-μm particles occurs quite rapidly. In view of the foregoing, this results can be explained as follows. Suspended particles with the size > 0.5 μm are floated by pulsating cavities, whereas particles with the size < 0.5 μm are repelled by them [66].

By varying the amount of a surfactant added, one can change the thickness of the solvate envelope and thereby the kinetics of coagulation. Experiments indicate that when the concentration of surfactant exceeds a particular value, the process of coagulation will terminate. Thus, the coagulation of latex ceases, if its particles are coated with more than two monolayers of the surfactant.

The mechanism of emulsion coalescence in an acoustic field does not virtually differ from the above mechanism of coagulation. This is because liquid droplets, especially those coated with the film of a surfactant, move in liquid in the same fashion as do solid particles.

The conditions, under which the rates of emulsification and coalescence become equal, affect the concentration and fineness of emulsions.

The rate of emulsification can be estimated analytically by taking into account two competitive processes – the proper emulsification and coalescence. The general expression, describing the kinetics of ultrasonic emulsification, can be written as

$$\frac{d(VC)}{dt} = \alpha S - \beta V C^n, \qquad (4.56)$$

where V and C are the emulsion volume and concentration, respectively; S is the interface area; α and β are parameters dependent on the characteristics of acoustic field, physicochemical properties of phases, and experimental conditions; n is the index of coalescence equal to the number of coalescing particles.

The first term in (4.56) describes the rate of dispersion, whereas the second term defines the rate of coagulation.

Bearing in mind that cavitation, in particular the motion of cavities, is decisive for emulsification, B. G. Novitskii [61] proposed a solution to the kinetic problem by assuming that it is cavities that are 'carriers' of a dispersed phase, and that the rate of coalescence is proportional to the concentration of emulsion. The derived expression for the ultimate concentration of emulsion does not depend on the number of cavities (or the intensity of ultrasound, if it exceeds the threshold of cavitation). At the same time, the rate of emulsification rises with the intensity of ultrasound.

The frequency of vibrations, all other parameters being equal, does not affect the rate of emulsification. But if the frequency rises, the dispersed particles become smaller, and the ultimate concentration of emulsion diminishes.

One should, however, emphasize that theoretical results do not always comply with experimental data. Thus, it was shown in a number of works [67–70] that the ultimate concentration of emulsion increases with the intensity of vibrations, and the rate of emulsification decreases with growing frequency.

Reduced viscosity of emulsion components favors their emulsification, whereas reduced surface tension leads to a more concentrated emulsion.

Feldman and Khavskii [70] suggested to promote emulsification by the simultaneous use of acoustic vibrations of different frequencies. Such an approach enables the dispersion of substances poorly soluble in water and makes it possible to control the fineness of the emulsions produced as well as their ultimate concentration and stability. Figure 4.27 illustrates some parameters of water-kerosene emulsions. It is seen that the simultaneous exposure of emulsions to two frequencies

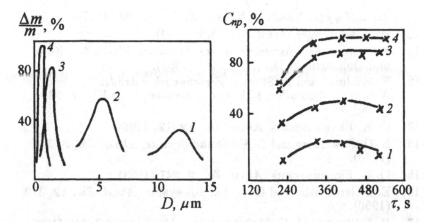

Figure 4.27. (a) Mass distribution of kerosene drops in kerosene-water emulsion and (b) dependence of its ultimate weight concentration on the time of emulsification: (1) mechanical emulsification in a blendor at 33 rps, $\tau = 360$ s; (2) ultrasonic emulsification at 20 kHz, $\tau = 380$ s; (3) ultrasonic emulsification by two frequences, $f_1 = 20$ kHz and $f_2 = 50$ Hz, $\tau = 340$ s; (4) ultrasonic emulsification by two frequences, $f_1 = 20$ kHz and $f_2 = 1$ MHz, $\tau = 300$ s [70].

favorably affects their quality, ultrasonic vibrations being more preferable.

Experiments performed by Agranat *et al.* [69] confirmed the enhanced stability of ultrasonically produced emulsions, especially those obtained by a combined action of kilo- and megahertz vibrations.

References

1. R. W. Wood and A. L. Loomis, *Phil. Mag.*, **4**, No. 22, 417 (1927).
2. K. Söllner, *Trans. Faraday Soc.*, **32**, 1532 (1936).
3. F. Streibl, *Inhalatiosterapie-ein neues anwendungsgebiet des Ultraschalls*, Diss. Erlangen, 1947.
4. T. K. McCubbin, *JASA*, **25**, 1013 (1953).
5. K. Bisa, K. Dirnage, and R. Eshe, *Siemens Z.*, **28**, 314 (1954).
6. K. Bisa, K. Dirnage, and R. Eshe, *Z. Aerosol-Forschung und Therapie*, **3**, 441 (1954).
7. K. Dirnage and R. Eshe, *Siemens Z.*, **29**, 382 (1955).
8. T. Kh. Sedel'nikov and O. V. Abramov, *Some Problems Related to Atomization of a Liquid Metal Jet* (in Russian), VINITI, Moscow (1982).
9. O. K. Eknadiosyants, In: *Physical Fundamentals of Ultrasonic*

Technology (in Russian), Nauka, Moscow, p. 337 (1970).

10. W. Eisenmenger, *Acustica*, **9**, 327 (1959).
11. K. Stamm, *Forschungsber dandes Nordheim-Westfallen*, Köln und Oplanden, W.D.V., No. 1480 (1965).
12. R. Pohlman and K. Stamm, *Forschungsber dandes Nordheim-Westfallen*, Köln und Oplanden, W.D.V., No. 933, (1960).
13. O. K. Eknadiosyants, *Akust. Zh.*, **12**, 127 (1966).
14. L. D. Rozenberg and O. K. Eknadiosyants, *Akust. Zh.*, **6**, 370 (1960).
15. O. K. Eknadiosyants, *Akust. Zh.*, **9**, 247 (1963).
16. E. L.Gershenzon and O. K. Eknadiosyants, *Akust. Zh.*, **12**, 310 (1966).
17. B. I. Il'in and O. K. Eknadiosyants, *Akust. Zh.*, **12**, 310 (1966).
18. O. K. Eknadiosyants, *Akust. Zh.*, **14**, 107 (1968).
19. Yu. Ya. Boguslavskii and O. K. Eknadiosyants, *Akust. Zh.*, **15**, 17 (1969).
20. O. K. Eknadiosyants, *Ultrazvuk. Tekhnika*, **4**, 8 (1966).
21. R. Pohlman and E. G. Lierke, *V Congress International d'Acoustique*, Liege, D35 (1965).
22. J. N. Antonewich, *Proc. Nat. Electronics Conf.*, Chicago, **13**, 798 (1957).
23. V. I. Sorokin, *Akust. Zh.*, **3**, 281 (1957).
24. L. D. Landau and E. M. Lifshits, *Mechanics of Continuum* (in Russian), Nauka, Moscow (1975).
25. P. L. Peskin and R. J. Raco, *JASA*, **35**, 1378 (1963).
26. O. S. Nechiporenko, Yu. M. Naida, and Medvedovskii, *Atomization of Metallic Powders* (in Russian), Naukova Dumka, Kiev, (1980).
27. D. G. Pazhi and V. S. Galustov, *Atomization of Liquids* (in Russian), Khimiya, Moscow (1979).
28. G. Lamb, *Hydrodynamics* (in Russian), GITL, Moscow–Leningrad (1948).
29. B. P. Konstantinov, *Hydrodynamic Sound Generation* (in Russian), Nauka, Moscow (1973).
30. R. G. Lang, *JASA*, **34**, 6 (1962).
31. M. A. Poger and O. K. Eknadiosyants, *Akust. Zh.*, **20**, 602 (1974).
32. O. B. Abramov, Yu. Ya. Borisov, and R. A. Oganyan, *Akust. Zh.*, **33**, 577 (1987).
33. O. B. Abramov, Yu. Ya. Borisov, and R. A. Oganyan, *Proc. XX Internat. Acoust. Conf.*, Praga (1981).

34. O. B. Abramov, V. S. Zuev, and V. S. Mebel', *Poroshkovaya Metallurgiya*, **6**, 6 (1981).
35. T. Kh. Sedel'nikov, *Oscillatory Noise of Issuing Gas Jets* (in Russian), Nauka, Moscow (1971).
00. R. Bogle and J. Tayller, *Trans. Roy. Soc. Canada*, **20**, 25 (1926).
37. A. Siverts, *Zh. Fiz. Khim.*, **88**, 109 (1914).
38. O. A. Kapustina, In: *Physical Fundamentals of Ultrasonic Technology* (in Russian), Nauka, Moscow, p. 253 (1970).
39. O. A. Kapustina, *Akust. Zh.*, **9**, 424 (1963).
40. O. A. Kapustina, *Akust. Zh.*, **10**, 440 (1964).
41. O. A. Kapustina, *Ultrazvuk. Tekhnika*, **5**, 40 (1964).
42. O. A. Kapustina, *Akust. Zh.*, **14**, 37 (1968).
43. O. A. Kapustina, *Akust. Zh.*, **14**, 399 (1967).
44. O. A. Kapustina, *Akust. Zh.*, **11**, 116 (1965).
45. V. G. Levich, *Physicochemical Hydrodynamics* (in Russian), GITTL, Moscow (1959).
46. S. A. Elder, *Acustica*, **31**, 54 (1959).
47. W. Nyberg, *JASA*, **30**, 329 (1958).
48. Yu. Ya. Boguslavskii, *Akust. Zh.*, **13**, 23 (1967).
49. F. G. Blake, *JASA*, **21**, 551 (1949).
50. V. F. Kazantsev, *Dokl. Akad. Nauk SSSR*, **129**, 64 (1959).
51. K. Yosioka and Y. Kawasima, *Acustica*, **5**, 167 (1955).
52. R. W. Boyle, G. B. Taylor, and D. K. Froman, *Trans. Roy. Soc. Canada*, **23**, 187 (1929).
53. O. Lindstrome, *JASA*, **27**, 654 (1955).
54. M. G. Sirotyuk, *Akust. Zh.*, **12**, 449 (1961).
55. R. Esche and P. Wenk, *Elektrotechn. Z.*, **5**, 97 (1960).
56. A. P. Kapustin, *Zh. Tekhn. Fiziki*, **24**, 1008 (1954).
57. A. P. Kapustin, In: *Application of Ultrasonics for Investigation of Substance* (in Russian), Izdat. MOPI, Moscow, p. 165 (1955).
58. F. Kruger, *Glastechn. Ber.*, **16**, 244 (1938).
59. *Entgasung Optischer Gläser mit Ultraschall*, Umschau (1953).
60. C. Sorensen, *Ann. Physik*, **26**, 121 (1936).
61. B. G. Novitskii, *Application of Acoustic Vibrations in Chemical Engineering* (in Russian), Khimiya, Moscow (1983).
62. N. L. Shirokova, In: *Physical Fundamentals of Ultrasonic Technology*, Nauka, Moscow, **641** (1970).
63. N. L. Shirokova and O. K. Eknadiosyants, *Akust. Zh.*, **1**, 3 (1961).
64. W. Konig, *Ann. Phys. und Chem.*, **42**, 353 (1991).

65. A. V. Il'in, V. P. Kuznetsov, B. G. Novitskii, and
     V. M. Fridman, *Akust. Zh.*, **18**, 537 (1972).
66. B. G. Novitskii, V. D. Petrushkin, and N. I. Chmarova,
     Ultrazvuk. Tekhnika, **3**, 66 (1968).
67. B. A. Agranat, O. D. Kirillov, and L. V. Preobrazhenskii,
     Ultrasound in Hydrometallurgy (in Russian), Metallurgiya,
     Moscow (1969).
68. B. A. Agranat, Yu. I. Bashkirov, and Yu. I. Kitaigorodskii,
     Ultrasonic Technology (in Russian), Metallurgiya, Moscow
     (1974).
69. B. A. Agranat, M. N. Dubrovin, and N. N. Khavskii, *Physical
     and Technical Fundamentals of Ultrasound* (in Russian),
     Vysshaya Shkola, Moscow (1987).
70. A. V. Feld'man and N. N. Khavskii, In: *Dynamic Effects of
     High-Power Ultrasound*, Izhevsk, p. 9 (1981).

Chapter 5

Nonlinear Effects at Liquid–Solid and Solid–Solid Interfaces

This chapter continues the analysis of nonlinear effects arising at liquid-solid and solid-solid interfaces.

These phenomena are, as a rule, the consequence of a combined action of several more simple nonlinear effects developing in a solid or liquid. The overall effect is determined by the intensity of ultrasound, properties and structure of treated substance, as well as by experimental conditions (in particular, the geometry of the system).

5.1 Liquid-Solid Interface

This section examines the phenomenon of liquid rise in a capillary tube in response to ultrasonic irradiation. The processes of wetting, disintegration, and crystallization of solids in an ultrasonic field are also considered.

5.1.1 The Sonocapillary Effect

The sonocapillary effect, or an anomalously high rise of a liquid in a capillary tube under the action of ultrasonic vibrations, was discovered in the 1960s [1–3]. Although the mechanism of this phenomenon has not yet been sufficiently understood, it is believed to be largely related to cavitation and acoustic streaming.

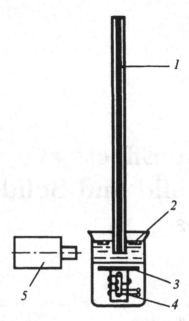

Figure 5.1. Setup for investigation of the sonocapillary effect [4]: (*1*) capillary; (*2*) liquid bath; (*3*) radiator; (*4*) magnetostrictive transducer; (*5*) microscope.

Some experiments with water, oils, and lacquers, aimed at studying the effect, were performed in [4, 5] on a setup schematically shown in Figure 5.1.

Capillary tube *1* was placed in liquid *2* at various distances from radiator *3*. Elastic vibrations with frequencies ~20 and 40 kHz were excited by magnetostrictive transducer *4*.

In the absence of ultrasound, distilled water raised in 300–400-μm capillaries to a height of 5–7 cm. Exposure of the system to ultrasonic vibrations of a precavitation intensity did not affect the height of water column (the displacement amplitude corresponding to the threshold of cavitation in water at a frequency of 18.5 kHz was 0.5 to 1.5 μm).

With the displacement amplitude of vibrations exceeding the cavitation threshold, the height of water column drastically increased (Figure 5.2), but diminished to an initial level (observed before the onset of vibrations) when the displacement amplitude was above 5 μm. The maximum water column height, observed at a fixed displacement amplitudes of about 2 μm, was also dependent on the distance from the capillary end to the radiator surface. The water column height was

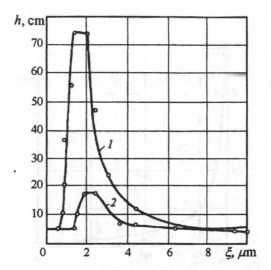

Figure 5.2.   The height h of water column in a capillary tube versus the displacement amplitude ξ_m of the radiator. The capillary end is positioned at sites of (1) minimum and (2) maximum acoustic pressure [4].

maximum when this distance was equal to the half wavelength of sound in the liquid, which corresponded to the position of the capillary end at the minimum of acoustic pressure. The maximum height of water column was as high as 100–120 cm, thus exceeding 15–20 times the column height observed before the onset of vibrations. The ultrasonically induced rise of water was minimum, when the capillary end was located at the maximum of acoustic pressure.

The velocity of water level diminished as the height of water column increased (Figure 5.3).

Higher-frequency vibrations (44 kHz as compared with 20 kHz) of equal intensity caused a lower and slower rise of liquids in capillary tubes (Figure 5.4). The rise of more viscous liquids (such as LBS-1 lacquer representing a solution of phenol-formaldehyde resin in ethanol) was lower than in the case of low-viscous liquids.

The velocity, at which a liquid raised, dropped with increasing viscosity of the liquid. Thus, in the case of K-47k lacquer (a solution of modified polymethylphenylsiloxane resin in xylene, $\eta \sim 150$ cP), the time of its steady-state level establishment was found to be about 30 min, whereas only 1–2 min in the case of water.

Investigation of the mechanism of the sonocapillary effect showed that it is accompanied by the occurrence of cavities near the capillary

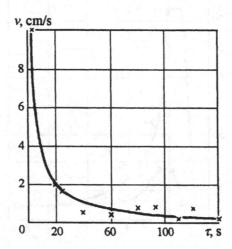

Figure 5.3. Water level velocity in a capillary tube as a function of time (or water column height) at the displacement amplitude of radiator $\xi_m = 1.5$ μm [4].

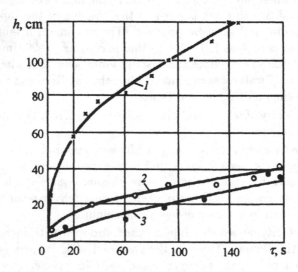

Figure 5.4. Dependence of the liquid column height on the frequency of vibrations and liquid viscosity at the intensity of vibrations 0.6 W/cm² [4]: (*1*) water ($\eta = 1$ cP), 18.5 kHz; (*2*) water, 44 kHz; (*3*) laquer LBS-1 ($\eta = 50$ cP), 18.5 kHz.

end. It was also found that the height of liquid column and the velocity of its increase are determined by the size of a cavitation region near the capillary end and by the character of cavity pulsations. The greater the number and the size of cavities near the capillary end, the higher the height of liquid column and the velocity of its increase. If there are no cavities near the capillary end, the height of liquid column corresponds to that provided for by only capillary forces.

Analysis of high-speed filming frames shows that the cavitation region near the capillary end is irregular, consists of cavities of different size, and is as long as 0.2–1.0 capillary diameter.

The dimensions of cavities at the periphery and at the center of the cavitation region are 300–400 and 20–200 μm, respectively. The growth time of cavities located near the capillary end is $\frac{2}{3}T$ ($\sim 40 \times 10^{-6}$ s at $f = 18.5$ kHz), and the collapse time is $\sim \frac{1}{3}T$ (T is the period of vibrations).

Apart from cavitation, high-intensity ultrasonic vibrations give rise to acoustic streamings. If their velocity and geometry are such that they dislodge the cavitation region from the capillary end, the sono-capillary effect takes no place.

Acoustic streaming was investigated in more detail in works [3, 6].

5.1.2 Wetting of Solids in Ultrasonic Fields

The problem of wetting of a solid substrate by a liquid phase in an ultrasonic field is of both theoretical and practical importance.

When being wetted, two free surfaces (those of a liquid and a solid) are substituted by its interface with a lower free surface energy of the system.

If a liquid drop is placed on the surface of a solid, this is a contact wetting with the number of interacting phases more than two* (Figure 5.5). Assuming that either of the surfaces possesses its own amount of energy per unit area (known as surface tension), σ_{LS}, σ_L, σ_S, and the contact angle is θ, Young's equation can be written as

$$\sigma_S - \sigma_{SL} - \sigma_L \cos \theta = 0. \qquad (5.1)$$

---

*This case differs from the wetting of a body totally immersed in a liquid, when the number of interacting phases is two.

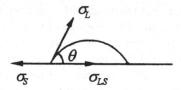

Figure 5.5. Illustration to the wetting of a solid substrate by a liquid.

In terms of the adhesional work W_a and the cohesional work W_c [7], equation (5.1) takes the form

$$W_a = \frac{1 + \cos \theta}{2} W_c, \qquad (5.2a)$$

$$W_a = \sigma_L (1 + \cos \theta). \qquad (5.2b)$$

The energy of interaction of the liquid drop with the solid substrate, E_i, can be written as

$$E_i = W_a S, \qquad (5.3)$$

where S is the area of the wetted spot.

Given experimental values for θ and S, one can estimate the energetic parameters of wetting.

For a perfectly wettable body, the contact angle is zero and the drop spreads over the body surface.

Various kinds of wetting can be grouped depending on which interaction forces between the liquid and solid phases are dominant:

– molecular forces, or

– forces of chemical interaction typical of ionic and molecular bonds.

In systems composed of liquid and solid metals, wetting is accompanied by the formation of chemical compounds and solid solutions. In such systems, the contact between phases may alter their physicochemical properties and consequently their wetting. The oxidation of solid-state surfaces or the formation of intermetallic compounds on them will reduce the contact angle. In some cases, liquid metal can penetrate through the structural defects of the oxide layer to the solid metal surface to spread over it.

In multi-component melts, the surface layer and the main body of melts differ in chemical composition.

In regard to specific features of wetting in an ultrasonic field, the main concern is how to supply vibrations to a liquid drop. This can be done through either solid or liquid phase. In the former case, there are two alternatives: the displacement velocity vector can be either normal or parallel to a vibrating substrate. In the latter case, there are also two alternatives: the drop can be placed on the surface of a sonotrode at the antinode of either displacement or strain.

The method of ultrasound feeding determines, to a certain degree, the character of vibrations in the drop-substrate system, the intensity of vibrations in the liquid phase, and the extent of manifestation of nonlinear effects.

The intensity of vibrations in the liquid phase is governed by their amplitude, frequency, and the ratio between the wave resistances of liquid (drop) and solid (substrate) phases. The extent to which nonlinear effects are manifested in the liquid bulk, on its surface, and at the liquid-solid interface is determined by the intensity of vibrations, properties of liquid phase (such as density, surface tension, and viscosity) and solid phase (such as strength, elasticity, grain and dislocation structure), as well as the physicochemical parameters of their interaction (interfacial energy, intersolubility, diffusion coefficients).

For general reasons, one may assume that the wetting of a solid subjected to vibrations must be affected by mechanical forces related to vibrational motion in the drop-substrate system. The magnitude of these forces must depend on the mass and geometry of the drop, the amplitude and frequency of vibrations, as well as on the mode of their supply. Under some conditions, resonance phenomena are possible.

It should be noted that mechanical forces and physicochemical differences between wetted and nonwetted regions of the substrate surface may be responsible for hysteresis phenomena.

Indeed, under the action of mechanical forces definitely directed over a half-cycle period, the drop spreads and wets the surface. Over the next half-cycle period, its motion on the wetted surface will obviously differ, in both the direction and character, from the motion it executed over the previous half-period.

Another group of effects involves nonlinear phenomena occurring in the liquid phase, on the free surface of the liquid, at the liquid-solid interface, and in the surface layers of the substrate. In the case of cavitation, a drastic rise in the local temperature and pressure near collapsing cavities will disturb the equilibrium of the system and thus will affect wetting. Acoustic streamings, including microjets arising near the substrate, act to reduce the thickness of the boundary layer, thereby contributing to liquid mixing. Shock waves and cumulative jets resulting from cavity collapse will erode the substrate surface.

One may also speculate as to the role of radiation pressure in maintaining equilibrium in the drop-substrate system.

Exposure of a drop to ultrasound may lead to generation of capillary waves on the free surface of the drop and even to its breakup, if the intensity of vibrations is above a particular value.

Nonlinear effects at the liquid-solid interface may be the reason for surface erosion, the process being especially intense, if the composition and structure of surface layers differ drastically from those in the substrate bulk. This is particularly true for oxide layers formed at the substrate surface.

If vibrations are supplied in such a manner that the drop is located near the strain (stress) antinode, analysis of wetting should obviously take into account a stressed state of the substrate.

It can also be anticipated that ultrasound will intensify the dissolving of solid phase in liquid and the penetration (in particular, diffusion) of liquid into the solid phase. The related effects that may be involved are the sonocapillary effect, grain boundary diffusion, and surface diffusion. In view of this, the question of the presence of surfactants in liquids appears to be essential.

Qualitative analysis of wetting in an ultrasonic field testifies to a great number of phenomena that may be involved. Although the number of publications dealing with particular aspects of wetting in an ultrasonic field is sufficiently large, none of them gave a comprehensive consideration of the relevant problems.

Below an attempt is made to view some of the works with a particular emphasis on those facets of the problem that have been touched above. In this respect, the mechanical forces are of prime interest; therefore, chemically inert materials will be considered.

Effects of the frequency and amplitude of elastic vibrations on the shape of a drop placed onto a solid-state substrate were studied in works [8–12]. Elastic vibrations were fed either via the substrate or directly to the drop.

Work [8] was undertaken to elucidate how the shape of the drop, placed onto a vertically vibrating fluoroplastic substrate, can change. The frequency of vibrations was varied from 20 to 400 Hz, the amplitude – from 0 to 100 μm.

Experiments revealed the existence of the frequencies that elicited a sharp increase in the amplitude of vibrations of drop's surface and the change of its shape (Figure 5.6). Above and below these critical

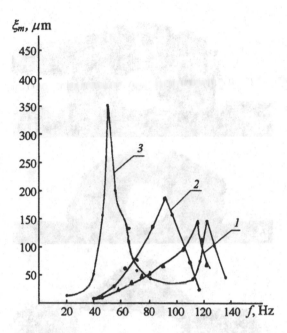

Figure 5.6. Amplitude–frequency dependences for the vibrating top of a liquid drop. The displacement amplitude of the substrate is 12 μm. The drop diameter (mm): (*1*) 2.75; (*2*) 3.3; (*3*) 4.72 [8].

frequencies, the drop vibrated in one piece with the substrate, and its shape did not vary. As the drop enlarged, its first critical frequency decreased. At critical frequencies, the vibrational motion of the drop dramatically changed: its radius and the displacement amplitude of its surface increased, whereas the contact angle decreased. The drop changed in its shape and finally spread over the substrate (Figure 5.7). Frame-to-frame analysis of film fragments showed that, at the first critical frequency, the drop alternately extended and contracted laterally, whereas higher critical frequencies gave rise to ridges on the drop surface, their number being increased with frequency.

At the first critical frequency, the drop was found to spread only over the first half-cycle period, being contracted during the second half-cycle period, but not to the initial size. At higher critical frequencies, the spreading also occurred for only part of the cycle. Any small deviations from the critical frequencies made the drop vibrate asymmetrically. Figure 5.7c illustrates a vibrating drop at the instant when its right edge is in the phase of contraction, while its left edge is in the phase of spreading.

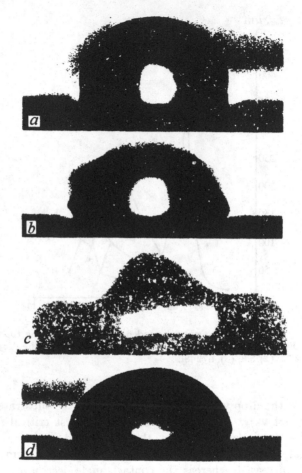

Figure 5.7.   Water drop on a substrate: (a) control (without vibrations); (b) substrate vibrates with frequency 170 Hz and amplitude 25 μm; (c) substrate vibrates with frequency 180 Hz and amplitude 25 μm; (d) after cessation of vibrations [8].

Comparison of above-mentioned critical frequencies with calculated natural resonant frequencies of the drop showed their satisfactory agreement (within an accuracy of up to 10–15%).

As the amplitude of vibrations increased, changes in the drop radius grew until, above a particular value of displacement amplitude, the drop broke down with the ejection of secondary droplets.

In another set of experiments, liquid drops were placed onto a horizontal polished face of a waveguide, in which ~20-kHz vibrations were

Table 5.1 Physicochemical properties of liquids used for wetting a steel substrate [13, 14].

Substance	Temperature t, °C	Density ρ_L, g/cm^3	Viscosity η, cP	Surface tension σ_L, dynes/cm	Wetting angle θ *
Castor oil	20	0.96	986	36	53
Water	20	0.99	1.00	72	80
Glycerol	20	1.26	1490	33	86
Gallium	35			358	115
Mercury	20	13.54	1.55	475	125

* experimental data of the author.

excited. The liquids used possessed different physicochemical properties (Table 5.1).

After the onset of vibrations, the shape of the drop was found to vary, being extended in the direction of the vibrational velocity vector. The contact angle diminished for all the liquids used, the decrease being inversely related to the amplitude of vibrations (Figure 5.8). It was also found that the degree of the contact angle changing was greater for water and organic liquids, which do not virtually interact with a steel substrate, than for molten metals.

V. A. Labunov *et al.* [12, 15] studied the effect of the frequency and amplitude of vibrations on the wetting of various solid substrate by various molten metals. Vibrations were supplied through the substrate in such a fashion that the vibrational velocity vector was normal to its surface. Experiments were performed in a vacuum, as well as in a protecting or reducing atmospheres. In [12], the authors studied the behavior of molten tin drops on the polished surface of CrI8Ni9Ti steel and found that the contact angle decreased with increasing amplitude of vibrations and temperature (Figure 5.9). The effect was more pronounced for larger drops.

Investigation of the wetting of a niobium substrate by molten tin allowed A. A. Shevchenko *et al.* [17] to obtain amplitude dependences of contact angle, wetted area, and time of drop spreading at 250°C (Figure 5.10). It is seen that at the displacement amplitude 1.5 μm, the drop is highly spread to be atomized at the amplitude 11–13 μm.

Similar results were obtained during the study of the wetting of a niobium-zirconium substrate by molten copper (Figure 5.11) [9]. The drop was placed onto the horizontal vibrating substrate near the antinode or node of the vibrational amplitude. A decrease in the contact

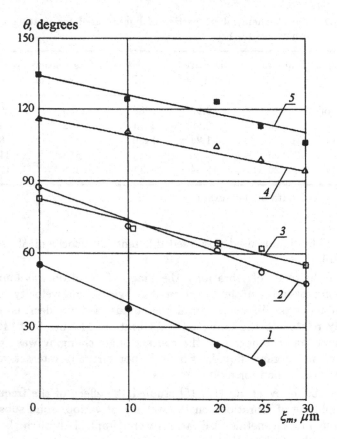

Figure 5.8.   Effect of ultrasound on the contact angle of drops of (*1*) castor oil; (*2*) water; (*3*) glycerol; (*4*) gallium; (*5*) mercury.

angle caused by ultrasound was more pronounced for the drop placed near the displacement antinode. For instance, vibrations with the amplitude 12-μm reduced the contact angle to 20 and to only 32 degrees, if the drop was placed at the displacement antinode or node, respectively.

Yu. A. Minaev *et al.* [9] investigated the effect of ultrasound on the kinetics of interaction of a copper melt with a niobium alloy (98.2% Nb + 1.8% Zr) substrate. A half-wave niobium sonotrode was immersed in the copper melt (1100°C) placed in a graphite crucible. The intensity of ultrasonic irradiation was sufficient to induce cavitation in the melt (the antinode displacement amplitude was 8 μm at a resonant frequency of 18.6 kHz).

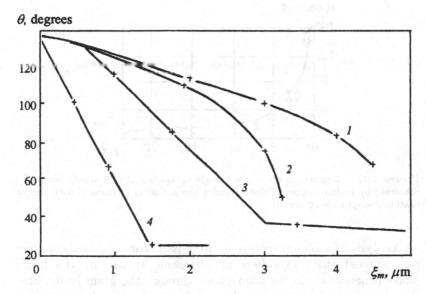

Figure 5.9. Amplitude dependences of the angle of wetting of Cr18Ni9Ti steel by molten tin at temperatures (°C): (1) 240; (2) 330; (3) 440: (4) 550 [12].

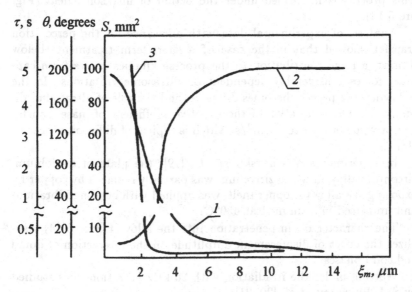

Figure 5.10. Amplitude dependences of (1) contact angle, (2) wetted area, and (3) time of wetting of niobium substrate by molten tin at 250°C [17].

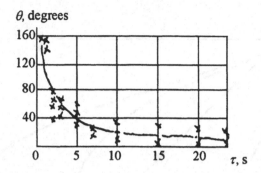

Figure 5.11. Changes in the contact angle of wetting of (1.8 mass% Nb + Zr) substrate by molten copper (1100°C) under the action of ultrasound with the vibrational amplitude 10 μm [9].

Analysis of niobium sonotrode after its occurrence in the copper melt showed that, regardless of ultrasonic treatment, the liquid phase penetrates into the solid phase through the grain boundaries, the rate of penetration being dependent on the boundary structure (Figure 5.12). In the course of time, the melt penetrates deeper into the solid phase to form a closed, branched system of intergrain layers. This process is intensified under the action of ultrasonic field (Figure 5.13).

Analysis of experimental data with allowance for the percolation transfer showed that in the case of a short-term treatment (below 6 min), a large contribution to the process kinetics comes from capillary forces considerably dependent on ultrasonic vibrations. In the following, the process becomes diffusive, and the role of ultrasound diminishes. The estimation of the effective coefficient of mass transfer gave a value of 3.0×10^{-5} cm^2/s, which is typical of diffusion processes [18].

In experiments of Khavskii et al. [19], the plate of a niobium-zirconium alloy, in which zirconium was partially replaced by copper by exposing the alloy to copper melt, was applied with bending vibrations and immersed in a tin melt at 450°C.

The character of tin penetration into the alloy (Figure 5.14) visualizes the effect of displacement amplitude on the interaction of liquid and solid phases.

Wetting under the irradiation with 20-kHz vibrations was studied by S. I. Pugachev et al. [20, 21].

The authors used titanium alloy VT5-1, aluminum alloy AMg6, ceramics TsTS-23 and TBK-3, as well as polymeric material MSN as

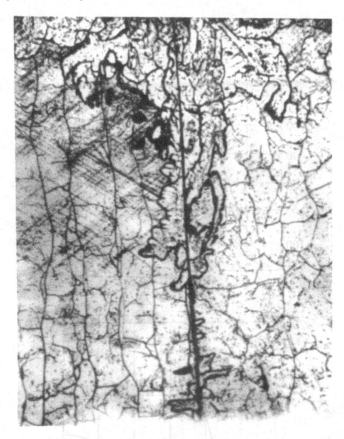

Figure 5.12. Typical structure of the alloy (1.8 mass% Nb + Zr) vibrating with the amplitude 10 μm in molten copper for 12 min. Magnification × 80 [9].

substrates. In the experiments with metallic and ceramic substrates, they used Sn–Zn melts (primarily, 90% Sn + 10% Zn), and in the case of polymeric substrate – Cd–Bi–Pb alloy with the melting point 92°C. In the absence of ultrasound, all of the substrates used showed a poor wettability; i.e., the contact angle was more than 90°.

However, after the onset of vibrations, the drop rapidly (for 0.01–0.04 s) spread over the surface of the substrate irrespective of the mode of vibrations feeding.

With the amplitude of vibrations exceeding a particular level, the drop was as a rule atomized into small secondary drops.

In [22], the ultrasonic wetting of a copper substrate by a Sn–Pb melt was studied.

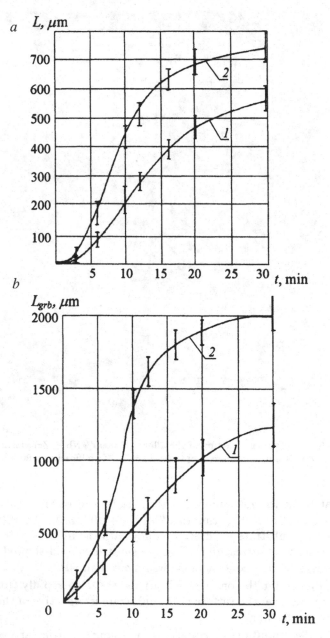

Figure 5.13. Time cources of penetration of molten copper into the solid phase (1.8 mass% Nb + Zr): (a) front of penetration; (b) penetration over grain boundaries; (1) penetration without vibrations; (2) penetration under the action of ultrasound [9].

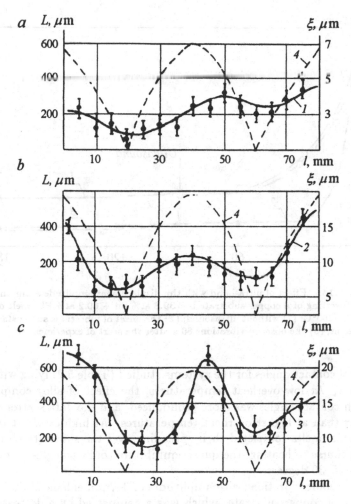

Figure 5.14. Distribution of penetration depth L of a substituting phase over the length l of Nb–Zr–Cu plate vibrated in tin melt for (min): (*1*) 5; (*2*) 10; (*3*) 25. Curve *4* shows the distribution of displacement amplitude over the plate length.

Before the onset of vibrations, the angle of wetting of copper substrate by melts of Sn, Pb, and their alloys was found to be strongly dependent on the overheat degree and the time of exposure. During the first 50–60 s of exposure, the contact angle decrease was maximum (Figure 5.15).

The quasi-equilibrium angle of copper wetting by Sn–Pb melts diminished with increasing overheat temperature (Figure 5.16), the entire

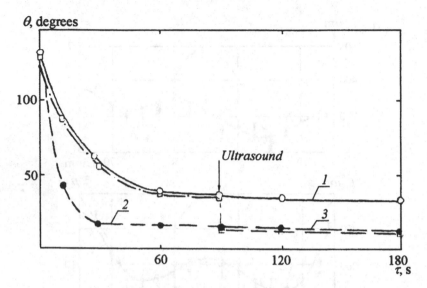

Figure 5.15.   Effect of ultrasound with the displacement amplitude 3 μm on the angle of wetting of a copper substrate by (39.8 at% Sn + 60.2 at% Pb) melt overheated by 200°C: (*1*) without ultrasound; (*2*) the onset of vibrations at the start of experiment; (*3*) the onset of vibrations 80 s after the start of experiment [22].

range of contact angles for the systems studied (Table 5.2) lying within region 1. At low overheat temperatures, the effect of alloy composition on contact angles was more pronounced, and the data scatter was greater than at higher overheat temperatures. At high overheat temperatures, melts completed their spreading over the substrate already in the course of heating, the quasi-equilibrium contact angles being as low as 25–30 degrees.

Ultrasonic vibrations with amplitudes of 3–7 μm enhanced the wettability of copper substrate, which was accompanied by a decrease in both the contact angle and the time of attainment of quasi-equilibrium state (Figures 5.15 and 5.16). At overheat temperatures of about 250–300°C, the quasi-equilibrium contact angles of binary alloys did not exceed 10–15 degrees and were independent of the intensity of ultrasound, if the displacement amplitude was not more than 3–7 μm. The melt drops could be atomized at higher displacement amplitudes (more than 8–10 μm) or overheat temperatures (more than 350°C).

After the onset of vibrations, the rate of contact angle reduction was so high that the time of attainment of quasi-equilibrium values for all investigated alloys and overheat temperatures up to 350°C was not more than 1 s.

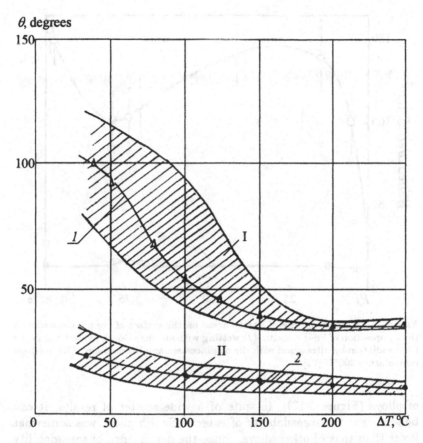

Figure 5.16.   Temperature dependences of the quasi-equilibrium angle of wetting of a copper substrate by (39.8 at% Sn + 60.2 at% Pb) melt: (1) without ultrasound; (2) irradiation by ultrasound with the displacement amplitude 3 μm. Hatched regions indicate the ranges of contact angle changing.

The ultrasonically induced spreading of pure tin and lead and their alloys differed. Thus, the extent of Sn and Pb spreading was insignificant, probably due to the formation of an intermetallic compound Cu_6Sn_5 and low solubility of Pb in copper. Oxide layers could also be the reason. Ultrasonic irradiation reduced the contact angle for Sn and Pb from 35–37 to 30–33 degrees.

At the same time, ultrasound enlarged the area of the spot wetted by alloys by a factor of 6–12, so that the resulting wetted areas appeared to be several times greater than those in the case of pure Sn and Pb. Wetted areas showed a nonmonotonic dependence on the composition

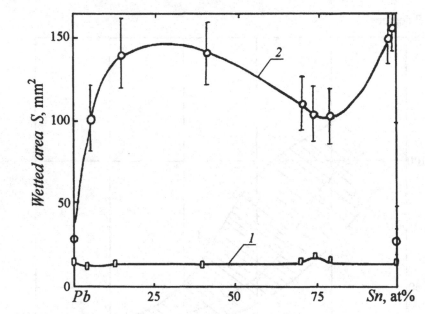

Figure 5.17. Dependence of wetted areas on the surface of copper substrate on the composition of Sn–Pb melts: (*1*) wetting without ultrasound; (*2*) wetting under the irradiation by ultrasound with the displacement amplitude 3 μm. The overheat temperature 200°C [22].

of alloys (Figure 5.17). In spite of a wide scatter of results, it can be seen that the spreadability of eutectic Sn–Pb alloys was somewhat lower than that of other alloys. Since the dependence of spreadability on contact angle is hyperbolic [27], the wide scatter in the wetted areas could be explained by their significant changes in response to slight changes in small contact angles. Curves in Figure 5.17 refer to the initial alloys rather than to alloys saturated with copper as a result of their contact. As follows from Young's equation (5.1), describing the spread of a liquid over a solid-state surface, the nonmonotonic behavior of the curve in Figure 5.17 cannot be related to changes in the surface tension σ_L, since the concentrational dependences of the surface tension of Sn–Pb alloys have no inflections [23]. Changes in the melt viscosity induced by ultrasound are insignificant as well [24]. It can be inferred that the nonmonotonic behavior of eutectic melts is associated with their specific structure with altered compressibility, which must affect the speed of ultrasound.

Based on experimental data, the adhesional work W_a and energy E of the interaction of Sn–Pb melts with solid copper were calculated

Table 5.2 Effect of ultrasonic vibrations ($\xi_m = 3$ μm) on the adhesional work W_a and the energy E_i of interaction of solid copper and Sn–Pb melts ($\Delta T = 200°$C).

Concentration of Sn, at.%	Surface tension $\sigma_{LG}\times10^3$, N/m*	Control				Ultrasonic irradiation			
		θ, degrees	$S\times10^4$, m^2	$W_a\times10^3$, J/m^2	$E\times10^7$, J	θ, degrees	$S\times10^4$, m^2	$W_a\times10^3$, J/m^2	$E\times10^7$, J
0	433	35	0.15	788	118	30	0.32	810	259
4.50	447	37	0.10	805	81	15	1.03	881	907
14.4	447	35	0.13	813	106	15	1.38	881	1216
39.8	457	35	0.13	832	108	12	1.40	905	1267
70.3	469	35	0.14	854	120	11	1.10	929	1022
73.8	476	34	0.17	871	148	10	1.05	942	989
78.4	488	35	0.16	888	142	11	1.02	966	985
99.3	530	35	0.13	965	125	12	1.48	1049	1552
99.8	530	35	0.13	965	125	13	1.57	1044	1639
100.0	527	35	0.11	959	105	33	0.27	970	262

* Data from work [23].

for control (nonirradiated) systems and those irradiated by ultrasound with the displacement amplitude 3-μm (overheat temperature 200°C). The results of calculations are summarized in Table 5.2. It can be seen that ultrasonic irradiation enhances W_a and E, the effect being especially pronounced for alloys rather than for pure metals.

As was already mentioned, the mechanism responsible for the effect of ultrasound on wetting in liquid–solid metal systems could be related to some additional forces arising due to vibrational motion in the substrate-drop system (mechanical factor) or to induced changes in the properties of the melt-solid phase system (physicochemical factor).

The degree of the manifestation of mechanical factor must correlate with the amplitude of vibrations, be dependent on the direction of the vibrational velocity vector relative to the melt-solid interface, and be virtually independent of the melt composition (acoustic properties of various binary Sn–Pb melts differ insignificantly).

The effect of physicochemical factor must depend on the melt composition and ultrasonic field parameters (this dependence may display a threshold behavior due to nonlinear acoustic effects).

Analysis of available experimental data suggests that the effect of mechanical factor is insignificant, as is evidenced by its independence of the amplitude of vibrations in the range 3–7 μm and by the radial symmetry of a spreading drop in an ultrasonic field (i.e., the absence of the expected drop expansion in the direction of the vibrational velocity vector).

At the same time, the strong dependence of ultrasonic wetting on the alloy composition in the range of small concentrations of the second component implies an essential part of physicochemical factor, the more so that this inference does not contradict the threshold dependence of the effect on the amplitude of vibrations. It can be suggested that with the amplitude of ultrasonic vibrations exceeding by a particular value (3 μm), the cavitation and high-intense microjets arising in the drop will activate physicochemical processes at the liquid–solid interface. These processes may be responsible for changes in the composition of surface layers due to dissolving and erosion of the solid phase, penetration of melt into the solid phase, narrowing of a diffusion layer in the melt, and the related intensification of convective mass transfer.

The strong dependence of ultrasonic effect on the addition of small amounts of the second component to a pure metal can be related to the action of this component as an interphase effector. The role of such effectors in the ultrasonic wetting of solid particles by metal melts was investigated in sufficient detail by V. I. Dobatkin and G. I. Eskin [25], who found that the addition of effectors considerably contributes to wetting.

Formally, the effect of physicochemical factor can be interpreted as a decrease in the liquid–solid surface tension.

If the amplitude of vibrations surpasses a certain level (7 μm), the surface–capillary waves arising on the drop surface will have amplitudes sufficient for the fragmentation of the wave crests and thus to atomization of the drop [26].

In the case of alloys containing $\sim$40–80 at.% Sn, there is a halo around the melt drop, which can be due to the melt propagation over the grain boundaries on the surface of the substrate (Figure 5.18).

Analysis of the vertical distribution of elements in the surface layer of substrate beneath the melt drop showed that it contained components of the melt (Figure 5.19). This indicates that ultrasonic vibrations induce an intense mass transfer through the melt-solid interface.

Similar results were obtained in the studies of the wetting of copper substrates by Pb–Sb and Sn–Sb melts.

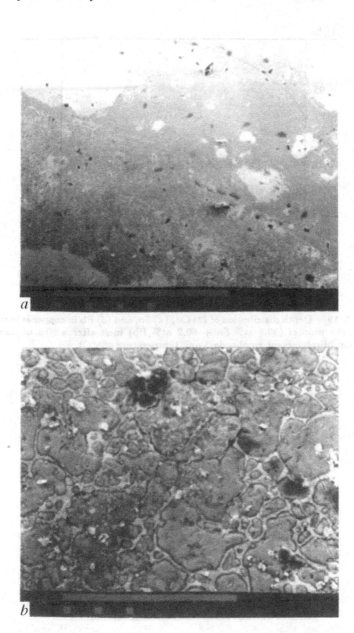

Figure 5.18. Effect of ultrasound on the wetting of copper substrate by (39.8 at% Sn + 60.2 at% Pb) melt (displacement amplitude 3 μm; temperature 435°C; exposure time 20 s): (a) halo around the melt drop; (b) penetration of melt over grain boundaries. Bars represent 100 μm.

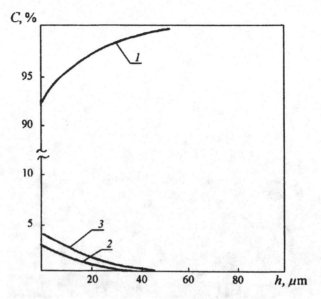

Figure 5.19.   Depth distribution of (*1*) Cu, (*2*) Sn, and (*3*) Pb in copper substrate beneath the drop of (39.8 at% Sn + 60.2 at% Pb) melt after a 20-s ultrasonic irradiation (displacement amplitude 3 μm; temperature 435°C).

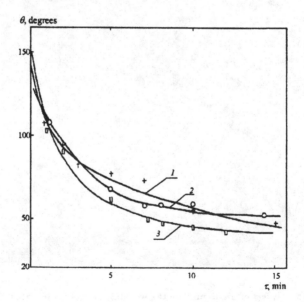

Figure 5.20.   Changes in the angle of wetting of aluminum by melts of (*1*) tin, (*2*) zinc, and (*3*) lead induced by ultrasonic irradiation (displacement amplitude 3-μm; overheat temperature 70°C; residual gas pressure 5.7×10^{-2} mm Hg.

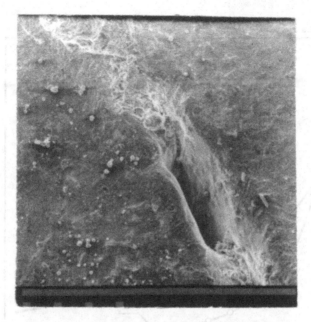

Figure 5.21. Erosion of aluminum plate in tin melt after a 5-min exposure to ultrasound (displacement amplitude 5-μm; temperature 300°C.

The problem of the ultrasonic wetting of oxide phases by metal melt was approached by N. V. Korchevskii,[*] who studied the wetting of alumina monocrystals by melts of low-melting-point metals (In, Sn, Bi, Pb, Zn, and Sb) and their alloys. Twenty kilohertz vibrations were supplied normally to the surface of the substrate. Raising the amplitude from 0 to ~10–12 μm did not virtually affect the contact angle, while a further increase in the amplitude led to atomization of the melt or brittle failure of the substrate.

Experiments with aluminum and titanium substrates (these materials form fairly stable surface oxide films) wetted by melts of low-melting-point metals revealed a decrease in the contact angle caused by ultrasonic vibrations (Figure 5.20), which was most likely due to the disinteration of surface oxide films as a result of ultrasonically induced cavitation in the melt (Figure 5.21).

The addition of a second component to melts promoted their ultrasonic spreading (Figure 5.22). The results were similar for Sn–Sb and Pb–Sb alloys.

---

[*]N. V. Korchevskii, Dissertation "Some Features of Ultrasonic Metallization of Aluminum and Copper", 1991.

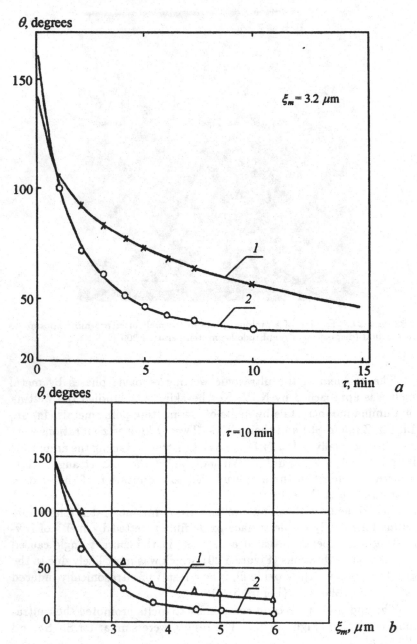

Figure 5.22. Angles of wetting of aluminum substrate by molten (*1*) tin and (*2*) tin alloy containing 0.7 at% Pb as functions of (*a*) time and (*b*) amplitude of vibrations (temperature 300°C).

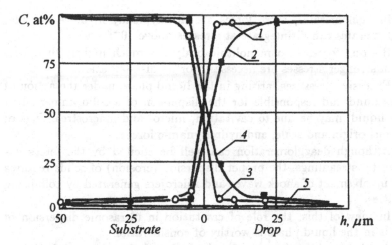

Figure 5.23. Depth distribution of (*1, 2*) Al, (*3, 4*) Sn, and (*5*) Pb across the substrate-drop interface after a 5-min exposure to ultrasound (displacement amplitude 5-μm; temperature 300°C). Curves *1* and *3* correspond to Sn drop; curves *2, 4*, and *5* correspond to a drop of (Sn + 0.7% Pb) alloy.

X-ray spectroscopy of the zone of contact of a solidifying drop and substrate showed that it was ~ 50 μm thick (Figure 5.23), i.e., thicker than in the case of wetting by pure tin.

The above consideration of some aspects of wetting in an ultrasonic field may appear useful for the development of ultrasonic technology of material metallization.

5.1.3 Ultrasonic Dispersion of Solids in Liquids

By dispersion is typically meant the breakup of a monolithic solid phase with the formation of small particles. However, this term is frequently referred to two different processes – proper dispersion in the above sense and deagglomeration, or the breakup of an agglomerate into constituting particles.

Both a particle and an extended body can be dispersed. In the latter case, one deals with erosion of a solid surface.

Dispersion will take place if pressures arising in an ultrasonic field are above a particular level of stresses in a solid, defined by its strength and plasticity. In the general case, mechanical properties of a material are determined by binding forces between the constituting particles.

In agglomerates, these forces are not typically high. As a rule, agglomerates can disintegrate at stresses above 10^{-2} Pa.

Binding forces in a monolithic body are much higher; therefore, much stronger stresses are necessary for its disintegration.

Excessive pressures, arising in the liquid phase under the action of ultrasound and responsible for the dispersion of a solid immersed in this liquid, may be due to cavitation, micro- and macrostreamings of various origin and scale, and hydrodynamic forces.

Although deagglomeration can well be elicited by the forces exerted by streamings, the proper dispersion (erosion) of solids requires the involvement of shock waves and microjets generated by collapsing cavities.

In view of this, the role of cavitation in ultrasonic dispersion of solids in the liquid phase is worthy of consideration.

As shown above (chapter 2), the collapse of cavities in the liquid phase gives rise to shock waves and cumulative jets. To estimate the energy of cavitation, B. A. Agranat and V. I. Bashkirov [28, 29] proposed a dimensionless criterion of its erosive activity, κ, representing the ratio of the energy stored by a cavity per unit time during its expansion to the mean specific power of the cavity in the phase of collapse. In other words, the criterion κ characterizes the transformation of power in a cavity. The greater κ, the higher is the energetic efficiency of the cavity. By expressing power through the cavity parameters, one can show that

$$\kappa \cong \frac{R_{\max}^3}{R_{\min}^3 \, \Delta t \, f}, \tag{5.4}$$

where $R_{\max}$ and $R_{\min}$ are the maximum and minimum radii of the cavity; Δt is the time of cavity collapse; f is the frequency of vibrations.

The values for $R_{\max}$, $R_{\min}$, and Δt can be derived from a numerical solution to the equation of cavity dynamics obtained by varying any of the parameters. By substituting the derived values into expression (5.4), one can determine the influence of this parameter on the energy liberated by a collapsing cavity.

In view of the foregoing, it is of interest to elucidate how the physicochemical properties of medium and the parameters of ultrasonic field influence the criterion of erosive activity.

The equation of cavity dynamics involves physicochemical parameters of liquids, such as vapor pressure, surface tension, sound speed, density, viscosity, and a nonlinear parameter from the equation of state.

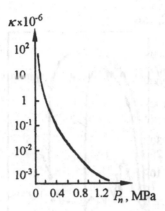

Figure 5.24. Dependence of the erosive activity criterion on the vapor pressure P_n [30].

When estimating the effect of vapor pressure on cavitation and its erosive activity, one should bear in mind that as vapor pressure decreases, the cavitation threshold rises, and the number of cavities formed falls. At the same time, the erosive activity of cavities greatly enhances (Figure 5.24).

Erosive activity of a liquid in low-intensity ultrasonic fields decreases with increasing viscosity of the liquid, which is partially due to the acoustic energy loss during the propagation of vibrations. On the other hand, high intensities of ultrasound in highly viscous liquids ($\nu \geq 5 \times 10^{-2}$ s/cm^2) appear beneficial for cavitation, which can be explained as follows. Viscous forces act to prevent the expansion of cavities; therefore, the beginning of their collapse coincides with the commencement of a compressional half-cycle of ultrasonic wave, which shortens the time of cavity collapse and may thus increase its erosive activity.

Cavity collapse is more intense at higher surface tension, although in this case the threshold of cavitation also rises. Changes in density, sound velocity, and nonlinear parameter of the equation of state in the limits typical of real liquids affect but little the erosive activity of cavities.

One should consider the effect of two ambient factors influencing the behavior of cavities – temperature and static pressure.

Temperature is an essential factor that may affect both cavitation and chemical activity of medium (the latter increases with temperature). The role of temperature in cavitation is more intricate, as

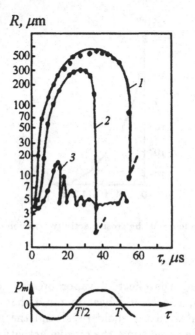

Figure 5.25.   Effect of static pressure on cavity pulsations at $p_m = 1$ MPa, $f =$ 20 kHz, and P_0 (MPa): (*1*) 0.1; (*2*) 0.5; (*3*) 1.0 [28].

it influences all of the physicochemical parameters of medium (vapor pressure, viscosity, surface tension, density, etc.), and it is difficult to estimate theoretically how changes in these parameters affect the erosive activity of cavities. However, experimentally it was revealed that the optimum temperature for cavitation in aqueous solutions lies between 60 and 80°C.

Excessive static pressures can drastically enhance the erosive effect of cavitation. Based on the numerical solution to the Noltingk–Neppiras equations obtained by varying static pressure, S. A. Nedyuzhii and B. G. Novitskii [31, 32] showed that the erosive activity of cavities increases with static pressure, unless it exceeds a particular value equal to acoustic pressure p_m (Figure 5.25). Once the static pressure surpasses this value, the erosive activity of cavities drops.

At atmospheric pressure $P_0 = 0.1$ MPa and acoustic pressure $p_m = 1$ MPa (Figure 5.25, curve *1*), the phase of cavity expansion lags behind the compressional half-cycle of acoustic wave, due to which acoustic field opposes the cavity collapse at its initial stage. As a result, the shock wave is dumped to reduce the erosive activity of cavities.

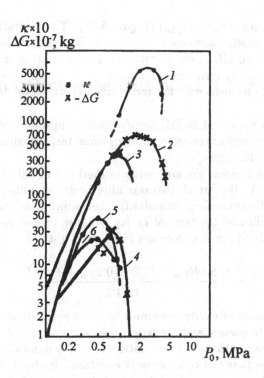

Figure 5.26. Dependence of the erosive activity criterion κ (curves *1, 3, 5*) and extent of cavitational erosion (curves *2, 4, 6*) on static pressure P_0 at different acoustic pressures p_m (MPa): (*1, 2*) 5; (*3, 4*) 2; (*5, 6*) 1 [28].

At a higher static pressure and the same acoustic pressure ($P_0 =$ 0.5 MPa, $p_m = 1$ MPa), the velocity of cavity collapse drastically increases (curve *2*), which is due to a shift in the commencement of cavity collapse. In this case, the cavity begins to collapse at the moment close to $0.75T$, i.e., when acoustic pressure, acting upon the cavity, is maximum. The combined effect of acoustic pressure and excessive static pressure drastically increases the velocity of cavity collapse, resulting in the generation of a high-intensity shock wave.

As the static pressure continues to rise ($P_0 = 1$ MPa, $p_m = 1$ MPa, curve *3*), the cavity, whose size remains virtually unchanged, executes complex nonharmonic vibrations. As a result, the erosive activity of cavitation becomes insignificant.

The erosive effect of cavitation can be largely enhanced by raising both acoustic and static pressures and maintaining the optimum ratio between them ($P_0 \sim 0.5p_m$) (Figure 5.26). The cavitational erosion was experimentally estimated from the mass loss ΔG of aluminum samples subjected to ultrasonic vibrations. The theoretical dependence of the erosive activity criterion κ on static pressure appeared to be in a satisfactory agreement with the respective experimental dependence of ΔG.

The data presented in [31] testify to the appropriateness of using the erosive activity criterion to select optimal technological parameters of ultrasonic cleaning.

Of certain interest are approximate methods of calculation of this criterion. A. V. Il'in *et al.* [33] was able to derive sufficiently correct and practically convenient formulas for determining the maximum and minimum radii and the time of cavity collapse in low-viscous liquids. By substituting them into formula (5.4), they got

$$\kappa = \frac{8.14(p_m - P_0)^{5/2} (0.2p_m + P_0)^{7/2}}{p_m^3 \, p_n^3}. \tag{5.5}$$

The equation of cavity dynamics involves such ultrasonic field parameters as frequency and acoustic pressure. For various frequencies, the respective solutions to the equation of cavity dynamics are similar, which implies that the criterion κ is constant. It should be, however, noted that the similarity of solutions can be disturbed at very low and very high frequencies, when the cavity becomes pulsating with a drastic decrease in the criterion κ.

At high frequencies, some decrease in $R_{\max}$ (resulting from the shortening of the time of action of tensile stresses) leads to an enhanced pressure of the vapor-gas mixture in a bubble at the beginning of the collapse phase. As a result, at $R_{\max}/R_0 \sim 2$, the bubble degenerates into pulsating one, which brings about to a drastic fall in the criterion κ.

At low frequencies, when the cavity changes relatively slow, the collapse conditions are no longer adiabatic, which again leads to pulsation of cavities. At $p_m = 1$ MPa, the lower and upper critical frequencies, corresponding to degeneration of collapsing cavity into pulsating one, are 10^3 and 10^6 Hz, respectively. In this case, κ is virtually constant in the frequency range 2×10^3–10^5 Hz.

Analysis of the cavity dynamics and the criterion κ as functions of the acoustic pressure amplitude p_m indicates that κ initially increases with p_m due to changes in $R_{\max}$, $R_{\min}$, and Δt; however, as p_m continues to increase, the cavity executes several pulsations before being collapsed (Figure 5.27).

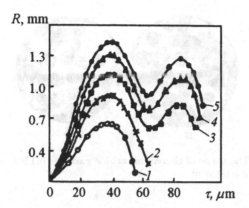

Figure 5.27. Motion of a cavity at different acoustic pressures p_m (MPa): (*1*) 1; (*2*) 2; (*3*) 3; (*4*) 4; (*5*) 5 [30].

Comparing the contributions from all the above factors to the erosive activity of cavitation allowed Agranat *et al.* [30] to rank them according to their increasing contributions to the criterion κ:

$$\rho \to \sigma \to \eta \to p_n \to f \to p_m \to P_0.$$

With regard to the role of shock waves and cumulative jets in dispersion, one should note that it is a scale factor that largely determines which mechanism will prevail. If a cavity is greater than a particle to be dispersed, the undistorted cavity will collapse to produce a shock wave. However, if the particle is greater than the cavity, the latter will no longer be spherical and will collapse with the formation of a cumulative jet.

Let us consider first the case when particles are smaller than the cavity. Due to the flotation effect of the pulsating cavity, particles will concentrate on its surface, where they experience the action of shock waves. The ledges on the surface of particles are the first to be destroyed; as a result, the particles become smooth (Figure 5.28).

The studies along this line led B. G. Novitskii [32] to an expression for the time during which treated particles attain a particular size. The presence of parameters and numerical coefficients that can hardly be determined experimentally makes this expression inappropriate for practical use, although the author emphasized a good agreement between the calculated and experimental data in the case of dispersion of gypsum particles.

Novitskii [32] treated also the problem of erosion of a solid surface in an ultrasonic field due to the action of cumulative jets resulting from

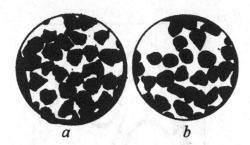

a   *b*

Figure 5.28. The shape of zirconium carbide particles (*a*) before and (*b*) after a 30-min ultrasonic treatment.

the cavity collapse and estimated the forces exerted by these jets on the solid surface.

Unfortunately, the expressions obtained are also virtually inapplicable because of the above reason.

5.1.4   Crystallization of Metals in Ultrasonic Fields

Crystallization of metals is the process that largely determines their quality. This especially holds for casting. The structure of a solidified casting completely determines its mechanical properties. This is also true for products produced from ingots. Many of the defects an ingot takes on during solidification are retained throughout further production stages to occur in a final product. This challenges the problem of improvement of metal structure in the process of crystallization.

For the first time the idea of metal conditioning by subjecting it to elastic vibrations was proposed by D. K. Chernov in the 19th century [34]. In regard to crystallization of a steel ingot he wrote that "if a steel melted in a crucible is subjected to severe shaking sufficient to set all its particles in a motion, the solidified ingot will have extremely fine crystals; but if the same steel is allowed to cool without any shaking, this very steel will have large, well-developed crystals". These ideas received confirmation in later works [35–37].

The pioneer investigations of the effect of ultrasonic vibrations on crystallization were performed in 1926–1927 by R. Wood and A. Loomis [38].

The studies along this line with the use of organic substances were undertaken in the 1930s by V. I. Danilov and co-workers [39–41] and later by A. P. Kapustin [42].

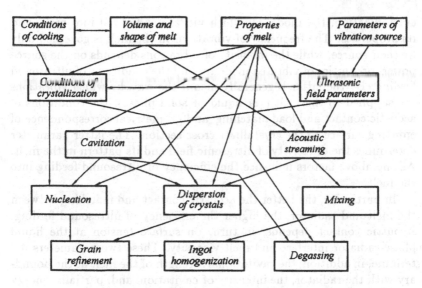

Figure 5.29. Diagram of the effect of ultrasound on a solidifying metal.

The works by S. Ya. Sokolov [43], I. I. Teumin [44], G. I. Pogodin-Alekseev [36], I. G. Polotskii [45], H. Seeman [46], G. I. Eskin [47], O. V. Abramov [48], and others contributed much to our understanding of the mechanism of metal crystallization in ultrasonic field.

A direct consequence of the effect of ultrasonic vibrations on a solidifying metal is structural alterations in it. Ultrasonic treatment metals eliminates their columnar structure, promotes the formation of fine equiaxial grains, enhances their homogeneity, and minimizes segregation. Ultrasonic irradiation of multi-phase alloys may sometimes alter their phase distribution.

Structural changes in a solidifying ingot are the results of the processes occurring in the melt and in the two-phase zone, namely, nucleation, crystal disintegration, phase mixing, which, in turn, are governed by cavitation, acoustic streaming, ambient conditions, and the properties of molten material. The degree of development of cavitation, the behavior and intensity of acoustic streaming are governed by the ultrasonic field parameters in the melt, as well as by its volume, configuration, and properties. Schematically, this relationship can be illustrated as shown in Figure 5.29.

Ultrasonic field parameters in the melt, such as the frequency of vibrations, vibrational velocity and pressure, and energy loss are determined by the parameters of ultrasonic source, the conditions of melt crystallization, the mode, by which vibrations are fed into melt, and its properties. The frequency of vibrations in the melt is governed only by their source, while the intensity of vibrations depends on the source power, properties of vibrated medium, and the conditions of ultrasound feeding. By the latter is generally meant the way, by which vibrations are supplied to the melt (via liquid or solid phase), the conditions of acoustic contact and load matching, and geometrical correspondence of emitting surface to the crystallizer cross section. The latter parameter determines the diffusivity of ultrasonic field and its pattern in the melt. All the above factors influence the efficiency of ultrasound feeding into the melt.

In particular, the better the acoustic contact and matching between the melt and radiator, the higher the efficiency of ultrasound feeding. Acoustic contact depends, in turn, on surface tension at the liquid phase-radiator interface and melt viscosity. These two parameters determine, in addition, the cavitational strength of the melt at the boundary with the radiator, the intensity of cavitation, and, partially, energy loss in the melt.

Matching depends on the proportion between the wave resistances of the radiator and the load, being the better the smaller is the resistance difference. The absence of experimental data concerning the ultrasonic velocity in melts of the majority of industrially essential metals and alloys makes the estimation of matching difficult. Moreover, as an ingot solidifies, its wave resistance varies due to changes in its density and sound speed.

Thus, the input resistance and ultrasonic energy loss in a melt are related to its properties, such as wave resistance, viscosity, and heat conductivity.

The conditions of metal crystallization - the velocity of crystal growth and temperature gradient in the liquid near the crystallization front – greatly affect the front structure, the two-phase zone width, and thereby the ultrasonic field parameters near the solid–liquid interface. It should be noted that analysis of the effect of ultrasound on the ingot structure has to take into account the phenomena that take place both in the liquid phase (the melt) and at the solid-liquid interfaces, i.e., at the crystallization front and in the two-phase zone.

Ultrasonic energy fed in the melt is responsible for excessive pressures and mixing, modifying the kinetics of crystallization. The magnitude of excessive pressures and the degree of mixing are functions of both the field and melt parameters (viscosity, heat conductivity, compressibility, liquid–vapor surface tension, densities of the liquid and solid phases). On the other hand, pressures required for dispersion of the liquid phase depend on the strength of growing crystals and their size, which, in turn, are determined by the conditions of crystallization (the rate of crystal growth and temperature gradient) and the properties of material (temperature range of crystallization and impurity distribution coefficient). It is also worth noting that the relationship (Figure 5.29) between the ultrasonic field parameters and the conditions of crystallization implies that ultrasound may alter these conditions themselves, that is, change the temperature gradient and the rate of crystal growth.

Primary parameters characterizing ultrasonic field – specific energy density, input resistance, and losses – determine the vibrational velocity and pressure in the melt. In turn, these two parameters are responsible for cavitation, acoustic streaming, radiation pressure, viscous forces, and other nonlinear effects in medium. As shown above, molten metals favor cavitation presumably due to a considerable saturation of melts with dissolved gases, which promotes the formation of cavitation bubbles.

Moreover, different solubility of gases in the solid and liquid phases may be responsible for excessive free gas near the crystallization front, which provides an additional source of cavities and must thereby decrease the cavitation threshold.

Cavitation, responsible for high-intensity shock waves near the crystallization front in the melt, should be taken into account, first, as one of the reasons for dispersion of growing crystals and, second, as a source of local disturbances in the melt homogeneity and thermodynamic equilibrium. Such disturbances can obviously change the conditions of the liquid-to-solid transition and consequently crystallization parameters.

Apart from being the cause of cavitation, vibrational pressure may also directly affect the process of crystallization. Calculations performed by J. Campbell [49] led him to nomograms relating the vibrational pressure in the melt to displacement amplitude and frequency (Figure 5.30).

As follows from Figure 5.30, at frequencies of about 2×10^4 Hz and a displacement amplitude of 2×10^{-6} m, the vibrational pressure in the melt is of order of 10 bar, which is too low to appreciably disturb the phase equilibrium in the melt-crystal system.

At a displacement amplitude of about 10^{-6} m and frequencies between 10^4 and 2×10^4 Hz, the vibrational pressure gradient in liquid metal is not steep enough to noticeably affect crystallization. At the same time, the vibrational pressure gradient across the solid-liquid interface can be sufficient for disintegration of growing crystals,

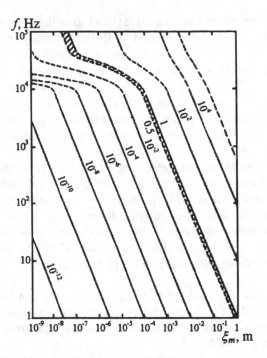

Figure 5.30. Acoustic pressure in iron melt as a function of amplitude and frequency of vibrations [49]. Hatched area corresponds to cavitation; dashed lines correspond to theoretical liquid strengths.

whose strength, at temperatures about the melting point, is quite low [44].

Acoustic streamings produced in the melt are responsible for energy loss. As was shown in chapter 2, these streamings mix the melt, alter its temperature field, promote convective diffusion, which all must affect the kinetics of crystallization defined, to a large extent, by the velocity of diffusion processes.

Vibrational velocity and acoustic streaming are responsible for viscous and inertial forces in the melt. Viscous forces are due to a difference in the vibrational velocities in liquid and solid phases, while inertial forces are due to a difference in their densities. Both kinds of forces are able to disintegrate growing crystals.

Campbell [49] estimated viscous and inertial forces resulting from the formation of a dendrite structure during crystallization of some alloys based on low-melting-point metals (bismuth and aluminum) for small (≤ 10) and large ($\geq 10^3$) Reynolds numbers. The estimations took into account the dependence of the growing dendrite geometry on the rate of alloy cooling.

Campbell also estimated shear and bending stresses imposed upon the melt by elastic vibrations of various amplitudes and frequencies (Figures 5.31 and 5.32). Comparison of these estimates with the strength characteristics of alloys and experimental data from [50] shows that the above-mentioned forces may disperse growing crystals.

Structural changes induced in the melt by an applied ultrasonic field largely depend on the conditions of crystallization and the mode of ultrasound feeding. In particular regions of a solidifying melt, the effect of one or another factor of ultrasonic field may be dominant. Thus, crystals are able to disintegrate only near crystallization front or in the two-phase zone, whereas acoustic streaming and mixing take place only in liquid phase.

In view of the foregoing, the effect of ultrasound on crystallization cannot be accounted for by the action of a single factor, for instance, mechanical disintegration of growing crystals, as it was assumed in earlier works.

If the elimination of a columnar structure and the refinement of grains may, to a certain degree, be explained by the direct ultrasonic dispersion of growing crystals, changes in the phase distribution and dendrite segregation can only be due to mixing and temperature gradients in the melt. In other words, the mechanism of the ultrasonic effect is rather sophisticated and manifests itself not only as the dispersion of growing crystals, but also as changes of the crystallization parameters of the substance.

The major ultrasonically induced structural alteration in solidifying metals is grain refinement, or a reduction in the mean size of grains. The coefficient of refinement, defined as the ratio of the mean sizes of grains in control (untreated) and ultrasonically vibrated ingots, can be used as a basic measure for ultrasonic processability of materials.

Various materials exhibit different ultrasonic processability: some of them can be worked by low-intensity vibrations, whereas others require high-intensity ultrasonic vibrations. Therefore, a second characteristic of ultrasonic processability of materials is the minimum intensity of ultrasound capable of producing structural changes.

Grain refinement takes place if an ultrasonic field is able to enhance the rate of nucleation or/and to disintegrate growing crystals.

As noted above, the disintegration of growing crystals may be due to cavitation, viscous forces, and vibrational pressure. Cavitation

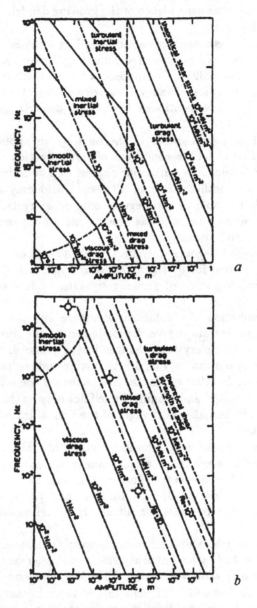

Figure 5.31. Diagrams illustrating the amplitude and frequency dependences of ultrasonically induced drag and inertial stresses arising near dendrites in the melts of bismuth-based alloys solidifying at (a) low and (b) high rates. Data from work [49] except for those represented by points, which are taken from work [50]. Dashed lines indicate the shear strengths of crystals.

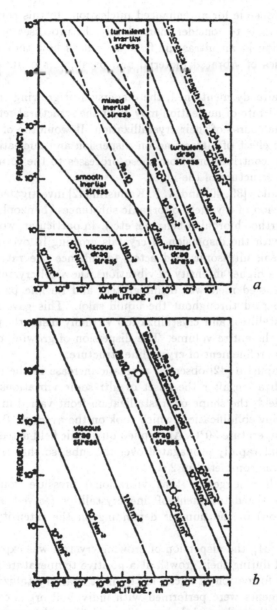

Figure 5.32. Diagrams illustrating the amplitude and frequency dependences of ultrasonically induced drag and inertial stresses arising near dendrites in the melts of aluminum-based alloys solidifying at (a) low and (b) high rates. Data from work [49] except for those represented by points, which are taken from work [50]. Dashed lines indicate the shear strengths of crystals.

may be responsible for an enhanced nucleation. In this connection, it seems reasonable to consider more closely the processes of dispersion and nucleation in an ultrasonic field as well as how they depend on the properties of vibrated material and parameters of its crystallization.

As was already mentioned, the dispersion of growing crystals and the enhanced rate of nucleation may elicit the structural refinement of metals in the course of their crystallization. It would be of interest to evaluate the effect of ultrasound on dispersion and nucleation, and to estimate the contributions from these processes to the formation of a fine-grained structure of the ingot.

V. I. Danilov [39, 40] and A. P. Kapustin [42] investigated the ultrasonic dispersion of a variety of organic substances (benzophenol, salol, piperonal, ortho-chloronitrobenzene, etc.). In particular, work [40] was concerned with the dispersion of crystals growing in overcooled salol melt. A weak ultrasonic field acted to enhance the rate of crystal growth. At a higher intensity of vibrations, the salol crystal immersed in the overcooled melt was disintegrated, and its fine particles were rapidly dispersed throughout the liquid salol. This gave rise to numerous crystallites, and crystallization virtually instantly propagated throughout the entire volume. The dispersion of growing crystals led to an intense refinement of crystalline structure.

A. P. Kapustin [42] observed a drastic increase in the rate of thymol crystallization after the onset of ultrasonic vibrations. Without ultrasonic field, the shape of crystallization front varied in a complex manner during solidification, which took on the average 30 min, if the overcool temperature 30°C. An applied ultrasonic field acted to smooth the front that rapidly propagated over the tube, so that thymol crystallization was completed in 2 s.

The author suggested that vibrational pressure could be responsible for the detachment of fine crystallites (served as nuclei of crystallization) in the number depending on the intensity of vibrations.

In work [51], the dispersion of growing crystals was experimentally investigated during their growth at a positive temperature gradient to exclude the formation of nuclei in the melt near crystallization front. The experiments were performed with individual organic substances and metals (thymol, naphthalene, tin, and bismuth), as well as with their alloys (naphthalene-azobenzene, tin-zinc, and tin-bismuth). All the substances used were chemically pure (99.99%).

A setup for the experiments with organic substances is shown in Figure 5.33, and that for the experiments with low-melting-point metals

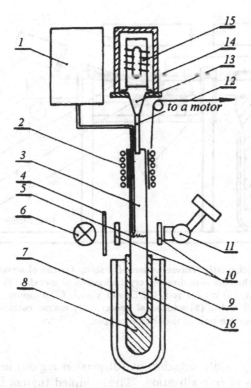

Figure 5.33. Schematic representation of a setup for the observation of crystallization front in the course of directed crystallization of organic substances: (*1*) potentiometer, (*2*) heater, (*3*) zone of molten substance, (*4*) ground glass, (*5*) crystallization front, (*6*) lamp, (*7*) Dewar vessel, (*8*) copper rod, (*9*) coolant, (*10*) polarizers, (*11*) microscope with a photographic camera, (*12*) 5-mm diameter waveguide, (*13*) horn, (*14*) cooling jacket of the transducer, (*15*) magnetostrictive transducer, (*16*) 7-mm diameter tube with a substance.

and their alloys is sketched in Figure 5.34. Graphite boat with a molten metal could be rotated around the A-axis to provide melt decantation and the exposure of crystallization front.

At a rate of thymol crystallization of $(5-16) \times 10^{-3}$ mm/s, the crystallization front remained smooth (Figure 5.35,*a*). Ultrasound of a precavitation intensity (1–20 W/cm^2) did not modify the shape of the crystallization front, but somewhat changed the temperature distribution in the melt by increasing its gradient near the crystallization front. Increasing the intensity of vibrations up to 25 W/cm^2 led to the formation of cavities in the melt, which made the front rough and caused detachment and removal of crystallites to the melt bulk (Figure 5.35,*b*).

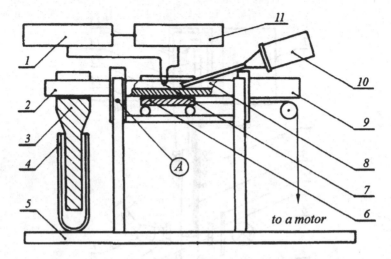

Figure 5.34. Schematic representation of a setup for the observation of crystallization front in the course of directed crystallization of metals: (*1*) potentiometer, (*2*) graphite boat, (*3*) copper rod, (*4*) Dewar vessel, (*5*) support, (*6*) heater, (*7*) thermocouple, (*8*) metal, (*9*) additional heater, (*10*) magnetostrictive transducer, (*11*) emf-compensation unit of the thermocouple.

The area of crystallite detachment (dispersion region) increased with decreasing rate of crystallization. The solidified thymol became more fine-grained.

With a further rise in ultrasonic intensity, the number of cavities and dispersed crystals increased, leading to enlargement of dispersion region and further refinement of crystalline structure.

It should be noted that the intensity of vibrations necessary to initiate cavitation in the melt depended neither on the rate of crystal growth nor the temperature gradient in the melt.

Analogous results were obtained in the experiments with naphthalene. One should only note that at a rate of naphthalene crystallization 16×10^{-3} mm/s, the crystallization front was irregular. The application of precavitation vibrations acted to smooth the front.

Similar experiments with low-melting-point metals showed that with the velocity of front migration being $(25–108) \times 10^{-3}$ mm/s and the temperature gradient in the melt ranging from 1 to 25°C/cm, the crystallization front in tin and bismuth was smooth without any evidence of roughness (Figure 5.36,*a*, ×20 magnification). The extent of front roughness was assessed from the longitudinal sections of samples by determining the interval that accommodated 90% of all top-to-base

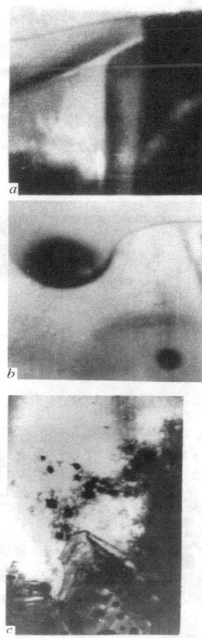

Figure 5.35.   Crystallization front of thymol crystals growing at a rate of 16 ×
10^{-3} mm/s in the ultrasonic field of intensity (W/cm^2): (a) 0; (b) 25; (c) 35.
Magnification × 100.

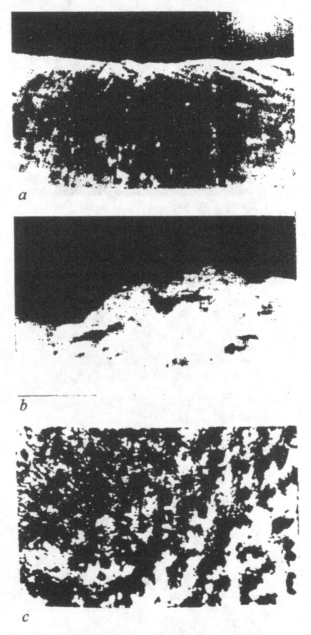

Figure 5.36. Crystallization front of tin crystals growing at a rate of 25 ×
10^{-3} mm/s in the ultrasonic field of intensity (W/cm^2): (*a*) 0; (*b*) 70 (side view);
(*c*) 70 (top view). Magnification × 20.

sizes of growing crystals. It is this interval that was a measure of front unevenness.

Low-intensity ultrasound (of approximately one third of the cavitation threshold) did not virtually modify the crystallization front but somewhat steepened temperature gradient.

The increase in ultrasound intensity led to cavitation in the melts. Cavitation thresholds for tin and bismuth were different (70 and 30 W/cm^2, respectively). Cavitation promoted the formation of depressions (craters) on crystallization front, which was probably due to the collapse of cavities and dispersion of crystals. Cavitation steepened temperature gradient in the melt.

To relate structural alterations in the crystallization front and the resulting structure of solidified specimens, the grain size and refinement coefficient were estimated from specimen sections.

In all cases, changes in the crystallization front were accompanied by changes in the grain size. The refinement coefficients of tin and bismuth were different: 8.5 (ultrasound intensity 70 W/cm^2) and 36 (ultrasound intensity 30 W/cm^2), respectively.

At higher intensities of ultrasound, the distortion of crystallization front and the extent of refinement were greater.

An higher rates of crystal growth and steeper temperature gradients, the intensity of vibrations being constant, the coefficient of grain refinement diminished, but cavitation occurred independently of these parameters.

The effect of ultrasound on dispersion and the form of crystallization front in melts was initially studied for naphthalene + 0.2 mass% azobenzene alloy.

When ultrasound was absent and the velocity of heater transposition was below 16×10^{-3} mm/s, the crystallization front remained smooth, while became irregular at a higher velocity (Figure 5.37,a). As the velocity of heater transposition increased, the length and radius of crystals, growing inward the melt, became greater. It is noteworthy that in the case of a rapid withdrawal of the heater from the melt, the velocity of crystal growth did not exceed $(50-80) \times 10^{-3}$ mm/s, the height of the needle protruding inward the melt being not more than 0.06–1 mm.

Ultrasonic irradiation of the melt with an intensity of 20 W/cm^2 smoothed the crystallization front in 2–3 s, if the velocity of heater withdrawal was below 30×10^{-3} mm/s (Figure 5.37,b). At higher heater velocities, the protruding needles became less acicular. Thus, the 20-W/cm^2 ultrasonic treatment influenced the growth of crystals rather than caused their disintegration.

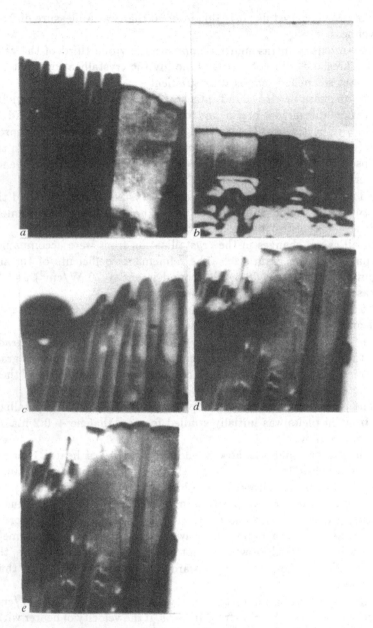

Figure 5.37. Crystallization front of (naphthalene + 0.2% azobenzene) crystals growing at a rate of 20×10^{-3} mm/s in the ultrasonic field of intensity (W/cm^2): (a) 0; (b) 20; (c, d, e) 35. In last cases, the photographs are taken (c) at the moment of cavity collapse as well as (d) 2 s and (e) 4 s later. Magnification × 100.

Raising the intensity of vibrations to 35 W/cm^2 gave rise to cavitation in the melt. Pulsating cavities produced depressions on the crystallization front (Figure 5.37,c), while collapsing cavities were responsible for the extensive growth of acicular crystals at the front that became uneven irrespective of the velocity of heater transposition (Figure 5.37,d,e).

A set of experiments was performed with naphthalene + 10 mass% azobenzene alloy, in which fairly slender and long crystal needles were found to grow (the rate of crystallization ranged from 5 to 100×10^{-3} mm/s). Precavitation ultrasound elicited both front smoothing and dispersion of growing crystals.

For protruding crystal of length l and radius r, the pressure P_d necessary for its destruction is given by [51]

$$P_d = \frac{1}{4}\left(\frac{r}{l}\right)^2 \sigma_{BT} \cong \gamma\sigma_{BT}, \qquad (5.6)$$

where σ_{BT} is the strength of the substance near the melting point.

The absence of experimental data on σ_{BT} and the size of growing crystals makes the verification of expression (5.6) difficult.

In experiments with a transparent naphthalene-based alloy, the dimensions of growing crystals (the length and radius of needles) were estimated to be then compared with ultrasound intensity required for their dispersion. The size of growing crystals was varied by changing the conditions of crystallization (the temperature and the velocity of a heater).

The data obtained showed that ultrasound intensity necessary for the dispersion of crystals rises with the parameter $\gamma = (r/l)^2$ (Table 5.3).

Table 5.3   Relationship between the parameters of growing crystals and ultrasound intensity necessary for their dispersion

Crystal growth rate $U \times 10^3$, mm/s	Growing crystal length l, mm	Growing crystal radius r, mm	$\gamma \times 10^2$	Ultrasound intensity I, W/cm^2
5	0.2	0.05	6.25	30
16	0.4	0.06	2.25	20
25	0.6	0.08	1.78	15
50	1.0	0.12	1.44	10
100	1.6	0.20	1.55	10

Experiments involving ultrasonic vibration of organic substances showed that when they solidify with the formation of irregular front, the dispersion of crystals takes place even if vibrations have precavitation intensity. The intensity of vibrations necessary for crystal dispersion falls with increasing length and decreasing radius of crystals.

Experiments with tin-bismuth alloys, in which the front structure and refinement coefficient were estimated versus the rate of crystallization and temperature gradient steepness in the melt confirmed these results.

At specified crystallization rates and temperature gradients (in the absence of ultrasound), the crystallization front was uneven. The steepening of temperature gradient or diminishing the crystallization rate led to a reduction in the two-phase zone width, estimated from inflections on the temperature curve. As the rate of crystallization decreased, the front became more rough.

Precavitation ultrasonic vibrations made the temperature gradient somewhat steeper, reduced the two-phase zone width, and increased the crystallization front roughness. All this led to a slight grain refinement. The coefficient of refinement increased with the two-phase zone width. Raising the intensity of ultrasound to a cavitation level contributed to both crystallization front roughness and grain refinement.

At ultrasonic intensities exceeding the cavitation threshold, the width of the two-phase zone could not be estimated because of the absence of a distinct inflection on the cooling curves near liquidus.

Thus, experiments show that ultrasound fed to a crystallizing melt causes a refinement of its structure even in the absence of cavitation, the refinement being the greater the wider is the two-phase zone. Steeper temperature gradients and lower crystallization rates act to reduce the refinement coefficient.

As for the reason for crystal dispersion in the absence of cavitation, a contribution to this process may come from vibrational and radiation pressures, as well as from viscous and dynamic forces produced by acoustic streaming.

Estimations show that peak vibrational pressure at the plane crystallization front in a solidifying steel is about 10 MPa (0.1 kg/mm^2), if the intensity and frequency of vibrations are 1 W/cm^2 and 20 kHz, respectively [44]. In the case of low-melting-point metals (tin, bismuth, lead, cadmium, and zinc), this value is 0.7 MPa under the same experimental conditions.

Viscous forces also contribute to dispersion of growing crystals, especially in alloys forming two-phase zone during solidification.

G. Schmid and A. Rolls [50] derived expression for the friction force F_f arising between a melt and a cylindrical crystal needle of length l and radius r:

$$F_f = \frac{2rl\xi_{mL}\omega}{\pi\left(\ln\dfrac{4r}{\rho_s\xi_{mL}\omega} - 0.0772\right)} \tag{5.7}$$

Experiments with low-melting-point metals and their alloys allowed the authors to confirm the role of viscous forces in the ultrasonic refinement of melts (Figures 5.31 and 5.32).

Later, I. I. Teumin [44] obtained expression for viscous forces exerted on a crystal of radius r, growing at the crystallization front

$$F_f = \frac{3}{2}\frac{\eta\omega\xi_{ms}}{r}\sqrt{(1-\beta r)^2 + \beta r^2\left(1+\frac{2\beta r}{9}\right)^2}, \tag{5.8}$$

where

$$\beta = \left(\frac{1}{2}\frac{\omega\rho_L}{9}\right)^{1/2}.$$

Calculations performed for a 0.1-cm crystal of a low-melting-point metal irradiated with ultrasound of intensity 1 W/cm^2 and frequency 20 kHz indicated that pressure on growing crystals due to viscous forces is 0.005 MPa. This implies that viscous forces will disperse growing crystals if their strengths at temperatures near melting points are rather low.

Radiation pressure is much lower than vibrational pressure and viscous forces. In the case of low-melting-point metals, radiation pressure exerted on crystallization front by a normally incident plane sound wave is described by formula

$$P_r = 2\rho_L\omega_L^2\left[\frac{w_{0L}^2 + w_{0S}^2 + 2\rho_L\rho_S c_L^2}{(w_{0L} + w_{0S})^2}\right] \tag{5.9}$$

that gives a value of 5 Pa. Therefore, the contribution from radiation pressure to crystal dispersion is apparently not great.

In addition to causing dispersion of growing crystals, ultrasonic field may also change the rate of nucleation.

As shown by V. I. Danilov [39, 41], A. P. Kapustin [42], R. E. Berlaga [52], and F. S. Gorsky [53] in experiments with transparent organic substances (salol, thymol, betol, piperin, etc.), the ultrasonic irradiation of overcooled melts enhances the rate of nucleation.

In these experiments, substances were heated above their melting points, kept for some time in the overheated state, and then overcooled to a particular temperature, at which they were held until nucleation began. After that, the melts were irradiated by ultrasound.

Under specially chosen experimental conditions, nucleation was not observed for 5–8 h, but started immediately after the onset of ultrasonic irradiation so that in a few seconds almost the entire volume of the melt was solidified. It should, however, be emphasized that organic substances used in these experiments were not purified.

In experiments with purified salol or betol, their melts could readily be overcooled to a glassy state, in which nucleation was not observed at any overcool temperature until the onset of ultrasonic vibrations. This allowed V. I. Danilov to infer that ultrasound enhances the rate of nucleation in overcooled liquids. However, other authors found that low-frequency (48 Hz) vibrations could sharply decrease the rate of nucleation in overcooled betol [54] and have no influence on the rate of nucleation in purified thymol and benzophenol [55]. Attempts to elucidate the influence of ultrasound on the rate of nucleation were also undertaken by Hunt, Frawley, and co-workers [56, 57].

In work [58], the effect of ultrasound on the rate of nucleation was estimated from metastability threshold and the expected time necessary for the induction of nucleation in transparent organic substances. Experiments were carried out with betol, naphthalene, azobenzene, salol and thymol. The time in which the first nucleus appeared was determined for melts either untreated or irradiated by ultrasound. In the last case, the ultrasound was applied after the melt holding at a given overcool temperature for 0.5–1.5 min.

All the substances studied exhibited a profound metastability threshold. A small decrease in the overcooling temperature by 1–2°C substantially increased the time of appearance of the first nucleus. Thus, for betol, this time was 40 min at a overcool temperature ΔT_c of 42.5°C, whereas no nucleation was observed for betol kept at $\Delta T_c = 41.5$°C even for 5 h (Figure 5.38). Analogous results were obtained for other substances.

Ultrasound with the intensity 100 W/cm^{2*} provoked nucleation at lower overcool temperatures ΔT_u that differed for various substances (Table 5.4). In this case, the time of appearance of the first nucleus substantially decreased. It should be noted that the applied ultrasonic field produced cavitation in overcooled melts, although this was not a sufficient condition for nucleation.

---

*Hereinafter the intensities of vibrations are given as they are measured in a gaging waveguide without allowance for losses in the primary waveguide, radiator, and water bath.

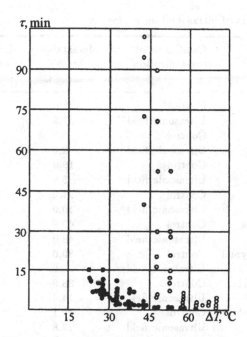

Figure 5.38. Effect of ultrasound on the metastability threshold of betol: τ is the expected time of appearance of the first nucleus in the melt overcooled by ΔT; (o) control experiments; (•) experiments with ultrasound.

Ultrasonically induced increase in the metastability threshold and decrease in the expected time of nucleation indicate that ultrasound enhances the rate of nucleation.

The effect of substance purity on metastability threshold was studied using commercially available chemically pure thymol, extra pure thymol, and thymol specially supplemented with insoluble impurities (powders of quartz, graphite, and boron carbide in an amount of 0.05 mass%). The influence of ultrasound on the metastability threshold of pure thymol was found to be insignificant, while the introduction of powdered quartz and graphite into thymol markedly reduced its metastability threshold.

Experiments with salol were performed to elucidate how the overheating of a melt influences its metastability threshold in an ultrasonic field. Increased overheat temperatures led to deactivation of insoluble impurities, which appeared just as a kind of substance purification bringing about a decrease in the metastability threshold. At the same time, the effect of ultrasound on the metastability threshold diminished with increasing purity of salol.

Table 5.4   Effect of ultrasound on nucleation.

Substance	Conditions of crystallization	Metastability threshold ΔT, °C	Expected time of nucleation τ, min
Betol	Control*	42.6	37
	Ultrasonic field**	22.6	6
Naphthalene	Control	8.0	6
	Ultrasonic field	2.0	0.1
Azobenzene	Control	16.0	2
	Ultrasonic field	5.0	0.1
Salol	Control	21.0	40
	Ultrasonic field	10.0	7
Purified thymol	Control	49.9	62
	Ultrasonic field	48.0	38
Nonpurified thymol	Control	30.0	54
	Ultrasonic field	20.0	12
Thymol + quartz	Control	38.9	55
	Ultrasonic field	14.4	2
Thymol + graphite	Control	39.1	58
	Ultrasonic field	13.8	1
Bismuth	Control	12.0	0.2
	Ultrasonic field	2.0	< 0.05
Antimony	Control	25.0	10
	Ultrasonic field	2.0	< 0.05

* Without ultrasonic irradiation.
** Ultrasonic field with an intensity sufficient to produce cavitation.

Since the visual observation of nucleation in metals is not possible because of their opaqueness, in [58] we studied the influence of ultrasound on the overcooling of bismuth and antimony melts that could be easily obtained in the overcooled state.

To obtain reproducible results, bismuth (240 g) was overheated by 130°C above the temperature of crystallization (271°C) and then overcooled by 12–32°C. The expected time of nucleation in bismuth, overcooled by 10°C, was 10–15 s. Irradiation of the melt with elastic vibrations (100 W/cm^2) was performed at temperatures 266 and 261°C. About 1–2 s after the onset of irradiation, the overcooled melt was found to crystallize irrespective of the extent of overcooling. If the bismuth melt was irradiated with ultrasound at the crystallization temperature and then was cooled without irradiation, no overcooling took place.

In experiments with antimony, 400 g of this metal was overheated to 700°C and then cothreshold oled in a furnace. The overcooling was found to be 25–40°C. In another experiment, 50 cm^3 of antimony melt, overcooled by 15°C, was kept in this state for 20 min. Common shaking, such as stirring with a porcelain rod, failed to cause crystallization of this overcooled melt, which, however, was crystallized in 1–2 s, when being exposed to ultrasonic irradiation (120 W/cm^2) at 625, 620, and 615°C. The time of melt holding in the overcooled state before the onset of ultrasonic vibration was varied from 0 to 15 min.

Thus, ultrasonic irradiation reduced the time of appearance of the first nucleus in overcooled antimony and bismuth by three and one order of magnitude, respectively. Like in the case of organic substances, this effect can be due to nucleation, promoted in overcooled metal melts by ultrasonic vibrations via cavitation.

The effect of cavitation can be explained as follows. Cavitation suggests the occurrence of a pulsating bubble in the melt. A drastic enlargement of this bubble during dilatation half-cycle period is accompanied by evaporation of liquid into the bubble. As a result, temperature inside it falls below the critical value, which will elicit overcooling of the melt at the bubble boundary. In turn, this will lead to the formation of a crystallization nucleus on the surface of the bubble. The nucleus may be detached from the bubble due to different velocities of liquid and solid phases. A shock wave resulting from the collapsing bubble will assist in removing the nucleus to the melt bulk.

The decrease in the temperature of a cavity caused by its expansion and evaporation of liquid was estimated by Cibula, Hunt, and Jackson [59, 60]. The temperature fall due to expansion can be found from the adiabatic conditions

$$\frac{\Delta T_s}{\Delta P} = \left(\frac{T}{C_p}\right)\left(\frac{\partial V}{\partial T}\right)_p = \frac{TV_m}{C_p}\alpha_V, \qquad (5.10)$$

where $\alpha_V = \frac{1}{V_m}\left(\frac{\partial V}{\partial T}\right)_p$ is the coefficient of thermal expansion.

A pressure drop induced by the expansion or collapse of the cavity can be estimated from the equation of cavity motion in an ultrasonic field. Hunt and Jackson [60] found that when pressure decreases to 10 Pa, the expansion of the cavity in water elicits a 20°C fall in its temperature; for benzene, this fall may be as large as 200°C.

Shock wave pressures may contribute to a shift in the melting point toward higher values for substances with $\Delta\rho = \rho_S - \rho_L > 0$ (Figure 5.39). Increase in pressure is equivalent to the overcooling of the melt and, hence, can enhance the rate of nucleation.

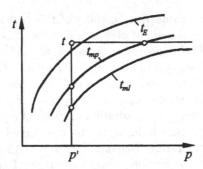

Figure 5.39.   Effect of ultrasound on the melting point and metastability level of metals with $\Delta\rho = \rho_S - \rho_L > 0$.

Of particular interest is the temperature, at which pressure begins to increase, and its rate. Even small pressures may be efficient, if the melt is cooled to a temperature close to T_0 [61]. Typically, the range of overcooling, in which the number of nuclei in the melt rises from unities to very large numbers, makes up some degrees. This is true for both homogeneous nucleation and nucleation on impurities. Therefore, if a melt is overcooled to the temperature at which nucleation may occur, the increase in the overcool temperature by some degrees (caused by elevated pressure due to cavity collapse) can substantially augment the number of nuclei.

Hickling [62] showed that cavity collapse in water may give rise to sufficiently large ice particles to serve as nuclei of crystallization.

It should be finally noted that part of the energy of a collapsing cavity is converted into thermal energy, which can heat the melt near the collapsing cavity and thus hinder nucleation.

It has been recognized by a number of authors that pure metals are more resistant to ultrasonic vibration than those containing impurities.

V. I. Danilov [39–41] was perhaps the first to emphasize the role of impurities in solidification of metals in an ultrasonic field. Investigations along this line were continued by A. P. Kapustin [42], Kh. S. Bagdasarov [55], I. I. Teumin *et al.* [63]. In particular, Teumin and co-workers were interested in how soluble impurities can affect the structural refinement of bismuth and zinc. In their experiments, sodium and magnesium were used as inoculants for bismuth and zinc, respectively. Vibrations (20 kHz) produced by a 3-kW ultrasonic generator were fed into a solidifying melt through a hole in the mould bottom. An inoculant was introduced into the melt directly before its casting

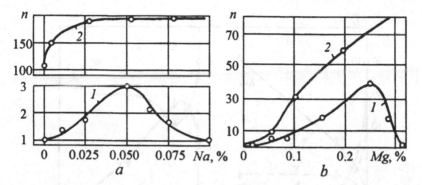

Figure 5.40. Effect of an inoculant and ultrasound on the grain refinement in (a) bismuth and (b) zinc either without (curves 1) or in the presence of ultrasound (curves 2).

performed at 330 and 480°C in cases of bismuth and zinc, respectively. Without ultrasound, the dependences of the grain number n on the concentration of impurity exhibited peaks (Figure 5.40,a,b, curves 1). The combined effect of the inoculant and ultrasound was many times greater than that produced by each of these factors (curve 2), which could be due to a more uniform distribution of the inoculant in the melt bulk and to the destruction of its layers adsorbed on the surface of growing crystals.

The mechanism of action of inoculants was studied by estimating the effect of soluble interfacially active impurities on the extent of metal refinement in an ultrasonic field [64, 65].

It is known that introduction of interfacially active substances into melts affects their overcooling [66]. For instance, if an inoculant reduces surface tension, it will diminish overcooling as well. In experiments with bismuth melts, sodium and potassium served as inoculants that reduced overcooling (Figure 5.41,a). The power of ultrasonic vibrations fed into a 100-cm^3 melt was 60 W. It was found that greater refinement coefficient K corresponds to smaller overcooling and, hence, to lower crystal–liquid surface tension (Figure 5.41,a).

Similar results were obtained in analogous experiments with an antimony melt and bismuth as a soluble impurity. Small concentrations of bismuth enhanced overcooling of antimony, but high concentrations reduced it. The refinement coefficient was found to change in the same concentration-dependent manner (Figure 5.41,b).

The mechanism of action of interfacially active inoculants on the structure of metals in an ultrasonic field was clarified by estimating

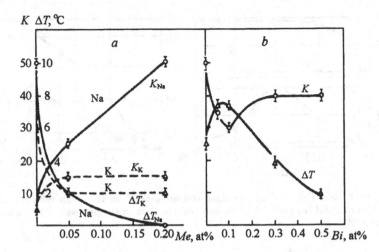

Figure 5.41.   Effect of interfacially active inoculants on the refinement coefficient K and overcool temperature ΔT for (a) bismuth and (b) antimony.

structural changes of the crystallization front in response to inoculant introduction. A setup shown in Figure 5.34 enabled a controlled crystallization of metals at specified velocity and temperature gradient in the liquid phase, as well as decantation of the crystallization front. Experiments were performed with molten chemically pure bismuth and sodium as an inoculant. The grain size and structural features of the crystallization front were estimated.

With the velocity of crystallization front in molten bismuth being $(25\text{–}175)\times10^{-3}$ mm/s and the temperature gradient in the liquid phase near the front ranging from 2.5 to 5°C/cm, the front was comparatively smooth with no evidence of surface roughness. The size of crystals protruding into the melt diminished with steepening temperature gradient and increasing front velocity.

Introduction of inoculant made the front rough, the roughness being increased with the inoculant concentration. Varying the front velocity within certain limits insignificantly influenced the size of crystals on the crystallization front (i.e., its roughness) and the dimensions of grains in the sample bulk. At the same time, steepening temperature gradient led to a noticeable refinement of grains and front crystals.

Irradiation of molten pure bismuth by ultrasonic vibrations of precavitation intensities (5 and 10 W/cm²) smoothed the crystallization front, without causing any changes in the grain size. Raising the intensity of vibrations reduced the size of crystals protruding inward the melt. Ultrasonic vibrations with intensity exceeding the cavitation threshold made the front more rough and the front crystals more large. The concurrent disintegration of crystals made the sample more fine-grained.

Similar structural changes were observed during the irradiation of inoculated melts. However, in this case, precavitation vibrations failed to smooth the crystallization front to the extent typical of pure bismuth.

Grains in the inoculated bismuth irradiated with postcavitation ultrasonic vibrations were much smaller than those in ultrasonically vibrated pure bismuth. Thus, the mean size of grains in inoculated bismuth (0.050% Na) solidified at a rate of 4 mm/min at a temperature gradient of 2.6°C/cm was 0.09 mm^2, whereas in pure bismuth solidified under the same conditions it comprised 0.15 mm^2.

Thus, the above experiments gave us an insight into the mechanism of action of ultrasound on grain refinement in metals conditioned by inoculants. The introduction of inoculants enhance the roughness of crystallization front, thereby facilitating the cavitation-mediated dispersion of crystals. This can explain the intense grain refinement in inoculated metals under the action of ultrasound reported by many authors.

Like soluble inoculants, insoluble inoculants may also significantly contribute to ultrasonic refinement of metals [58]. It was found that the threshold intensity of ultrasound eliciting grain refinement in metals inoculated with insoluble impurities was much lower than that for pure metals.

The grain refinement of inoculated metals treated with ultrasound was greater than in the case of pure metals irradiated with the same or even higher ultrasonic intensities. Thus, the refinement coefficient of pure tin irradiated with 120-W ultrasonic vibrations was approximately 20. At the same time, in the case of tin inoculated with 0.5 mass% SiO$_2$ and irradiated with 90-W vibrations, it ranged from 50 to 70 [67].

As for the mechanism of action of insoluble inoculants on the threshold intensity of ultrasound and refinement coefficient, two assumptions are possible:

(1) Inoculant may reduce the threshold of cavitation involved in the structural refinement of metals.

(2) Elastic vibrations activate the inoculant, due to which its particles become nucleation centers.

The first assumption can be proved by the following data. The introduction of 0.6 mass% of Al_2O_3 into molten aluminum reduced the threshold acoustic pressure of cavitation from 0.8 to 0.5 MPa [68].

In work [67], the cavitation threshold was measured for molten tin inoculated with 0.2 mass% of Al_2O_3, SiO_2, and NiO. To compare the thresholds of cavitation and the wettabilities of inoculants, their contact angles were measured. Cavitation threshold was found to be reciprocally related to contact angle. Thus, inoculates reduce the cavitation threshold of melts, thereby promoting grain refinement.

As for the activation of insoluble impurities in an ultrasonic field, one should dwell on the problem of ultrasonic metallization of solid particles by melts that fail to do this in the absence of ultrasound (see section 5.1.2).

Experiments were performed with molten tin inoculated with Al_2O_3, SiO_2, and W particles unable to affect the structure of tin in the absence of ultrasound.

It was found that ultrasonic field promotes the metallization of particles and their transformation into nucleation centers. The activation of impurities was indirectly confirmed in experiments on melt overheating. Thus, the overheating of ultrasonically treated molten tin inoculated with Al_2O_3 by 1–2°C above its crystallization temperature caused no deactivation of the inoculant, i.e., the solidified ingot remained fine-grained (Figure 5.42,a). At the same time, the identical overheating of the inoculated melt without ultrasound led to the ingot structure with larger grains (Figure 5.42,b).

The overheating of inoculated metal in the presence of ultrasound by 5–7°C above its crystallization point resulted in the coarse-grained structure. On the other hand, the irradiation of this melt with 45-W ultrasonic vibrations at temperature 1–2°C above its crystallization point caused the refinement of grains to the extent which was lower than that in the case of the melt ultrasonically irradiated until it completely solidified. These results confirm the assumption that ultrasonic fields favor the metallization of inoculant particles to become nucleation centers.

Speculating as to the possible mechanism of ultrasonically induced activation of inoculant particles, it can be suggested that elastic vibrations assist in removing various contaminations from particle's surface, whereas an intense motion of molten metal (due to microjets and difference in the vibrational velocities of liquid phase and particles) favors the filling of cracks on particle's surface with the liquid. In this case, the probability of an inoculant particle to become a

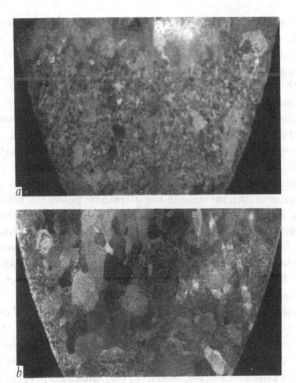

Figure 5.42. Macrostructure of (a) ultrasonically treated and (b) control ingots of tin inoculated with 0.5 mass% Al_2O_3 and remelted with 1–2°C overheat. Magnification × 2.

nucleation center is determined by its isomorphism with a solidifying metal. The cleaning of particles and the motion of liquid phase are associated with both cavitation and viscous forces. The involvement of these forces in the inoculant activation was demonstrated experimentally [67].

Depending on the conditions of crystallization, ultrasonic field may disperse growing crystals and increase the rate of nucleation. When temperature gradient in a melt is positive, the major effect of ultrasound is the dispersion of growing crystals. On the other hand, the ultrasonic irradiation of the overcooled melt enhances the rate of nucleation and apparently the dispersion of growing crystals. To estimate the contributions from dispersion and nucleation to the overall effect of ultrasound, molten organic substances and metals were crystallized at the negative temperature gradient in an ultrasonic field [51].

Molten substance was overcooled to a particular temperature exceeding the metastability threshold and then supplemented with seeds capable of inducing crystallization. Ultrasonic vibrations of a certain intensity were fed into the melt during its crystallization. In experiments with organic substances, the growth of crystals were observed under a microscope.

The rate of growth of purified thymol crystals was virtually independent of overcool, the crystallization front being smooth. Exposure of melts overcooled by 10, 20, and 30°C to ultrasonic vibrations with the intensity varying from 1 to 20 W/cm^2 affected neither nucleation not the front of crystallization, although somewhat increased the rate of crystal growth. An increase in the ultrasonic power fed into the melts overcooled by 10, 20, and 30°C initiated cavitation and thus elicited the dispersion of crystallization front. Exposure to ultrasound with an intensity of 25 W/cm^2 induced no nucleation in the melt overcooled by 10°C, whereas in the melt overcooled by 20°C, the formation of nuclei ahead the crystallization front was observed. The number of nuclei, arising near pulsating cavities, increased with overcool.

More intense ultrasonic vibrations at the same extent of overcooling augmented the relative number of crystals detached from the crystallization front in comparison with the number of arising nuclei.

With the intensity of vibrations lower than the cavitation threshold, the grains of pure metals (tin, bismuth, and antimony) crystallized from overcooled melts tended to enlarge. However, above the cavitation threshold, the grains became smaller with increasing overcool.

The fineness of grains in these experiments was higher than in the case of the positive temperature gradient in melts, which gave grounds for the supposition that ultrasonic treatment at negative temperature gradients not only disperses growing crystals, but also promotes nucleation.

Depending on substance and the conditions of its crystallization, either the dispersion of growing crystals or the intensification of nucleation may be determining in grain refinement. Regardless of the crystallization conditions, the grain refinement in pure metals can occur only in the presence of cavitation. If pure metals are crystallized at the positive temperature gradient, the major contribution to grain refinement comes from dispersion. In this case, grains become smaller at higher rates of crystal growth and steeper temperature gradients. If overcooled melts are crystallized, the considerable contribution to grain refinement comes from an increased rate of nucleation. In this case, the grain size decreases with an increase in overcool and temperature gradient.

It should be noted that the ultrasonic irradiation of alloys brings about grain refinement even in the absence of cavitation, although the extent of refinement in this case is not high. At the same time, raising ultrasonic power to the cavitation threshold drastically reduces the size of grains. A steeper temperature gradient and lower rate of crystallization shorten the two-phase zone and make the crystalline structure less fine. If overcooled alloys are crystallized, the major contribution to grain refinement comes from the increased rate of nucleation. As in the case of crystallization of pure overcooled metals, the increased rate of nucleation may be due to the activation of insoluble impurities in molten alloys.

Ultrasonic refinement is obviously determined not only by excessive pressures, but also by material strength and dimensions of growing crystals.

The elucidate how the ultrasonic treatability of materials is related to their properties, ultrasound efficiency threshold, and the refinement coefficient were estimated for some chemically pure low-melting-point metals (aluminum, bismuth, cadmium, tin, lead, antimony, and zinc) irradiated at an acoustic power of 600 W [69].

The comparison of ultrasound efficiency thresholds and the refinement coefficients of metals with their strengths and some other properties that might influence cavitation showed that the refinement coefficient is mainly determined by material strength. Its viscosity, surface tension, sound velocity, vapor pressure, and heat conductivity are of no great importance.

For more correct analysis of the results obtained, it is necessary to know the mechanical properties of metals near their melting points. Since no reliable relevant data are available, the mechanical properties of investigated metals at high temperatures were tentatively estimated by ball indentation into metals heated to the temperature 2–5°C below their melting points. The results obtained (Table 5.5) showed that stronger metals possess higher ultrasound efficiency thresholds and lower coefficient of refinement.

It should be noted that ultrasonic treatability of metals depends also on the intensity of cavitation in the melt: the coefficient of refinement grows with the pressure $P_{s,\max}$ produced by collapsing cavities and, hence, with the parameter $\beta = P_{s,\max}/H_T$.

In alloys, the dispersion of growing crystals may be governed not only by the strength of alloy, but also by the geometrical sizes of crystals. The latter are known to be related to the two-phase zone width and, hence, to the phase diagram of alloys (i.e., to the temperature range of crystallization and the coefficient of impurity distribution)

Table 5.5   Strength of materials and parameters of their ultrasonic treatability.

Metal	Strength σ_B at $T=20°C$, MPa *	Dent diameter d, at $T\sim T_m$, mm	Hardness H_T at $T\sim T_m$, MPa **	$\beta = \frac{P_{s,max}}{H_T}$	Ultrasound threshold N_t, W	Coefficient of refinement, K
Tin	0.12	1.75	0.042	27.9	350	30
Bismuth	0.05–0.2	1.8	0.039	27.2	60	70
Lead	0.15	2.20	0.026	35.8	250	20
Cadmium	0.63	0.75	0.227	4.3	400	10
Zinc	1.18	0.5	–	–	500	4
Antimony	0.86	1.7	–	–	300	8
Aluminum	1.27	1.6	–	–	400	15
Indium	–	1.72	0.045	15.0	200	24

* Data from [90].
** Bearing stress calculated by formula $H_T = 4F/\pi d^2$ at $F = 10$ N was taken as hardness.

and to the conditions of crystallization (the rate of crystal growth and temperature gradient steepness).

To substantiate the above supposition about the relationship between the ultrasonic treatability of a material and its properties, the ultrasonic treatabilities of carbon steels, aluminum-silicon alloys, and aluminum-copper alloys (determined from the ultrasound efficiency thresholds and grain sizes) were compared with their mechanical properties at temperatures close to melting points as well as with the two-phase zone widths calculated by formula [70]

$$\Delta l = \frac{\left(T_L - T_S - \frac{DG_L}{U}\right)\left(1 - S_L^{k-1}\right)k}{(1-k)G_L}, \qquad (5.11)$$

where D is the coefficient of diffusion; G_L is the temperature gradient near the crystallization front; k is the equilibrium coefficient of impurity distribution; S_L is the cross-sectional area of the liquid phase in the two-phase zone (in further calculations, $S_L = 0.1$).

Strengths of aluminum-based alloys near their melting points were taken from [71]. To compare the structural changes in steels with their mechanical properties, the latter were determined at temperatures close to solidus. In calculations, it was assumed that $G_L = 50°/cm$, $D = 5 \times 10^{-4}$ cm²/s, $U = 0.05$ cm/s. The results are summarized in Table 5.6.

Table 5.6 Ultrasonic treatability and related properties of steels and aluminum-based alloys.

Alloy	Strength σ_B at $T \sim T_s$, MPa	Impurity distribution factor, k_0	Temperature interval of crystallization δT, °C	Two-phase zone width Δl, cm	Coefficient $\alpha' = \Delta l / \sigma_{BT}$	Refinement coefficient K
Al–1% Cu	0.016	0.20	15	0.40	2.5	15
Al–2% Cu	0.030	0.50	65	2.80	9.3	27
Al–3% Cu	0.020	0.33	92	3.40	15.9	30
Al–5% Cu	0.018	0.30	82	2.80	15.0	25
Al–0.4% Si	0.008	0.12	26	0.46	5.8	20
Al–1% Si	0.012	0.10	50	0.77	6.4	25
Al–2% Si	0.006	0.07	70	1.13	18.8	27
Al–5% Si	0.005	0.20	50	1.32	26.4	32
Al–0.2% C	0.054	0.25	35	1.07	1.9	2
Al–0.4% C	0.056	0.25	47	1.44	2.6	3
Al–0.8% C	0.060	0.40	90	3.60	6.0	5
Al–1.0% C	0.057	0.40	110	4.35	7.8	6

It can be seen that aluminum-based alloys with a lower strength of growing crystals and wider two-phase zone (i.e. with a greater coefficient $\alpha' = \Delta l / \sigma_{BT}$) are more ultrasonically treatable as have greater refinement coefficients.

Analogous results were obtained for carbon steels. In particular, it was shown that the temperature interval of crystallization (and the two-phase zone width) increases with the carbon content of steels. Of two alloys with close strengths of growing crystals, an alloy with a wider two-phase zone possesses a greater coefficient of refinement. This fact explains a high treatability of single-phase alloys with a wide temperature interval of crystallization. Ultrasonic treatment produces a more significant grain refinement in aluminum-magnesium alloys and tin bronze than in brasses [72].

To verify the supposition as to the relationship between the treatability of alloys and their mechanical properties and temperature interval of crystallization, the authors of the work [73] made and ultrasonically processed, under identical conditions, steels and iron-based nickel–chromium alloys Cr18, Cr25, Cr18Ni4, Cr18Ni9Ti, Cr23Ni18, and Cr25Ni20, as well as nickel-based alloys Cr20Ni60, Cr20Ni80,

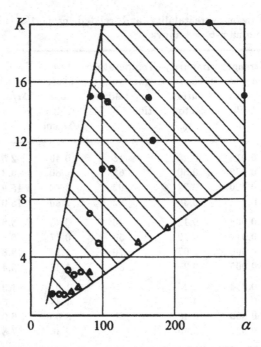

Figure 5.43. Relationship between material properties and the grain refinement coefficient K of (*1*) carbon steels, (*2*) austenitic and ferritic steels, and (*3*) nickel-based alloys.

CrNi77TiAlB, CrNi62VMoCoAl, CrNi70MoAlB, CrNi58VNiCoAl, CrNi5VMoCoAl, and CrNi51VMoTiAlCoW. The alloys were tested for their mechanical properties at high temperatures, and their temperature intervals of crystallization were determined.

As was already mentioned above, the size of crystals is proportional to the temperature interval of crystallization, $\delta T = T_L - T_s$. To account for both the strength and dimension of growing crystals, the authors introduced a parameter $\alpha = \delta T / \sigma_{BT}$ and compared it with the coefficient of structural refinement.

In spite of rather conventional character of such comparison, it could be inferred that the refinement coefficient increases with the parameter α (Figure 5.43), which testifies to an applicability of this approach for the qualitative estimation of ultrasonic treatability of materials.

It can be concluded that the ultrasonic treatability of materials strongly depends on their properties.

5.2   Ultrasonically Induced Phenomena at the Solid–Solid Interface

High-power ultrasound passing through the interface between two solid phases gives rise to certain phenomena at the interface and near it. In particular, the excitation of vibrations in one of the phases leads to its heating and plastic deformation. This, in turn, may affect the state of the interface.

Of great practical and theoretical interest are cases, when an interface is subjected to a combined effect of ultrasound and some other factors, such as static pressure, heating, evacuation, and external force. As this takes place, interfacial phenomena are strongly intensified, which underlies various technological applications of ultrasound, including ultrasonic welding, machining, impact hardening, softening, pressing, etc.

Ultrasonically induced interfacial physical processes were investigated at length by A. M. Mitskevich [74–77] and, as applied to metal processing, by V. P. Severdenko et al. [78], V. I. Petukhov et al. [79], and other authors [80, 81]. Ultrasonic welding, in which contact phenomena are essential, was studied by A. M. Mitskevich [74–77], Yu. V. Kholopov [82], Sh. Shvegla [93], and others [84–86].

This section deals with the effect of ultrasound on the structure of surface layers of contacting solid bodies, their attachment, and friction forces.

Let us consider some processes occurring in the zone of contact of two solid plates, one of which is applied with ultrasonic vibrations of the amplitude ξ_1 (Figure 5.44). The plates are pressed to each other by a constant force N. As this takes place, two regimes are possible – slip ($\xi_1 > \xi_{th}$) and no-slip ($\xi_1 < \xi_{th}$, where ξ_{th} is the ultimate permissible material microdisplacement in the contacting surfaces without their fracture). If there is no slip between the contacting surfaces, the initial interaction between them is defined by their microgeometry.

Figure 5.45,a gives the surface profile of one of the copper plates before ultrasonic irradiation. The pressing of such surfaces to each other brings about the contact and plastic deformation of the greatest unevennesses, since stresses at the sites of contact can be several times as high as the plastic strength of metals.

Ultrasonic vibrations (the direction of displacement is indicated by arrows in Figure 5.44) produce cyclic tangential stresses on contacting microunevennesses whose action, in combination with a static stress

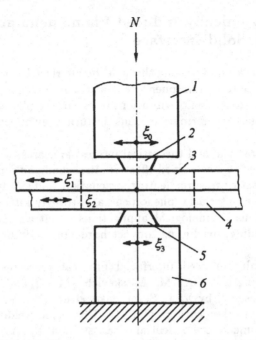

Figure 5.44.   Diagram illustrating the excitation of vibrations in the contact zone of two solids: (*1*) sonotrode, (*2*) radiator, (*3*) upper solid, (*4*) lower solid, (*5*) reflector.

$\sigma_u = N/S_u$ (S_u is the total contact area of microunevennesses), accelerates the plastic deformation of contacting microunevennesses, thereby promoting the approach of the surfaces and increasing the number of contacting microunevennesses. As a result, the contacting surfaces become more smooth. Figure 5.45,*b* presents a surface profile of one of the contacting copper plates after a short-term ultrasonic loading. It is seen that the surface in the region of contact became more smooth with the mean height of microunevennesses having been reduced from 2 to 0.2 μm and their mean radius having been increased from 30 to 230 μm [74].

In some works, the problem was treated theoretically. Thus, R. D. Mindlin [87] considered the problem of the contact of two equal-diameter spheres with radius R pressed by a central force N and subjected to a tangential, direction-variable force $F_\tau \leq f_s N$, where f_s is the coefficient of rest friction (Figure 5.46).

It can be shown that the contacting spheres are nonelastically displaced relative to each other in the limits of a ring, inside which there is no shift, although the specific tangential force f_τ is not zero.

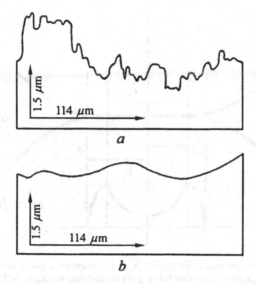

Figure 5.45. Surface profiles of copper plates (*a*) before and (*b*) after a 1-s ultrasonic irradiation with frequency 20 kHz and amplitude 10 μm [74].

Figure 5.46,*b* shows the distribution of forces f_τ, for which the following expressions are valid

$$a = 1.11 \left(\frac{NR}{E} \right)^{1/3}, \tag{5.12a}$$

$$\frac{a'}{a} = \left(1 - \frac{f_\tau}{f_s} \right)^{1/2}. \tag{5.12b}$$

Inside a ring of radius a', there occurs a shift. The force f_τ is maximum at $r = a'$.

The stronger the force F_τ, the smaller is the ring, inside of which inelastic displacement is absent.

Experiments, performed with ball-plane pairs possessing various ball radii, axial loads, and tangential forces $F \sin \omega t$, showed a good agreement with theory [91]. The author observed the formation of a wearing ring, narrow at small $f_s = F_\tau / N$. The internal radius of this ring diminished with increasing F_τ and f_τ in accordance with expression (5.12b). The surface inside the ring appeared to be polished, which was ascribed to plastic deformation of microunevennesses.

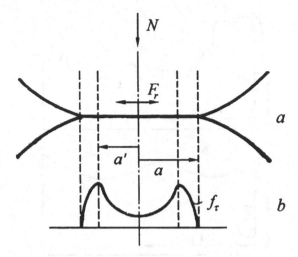

Figure 5.46. Schematic representation of (a) contact zone of two balls and (b) distribution of specific tangential force f a) in the contact zone [87].

Ultrasound was also found to modify the metallophysical properties of contacting surfaces. In particular, the microhardness of copper inside the contact ring increased by 10–70% after ultrasonic exposure [74]. Structural analysis of the surface in the contact zone showed that it represents an amorphous layer 10^{-2} to 10^{-3} μm in thickness, known as Beilby layer, with a great number of point and linear defects. Such a surface layer, containing a considerable amount of active centers, should favor the adhesion of contacting surfaces.

Thermal events are also essential in the deformation and adhesion of contacting surfaces. Exposure to ultrasound warms the contact zone (Figure 5.47), the extent of warming-up being determined by experimental conditions, in particular, by vibrational amplitude. Raising the pressing force leads to higher warming-up temperatures only to certain limits, above which warming would diminish.

Theoretical consideration of this phenomenon [76, 84] showed a good agreement with experiment, especially at the early stages of the process (a dashed line in Figure 5.47)

$$T(\tau) = \frac{4N f_s \omega \xi}{j\left[\dfrac{S_u \lambda_1}{(a_1 \tau/\pi)^{1/2}} + \dfrac{2S_1 \lambda_2}{r_0 \ln(4a_2 \tau/Cr_0^2)}\right]}, \qquad (5.13)$$

here, j is the mechanical equivalent of heat; $\xi = \xi_1 - \xi_2$ (see Figure 5.44); λ_i and a_i are the coefficients of heat and temperature con-

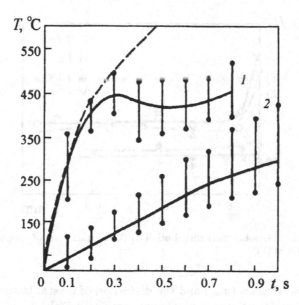

Figure 5.47. Time course of temperature in the contact zone of two 1-mm thick copper plates vibrated with frequency 20 kHz and amplitude (*1*) 16–17 and (*2*) 12–13 μm. Pressing force $N = 4400$ N [74].

ductivity of contacting materials; subscripts 1 and 2 correspond to upper and lower contacting plates, respectively (Figure 5.44); r_0 is the radius of the contact zone; S_1 is the cylindrical surface, through which heat propagates into the plates; $S_u = \pi r_0^2$; ln c is the Euler constant.

It follows from (5.13) that temperature in the contact zone rises with N and ξ, which agrees with the above-mentioned experimental data.

Analysis of processes occurring in the contact zone indicates the appearance of pickup zones on a polished surface, their number being increased with time. It was shown that pickup zones are oriented in a direction coinciding with the direction of vibrations [74].

As the number of pickup zones increases, they fuse to form a join region, in which various pickup zones have common grains. The occurrence of a specific strain texture may perhaps be observed here, and sometimes turbulent plastic flows may take place. The strength of material in this region is close to that in the material bulk, but prolonged ultrasonic treatment may reduce the microhardness of the metal near the interface (Figure 5.48).

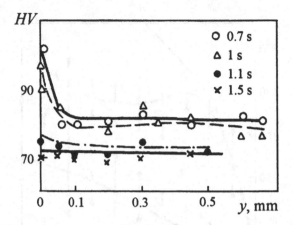

Figure 5.48. Microhardness distribution in the contact zone of copper vibrated for various times [74].

Residual stresses (σ_{res}) and the distortion of crystal lattice ($\Delta a/a$) in the join region of copper plates are significant [74]:

$$\sigma_{\text{res}} \sim 100 \pm 20 \text{ N}$$

$$\frac{\Delta a}{a} = 6.6 \times 10^{-4}.$$

At $\xi_1 \sim \xi_{th}$, the contact between metals may markedly affect diffusion processes, the effect being most pronounced when contacting metals are different. X-ray spectral analysis of the bond region of Cu and Ti plates indicated an interdiffusion of these metals [88]. Similar results were obtained during investigation of Cu–Al, steel-iron, and other pairs. If the contact of materials occurs at $\xi_1 > \xi_{th}$, the processes of polishing, adherence, and the formation of joint zone are not separated in time but proceed virtually simultaneously.

The interaction of contacting surfaces accompanied by friction is rather complex and has a dual, molecular and mechanical, nature. Molecular interaction is due to a mutual attraction of two solids, or their adhesion, whereas mechanical interaction is due to the interpenetration of the microunevennesses of contacting surfaces.

By now, the effect of ultrasound on contact friction is quite fully understood in the respect of slip kinematics.

Experiments were performed on a setup (Figure 5.49) consisting of a rotary disk and a waveguide-radiator system rotatable through any specified angle in both, vertical (positions I–II) and horizontal (positions III-IV), planes of contact of friction pairs. Waveguide (2) was

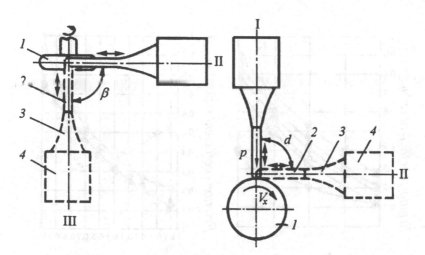

Figure 5.49. Schematic representation of a setup for investigating the effect of ultrasound on contact friction. The direction of ultrasonic wave propagation may be parallel to friction force and contact surface (position I), perpendicular to friction force and contact surface (position II), perpendicular to friction force and parallel to contact surface (position III): (*1*) disk, (*2*) sonotrode, (*3*) horn, (*4*) transducer.

pressed to the disk with a definite force, which enabled the normal pressure P across the contact surface to be determined. The friction force F_f, arising between the friction pairs, was estimated from the torque M on the disk shaft. Given the arm (the disk radius R), one could determine the friction force and the coefficient of friction

$$F_f = \frac{M}{R}, \quad f = \frac{F_f}{P}. \tag{5.14}$$

The ultrasonic frequency was 19.5 kHz. The displacement amplitude at the radiator, ξ_{m0}, was varied from 2 to 10 μm. Steel–lead, steel–copper, steel–aluminum, and steel–steel pairs were investigated.

The results are presented in Figure 5.50. The friction coefficient was most drastically reduced (by 75–90%), if the directions of vectors of vibrational velocity and friction force coincided ($\alpha = \beta = 0$, position II, Figure 5.49). With the angle α increasing from 0 to 90°, the reduction in friction coefficient, f_u/f, gradually diminished to comprise 0.7–0.9 at $\alpha = 90°$ (position I).

The quantity f_u/f behaved similarly when the angle β varied from 0 to 90°. The reduction in friction coefficient was 47–60% at $\beta = 90°$ (position III).

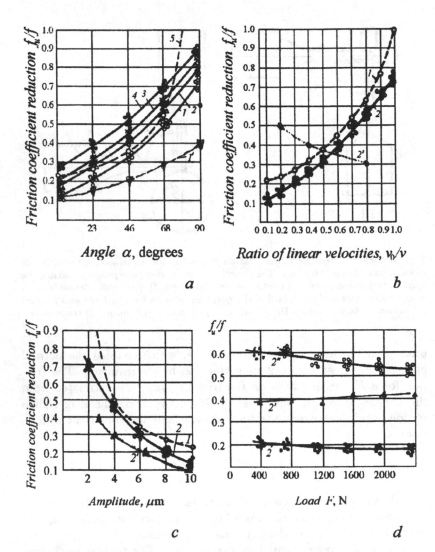

Figure 5.50. Reduction in the contact friction coefficient as a result of variation of: (*a*) angle between the vectors of vibrational velocity and friction force ($F = 140$ N, $\xi_m = 10$ μm, $v_m = 0.1$ m/s, (*1*)–(*5*) angle α, (*1′*) angle β); (*b*) ratio v_0/v_m ($\beta = 0$, (*1*) $a = 0$, $F = 140$ N; (*2*) $\alpha = 0$, $F = 400$ N; (*2′*) $\alpha = 90°$, $F = 400$ N); (*c*) displacement amplitude ξ_m ((*1*) $\alpha = \beta = 0$, $v_0 = 0.1$ m/s, $F = 140$ N; (*2*) $\alpha = \beta = 0$, $v_0 = 0.23$ m/s, $F = 400$ N; (*2′*) the same as (*2*) but with a lubricant); (*d*) load ($v_0 = 2$ m/s, $\xi_m = 7$ μm, (*2*) $\alpha = \beta = 0$; (*2′*) $\alpha = 0$, $\beta = 90°$; (*2″*) $\alpha = 90°$, $\beta = 0$). Contacting pairs: (*1*), (*1′*) steel–lead; (*2*), (*2′*), (*2″*) steel–copper; (*3*) steel–steel; (*4*) steel–aluminum; (*5*) steel–lead (calculations).

The effect of ultrasound on the friction coefficient was the more severe, the lower was this coefficient.

The amplitude ξ_{m0} and the ratio v_0/v_m (v_m is the amplitude of vibrational velocity and v_0 is the linear velocity on the disk surface) were varied (the ratio v_0/v_m was varied by changing both v_m and v_0). It was found that the effect of ultrasound on the contact friction increased with the displacement amplitude and diminished with increasing ratio v_0/v_m at $\alpha = \beta = 0$ and at $\alpha = 0$, $\beta = 90°$. At $\alpha = 90°$ and $\beta = 0$ (position I), the efficiency of ultrasound increased with the slip rate (Figure 5.50, curve $2'$) [78].

With the amplitude of vibrations above 18 μm, the seizure of contacting steel–copper and steel–steel pairs occurred at lower slip rates and amplitude of vibrations than in the case of position III (Figure 5.49), i.e., when the vibrational velocity vector was perpendicular to friction force and parallel to contact surface. Seizure drastically increased the coefficient of friction and reduced the efficiency of ultrasound.

The efficiency of ultrasound was high even if contacting surfaces were lubricated with a mineral oil (Figure 5.50,c).

Stronger forces F reduced the coefficient of friction, whether or not ultrasound was applied [78, 85], although the ratio f_u/f was virtually constant for F ranging from 50 to 2000 N. In other words, the effect of elastic vibrations on contact friction was independent of the pressing force in the range mentioned (Figure 5.50,d).

The effect of ultrasound on contact friction was found to be determined by slip kinematic, the state of contacting surfaces, characteristics of their interaction, and the temperature in contact zone (which, in turn, is influenced by ultrasound). The effect of kinematic factor on the ultrasonically induced reduction in friction coefficient was basically explained by Verderevskii et al. [80] who derived the following expression for the friction coefficient reduction

$$\frac{f_u}{f} = \frac{\pi}{2} \frac{1}{\arcsin \dfrac{v_0}{\omega \xi_{m0}}}. \tag{5.15}$$

It can be seen that f_u/f does not depend on the friction coefficient f, i.e., on the properties of contacting surfaces, but is dependent on the linear velocity v_0, displacement amplitude ξ_m, and frequency ω. The calculated and experimental values for f_u/f show a substantial disagreement, probably because in calculations, friction force was assumed to be independent of slip velocity [80].

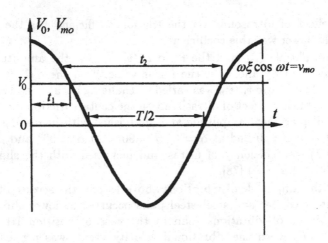

Figure 5.51.   Time course of slip rate during a cycle period.

The relationship of these parameters was taken into account in calculations performed by Petukhov *et al.* [79].

Based on experimental data, Kragel'skii [89] derived the following empirical formula

$$f = (l + bv)e^{-cv} + d, \qquad (5.16)$$

where v is the slip velocity; l, b, c, and d are constants.

At relatively small slip velocities

$$f = f_s - \alpha|v|, \qquad (5.17)$$

where f_s is the friction coefficient at $v \sim 0$; α is the coefficient of proportionality; $v = \omega\xi_m \cos \omega t - v_0 \cos \alpha$.

Let us consider how v changes during a vibrational period provided that $v_m > v_0$ (Figure 5.51). According to the diagram in this figure, the work done by a friction force over the time interval $0 - t_1$ is given by

$$A_1 = -\int_0^{t_1} Fvf \, d\tau. \qquad (5.18a)$$

The sign minus indicates that the friction force and the direction of motion are opposite.

Over the time interval $t_2 - t_1$, the directions of force and motion coincide; therefore

$$A_2 = \int_{t_1}^{t_2} Fvf \, d\tau. \qquad (5.18b)$$

At $t_2 < \tau < T$

$$A_3 = - \int_{t_2}^{T} Fvf \, d\tau. \qquad (5.18c)$$

Without ultrasound, the work of the friction force is

$$A = - \int_{0}^{T} Fv_0(f_s - \alpha v_0) \, d\tau. \qquad (5.19)$$

The ratio $(A_1 + A_2 + A_3)/A$ determines the extent to which ultrasound reduces friction coefficient.

Taking into account the behavior of v over the vibrational cycle period, one can write

$$\frac{f_u}{f} = 1 - \frac{\dfrac{2}{\pi} f_s \arccos \left(\dfrac{v_0 \cos \alpha}{\omega \xi_m} \right)}{f_s - \alpha v_0} +$$

$$+ \frac{\dfrac{\omega}{\pi} \alpha \xi_m \left(1 - \dfrac{v_0 \cos \alpha}{\omega \xi_m} \right) \sqrt{1 - \left(\dfrac{v_0 \cos \alpha}{\omega \xi_m} \right)^2}}{f_s - \alpha v_0}. \qquad (5.20)$$

Based on formula (5.17), the ratio f_u/f was calculated for the steel-lead pair (Figure 5.50, dashed lines). It can be easily seen that the calculated values agree well with experimental results.

It can be inferred from expression (5.20) that ultrasonically induced reduction in the friction coefficient diminishes with increasing friction coefficient f_0, which is also in agreement with experimental data.

Ones the vibrational velocity vector happens to be perpendicular to the friction force vector and the surface of contact (Figure 5.49, position III), the interaction of contacting plates may be accompanied by regular breakings of the contact. As this takes place, stresses in the contact zone change periodically from the maximum values (at the time of the closest contact) to zero (when there is a gap between contacting surfaces). It was shown [81] that the maximum stress occurs at $t_k \sim 0.75$ T, being twice as high as the static stress.

Analysis of this situation allowed Severdenko *et al.* [78] to derive expressions providing a good description of the effect of ultrasound on friction forces.

References

1. E. G. Konovalov and I. N. Germanovich, *Dokl. Akad. Nauk Belorus. SSR*, 6, 492 (1962).

2. Yu. P. Rozin and V. S. Tikhonova, *Kolloid. Zh.*, **31**, 568 (1969).

3. Yu. P. Rozin and V. S. Tikhonova, *Ultrazvuk. Tekhnika*, no. 3, p. 76 (1969).

4. V. I. Drozhalova and B. A. Artamonov, *Ultrasonic Impregnation of Workpieces* (in Russian), Mashinostroenie, Moscow, (1980).

5. Yu. I. Kitaigorodskii and V. I. Drozhalova, In: *Advances in Ultrasonic Engineering and Technology* (in Russian), Mashprom, Moscow, p. 21 (1974).

6. P. P. Prokhorenko, N. V. Dezhkunov, and E. G. Konovalov, *Ultrasonic Capillary Effect* (in Russian), Nauka & Tekhnika, Minsk, (1981).

7. F. A. Bronin, In: *Intensification of Metal Recovery from Ores in an Ultrasonic Field* (in Russian), Metallurgiya, Moscow, p. 31 (1969).

8. V. G. Churin, N. N. Dubrovin, and O. V. Abramov, In: *Physical and Physicochemical Methods in Technological Processes* (in Russian), Metallurgiya, Moscow, p. 35 (1986).

9. Yu. A. Minaev, A. A. Shevchenok, and O. V. Abramov, *Izv. Vuzov. Ser. Chernaya Metallurgiya*, no. 5, 141 (1986).

10. M. P. Alieva, E. E. Glikman, and Yu. V. Goryunov, *Fiz. Khim. Obrab. Mater.*, no. 1, 130 (1973).

11. O. A. Shaposhnikov, V. G. Dvoryanchikov, and Yu. A. Kudrin, *Elektron. Tekhnika*, Ser. 7, no. 1 (92), 164 (1979).

12. V. A. Labunov and N. I. Danilovich, *Fiz. Khim. Obrab. Mater.*, no. 4, 56 (1976).

13. *Handbook of Chemistry and Physics*, (1981).

14. *Handbook of Chemistry*, Izdat. Khim. Literatury, Moscow–Leningrad, (1962).

15. V. A. Labunov, N. I. Danilovich, and N. N. Leshchenko, *Zavodskaya Lab.*, no. 7, 31 (1976).

16. V. V. Klubovich, M. D. Tyavlovskii, and V. L. Lagin, *Ultrasonic Soldering in Radioengineering and Instrument Making* (in Russian), Nauka & Tekhnika, Minsk (1985).

17. A. A. Shevchenok, Yu. A. Minaev, and N. T. Konovalov, *Izv. Vuzov, Ser. Chernaya Metallurgiya*, no. 7, 35 (1986).

18. B. S. Bokshtein, *Diffusion in Metals* (in Russian), Metallurgiya, Moscow (1978).

19. N. N. Khavskii, A. A. Shevchenok, and O. V. Abramov, *Izv. Vuzov, Ser. Tsvetnaya Metallurgiya*, **3**, 117 (1986).

20. S. I. Pugachev and N. G. Semenova, In: *Interfacial Effects of High-Power Ultrasound in Metals* (in Russian), Nauka, Moscow, p. 72 (1986).

21. A. A. Prokhorenko, S. I. Pugachev, and N. G. Semenova, *Ultrasonic Metallization of Metals* (in Russian), Nauka & Tekhnika, Minsk (1984).

22. N. V. Korchevskii and O. V. Abramov, *Zh. Neorg. Khim.* (in press).

23. V. K. Semenchenko, *Superficial Phenomena in Metals and Alloys* (in Russian), GITL, Moscow (1957).

24. E. G. Konovalov and Zh. S. Vorob'eva, *Dokl. Akad. Nauk Belorus. SSR*, **18**, 710 (1974).

25. V. I. Dobatkin and G. I. Eskin, In: *Interfacial Effects of High-Power Ultrasound in Metals* (in Russian), Nauka, Moscow, p. 6 (1986).

26. O. K. Eknadiosyants, In: *Physical Fundamentals of Ultrasonic Technology* (in Russian), Nauka, Moscow, p. 337 (1970).

27. S. V. Lashko and N. F. Lashko, *Soldering of Metals* (in Russian), Mashinostroenie, Moscow (1988).

28. B. A. Agranat, V. I. Bashkirov, and Yu. I. Kitaigorodskii, In: *Physical Fundamentals of Ultrasonic Technology* (in Russian), Nauka, Moscow, p. 166 (1970).

29. B. A. Agranat and V. I. Bashkirov, *Dokl. Akad. Nauk SSSR*, **179**, 72 (1968).

30. B. A. Agranat, M. N. Dubrovin, and N. N. Khavskii, *Physical and Technical Aspects of Ultrasound* (in Russian), Vysshaya Shkola, Moscow (1987).

31. S. A. Nedyuzhii, *Akust. Zh.*, **10** (1964).

32. B. G. Novitskii, *Employment of Acoustic Vibrations in Chemical Engineering Processes* (in Russian), Khimiya, Moscow, p. 191 (1983).

33. A. V. Il'in, V. P. Kuznetsov, B. G. Novitskii, and V. M. Fridman, *Akust. Zh.*, **18**, 37 (1972).

34. D. K. Chernov, *Physical Metallurgy* (in Russian), Metallurgizdat, Moscow (1960).

35. G. F. Balandin, *Crystalline Structure of Castings* (in Russian), Mashinostroenie, Moscow (1973).

36. G. I. Pogodin-Alekseev, *Ultrasound and Low-Frequency Vibration in Production of Alloys* (in Russian), Mashgiz, Moscow (1961).

37. A. A. Romanov, *Vibrating-Mould Casting* (in Russian), Mashgiz, Moscow (1959).

38. R. Wood and A. Loomis, *Phyl. Mag.*, **4**, 417 (1927).

39. V. I. Danilov, E. E. Pluzhnik, and B. M. Teverovskii, *Zh. Eksper. Teor. Fiziki*, **9**, 66 (1939).

40. V. I. Danilov and B. M. Teverovskii, *Zh. Eksp. Teor. Fiziki*, **10**, 1305 (1940).

41. V. I. Danilov and G. Kh. Chedzhemov, In: *Problems of Metallurgy and Physics of Metals*, no. 4 (in Russian), Metallurgizdat, Moscow, p. 34 (1955).

42. A. P. Kapustin, *Effect of Ultrasound on Kinetics of Crystallization*, Izdat. AN SSSR, Moscow, (1962).

43. S. Ya. Sokolov, *Zh. Tekh. Fiziki*, **3**, 81 (1936).

44. I. I. Teumin, In: *Problems of Metallurgy and Physics of Metals*, no. 7 (in Russian), Metallurgizdat, Moscow, p. 376 (1962).

45. I. G. Polotskii, *Application of Ultrasonic Vibrations for Investigation, Control, and Treatment of Metals and Alloys* (in Russian), Izdat. AN Ukrain. SSR (1960).

46. H. Seeman, H. Staats, and K. G. Pretor, *Arch. F. Eisenhutten wessen*, **38**, 257 (1967).

47. G. I. Eskin, *Ultrasonic Treatment of Molten Aluminum* (in Russian), Metallurgiya, Moscow (1972).

48. O. V. Abramov, *Crystallization of Metals in an Ultrasonic Field* (in Russian), Metallurgiya, Moscow (1972).

49. J. Campbell, *Int. Metals Reviews*, **2**, 71 (1981).

50. G. Schmid and A. Rolls, *Z. Electrochem.*, **46**, 653 (1940).

51. O. V. Abramov, In: *Growth and Imperfections of Metal Crystals* (in Russian), Metallurgiya, Moscow, p. 326 (1966).

52. R. Ya. Berlaga, *Zh. Eksper. Teor. Fiziki*, **16**, 647 (1946).

53. F. S. Gorskii, *Izv. AN Belorus. SSR*, **6**, 117 (1950).

54. G. L. Mikhnevich and P. I. Dombrovskii, *Zh. Eksper. Teor. Fiziki*, **10**, 252 (1940).

55. Kh. S. Bagdasarov, *Kristallografiya*, **1**, 61 (1956).

56. J. D. Hunt and K. A. Jackson, *J. Appl. Phys.*, **37**, 254 (1966).

57. I. I. Frawley and W. J. Childs, *Trans. Metal Soc. of AIME*, **242**, 736 (1968).

58. O. V. Abramov, In: *Crystallization and Phase Transitions* (in Russian), Izdat. AN Belorus. SSR, p. 358 (1962).

59. A. Cibula, *J. Inst. Metals*, **95**, 159 (1967).

60. J. D. Hunt and K. A. Jackson, *J. Appl. Phys.*, **37**, 254 (1966).

61. I. L. Aptekar' and D. S. Kamenetskaya, *Fiz. Mekh. Matem.*, **14**, 316 (1962).

62. R. Hickling, *Nature*, **206**, 915 (1965).
63. I. I. Teumin, V. E. Neimark, and M. Ya. Fishkis, *Liteinoe Proizvodstvo*, **6**, 31 (1962).
64. O. V. Abramov, In: *Interfacial Effects of High-Power Ultrasound in Metals* (in Russian), Nauka, Moscow, p. 52 (1986).
65. O. V. Abramov and I. I. Teumin, In: *Physical Fundamentals of Ultrasonic Technology* (in Russian), Nauka, Moscow, p. 327 (1970).
66. P. P. Pugachev and I. P. Altynov, *Dokl. Akad. Nauk SSSR*, **86**, 117 (1952).
67. O. V. Abramov and I. I. Teumin, In: *The Mechanism and Kinetics of Crystallization* (in Russian), Nauka & Tekhnika, Minsk, p. 258 (1964).
68. V. I. Dobatkin and G. I. Eskin, In: *Interfacial Effects of High-Power Ultrasound in Metals* (in Russian), Nauka, Moscow, p. 3 (1986).
69. O. V. Abramov and I. I. Teumin, *Fiz. Met. Metallov.*, **18**, 87 (1964).
70. V. T. Borisov, *Dokl. Akad. Nauk SSSR*, **136**, 583 (1960).
71. G. F. Balandin and Yu. P. Yakovlev, *Fiz. Met. Metallov.*, **13**, 436 (1962).
72. O. V. Abramov, In: *Employment of Ultrasound in Engineering* (in Russian), Nauka & Tekhnika, Minsk, p. 10 (1964).
73. N. N. Dmitriev, O. V. Abramov, and V. P. Kudel'lkin, *Izv. Akad. Nauk SSSR, Ser. Metally*, no. 4, 102 (1977).
74. A. M. Mitskevich, In: *Physical Fundamentals of Ultrasonic Technology* (in Russian), Nauka. Moscow, p. 70 (1970).
75. A. M. Mitskevich and Yu. V. Kholopov, *Ultrazvuk. Tekhnika*, no. 5, 22 (1964).
76. A. V. Kulemin and A. M. Mitskevich, *Ultrazvuk. Tekhnika*, no. 2, 7 (1966).
77. A. M. Mitskevich, S. K. Ginzburg, and Yu. V. Kholopov, *Ultrazvuk. Tekhnika*, no. 2, 15 (1966).
78. V. P. Severdenko, V. V. Klubovich, and A. V. Stepanenko, *Ultrasonic Shaping of Metals* (in Russian), Nauka & Tekhnika, Minsk (1973).
79. V. I. Petukhov, O. V. Abramov, and A. V. Kulemin, In: *Application of Novel Physical Methods for Intensification of Metallurgic Processes* (in Russian), Metallurgiya, Moscow, p. 211 (1974).
80. V. A. Verderevskii, V. V. Nosal', and O. M. Rymsha, *Ultrazvuk. Tekhnika*, no. 5, 18 (1964).
81. A. I. Markov and T. G. Suvorova, In: *Machining of New Construction Materials and Diamond Instruments* (in Russian),

Proc. of Moscow Institute of Aviation, no. 402, p. 45 (1977).

82. Yu. V. Kholopov, *Ultrasonic Welding* (in Russian),
     Mashinostroenie, Moscow (1972).

83. Š. Švegla, In: *The World Electrotechnical Congress*, Moscow,
     section 4B, report 27 (1977).

84. L. L. Silin, G. F. Balandin, and M. G. Kochan, *Ultrasonic
     Welding* (in Russian), Energiya, Moscow (1962).

85. V. M. Koloshenko, *Ultrasonic Microwelding* (in Russian), Nauka
     & Tekhnika, Minsk (1974).

86. S. S. Volkov, Ya. N. Orlov, and B. Ya. Cherpak, *Ultrasonic
     Welding of Plastics* (in Russian), Energiya, Moscow (1974).

87. R. D. Mindlin, *J. Appl. Mech.*, **19**, 259 (1949).

88. M. Okada, *J. Jap. Inst. Metals*, **26**, 585 (1962).

89. I. V. Kragel'skii, *Friction and Wear* (in Russian),
     Mashinostroenie, Moscow (1968).

90. *Physicochemical Properties of Elements*, ed. G. V. Samsonov,
     Naukova Dumka, Kiev (1965).

91. K. L. Johnston, *Proc. Roy. Soc.*, **A230**, 531 (1955).

Part II

Ultrasonic Technology

The technical aspects of ultrasound – creation of new generators, development of methods for calculating complex oscillatory systems, and the ways for efficient transmission of vibrations to a varying load – are now a matter of extensive research. The extent of their development largely determines the efficiency of high-intensity ultrasound applications in production and treatment of metals and in other industrial processes.

Part 2 of the book is concerned with the design principles of ultrasonic installations, mechanoacoustic radiators, and other vibratory systems, as well as the control of acoustic parameters.

Chapter 6

Sources of Ultrasonic Energy

Practical applications of high-intensity ultrasound are largely determined by technical feasibilities for generation and transmission of vibrations of a certain intensity and frequency.

Principally, the industrial ultrasonic installation involves a transducer converting a particular type of energy (electric or mechanical) into acoustic energy, power supply, radiator, and various systems for providing measurement and control of acoustic parameters. The processed medium serves as a load (Figure 6.1).

At present, electroacoustic transducers are most widely used, in particular, those based on magnetostrictive or piezoelectric effects.

Magnetostriction was discovered in 1842 by J. Joule who observed changes in the dimensions of a ferromagnetic body elicited by changes in its magnetic state.

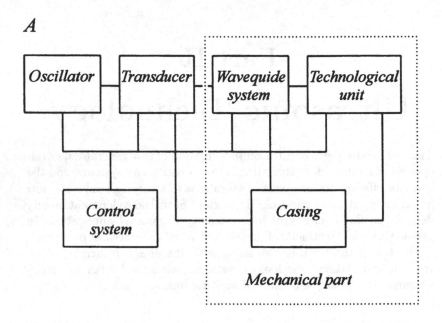

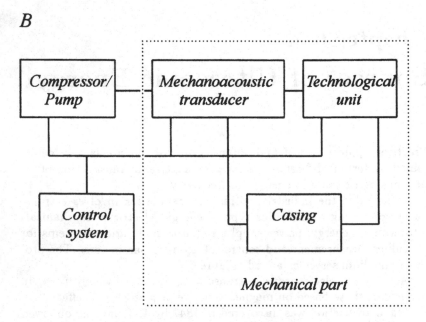

Figure 6.1. Block diagrams of industrial ultrasonic installations: (*a*) installation with an electroacoustic transducer; (*b*) installation with a mechanoacoustic transducer.

The piezoelectric effect lies in the generation of electric polarization as a result of the application of mechanical stresses; the reverse effect lies in the production of strains in some anisotropic insulators and semiconductors applied with an electric field. This effect was discovered in 1880 by P. Curie for quartz crystals.

To date, a diversity of magnetostriction and piezoelectric materials have been obtained, which are highly efficient in converting one type of energy into another.

Electromagnetic and electrodynamic transducers are comparatively uncommon in ultrasonic technology.

Other physical principles underlie the operation of mechanoacoustic transducers that convert mechanical energy (e.g., the energy of a compressed gas) into acoustic energy. In this case, vibrations are generated as a result of the collision of gaseous or liquid flows with an obstacle.

Electric generators, as well as various pumps and compressors, serve as power supplies to drive ultrasonic transducers.

6.1   Power Supplies for Ultrasonic Transducers

According to desired parameters of ultrasonic vibrations in gaseous and liquid media, mechanoacoustic transducers can be driven by pumps and compressors operated at various pressures and powers. As a rule, there are no specific "ultrasonic" demands that would imposed on these pumps and compressors.

Electroacoustic transducers can be energized by high-frequency motor-driven, vacuum-tube (going out of current use), or semiconductor oscillators, which feature their own electric, technical, and performance parameters.

The motor-driven oscillator is a device consisting of a power-line-frequency motor and a high-frequency generator with a single shaft. Such oscillators are coupled to electroacoustic transducers either directly or via special frequency multipliers. Motor-driven oscillators are rather efficient, reliable, and easy-to-use devices that have, however, shortcomings associated with the difficulty of operation frequency control.

The design principles of ultrasonic vacuum-tube and semiconductor oscillators are analogous to those employed in radio engineering.

Relatively low-frequency (up to 50 kHz) oscillators with an output of 0.025 to 10 kW are most common in ultrasonic technology.

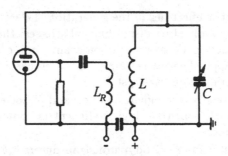

Figure 6.2.   Circuit diagram of a self-oscillator with inductive feedback.

The loading of oscillators by magnetostrictive or piezoelectric transducers specifies some requirements for the oscillators. Thus, designers
try to obtain their maximum efficiency and stable operation over a wide
range of operation frequencies, as well as to provide the possibility of a
continuous control of operation frequency and power output. Besides,
the coupling of an oscillator to a transducer should be easy.

Vacuum-tube oscillators can be easily tuned to provide a wide range
of operation parameters; however, these power supplies possess a substantial disadvantage related to their relatively low efficiency.

A one-stage tube self-oscillator involves one or two tubes connected
in push-pull circuit. The simplest scheme of such oscillator is that with
an inductive feedback (Figure 6.2). An oscillatory circuit, consisting
of inductor L and capacitor C, is connected to anode and, through
feedback coil L_R, to the tube grid. Part of the energy of the oscillatory circuit is continuously delivered to the feedback coil that controls
the tube. The tube, in turn, delivers to the oscillatory circuit such
an amount of energy that compensates for the energy consumed by a
loading transducer.

In multistage oscillators, the first, master stage generates high-
frequency, low-intensity vibrations that are gradually amplified by gain
stages to reach, in the output stage, an amplitude sufficient for the
loading transducer to be driven.

Injection-driven oscillators are convenient as providing a high frequency stability and continuous output tuning. The loading transducer
would not alter the operation frequency of such oscillators, since it is
controlled by the master self-oscillator.

Recent trend toward increasing use of transistor and thyristor driving oscillators is due to their efficiency, reliability, compactness, and
low weight [1].

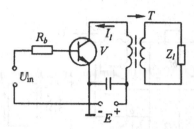

Figure 6.3. Circuit diagram of a transistor injection-driven oscillator [1].

The circuit diagram of an injection-driven transistor oscillator is shown in Figure 6.3. Transistor V is controlled by the voltage U_{in} across its base. The load Z_l is connected to a collector through transformer T. The values for load resistance, supply voltage E, and control voltage U_{in} are chosen so as to provide the oscillator operation in a flip-flop mode (regime D). In this case, the transistor operates in the pinch-off and cut-off portions of its characteristic, when energy dissipation in the collector circuit is minimum, whereas the efficiency of operation is maximum (85–98%).

Transistor oscillators with the powers 50–5000 W are usually designed in push-pull circuit, and those with the powers 100–1000 W – in half-bridge circuit (Figure 6.4).

Power supply E is connected to a half-bridge, whereas transistors $V1$, $V2$, and output transformer $T2$ are connected between junction points cd and ab, respectively. The exciting voltage from transformer $T1$ is supplied to the transistors in antiphase. The load Z_l is connected to transformer $T2$ through correction filter CF.

High-power oscillators are usually designed in circuits allowing the summation of the powers of constituting stages.

Apart from transistor oscillators, the present-day technologies make a wide use of thyristor oscillators.

The ultrasonic thyristor generator (Figure 6.5) involves a supply voltage filter, controlled rectifier, direct-current filter, autoinverter, matching high-frequency transformer, compensating ultrasonic-frequency filter, as well as a control system and a circuit for the magnetic biasing of a magnetostrictive transducer. The transducer should be expediently designed in a plug-in mode to provide its exchange by a piezoceramic transducer. The filters of supply voltage and ultrasonic frequency are necessary to match the generator with a load and supply line. The controlled rectifier serves as a power regulator, because of considerable differences in the transients occurring in the autoinverter

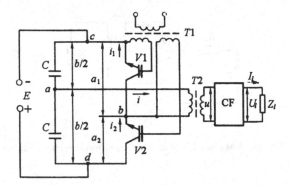

Figure 6.4. Circuit diagram of a half-bridge push-pull transistor oscillator [1].

and the load. Simultaneously, together with direct-current filter chokes, the rectifier protects the generator in cases of the inverter failure and other troubles.

Control systems in industrial ultrasonic generators should provide normal operation of transducers, maintenance of their nominal electric parameters, and high efficiency. The functional capabilities of present-day generators can be improved by employing mean-scale and large-scale integrated circuits. In particular, industrial ultrasonic installations should be programmable and capable of recognizing malfunctions, automatic switching off and on, actuating a reserve, evaluating and memorizing technological parameters, and so forth. This can well be approached by using microprocessors and integrating such ultrasonic installations into computer-aided production lines.

At present, ultrasonic thyristor generators are mainly controlled with respect to two parameters. First, driving pulses for the inverter thyristors must have the intensity such as to ensure their reliable triggering. Second, the control must provide an automatic tuning of the operation frequency of the generator to the value corresponding to its maximum output, i.e., to the frequency of its electromechanical resonance.

The main part of multipurpose generators is an autoinverter with an output transformer, since it is these units that match electric and mechanical vibratory systems and perform the basic function of ultrasonic generators – the generation of electric oscillations of ultrasonic frequency.

The most promising in this respect are resonant autoinverters, since they allow a smooth change of currents and voltages across the el-

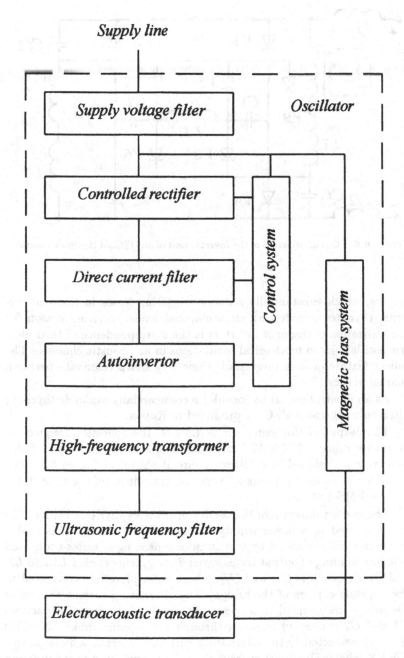

Figure 6.5. Block diagram of an industrial ultrasonic thyristor generator.

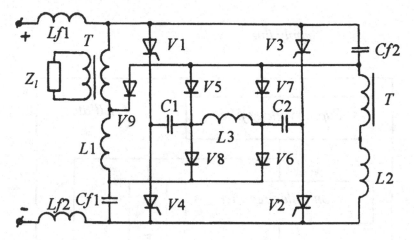

Figure 6.6.   Circuit diagram of the inverter unit of an UZG3-4 thyristor generator [5].

ements, which substantially reduces triggering losses in resonant circuits, thyristors, protecting elements, and loads. Another reason for the utilization of resonant inverters is the correspondence of their electric oscillations to mechanical oscillations in an acoustic channel. The salient feature of such inverters is their very sharp resonant operation characteristics.

As an illustration, let us consider a commercially available thyristor ultrasonic generator UZG3-4 produced in Russia.

The output of this generator is 4.5 kW, the operation frequencies lie within ranges of 17.5–19.3 and 20.5–23.5 kHz. The inverter of this generator is designed in a full-wave circuit shown in Figure 6.6. The oscillator is loaded by a magnetostrictive transducer of the type PMS-6-22 or PMS-15A-18.

The curcuit diagram of the oscillator involves thyristor bridge $V1$–$V4$ connected to a power supply through a choke filter. This bridge is direct-current shunted by two circuits consisting of series-connected primary winding of output transformer T, triggering chokes $L1$ and $L2$, and filter capacitors C_{f1} and C_{f2}, whose outer plates are connected to the opposite corners of the bridge. The alternating-current corners of the bridge are coupled to a triggering circuit consisting of capacitors $C1$ and $C2$ connected in series through a triggering choke $L3$. This choke is connected to the alternating-current corners of a diode bridge $V5$–$V8$, whereas its direct-current corners are connected to the junction points of filter capacitors C_{f1} and C_{f2} and triggering chokes $L1$ and

$L2$. Important elements of the circuit is diode $V9$, which shunts the respective primary windings of the transformer and filter capacitor C_{f2}. The essential elements of this type of inverters are the diode bridge $V5$–$V8$ (with a central triggering choke $L3$) and the capacitor filter (consisting of C_{f1} and C_{f2} connected in series with transformer T and triggering chokes $L1$ and $L2$).

The inverter operates as follows. Being supplied with voltage, capacitors C_{f1} and C_{f2} are charged through filter chokes L_{f1} and L_{f2} to the supply voltage (the upper plates are charged positively). The driving pulses trigger thyristors $V1$ and $V2$, and the transient current of triggering capacitors $C1$ and $C2$ flows over two circuits: C_{f1}–$L1$–T–$V1$–$C1$–$L3$–$C2$–$V2$–C_{f1} and C_{f2}–V–$C1$–$L3$–$C2$–$V2$–$L2$–T–C_{f2}.

At this stage, voltages across thyristors $V3$ and $V4$ gradually change from negative to positive, terminating the stage of their cutoff, whereas diodes $V5$ and $V6$ are applied with a peak negative voltage that falls to zero by the end of this stage. The current through thyristors $V1$ and $V2$ will cease ones the filter capacitors are recharged (the upper plates being negatively). By that time, magnetic energy accumulated in the triggering chokes may not be totally consumed by a load and for recharging capacitors $C1$ and $C2$; therefore the voltage across the chokes $L1$–$L3$ will turn on diodes $V5$ and $V6$. In this case, the current flows over circuits C_{f1}–$L2$–T–$V5$–$L3$–$V6$–C_{f1} and C_{f2}–$V5$–$L3$–$V6$–$L3$–T–C_{f2} in the direction opposite to the voltage across C_{f1} and C_{f2}, charging them. The excessive energy accumulated over the operation period of the main thyristors $V1$ and $V2$ returns to the power supply of the inverter and produces useful work in a load. As the current falls to zero, diodes $V5$ and $V6$ are blocked to remain in this state until a new operation cycle. Ones diodes $V5$ and $V6$ are in conducting state, thyristors $V1$ and $V2$ are applied with an inverse voltage equal to the difference in the voltages of triggering and filter capacitors. In this case, the direct voltage across antiphase thyristors $V3$ and $V4$ is maximum, and the voltage across diodes $V7$ and $V8$ is equal to that across choke $L3$.

Diode $V9$ is activated when the voltage across the primary windings of the transformer exceeds the voltage across capacitor C_{f2}. In this case, current will flow over the circuit $V9$–T–C_{f2}–$V9$. As the winding voltage is opposite to the voltage across C_{f2}, part of the load energy is transferred to the power supply. At this stage, the output voltage of the inverter is not higher than the supply voltage. Diode $V9$, which is activated by the output voltage of negative polarity, stabilizes the amplitude of the output voltage as the load of the generator changes from nominal to idle. Now, the network elements are charged by the voltages of the power supply, filter, and triggering capacitors, which are virtually independent of the load. This leads to constant voltages across the inverter elements, including the thyristors.

By this time, one period of output frequency is completed, since this network works as a frequency doubler. The second period is determined by analogous processes that arise upon the triggering of antiphase thyristors $V3$ and $V4$, bridge diodes $V7$ and $V8$, and diode $V9$ (this diode is conducting only when the load is small).

This circuit features a high triggering stability when operating with a varying load. With identical parameters of elements of the circuits shunting the main bridge, the distribution of current in the primary windings of the output transformer T, as well as in the triggering chokes $L1$ and $L2$ and filter capacitors C_{f1} and C_{f2}, is uniform. This is very important, because the skin-effect at working frequencies of the generator (18–22 kHz) is fairly profound. Another advantage of this scheme that its triggering elements are separated. Thus, three triggering chokes $L1$, $L2$, and $L3$ with inductances $L1 = L2 = L3$ are connected into the constant-current and alternating-current branches of the thyristor bridge.

The shortcoming of this device is a unidirectional inverter current through the load, which requires an increased power of the ultrasonic-frequency filter to retain the necessary harmonic composition of output voltage.

6.2   Mechanoacoustic Transducers

As already mentioned, the excitation of vibrations in mechanoacoustic transducers results from the collision of a gaseous or liquid flow with an obstacle. Aerodynamic and hydrodynamic transducers are used to produce vibrations in gases and liquids, respectively.

Depending on the principle of ultrasound generation, aerodynamic transducers can be divided into two groups, namely, sirens, in which a gas flow is interrupted by movable (usually rotating) surfaces, and gas-jet generators (whistles), in which gas passes through a nozzle-resonator system.

In turn, sirens can be subdivided into dynamic (rotating) and pulsating. In pulsating sirens, a gas flow is interrupted by a shutter. In dynamic sirens, which are most frequently used, a gas flow is interrupted by a perforated or slotted rotor. The working frequency of dynamic sirens (usually, 40–50 kHz) is determined by the number of holes in the rotor and the rate of its rotation. The acoustic power of sirens depends on the pressure and flow rate of working gas, and may vary from hundreds of watts to tens of kilowatts. The efficiency of dynamic sirens can be as high as 50–60%.

The shortcoming of dynamic sirens is that they involve rotating parts, which makes their production and maintenance more complicated, especially in the case of siren working at high temperatures and in corrosive media.

The Hartmann-type gas-jet generators are the most promising devices for industrial applications. They are compact, simple, and allow high-duty operation.

Analysis of a gas jet issuing from a vessel through a converging nozzle shows that with the pressure in the vessel exceeding a particular, critical value, the velocity of the gas in the nozzle near its edge becomes equal to the speed of sound, whereas just outside the nozzle (close to its edge) the gas jet will be supersonic. Such gas flow is characterized by perturbations near the nozzle edge, which propagate as waves of compressions and rarefactions. As a result, the gas jet becomes inhomogeneous with regularly varying local pressure, density, and velocity (Figure 6.7).

Analysis of the issuing gas flow allows one to derive formula for the wavelength of jet oscillations

$$\Delta = A d_c \sqrt{P_a - 0.9}, \qquad (6.1)$$

where P_a is the excessive gas pressure (atm), d_c is the nozzle edge diameter (cm), A is the coefficient varying from 0.77 to 1.22 for different nozzles.

If a resonant cavity is placed some distance from but coaxially with the nozzle, the jet becomes a source of high-intensity acoustic oscillations. The resonator should be expediently placed in the region of an increased jet pressure (Figure 6.7,b, region $a_1 b_1$), called by Hartmann the region of instability.

Although as many as 70 years have passed since the first Hartmann's work, we are now making only the first attempts to the theoretical consideration of gas-jet generators.

Ones the resonator is placed in the first region of instability (Figure 6.7,c), the distribution of averaged (with respect to time) static pressure over the jet has the form illustrated by curve P_a (Figure 6.7,d). The jet entering into the resonator undergoes retardation, due to which its velocity becomes subsonic, whereas pressure and gas density in-

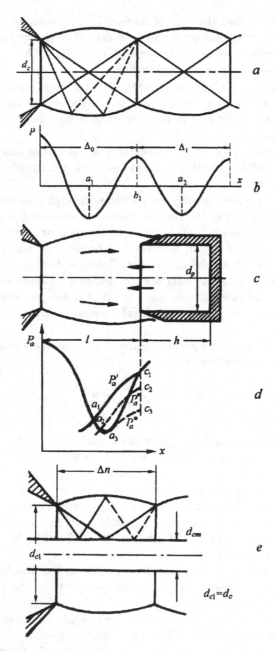

Figure 6.7.   Structure of a gas jet (a) without resonator, (c) with a resonator, and (e) in the presence of a rod. Pressure distribution over the jet (b) without and (d) with the resonator [2].

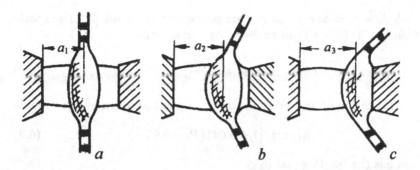

Figure 6.8. Distribution of shock pressures in the jet during the resonator discharge [2].

crease. The kinetic energy of the jet is converted into the energy of gas compression. When pressure inside the resonator rises to a value equal to inlet pressure (point c_1), the gas inside the resonator becomes unstable and gives rise to a backward jet. The distribution of pressure in the backward jet is reciprocal of that in the primary, forward jet (curve P_a').

At the initial stage of resonator discharge, pressure in the portion (up to point a_1) of the backward jet is higher than in the primary jet, so that two counter jets collide approximately midway between the nozzle and resonator (Figure 6.8,a). The air flowing out of the region of jet collision moves laterally to the nozzle–resonator axis.

As the gas issues from the resonator, the inside pressure falls and the backward jet diminishes so that the point where jets will collide shifts toward the resonator (to point a_2). By the end of the discharge stage, the resonator pressure drops to the value corresponding to the minimum pressure in the forward jet (curve P_1'', point a_3). The point of the jets collision displaces to the ultimate position (Figure 6.8,c), and the exhaust gas flows around the resonator. The discharge stage is terminated to be replaced by a new stage of filling.

Thus, the interaction of the permanent forward and cyclic backward jets gives rise to gas pulsations.

As mentioned above, acoustic oscillations in the gas-jet generator arise when the distance between the nozzle and resonator lies within the region of instability a_1b_1 (Figure 6.7,b). The frequency and intensity of these oscillations depend largely on the dimensions of the system, d_c, l, h, d_p (for these designations see Figure 6.7).

Given a desired frequency, one can determine the sizes and acoustic characteristics of a transducer. In the case of the Hartmann generator with $d_c = d_p = h$, calculations are performed as follows [2]. The nozzle diameter d_c (in cm) can be found by the formula

$$d_c = \frac{5860}{f}. \tag{6.2}$$

The region of instability begins from point a_1 given by the formula

$$a_1 = d_c \left[1 + 0.041(P_a - 0.93)^2\right], \tag{6.3}$$

and has the length equal to Δ.

Inasmuch as the preferable position of the nozzle edge is in the second third of the interval $a_1 b_1$, parameter l is given by

$$l = a_1 + 0.6\,(b_1 - a_1). \tag{6.4}$$

The acoustic power of a vibrating system with parameters l and h can be calculated by the formula

$$N_a = 29\,d_c^2\sqrt{P_a - 0.9},\ \text{W}. \tag{6.5}$$

Power necessary for the maintenance of a required air discharge is

$$N_n = 5250\,(P_a + 1.03)\left[(P_{01} + 1.03)^{0.29} - 1\right] d_c^2,\ \text{W}. \tag{6.6}$$

Then, the generator has the efficiency

$$\eta = \frac{N_a}{N_n} \times 100. \tag{6.7}$$

At a given pressure P_a, the air consumption Q is

$$Q = 0.852\,d_c^2(P_a + 1.03),\ \text{m}^3/\text{min}. \tag{6.8}$$

The Hartmann generator has limitations associated with its low efficiency and stability with respect to small variations in pressure and the nozzle-to-resonator distance.

A modified Hartmann generator, which is largely free of above limitations, contains an axial metal rod often used also for supporting a resonator. This type of ultrasonic generators is especially efficient when the rod is slender, and pressure is low.

Figure 6.7,e illustrates how the rod affects the jet parameters.

By now, the general design principles of rod-type gas-jet generators has not been developed. Nor are there general formulas relating their acoustic parameters to dimensions and blowing pressures. However, the

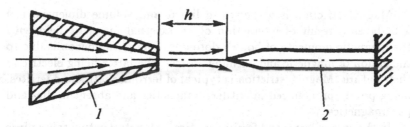

Figure 6.9. Schematic representation of a blade hydrodynamic transducer: (*1*) nozzle, (*2*) blade [3].

accumulated relevant experimental evidence enabled Yu. Ya. Borisov [2] to derive an empirical formula for the frequency of the rod-type gas-jet generator

$$f = \frac{c}{4\left[h + 0.4l + (d_p - d_{cm})(0.4 - 0.2h/d_c)\right]},\qquad(6.9)$$

where c is the sound velocity in the working gas, m/s; d_{cm} is the rod diameter; parameters h, l, d_c, and d_{cm} are expressed in mm.

Problems associated with the employment of gas-jet generators for liquid atomization will be considered below.

Ultrasonic vibrations in liquids can be generated with the aid of hydrodynamic transducers, the most frequent of which are those containing a rectangular slit-like nozzle and a blade fixed at a distance h apart (Figure 6.9). The blade can be fixed either as a console or at two nodal points.

Bending vibrations are excited in the blade as a result of its collision with an inflowing liquid. Blade hydrodynamic transducers can generate oscillations with frequencies of 2 to 35 kHz and intensities varying from 1.5 to 2.5 W/cm^2.

At present, a diversity of modifications of blade and rod-type transducers are known [3, 4, 6].

Apart from blade transducers, there also exist rotor and turbine hydrodynamic transducers based on either rotation of a working liquid or the regular interruption of its flow [7, 8].

6.3   Magnetostrictive Transducers

Electroacoustic transducers are based on the magnetostriction or electrostriction effects when a magnetic or electric field induces elastic strains in electromechanically active substances with corresponding changes in their shape and dimensions.

Magnetostriction is a change in linear and volume dimensions of a body as a result of interaction of an external magnetic field with the magnetic momenta of body's domains. The phenomenon is due to the action of magnetic and exchange forces, the latter being electric in their nature. Magnetostriction is typical of ferromagnetics and ferrites, being poorly pronounced in antiferromagnetics and absent in dia- and paramagnetics.

In the case of ferro- and ferrimagnetics, magnetostriction takes place if the magnetizing field strength does not exceed a value corresponding to the saturation magnetization of the substance (the so-called region of technical magnetization). In this region, the magnetic field elicits a displacement of domain boundaries and rotation of the magnetic momenta of domains, thereby changing the energy state of the crystalline lattice and, consequently, steady-state distances between its atoms. In this case, magnetostriction is anisotropic and affects the shape of the crystal rather than its volume (the so-called linear magnetostriction). The strain produced by linear magnetostriction is characterized by a relative change Δl in the length l of a sample subject to a magnetic field

$$\lambda = \frac{\Delta l}{l}. \tag{6.10}$$

This strain is typically $\sim 10^{-5}$–10^{-3}; however, in some rare-earth elements, it may be as high as $\sim 10^{-2}$. The elongation of the specimen may be either in the direction of the magnetizing field (the longitudinal magnetostriction) or perpendicular to it (the transverse magnetostriction). In the region of technical magnetization, transverse and longitudinal magnetostrictions have, as a rule, opposite signs, the magnitude of the former being smaller than that of the latter.

Magnetostriction due to exchange forces is usually observed in magnetic fields above the region of technical magnetization when the magnetic momenta of domains are already aligned by the field and can only grow in magnitude.

Magnetostriction curves for some metals and alloys, representing a specific elongation versus the magnetizing field strength, are shown in Figure 6.10. At a particular strength of the field, magnetostriction reaches a maximum value called the saturation magnetostriction λ_s. The values of magnetic induction and magnetization intensity, at which magnetostriction is maximum, are known as the saturation magnetic induction B_s and saturation magnetization intensity I_s.

Magnetostriction due to exchange forces manifests itself as a change in body's volume. The sign, value, and behavior of this type of mag-

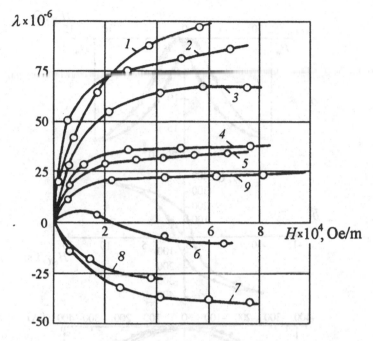

Figure 6.10. Static magnetostriction of some materials in relation to the magnetic field strength: (*1*) 54% Pt + 46% Fe; (*2*) 65% Co + 35% Fe; (*3*) 49% Co + 49% Fe + 2% V; (*4*) 50% Ni + 50% Fe; (*5*) 14% Al + 86% Fe; (*6*) Fe; (*7*) Ni; (*8*) nickel-zinc ferrite; (*9*) 14% Cr + 86% Fe [11].

netostriction versus the magnetizing field strength depend on both the material nature and its structural characteristics, which are often determined by the conditions of heat treatment and machining. Being subjected to a magnetic field, some substances would contract (the negative magnetostriction), but others would elongate (the positive magnetostriction). In some cases, the sign of magnetostriction transforms with increasing magnetizing field. Thus, either elongation or contraction of iron samples can be observed in weak and intense magnetic fields, respectively.

Neither the sign nor the value of magnetostriction depend on the direction of magnetizing field; therefore, it is even parity effect. (Figure 6.11,*a*). The dependence of both magnetization and magnetostriction on the magnetizing field strength is characterized by hysteresis. However, the transformation of the dependence $\lambda(H)$ by invoking the magnetization curve (Figure 6.11,*b*) gives the dependence $\lambda(I)$ almost without hysteresis (Figure 6.11,*c*) [9].

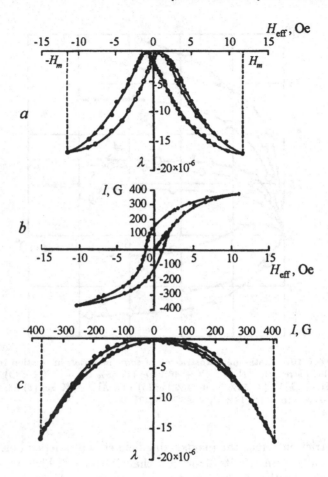

Figure 6.11. Static magnetostriction hysteresis in nickel: (a) magnetostriction versus the magnetic field strength; (b) magnetization versus the magnetic field strength; (c) magnetostriction versus magnetization [9].

Magnetostriction can also be strongly affected by temperature, elastic stresses, and other factors. It is reversible. In other words, if a ferromagnetic material is subjected to a mechanical stress, its magnetization will change (the so-called magnetoelastic effect).

The relationship between the elastic and magnetic properties of a ferromagnetic material is described by the static equations of coupling whose number depends on the set of independent variables. In a linear approximation for a rod-type system, the following four equations are sufficient for determining the main magnetomechanical parameters:

$$\sigma = E_H \mathcal{E} - eH, \qquad (6.11a)$$

$$B = e\mathcal{E} + \mu_\mathcal{E} H, \qquad (6.11b)$$

$$\sigma = E_B \mathcal{E} + aB, \qquad (6.11c)$$

$$H = a\mathcal{E} + \frac{1}{\mu_\mathcal{E}} B, \qquad (6.11d)$$

where H is the magnetic field strength; B is the magnetic induction; $e = (\partial B/\partial \mathcal{E})_H$ is the magnetostriction constant characterizing the reverse magnetostriction effect*; $a = -(\partial \sigma/\partial B)_\mathcal{E}$ is the magnetostriction constant characterizing the direct magnetostriction effect; $E_H = (\partial \sigma/\partial \mathcal{E})_H$ is the module of elasticity at a constant strength of magnetic field; $E_B = (\partial \sigma/\partial \mathcal{E})_B$ is the module of elasticity at a constant magnetic induction; $\mu_\mathcal{E} = (\partial B/\partial H)_\mathcal{E}$ is the magnetic permittivity at a constant strain.

Equations (6.11a, c) and (6.11b, d) describe direct and reverse magnetostriction effects, respectively. Parameters e, a, E_H, E_B, and $\mu_\mathcal{E}$ characterize a magnetostrictive material with respect to its efficiency in transforming magnetic (or electric) energy into mechanical energy and vice versa.

Applied to magnetostrictive material, an alternating magnetic field generates mechanical vibrations with a frequency twice as large as that of the magnetizing field (due to magnetostriction effect parity). This can be illustrated by Figure 6.12, which shows magnetostriction curves and elongation of the specimen subjected to an alternating magnetic field. In practice, the even parity of magnetostriction is undesirable, since the arising strains have small amplitudes because of a low slope of magnetostriction curves near a zero point. But if, in addition to the alternating magnetic field, the material is applied with a permanent magnetic field H_0 producing a constant strain, the operation region shifts to a linear portion of the magnetostriction curve, due to which the frequency of generated vibrations will be equal to the frequency of alternating magnetic field.

These features of magnetostrictive materials make them suitable for the manufacturing of magnetostrictive transducer cores. For this, soft magnetic materials are commonly used, which possess high mechanical durability, magnetostriction, saturation magnetization, and such favorable technological properties as high resistance to corrosion and ductility.

---

*Instead, the magnetostriction sensitivity constant $\lambda = (\partial B/\partial \sigma)_H$ is sometimes used.

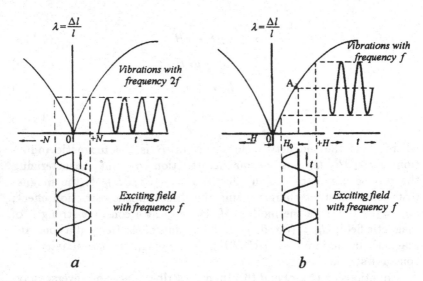

Figure 6.12.   Magnetostriction in an alternating magnetic field: (*a*) unbiased and (*b*) biased [11].

Important characteristics of magnetostrictive materials to be used in transducers are the magnetostriction constant a, sensitivity λ, and the magnetomechanical coefficient k. The last parameter characterizes the efficiency of the transformation of consumed magnetic energy into the mechanical energy of oscillations; in the case of longitudinal vibrations, it is

$$k = \sqrt{\frac{4\pi a^2 \mu_\varepsilon}{E_B}}. \tag{6.12}$$

For magnetostrictive materials to be used in transducers, the above parameters should be as great as possible.

The magnetomechanical coefficient is a valid parameter if energy loss in material are negligibly small. However in real materials, there are always two types of energy losses, mechanical and magnetic. The mechanical loss is determined by the mechanical quality of the transducer as an oscillatory system, whereas the magnetic loss is due to hysteresis and eddy currents.

Due to hysteresis, the magnetic flux lags behind the winding current and the field strength [9]

$$\Phi = \mu_\varepsilon H S e^{-j\varphi_1}, \tag{6.13a}$$

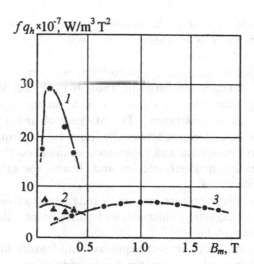

Figure 6.13. Dependence of the magnetic loss rate on induction amplitude: (*1*) nickel, (*2*) 21SPA ferrite, (*3*) 50CoW alloy [10].

where Φ is the magnetic flux; S is the cross-sectional area of the specimen; φ_1 is the phase angle of the magnetic field due to hysteresis.

With allowance for eddy currents, the equation (6.13a) takes the form

$$\Phi = \mu_\varepsilon \chi H S e^{-j\varphi_1} = \mu_\varepsilon H S \chi_0 e^{-j(\varphi_1+\varphi_2)}, \qquad (6.13b)$$

where χ is the parameter accounting for eddy currents: $\chi = \chi_0 e^{-j\varphi_2}$; φ_2 is the phase angle of a magnetic flux caused by eddy currents.

In practice, the magnetic loss of a transducer is characterized by the quantity $\operatorname{tg}\beta$, where $\beta = \pi/2 - (\varphi_1 + \varphi_2)$.

Energy losses in magnetostrictive materials due to hysteresis and eddy currents can also be characterized by the coercive force H_c and specific ohmic resistance ρ_e, respectively. The specific loss rate due to magnetic hysteresis can be described by formula from [10]:

$$N_h = f q_h B_m^n, \qquad (6.14)$$

where q_h is the coefficient of magnetic loss dependent on the magnetic induction amplitude and the strength of a permanent magnetic field (Figure 6.13); $n = 1.5\text{--}2$.

The specific loss rate due to eddy currents can be written as

$$N_e = q_e f^2 B_m^2, \qquad (6.15)$$

where $q_e = 2/3\pi^2 d^2 \rho_e^{-1}$; d is the plate thickness.

Energy losses, as well as the magnetomechanical coefficient, characterize the efficiency of magnetostrictive materials in converting one form of energy into another.

The main characteristics of magnetostrictive materials that are commonly used in practice are listed in Table 6.1. Nickel and its alloys have high magnetostrictive, mechanical, and anticorrosive characteristics, but low ohmic resistance. The alloying of nickel with cobalt, chromium, silicon, and iron substantially improves its ohmic resistance, magnetostrictive constants, and magnetomechanical coefficient, but diminishes saturation magnetostriction and, hence, the rating intensity of ultrasonic vibrations.

Iron–cobalt alloys possess very high magnetostrictive characteristics, saturation magnetic induction, and Curie point. However, they are relatively corrodable.

Iron-aluminum alloys are not expensive and feature high ohmic resistance (in other words, energy loss due to eddy currents is small). At the same time, they are corrodable and poorly workable.

Ferrites exhibit high magnetostrictive characteristics, mechanical quality factor, and ohmic resistance (due to which eddy currents are virtually absent), but their mechanical properties are poor.

The simplest transducer represents a ferro- or ferrimagnetic core inside a coil through which an alternating current flows to create a magnetic field. Magnetization of the core gives rise to elastic stresses and strains directed axially and varying with the alternating magnetizing field; that is, the core begins to vibrate longitudinally. The efficiency of transformation of electric energy into acoustic energy is, however, very low in such a device.

In the general case, the electroacoustic efficiency of a transducer, η_{ea}, is defined as the ratio of the radiated acoustic power N_a to the consumed electric power N_e and can be presented as follows

$$\eta_{ea} = \frac{N_a}{N_e} = \eta_{em}\,\eta_{ma}, \qquad (6.16)$$

where η_{em} is the electromechanical efficiency; η_{ma} is the mechanoacoustic efficiency.

The electromechanical efficiency of a transducer is determined by energy losses that occur during the conversion of electric energy into the energy of mechanical vibrations. These losses are mainly due to magnetic hysteresis and eddy currents in the transducer. Thus, electromechanical efficiency is eventually determined by the properties of magnetostrictive material and transducer design.

Table 6.1 The main characteristics of magnetostrictive materials.

Material	Chemical composition	Density ρ_s, g/cm³	Velocity of sound c_t, 10³ m/s	Young's modulus E, 10¹¹ N/m²	Magnetic permittivity $(\mu_r)^*$, 10⁶ Tm/A	Coercive force H_c, A/m	Saturation induction B_s, 10⁻¹ g	Saturation magnetostriction λ_s, 10⁻⁶
Nickel	99% Ni	8.9	4.9	2.15	20–50	1.7	6.1	−37
Nicosi	4% Co, 2% Si, 94% Ni	8.8	4.8	1.90	–	0.17–0.25	6.2	(−25)–(−27)
Permendur	49% Co, 2% V, 49% Fe	8.2	5.2	2.05	50–80	1.4	24.0	70
Permendur	65% Co, 35% Fe	8.2	5.2	2.05	–	1.0	22.0	90
Alfer	12.4% Al, 87.6% Fe	6.7	4.8	1.58	60–100	0.1	16.0	40
Hypernic	50% Ni, 50% Fe	8.3	–	1.4	–	0.05	–	25
Permalloy	40% Ni, 60% Fe	8.2	–	1.3	–	0.1	–	25
Hyperco	35% Co, 0.45% Cr, 64.55% Fe	8.1	–	2.1	120	0.2	–	55
Nickel ferrite	NiO, Fe_2O_3	5.2	5.9	1.8	34	2.7	3.2	−26
Nickel-zinc Ferrite	$NiO_{0.5}, ZnO_{0.5}, Fe_2O_3$	5.3	–	1.8	270	0.35	–	−9

Magnetostriction constant a^*, 10^7 N/m² T	Constant of sensitivity ν^*, 10^{-9} T m²/N	Magneto-mechanical coefficient k^*	Magnetic loss tangent, $tg\beta \times 10^{-3}$	Ohmic resistance ρ_e, 10^{-7} Ohm m	Curie point θ, °C	Mechanical Q-factor	Optimal magnetizing field H_{opt}^{**}, 10^3 A/m	Yield limit σ_T, MPa	Tensile limit σ_B, MPa	Fatigue limit σ_{-1}, MPa
2.3	4.2	0.26–0.30	60	0.7	360	750	1–2	100	360	100
1.8	28	0.49	–	1.8	350	–	0.4–0.6	85	420	–
2.2	27	0.48–0.54	400	3.4	980	600	0.4–0.6	350	500	110
1.5	7	0.27–0.30	500	0.8	980	400	1.0–1.7	450	670	–
0.8	0	0.30	30	16	600	400	0.3–0.6	600	800	30
0.4	–	0.20–0.32	–	4.6	–	–	–	–	–	–
0.3	–	0.20–0.35	–	7.5	–	–	–	–	–	–
0.8	3.8	0.14	–	1.4	–	–	–	–	–	–
2.4	2.8	0.21	30	10^7	590	2000	1.0–1.7	–	60–80	20–25
2.3–5.5	3.9	0.10–0.14	–	10^7	–	–	–	–	–	–

* These parameters significantly depend on the constant magnetizing field strength H_o and, therefore, on magnetic induction B_o.

** The optimum value H_{opt} corresponds to the maximum k.

Mechanoacoustic efficiency depends on the losses due to mechanical hysteresis and internal friction in transducer material, on the losses in a fixture and passive elements, as well as on the losses during the emission of vibrations into a load. This efficiency is also dependent on the properties of magnetostrictive material, vibratory system design, and characteristics of the load.

Although simple in design, the magnetostrictive transducer with a solid core has some disadvantages that are reduced in the final analysis to its low electroacoustic efficiency. Because of a significant dissipation of magnetic flow, vibrations in the transducers with such cores can be excited only by relatively strong magnetic field. Therefore, it would be expedient to use, instead of open core, a closed magnetic core, just as in electric transformers. Such cores provide a desired magnetization at a significantly lower strength of the magnetizing field. Transducers with closed magnetic cores can be either straight or circular.

Magnetostrictive transducer cores should not be fabricated from a solid piece of a material (except ferrites), because of a great eddy-current loss. This loss can, however, be easily reduced by building up the transducer core from separate laminations 0.1–0.3 mm thick, as in electric transformers. These laminations are commonly stamped from a sheet of a magnetostrictive material subjected to heat treatment. If the treated material fails to produce a firm insulating oxide layer (an example of such material is 65Co alloy), the core can be built from laminations insulated by a special paste.

Figure 6.14 illustrates the form of laminations for the simplest axial and annular transducers.

Depending on the transducer power, laminations can be made with one or several windows. Coils are mounted on the core legs, while a yoke, closing the magnetic circuit, serves as a radiator. Such a construction is most appropriate for high-power magnetostrictive transducers, as it is compact and provides lower energy losses and better cooling of the device in comparison with other transducer designs.

The electroacoustic efficiency of core transducers operated at a frequency of ~20 kHz is 40–50% and diminishes with increasing frequency.*

The design calculation of magnetostrictive transducers involves the determination of its dimensions, resonant frequency, consumed and acoustic powers, as well as input resistance.

---

*Calculation of transducer efficiency will be considered in more detail in section 8.1.

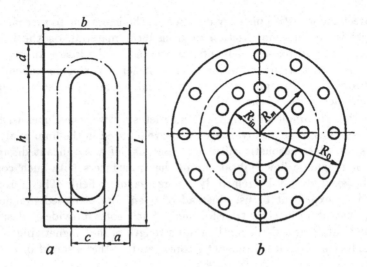

Figure 6.14. Laminations for (a) axial and (b) annular magnetostrictive transducers.

Exact calculation of electric and acoustic parameters of transducers is quite sophisticated; therefore, it is a common practice to use approximated methods.

When designing a magnetostrictive transducer of a desired frequency and power, the dimensions of its radiating surface are estimated by taking into account the properties of a magnetostrictive material to be used as well as the permissible power from unit radiating surface (a specific acoustic power n_a). In a linear approximation, i.e., when force and velocity change harmonically and their amplitudes are proportional, the acoustic power transmitted to a load is

$$N_a = \frac{1}{2} R_l v_{mr}^2, \qquad (6.17)$$

where R_l is the load resistance; v_{mr} is the vibrational velocity amplitude on the radiating surface of the transducer.

By a load is meant the medium to which ultrasonic vibrations are emitted. The load impedance is defined as the ratio of the instantaneous values for force and velocity (see expression 1.34). In the general case, this ratio has real and imaginary components

$$Z_l = R_l + jX_l. \qquad (6.18)$$

Nonlinear phenomena in the load (for example, cavitation in liquid) affect the relationship between force and velocity, which can be taken into account by assuming that R_l depends on the amplitude of vibrations ξ_{mr}

$$R_l = r_l(\xi_{mr})S_r, \tag{6.19}$$

where S_r is the area of the radiating surface of the transducer, r_l is the specific resistance of the load.

Taking into account (6.17) and (6.19), as well as the relationship between the vibrational velocity and displacement, one can write

$$N_a = \frac{1}{2}r_l(\xi_{mr})\omega^2\xi_{mr}^2 S_r = n_a S_r. \tag{6.20}$$

Thus, with the electric power and efficiency being given, the acoustic power of the transducer is determined by its displacement amplitude.

Actually, the displacement amplitude of a transducer cannot be obtained infinitely great, since it is limited by the fatigue strength and saturation magnetostriction of transducer material. For axial magnetostriction transducers, the expression relating the stress amplitude at the strain antinode, σ_m, to the displacement amplitude has the form

$$\xi_{mr} = \frac{\varphi_i \sigma_m c_0}{\omega E}, \tag{6.21}$$

where φ_i is the shape factor dependent on the design and dimensions of transducer.

For a symmetric three-leg core, the shape factor is given by

$$\varphi_3 = \frac{1}{\sqrt{\cos^2 \alpha_1 + q^2 \sin^2 \alpha_1}}, \tag{6.22}$$

where $\alpha_1 = \omega/c \cdot d$ is the wave length of the yoke; $q = b/2a$ is the ratio of the cross-sectional areas of yoke and legs; for designation of a, b, and d see Figure 6.14.

Since the maximum value for σ_m cannot exceed the fatigue strength of material, then, by substituting σ_{-1} for σ_m, one can estimate the ultimate displacement amplitude, at which the lifetime of magnetostrictive transducer, ξ_{mr}^σ, is still great.

Ultimate displacement amplitude is limited not only by fatigue strength, but also by magnetic saturation. At low magnetic inductions, the amplitude of alternating magnetostriction stress is proportional to B_m, being maximal at $B_m \sim 0.5B_s$ (Figure 6.15) [9]. The greatest attainable amplitude of displacement, ξ_{mr}^B, can be estimated by formula

$$\xi_{mr}^B = \frac{\sigma_{m\,max} S_c A}{\omega(R_l + R_{ml})}, \tag{6.23}$$

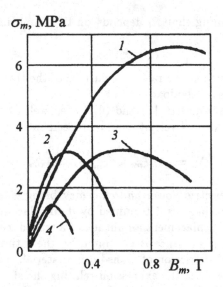

Figure 6.15. Dependence of the mechanical stress amplitude on induction amplitude: (*1*) 50CoW alloy, (*2*) 14Al alloy, (*3*) nickel, (*4*) 21SPA ferrite [10].

where $\sigma_{m\ max}$ is the ultimate amplitude of magnetostriction stress; S_c is the cross-sectional area of the magnetostrictive transducer core; A is the design coefficient (for axial transducer, $A = \cos\alpha_1$); R_{ml} is the mechanical loss resistance of transducer.

The internal resistance of transducer is a complex value known as internal impedance Z_i. At resonance, its imaginary part is zero, whereas its real part is equal to the mechanical loss resistance R_{ml}.

Expression for the mechanical loss resistance will be given below. The values for the greatest admittable and attainable displacement amplitudes of transducer enable its specific acoustic power n_a to be estimated provided that the expression $r_l = \rho_l c_l$ is independent of displacement amplitude. It should be noted that r_l decreases when load behaves nonlinearly. With allowance for (6.20) and (6.23), one can obtain expression for the power of a rod-type transducer

$$n_a = \frac{1}{2}\frac{\sigma_{m\ max}^2}{r_l}\cos kl\,\eta_{ma},\qquad(6.24)$$

where l is the rod length.

The acoustic power emitted from the square centimeter of nickel or alfer radiator may reach 30 W [10]. At the same time, the acoustic power of iron-cobalt transducers may be significantly higher.

Based on the specific acoustic power n_a and consumption power N_a, one can calculate the area of radiating surface

$$S_r = \frac{N_a}{n_a} = \frac{N_e\,\eta_{ea}}{n_a}. \tag{6.25}$$

The radiating surface of transducers is usually square; hence, the radiator width b is (Figure 6.14)

$$b = \sqrt{S_r}. \tag{6.26}$$

The condition $b < \lambda/2$ should be fulfilled to avoid bending vibrations [12].

To obtain maximum acoustic power, the dimensions of transducer should be chosen so as to satisfy the following conditions

- the ratio of the cross-sectional area of core legs to the radiating area should be not less than 0.65–0.75;

- the thickness d of joke should comprise 0.9–1.2 of the leg width a.

Under these conditions, the mechanical loss of transducer will be minimum. The window should be as wide as possible to provide a convenient mounting of windings and efficient cooling of the transducer. The number of legs is also dictated by the condition of efficient cooling, inner legs being taken twice as wide as outer legs. The width of any leg should not exceed 50 mm, otherwise the arising temperature gradient will reach a critical value and impair the magnetostrictive properties of material.

The next stage of transducer design is the selection of its resonant length l, which is dependent on desired frequency. The natural frequency of a multi-leg core is less than that of a single-leg core of the same length, the difference being increased with the width and height of yoke.

The resonant frequency of core can be obtained from the equation

$$\mathrm{tg}\,\alpha_1\,\mathrm{tg}\,\frac{\alpha_2}{2} - \frac{1}{q} = 0, \tag{6.27}$$

where $\alpha_2 = \omega/c \cdot h$ is the wave length of the leg; other designations are as for expression (6.22).

A nomogram for estimating the wave dimensions of yoke and the total length of transducer is presented in Figure 6.16.

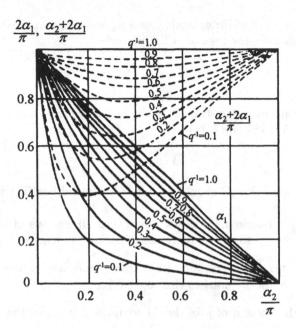

Figure 6.16. Nomogram for estimating the resonant sizes of yokes $(2\alpha_1/\pi)$ and the total length $(\alpha_2 + 2\alpha_1/\pi)$ of a symmetric vibratory system at various q [9].

The design calculation of annular transducer was given in [9, 10]. The natural frequency of such transducer is

$$f_r = \frac{1}{2\pi R_c}\sqrt{\frac{E}{\rho}[1 + (1-n)^2]}, \qquad (6.28)$$

where n is the serial number of natural harmonics $(n = 1, 2, 3)$.

The ring width $R_{ex} - R_{in}$ should not exceed $\lambda/2$. Further calculations aim at establishing the relationship between displacement amplitude, coil voltage, load resistance, and mechanical loss.

As mentioned above, for transducer operating in a linear mode, the magnetostriction stress amplitude is proportional to induction amplitude $(\sigma_m = \sigma_M = aB_m)$, which is related to the coil voltage as follows

$$B_m = \frac{1}{\pi f N S_c} U_m. \qquad (6.29)$$

With allowance for (6.23), one can obtain an expression relating the amplitude of vibrations to the input voltage of transducer

$$\xi_m = \frac{aA}{2\pi^2 f^2 N(R_l + R_{ml})} U_m = \frac{gA}{2\pi f(R_l + R_{ml})} U_m, \qquad (6.30)$$

where
$$g = \frac{a}{\pi f N}.$$

By introducing the idle sensitivity of transducer

$$\nu_U = \frac{\xi_m}{U_m} \quad \text{at } R_l = 0, \tag{6.31}$$

one can obtain from formula (6.30)

$$\xi_m = \frac{\nu_U U_m}{1} + \frac{R_l}{R_{ml}}, \tag{6.32}$$

where

$$\nu_U = \frac{9A}{2\pi f R_{ml}}.$$

Expression for the mechanical loss resistance was obtained in [10] by estimating the mechanical loss rate through the Q-factor of material

$$R_{ml} = \frac{\rho c S_c}{2Q} \left\{ 2\alpha_1 + \left[\pi - 2\arctg\left(\frac{S_r}{S_c}\tg\alpha_1\right)\right] \frac{S_c}{S_r} \right.$$
$$\left. \times \left(\cos^2\alpha_1 + \frac{S_r^2}{S_c^2}\sin^2\alpha_1\right)\right\}. \tag{6.33}$$

Taking into account relation (6.30), one can obtain the resulting expressions for the sensitivity and displacement amplitude of idle-operated transducer and one loaded by R_l. It should be noted that there are no stringent data on the load resistances of various technological processes; therefoe, Table 6.2 lists approximate values for acoustic load parameters taken from [10].

In design calculations of magnetostrictive transducers one should bear in mind that energy loss

$$N_n = N_e(1 - \eta_{ea}) \tag{6.34}$$

is dissipated as heat, due to which the transducer warms up. The temperature, at which the magnetostrictive properties of material begin to be drastically impared, lies far below its Curie point. In particular, the permissible surface temperature of core made of 65Co alloy is 70°C, the bulk temperature being as high as 150–200°C. Therefore, magnetostrictive transducers should be cooled to ensure a normal thermal

Table 6.2   Load impedances and their components for some technological processes [10].

Process	Frequency of vibrations f, kHz	Amplitude of vibrations ξ_m, μm	Load impedance $\times 10^{-3}$, kg/m^2 s	
			Load resistance, r_l	Load reactance, X_l
Cleaning	18.7	6	200	40
	18.7	18	140	–
	19	30	30	1
Impregnation	27	6	75	–
Welding of metals	22	5	20	–
Cutting of metals	23	10	60	40
Surface treatment	20	12	20	10

regime. The efficiency of cooling depends on the coolant temperature T_c, the area of cooled surface, S_c, and the heat transfer coefficient α

$$N_n = S_c\,\alpha(T_S - T_c). \tag{6.35}$$

The area of the cooled surface of core transducer can be quite accurately determined by using the expression

$$S_c = b(b + 4l + 2nh), \tag{6.36}$$

where n is the number of core legs.

The heat transfer coefficient largely depends on the coolant speed.

An exact calculation of the electric parameters of magnetostrictive transducer is rather sophisticated, which is because magnetostriction parameters, such as magnetic permittivity μ, magnetostriction constant a, the coefficient of magnetic loss, and Q-factor, are not actually constant. That is why the results of calculations have to be verified experimentally. Since magnetostriction characteristics depend on the magnetic field strength, calculations should involve the estimation of the optimum strength H_0 of a biasing magnetic field and corresponding magnetic permittivity.

Estimations of the alternating magnetic field strength H_m, should be done with account for the following relations

$$H_0 \geq \frac{H_{\max}}{2}; \quad H_m = H_{\max} - H_0, \tag{6.37}$$

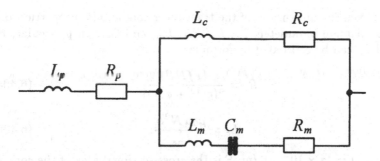

Figure 6.17. Equivalent circuit of a magnetostrictive transducer.

where $H_{\max}$ is the alternating field strength, at which the amplitude of the alternating magnetostriction component λ_m is maximum.

Direct and alternating currents necessary for the creation of biasing and alternating fields H_0 and H_m can be determined by formulas

$$H_0 = \frac{0.8\pi N\, I_o}{l_{me}}, \tag{6.38a}$$

$$H_m = \frac{\sqrt{2}\,0.8\pi N\, I_m}{2 l_{me}}, \tag{6.38b}$$

where N is the number of the winding turns; l_{me} is the mean length of magnetic core.

The calculation of electric parameters of transducer should begin with the determination of its input impedance, which can be done by analyzing the equivalent circuit of the transducer (Figure 6.17)*.

Transducer impedance can be presented as the electric impedance of core, Z_c, connected in parallel to the mechanical impedance of transducer recalculated into the equivalent electric impedance Z_m:

$$Z_c = R_c + j\omega L_c, \tag{6.39a}$$

$$Z_m = R_m + j\left(\omega L_m - \frac{1}{\omega C_m}\right), \tag{6.39b}$$

where $R_c = \omega L_0 \chi_o \sin(\varphi_1 + \varphi_2)$ is the resistance due to hysteresis and eddy-current losses; $L_c = L_0 \chi_o \cos(\varphi_1 + \varphi_2)$ is the inductance of the coil with a core; L_0 is the inductance of the coil with no core; R_m is the mechanical loss resistance; L_m is the inductance due to transducer mass; C_m is the capacitance due to elasticity of transducer material.

---

*Along with series-parallel connection, parallel-series and parallel connections may be used.

Given the dimensions of the transducer core and its properties, one can calculate parameters R_c, L_c, R_m, L_m, and C_m. In particular, R_c and L_c can be estimated by formulas

$$R_c = \frac{(\pi f N S)^2}{2(q_l f^2 + q_h f)V}, \tag{6.40a}$$

$$L_c = \frac{\mu_0 \mu N^2 S}{l_{me}}, \tag{6.40b}$$

where $\mu = 4\pi \times 10^{-7}$, H/m; S is the cross-sectional area of the core; V is the core volume, $V \cong (lb - hc)b$ (Figure 6.14).

The mechanical loss resistance R_{ml} can be recalculated into the equivalent electric resistance

$$R_m = \frac{U_m^2}{2N_{ml}}, \tag{6.41}$$

where $N_{ml} = 1/2 R_{ml} v_{mr}^2$ is the mechanical loss rate; v_{mr} is the amplitude of vibrational velocity at the radiator.

With allowance for the scattered flux L_p and the ohmic resistance of coil, R_p, the total impedance of transducer can be written as

$$Z_t = Z_p + \frac{Z_m Z_c}{Z_m + Z_c} = R_p + jL_p + |Z_c|e^{j\Phi_c} - \frac{|Z_c|^2 e^{2j\Phi_c}}{R^1 + jX^1}, \tag{6.42}$$

where

$$R^1 = R_m + R_c; \quad X^1 = X_m + X_c = \omega(L_m + L_c) - \frac{1}{\omega C_m};$$

Φ_c is the phase angle of electric impedance.

The last term on the right-hand side of expression (6.42) is maximal at $X_1 = 0$, i.e., at frequency $\omega_a = 1/\sqrt{(L_m + L_c)/C_m}$ known as the antiresonant frequency in contrast to the mechanical resonant frequency $\omega_r = 1/\sqrt{L_m C_m}$. The approximate ratio of these frequencies is

$$\frac{\omega_r}{\omega_a} = \sqrt{1 + \frac{L_c}{L_m}} \cong 1 + \frac{1}{2}\frac{L_c}{L_m}. \tag{6.43}$$

It should be noted that in calculations of the input impedance of transducer, the quantities L_p, L_m, and C_m may be neglected, and the amplitude of the coil current can be rather accurately calculated by formula

$$I_m = \frac{U_m}{Z_t} = \frac{U_m}{R_p + \dfrac{Z_c R_m}{Z_c + R_m}}. \tag{6.44}$$

The coil resistance is given by

$$R_p = \rho'_l \frac{l_{tm}}{S_0} N, \qquad (6.45)$$

where l_{tm} is the mean length of the coil turn; S_0 is the cross-sectional area of the coil wire, determined from the net (alternating plus direct) current flowing across it.

If vibrations are emitted to a load, its resistance R_l should be introduced into the equivalent circuit of the transducer.

6.4   Piezoceramic Transducers

Piezoelectric transducers have also found a wide current use in ultrasonic technology. The piezoelectric effect lies in the electric polarization of some anisotropic insulators and semiconductors produced by a mechanical stress (it is the so-called direct piezoelectric effect). The reverse piezoelectric effect consists in a strain imposed on these materials by external magnetic field.

The displacement of lattice atoms (or ions) and the disturbance of their electron shells in a crystal subject to an electric field give rise to strain*. As in the case of the magnetostriction effect, four coupling equations with their own sets of independent variables are used to describe the elastic, piezoelectric, and dielectric properties of a crystal. In a linear approximation, these four equations can be written in the form of a matrix

$$\sigma = c\mathcal{E}_1 + e_t E, \qquad (6.46a)$$

$$D = c\mathcal{E}_1 + \mathcal{E}E, \qquad (6.46b)$$

$$\mathcal{E}_m = S\sigma + d_t E, \qquad (6.46c)$$

$$E = -g\sigma + \frac{1}{\chi}P, \qquad (6.46d)$$

where E, D, P are, respectively, the components of the stress, induction, and polarization vectors of the electric field.

Matrix components have the following meanings: c_{ik} are the coefficients of elasticity; s_{ik} are the coefficients of elastic compliance; $\mathcal{E}_{ik}$ is the permittivity; χ_{ik} is the dielectric susceptibility; e_{ik}, d_{ik}, and g_{ik}

---

*In some complex crystals, known as pyroelectric crystals, the distribution of charges may be so asymmetric that these crystals will be self-polarized. If the direction of polarization of such crystals depends on external field, they are known as ferroelectrics.

are the piezoelectric constants (d_{ik} are also known as piezoelectric moduli). The subscript t indicates the transpose of the matrix. The tensile mechanical stresses and strains have a positive sign.

Parameters e, $\mathcal{E}$, s, d, and χ characterize piezoelectric material in terms of its efficiency in converting electric energy into mechanical one and vice versa. Piezoelectric material is also characterized by sound velocity c_s), dielectric loss tangent $\operatorname{tg}\delta$, mechanical Q-factor, and tolerable temperature θ (for ferroelectrics, it corresponds to the Curie point). Sometimes, piezoelectric material can be conveniently characterized by derived quantities, such as the electromechanical coupling coefficient k_{ik}, whose quasi-static value is given by

$$k_{ik} = \frac{d_{ik}}{\sqrt{\mathcal{E}_{ii}\mathcal{E}_0 s_{kk'}}}, \tag{6.47}$$

where $\mathcal{E}_0$ is the dielectric constant.

An essential parameter of industrial ultrasonic transducers is rating strain related to the tensile strength of material.

In the case of high-power transducers, of great importance is the amplitude dependence of material properties. In this connection, one should note that $\operatorname{tg}\delta$ grows and Q-factor drops with increasing electric field strength, i.e., with increasing alternating stress amplitude (Figure 6.18).

At present, some monocrystalline substances (quartz, ammonium dihydrophosphate, lithium niobate) and ferroelectric ceramics (piezoceramics) have found extensive practical applications as piezoelectric materials. Among piezoceramics with a strong piezoeffect, solid solutions of Pb zirconate–titanate (PZT) are most frequently used. The relevant parameters of piezoceramics can be widely varied by introducing traces of certain chemical elements (Table 6.3).

In their electromechanical properties, piezoceramic materials can be either segneto-soft or segneto-hard. The former (such as PZT-5H or PZTMB-1) exhibit high values of piezoelectric modulus and permittivity, their disadvantage being great electric losses and poor mechanical quality. Conversely, the latter (such as PZT-8 or PZT-23) have small electric and mechanical losses and nonlinearity, but their piezoelectric moduli are relatively low. Transition materials, such as PZTB-3, are also widely used.

Piezoceramic materials are usually produced from powdered components that are roasted at 1300–1400°C, compacted, sintered, and then subjected to machining. After coating the electrodes, piezoelements are polarized in a permanent electric field.

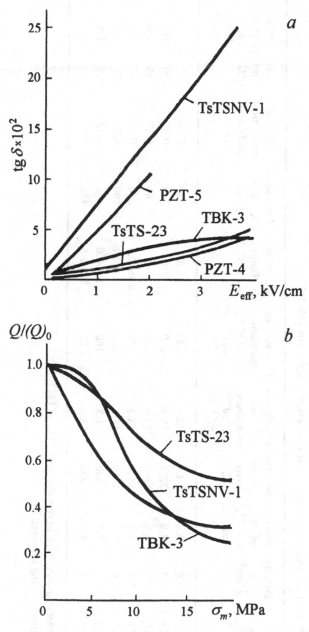

Figure 6.18. (a) Dielectric loss tangent versus the exciting electric field strength and (b) mechanical quality versus the mechanical stress amplitude for some piezoceramic materials [10].

Table 6.3 The main characteristics of piezoelectric materials [13–15].

Piezo-electric	Density ρ_s, g/cm^3	Sound velocity $c_s \times 10^3$, m/s	Relative permittivity ε_{ii}^σ	Coefficient of elastic compliance $S_{kk} \times 10^{-12}$, m^2/N	Piezoelectric modulus $d_{ik} \times 10^{-12}$, C/N	Dielectric loss tangent $\mathrm{tg}\,\delta \times 10^{-2}$	Mechanical Q-factor	Electro-mechanical coefficient k_{ik}	Curie point θ,°C	Tensile strength σ_s, MPa
Quartz	2.6	5.47	4.5	12.8	2.31	0.9	10^4	0.095	–	–
PZTMB-1	7.3	2.9	2200	16.30	200	2–9	60	0.35	250	196
PZT-5H	7.5	2.8	3400	17.0	274	2–9	65	0.39	193	100
PZT-23	7.4	3.2	1100	13.2	75	0.75–2.0	300	0.20	285	182
PZT-8	7.6	3.4	1000	11.4	93	0.4–0.7	1000	0.29	300	700*
PZTB-3	7.2	3.5	2300	11.3	160	1.2–2.0	350	0.32	180	196
TBK-3	5.4	4.7	1200	8.4	51	1.3–2.5	450	0.17	105	182

Piezoceramic parameters have subscripts (11) or (31).

* indicates bending strength.

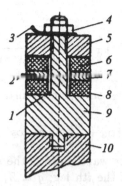

Figure 6.19. Cross-sectional view of a compound piezoelectric transducer: (*1*) insulating bush, (*2*) electrode, (*3*) grounding electrode, (*4*) nut, (*5*) upper metal plating, (*6–8*) piezoceramic elements, (*7*) metal washer, (*9*) lower metal plating, (*10*) waveguide.

The design of piezoceramic transducers is defined by their destination and operation frequency range. In industrial transducers, piezoelements are often made in the form of plates, rods, and rings. To reduce the resonant frequency, plates and rods may be coated with metals.

Figure 6.19 shows a commonly used industrial vibratory system consisting of a compound piezoelectric transducer with a uniform cutoff waveguide at its edge. Polarized piezoceramic elements (*6*, *8*) with electrodes on their end surfaces are fixed with screw (*1*) between metal platings (*5*, *9*). If the neighboring
piezoceramic plates are polarized oppositely, the outer electrodes contacting metal platings are grounded, whereas central washer (*7*) is charged. The tightening screw imposes a compressive stress on the piezoelement, thus enhancing its dynamic strength.

As in the case of magnetostrictive transducers, the designing of piezoelectric transducers involves the calculation of their dimensions and the relationship between electric and acoustic parameters (input voltage U, displacement ξ, vibrational velocity v).

The calculation of piezoelectric transducers to be operated at a resonant frequency is commonly performed by invoking equations describing the vibrations of an asymmetric system [10].

The material and cross-sectional areas of outer core legs are usually the same (therefore, $Z_1 = Z_3$), but their wavelengths are different ($\alpha_1 \neq \alpha_3$). In this case, equations for wave dimensions are as follows

$$\operatorname{tg}(\alpha_1 + \alpha_2)\operatorname{ctg}\alpha_2 = -\left[\frac{1}{2}\left(q + \frac{1}{q}\right) + \frac{\cos(\alpha_1 - \alpha_3)}{\cos(\alpha_1 + \alpha_3)}\left(\frac{1}{q} - q\right)\right],$$

$$\tag{6.48a}$$

$$\cos\alpha_1\cos\alpha_2\sin\alpha_3 - \frac{Z_1}{Z_2}\sin\alpha_1\sin\alpha_2\sin\alpha_3 +$$

$$+\frac{Z_2}{Z_3}\cos\alpha_1\sin\alpha_2\cos\alpha_3 + \frac{Z_1}{Z_3}\sin\alpha_1\cos\alpha_2\cos\alpha_3 = 0, \tag{6.48b}$$

here $\alpha_1 = (2\pi f/c_i)l_i$ is the wavelength of the ith leg; $Z_i = \rho_i c_i s_i$ is the mechanical impedance of the ith leg; $q = Z_1/Z_2$; c_i is the velocity of propagation of vibrations in the ith rod; l_i is the length of the ith rod.

Taking into account that ξ_m is proportional to the applied force amplitude $F_m = \sigma_m S_2$ and that $\sigma_m = d_{ik}E_m$, where d_{ik} is the piezoelectric modulus, one can determine the amplitude of vibrations at the radiating surface by using expressions (6.23).

For a piezoelectric transducer,

$$E_m = \frac{U_m}{l_2}, \tag{6.49}$$

where l_2 is the thickness of piezoelement.

In the case of a compound piezoelectric transducer, the expression for the mechanical loss resistance has the form

$$R_{ml} = \rho_1 c_1 S_1\left[\frac{\pi - \alpha_2}{2Q_1} + \frac{\alpha_2 Z_1}{Q_2 Z_2}\sin^2\alpha_1\right], \tag{6.50}$$

and that for the driving force amplitude is given by [10]

$$F_m = 2\pi f d_{ik} Z_2 \sin\alpha_1 U_m. \tag{6.51}$$

By substituting the displacement amplitude into formula (6.23), one can estimate the sensitivity of piezoelectric transducer with an arbitrary load resistance

$$\nu_U = \frac{d_{ik} Z_2 \sin\alpha_1}{R_m + Z_2\left[\dfrac{\pi - \alpha_2}{2Q_1} + \dfrac{\alpha_2 Z_1}{Q_2 Z_2}\sin^2\alpha_1\right]}. \tag{6.52}$$

Analysis of this expression indicates that for a resonant transducers, there exists such an optimal position of the piezoelement where its sensitivity is maximum. This position can be defined by a parameter p

$$p = \frac{\pi Q_2 Z_2}{2\alpha_2 Q_1 Z_1}\left(1 + \frac{2Q_1 R_l}{\pi Z_1}\right). \tag{6.53}$$

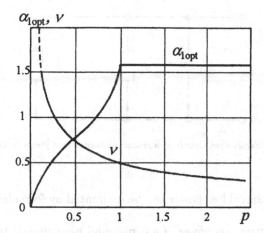

Figure 6.20. Dependence of $\alpha_{1\text{opt}}$ and the maximum sensitivity ν on parameter p for a compound piezoelectric transducer [10].

For small p

$$\sin \alpha_{1\text{opt}} = \sqrt{p},$$

$$\nu_{\text{opt}} = \frac{d_{ik}Q_2 Z_2^2}{2Z_1^2 \alpha_2}\frac{1}{\sqrt{p}}. \tag{6.54}$$

As the parameter p increases, the optimum position of the piezoele-ment shifts toward the center of the system (in this case the maximum sensitivity of transducer falls). At $p \geq 1$, which corresponds to the central position of the piezoelement,

$$\alpha_{1\text{opt}} = \frac{\pi}{2},$$

$$\nu_{\text{opt}} = \frac{d_{ik}Q_2 Z_2^2}{Z_1^2 \alpha_2}\frac{1}{1+p}. \tag{6.55}$$

Figure 6.20 shows the dependences of $\alpha_{1\text{opt}}$ and relative sensitivity on parameter p.

Much like the calculation of magnetostrictive transducers, the de-sign of piezoceramic transducers require the determination of their in-put powers and impedances.

The total power of a compound piezoceramic transducer is the sum of its acoustic power, mechanical loss power, and dielectric loss power.

Acoustic power is a function of the load. When the piezoceramic transducer operates under common conditions, its electroacoustic effi-ciency ranges from 40 to 70%.

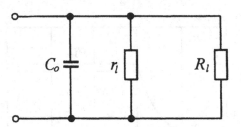

Figure 6.21. Equivalent circuit of a resonant compound piezoelectric transducer [10].

The mechanical loss power can be estimated by formulas (6.41) and (6.52).

The dielectric loss power of a compound piezoelectric transducer is given by

$$N_l = 2\pi f \mathcal{E}\mathcal{E}_0 \operatorname{tg}\delta \frac{U_m^2}{l} Sn, \qquad (6.56)$$

where $\mathcal{E}_0 = 8.55 \times 10^{-12}$, F/m; n is the number of piezoelectric plates.

The input power of the transducer can be calculated from the equivalent circuit, which is shown in Figure 6.21 for the case of a compound resonant piezoelectric transducer.

The capacitance of the piezoelectric plate determines the input reactance of the transducer, which can be calculated for the case presented in Figure 6.19 by formula

$$C_o = \frac{\mathcal{E}\mathcal{E}_0 S_2 (1 - k^2) n}{l_2}, \qquad (6.57)$$

where k is the electromechanical coefficient.

The electric loss resistance due to dielectric hysteresis and insulation currents can be estimated by formula

$$r_l = \frac{U_m^2}{2N_l} = \frac{l_2}{8\pi f \mathcal{E}\mathcal{E}_0 \operatorname{tg}\delta Sn}, \qquad (6.58)$$

whereas the mechanical loss resistance R_{ml} can be recalculated into the equivalent electric resistance R_m by formula (6.41).

References

1. O. K. Keller, *Ultrasonic Transistor and Thyristor Generators* (in Russian), Mashinostroenie, Moscow (1978).

2. Yu. Ya. Borisov, In: *Sources of High-Intensity Ultrasound* (in Russian), Nauka, Moscow, p. 8 (1967).

3. *Ultrasonic Technology* (in Russian), ed. B. A. Agranat, Metallurgiya, Moscow (1974).

4. D. A. Gershgal and V. M. Fridman, *Industrial Ultrasonic Tools* (in Russian), Energiya, Moscow (1976).

5. A. A. Rukhman, *Advanced Design of Thyristor Generators* (in Russian), Cand. Sci. (Techn.) Dissertation (1982).

6. A. F. Nazarenko, *Design of Hydrodynamic Axial Transducers*, Doctoral (Phys.-Math.) Dissertation (in Russian) (1981).

7. R. G. Sarukhanov, *Design of Hydrodynamic Vortex Transducers*, Cand. Sci. (Techn.) Dissertation (in Russian) (1977).

8. A. I. Zimin, *Impulse Excitation of Cavitation*, Teor. Osn. Khim. Tekhn., 1996, no. 2, p. 57.

9. E. Kikuchi, *Ultrasonic Transducers* (Russian translation), Mir, Moscow (1972).

10. V. F. Kazantsev, *Calculation of Industrial Ultrasonic Transducers* (in Russian), Mashinostroenie, Moscow (1980).

11. O. V. Abramov, *Crystallization of Metals in an Ultrasonic Field* (in Russian), Metallurgiya, Moscow (1972).

12. I. I. Teumin, *Ultrasonic Vibratory Systems* (in Russian), Mashgiz, Moscow (1959).

13. I. A. Glozman, *Piezoceramics* (in Russian), Energiya, Moscow (1972).

14. E. G. Smarzhevskaya and N. B. Fel'dman, *Piezoelectric Ceramics* (in Russian), Sovetskoe Radio, Moscow (1971).

15. B. Jaffe, W. Cook, and G. Jaffe, *Piezoelectric Ceramics* (Russian translation), Mir, Moscow (1974).

Chapter 7

Ultrasonic Stacks

Current technological applications of ultrasound with the use of electroacoustic transducers often requires a constructive separation of devices that convert electric energy into the energy of elastic vibrations and those radiating these vibrations to a processed medium, or load. This is dictated by the need to space the transducer from a technological unit or to match the load with the transducer. Technically, it is implemented by using ultrasonic stacks incorporating, in particular, waveguide sections and a radiator.

By now, L. G. Merkulov [1, 2], I. I. Teumin [3–5], E. A. Neppiras [6, 7], E. Eisner [8], A. E. Crawford [9], V. F. Kazantsev [10], Yu. I. Kitaigorodskii [11], R. D. Stafford [12], and others have developed theory and suggested methods for design calculation of ultrasonic stacks guiding and radiating high-intensity vibrations under conditions when the cross-sectional dimensions of waveguide elements is much less than or comparable with the half-wavelength of these vibrations in construction materials. Developed installations enable ultrasonic vibrations be efficiently transmitted to processed media.

7.1 Vibrations in Finite-Size Solids: Cylindrical Sonotrodes

Technological ultrasonic vibratory system incorporates transducer 1, sonotrodes 4 and 5, radiator 6, and fixture 3 (Figure 7.1). Cyclic mechanical stresses arising in transducer excite vibrations in the sonotrodes, which transform vibrations and match the mechanical resistance of an external load and the internal resistance of transducer.

By sonotrode is meant a confined elastic medium capable of transmitting vibrations of a certain type (longitudinal, bending, etc.) from

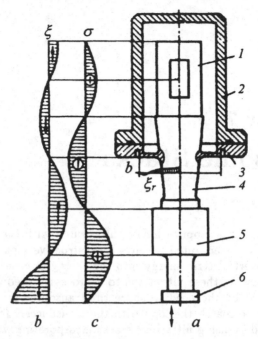

Figure 7.1. (a) Cross-sectional view of an ultrasonic stack and the longitudinal distribution patterns of (b) displacement and (c) stress: (1) magnetostrictive transducer; (2) housing; (3) fixture; (4) horn; (5) sonotrode; (6) radiator. Arrows indicate the direction of wave propagation.

their source to a load. To operate as a guide of certain waves, sonotrode must satisfy some requirements. For instance, longitudinal vibrations in a waveguide representing a uniform-section rod will be excited if [5]

$$0.05 < \frac{d}{\lambda} < 0.5,$$

where d is the maximum linear dimension across a waveguide and λ is the wavelength in the waveguide material.

Waveguides are systems with distributed parameters, i.e. any arbitrarily small parts of a waveguide have mass, elasticity, and energy loss. Generally, a wave guiding system represents an aligned set of uniform and nonuniform waveguide sections (also known as sonotrodes), lumped masses, and lumped elasticities. Uniform waveguide is a rod, whose physical parameters (E or ρ) and cross section do not change lengthwise. The waveguide cross section may be arbitrary; however, tubular and rectangular waveguides are most frequently used. In a nonuniform waveguide, physical parameters or cross section vary longitudinally.

Horn, or velocity transformer, is a sonotrode whose cross section varies longitudinally according to a certain law. Horn serves for the transformation (in particular, amplification) of vibration amplitude and velocity (the type and intensity of vibrations are retained) and for matching a load with the transducer.

Radiator, is an element of an ultrasonic stack that is in a direct contact with processed media, or loads. Radiator is designed so as to provide the most efficient delivery of elastic vibrations to a load at given processing parameters – vibration stress and velocity.

Fixture ensures a rather rigid mounting of a vibratory system to a stationary base with a minimum acoustic energy loss.

Dimensions of a vibratory system should be chosen such that the frequency of its mechanical resonance is close to the electric resonant frequency of the oscillator-transducer pair. When the vibratory system operates at a resonant frequency, vibrations in the sonotrodes attain their maximum intensity. In this case, the efficiency of ultrasonic treatment is also maximum.

When the vibratory system operates in a no-load mode, the longitudinal distribution of displacements and stresses resembles their distribution in standing waves, provided the energy loss in the sonotrodes and radiator is negligibly small (see distribution diagrams of displacement ξ and stress σ in Figure 7.1).

Another important characteristic of vibratory systems is the so-called quality factor, which determines the resonant peak sharpness, the amplitude-frequency dependence, and the operation frequency range (this is because the vibratory system can operate with a varying load).

Connectors, providing a mechanical coupling of the mating elements of the vibratory system, may be either detachable or nondetachable. In any case, they should be reliable, durable, and designed to minimize energy loss in joints.

In addition to acoustic requirements, there are also special requirements to vibratory systems. In particular, radiators for liquid treatment must be resistant to corrosion and cavitation, whereas those intended for the irradiation of molten metals should be refractory.

Depending on technological demands, vibratory systems may differ in their design. The central problem of designing ultrasonic stacks is the calculation of resonant dimensions of their constituting parts.

Generally, oscillations of a solid are described by a set of three differential equations, in which displacement components are interrelated. An exact solution of this set presents a problem; therefore, it is a common practice to simplify equations by taking into account only some of the relations between displacement components.

In practice, the most frequent vibrations are longitudinal vibrations in axial ultrasonic stacks, whereas bending and torsional vibrations are rare.

In ultrasonic technology, the following sonotrodes are used:

- a uniform cylinder (Figure 7.2,*a*);

- a cylinder with an apparent additional mass or elasticity (Figure 7.2,*b*);

- a tubular emitting waveguide (Figure 7.2,*c*);

- a rod with a cross section varying according to a certain law, that is, horn or velocity transformer (Figure 7.2,*d*);

- a blade-like sonotrode (Figure 7.2,*e*);

- a block-like sonotrode (Figure 7.2,*f*);

- a membrane sonotrode(Figure 7.2,*g*).

When longitudinal vibrations are excited in sonotrodes, their lateral dimensions are either much less than, or comparable with, the half-wavelength in the sonotrode material.

Allowance for only longitudinal displacements transforms the above-mentioned set of equations into a differential equation of longitudinal vibrations in a rod. If the rod has arbitrary cross section $S(x)$, the equation can be written as

$$E \frac{\partial}{\partial x}\left[S(x)\frac{\partial \xi}{\partial x}\right] = \rho S(x)\frac{\partial^2 \xi}{\partial t^2}. \tag{7.1a}$$

This equation is valid under the condition that the rod changes so smoothly that stress across any of its cross sections is uniform. Assuming that vibrations follow the harmonic law $\xi = \xi_m e^{j\omega t}$, equation (7.1a) can be simplified

$$\frac{d^2 \xi_m}{dx^2} + \frac{d}{dx}\ln S(x)\frac{d\xi_m}{dx} + \omega^2 \frac{\rho}{E}\xi_m = 0. \tag{7.1b}$$

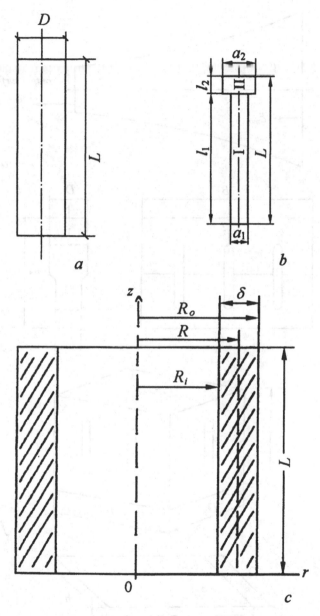

Figure 7.2. Sonotrode shapes: (a) a uniform, circular cylinder; (b) cylinder I with apparent additional mass II; (c) tube; (d) a body of revolution with a cross section varying by a certain law; (e) blade; (f) block; (g) membrane. Input and output amplitudes are ξ_{m0} and ξ_{ml}, respectively.

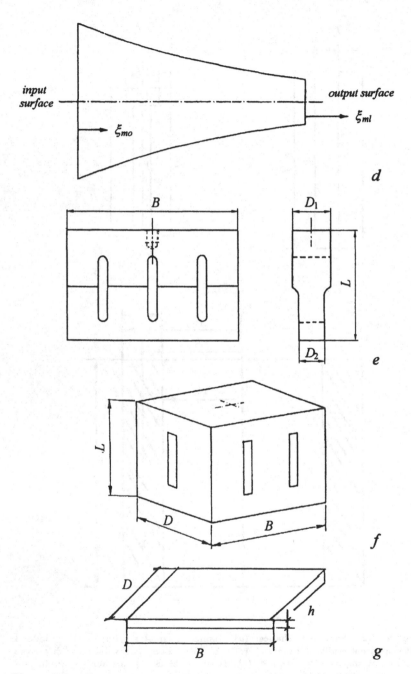

Figure 7.2.   Continued.

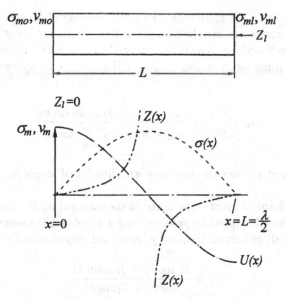

Figure 7.3.   (a) Uniform sonotrode and (b) longitudinal distribution of vibration stress, velocity, and mechanical resistance in the absence of load.

In the case of a uniform rod and absence of energy loss, this equation takes the form

$$\frac{d^2\xi_m}{dx^2} + k^2\xi_m = 0, \qquad (7.1c)$$

where $k = \omega/\sqrt{E/\rho} = \omega/c$ is the wave number.

The solution to equation (7.1c) can be written as

$$\xi_m = A\cos kx + jB\sin kx, \qquad (7.2)$$

where A and B are constants derived from the boundary conditions.

Taking into account a relationship between the vibration velocity and stress

$$\sigma_m = -\omega\rho c\,\xi_m\sin\left(kx\right) \qquad (7.3)$$

and designating peak stress and velocity at the waveguide end adjacent to the load (Figure 7.3) as σ_{ml} and v_{ml}, one can obtain for the amplitudes of stress and velocity at any point of vibratory system

$$\left.\begin{aligned}
\sigma_m &= \sigma_{ml}\cos kx + jv_{ml}w_0\sin kx \\
v_m &= v_{ml}\cos kx + j\frac{\sigma_{ml}}{w_0}\sin kx
\end{aligned}\right\}, \qquad (7.4)$$

where w_0 is the wave resistance. The distribution of vibration parameters lengthwise a uniform free-end sonotrode with the length equal to the half-wavelength in sonotrode material is shown in Figure 7.3. The mechanical impedance at any point of the waveguide system is given by

$$Z_m = \frac{\sigma_m}{u_m} = \frac{\sigma_{ml} \cos kx + j v_{ml} w_0 \sin kx}{v_{ml} \cos kx + j \dfrac{\sigma_{ml}}{w_0} \sin kx}, \tag{7.5}$$

where σ_m and v_m are the complex amplitudes of stress and velocity, respectively.

The mechanical input impedance of the waveguide, $Z_{in} = \sigma_{m0}/v_{m0} = R_{in} + jx_{in}$, is called input impedance and depends on the waveguide parameters, such as length, frequency, and load impedance $Z_l = \sigma_{ml}/v_{ml}$

$$Z_{in} = \frac{Z_l \cos kl + j w_0 \sin kl}{\cos kl + Z_l \sin kl}, \tag{7.6}$$

where l is the waveguide length.

Given the input impedance, it is possible to determine vibration parameters of the waveguide system and resonant frequencies of its constituents. Figure 7.4 gives the distribution curves of vibration displacement over a uniform waveguide for various types of load. Consider some particular cases.

(a) The load is absent, $Z_l = 0\,(\sigma_{ml} = 0)$. In this case, the wave, traveling along the waveguide, is totally reflected from its free end to produce a standing wave. If the waveguide length is multiple to $\lambda/2$, a displacement antinode is at the waveguide end, and a displacement node is at distance $x_0 = \lambda/4$ from the end. The antinode displacement amplitude will be maximum (in Figure 7.4 it is taken to be unity) at a given driving force. Based on relation (7.6), it is possible to estimate the input impedance of the free-end waveguide

$$Z_{in} = X_{in} = j w_0 \operatorname{tg} kl. \tag{7.7}$$

In this case, the input impedance is purely imaginary (there is no energy absorption in the absence of resistive load at the waveguide end).

(b) The waveguide end is fixed, $Z_l = \infty (v_{ml} = 0)$. Now, the distribution of amplitudes over the waveguide is reciprocal of that in the previous case (at $Z_l = 0$), i.e., there is a displacement node and stress antinode at the waveguide end (displacement and stress in the waveguide are out of phase by 90°) (Figure 7.4,b). The input

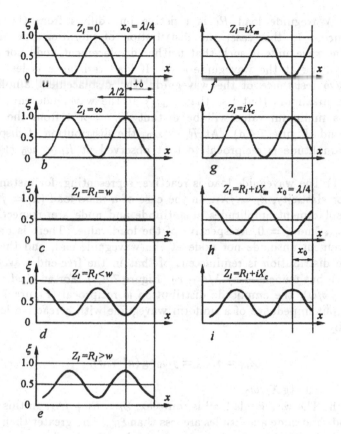

Figure 7.4. Longitudinal distribution of vibration displacements in a uniform sonotrode as a function of load type.

impedance of the waveguide with a fixed end is again purely imaginary

$$Z_{\text{in}} = X_{\text{in}} = -jw_0 \operatorname{ctg} kl. \tag{7.8}$$

(c) The waveguide load is equal to the wave resistance of the waveguide,

$$Z_l = R_l = w_0.$$

In this case, the input impedance of the waveguide is also equal to the wave resistance, $Z_{\text{in}} = R_{\text{in}} = w_0$, which gives rise to a traveling wave. The elastic wave is totally absorbed in the load, called now matched load. The displacement amplitude of the traveling wave is equal to $\xi = 1/2\,\xi_{\text{max}}$ at any point of the waveguide (Figure 7.4,c).

(d) Waveguide load R_l is resistive but differs from the wave resistance. In this case, the distribution of vibration parameters over the waveguide is such that neither displacement node nor anti-node will be at the waveguide end. If load resistance is less than the wave resistance of the waveguide, the displacement amplitude will be maximum (but less than ξ_{max}) at the waveguide end, whereas its minimum value will be distant $X_0 = \lambda/4$ from the wave guide end (Figure 7.4,d). At $R_l > w_0$, the distribution of displacement amplitude is reciprocal to that observed at $R_l < w_0$ (Figure 7.4,e).

(e, f) The waveguide load is reactive, representing, for instance, a mass or elasticity, $Z_l = jX_l$. In the case of a mass load ($Z_l = jm\omega$), the displacement amplitudes at antinode and node are, respectively, $\xi = \xi_{max}$ and $\xi = 0$, irrespective of the load value. There is neither displacement antinode nor node at the waveguide end, and the amplitude distribution is reminiscent of that in the free-end waveguide (case a), but truncated by $\lambda/4 - x_0$ (Figure 7.4,e). For an elastic load ($Z_l = j\kappa/\omega$), the amplitude distribution is reciprocal (Figure 7.4,f). The input impedance of a uniform waveguide with a reactive load is given by

$$Z_{in} = jX_{in} = jw_0 \, \text{tg} \, (kl + \varphi), \qquad (7.9)$$

where $\varphi = \text{arctg} \, X_l/w_0$.

(g, h) The waveguide load is complex, $Z_l = R_l + jX_l$. In this case, antinode and node amplitudes are less than ξ_{max} and greater than zero, respectively. The disposition of nodes and antinodes depends on the sign of the reactive component of the load (Figure 7.4,g, h)

The case, when $l \ll \lambda/2$ (i.e., the sonotrode is short), is of practical interest. Now, $kl \ll 1$ and $\cos kl \sim 1$, $\sin kl \sim kl$, and the expression for the input impedance takes the form

$$Z_{in} \cong \frac{Z_l - w_0 kl}{1 + \dfrac{Z_l}{w_0} kl} = \frac{Z_l - w_m}{1 + \dfrac{Z_l}{w_0} kl} \cong Z_l + X_m. \qquad (7.10)$$

For real loads and metal rods, $Z_l/w_0 \ll 1$ and, hence, $Z_l/w_0 kl \ll 1$. At $Z_l = 0$, the impedance $Z_{in} = X_m$; therefore, short rod can be considered as an apparent additional mass.

Proceeding from (7.7–7.9) and taking into account that at resonance, the imaginary part of the input impedance is equal to zero, one can derive formula relating the resonant frequency of the waveguide to

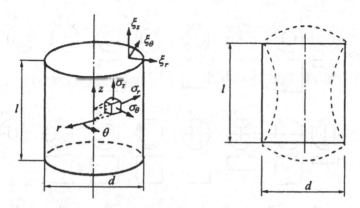

Figure 7.5. Uniform cylindrical sonotrode with length l and diameter d: (*a*) designations; (*b*) longitudinal vibration mode in the sonotrode.

its length l. Resonant frequency f_0 of the waveguide with a free end is given by

$$f_0 = \frac{c_l}{2l} n, \tag{7.11a}$$

where $n = 1, 2, 3$; and c_l is the sound velocity in the waveguide material.

In such a waveguide, distance to the nodal plane is $x_0 = \lambda/4$.

Resonant frequency of the waveguide with a fixed end is given by

$$f_0 = \frac{c_l}{4l} n, \tag{7.11b}$$

where $n = 1, 3, 5$. The first nodal plane coincides with the fixed end.

At certain wavelength-to-diameter ratios, vibrations in the waveguide are more complex than simply longitudinal. Thus, G. McMagon [12] experimentally investigated vibrations in aluminum and steel cylindrical sonotrodes with the length-to-diameter ratio $0 < l/d < 1.7$ and frequency parameters such that $1.2 < kd < 6.2$ (these ranges are of practical interest for the design of ultrasonic cylindrical resonators). More recently, J. Hutchinson [15] performed computer calculations of the resonant frequencies of cylindrical sonotrodes.

Figure 7.5 depicts a cylindrical sonotrode and the longitudinal vibration mode on an expanded scale. Vibration modes observed by McMagon are presented in Figure 7.6. The modes were classified according to the circumferential and longitudinal symmetries of vibra-

Chapter 7. Ultrasonic Stacks

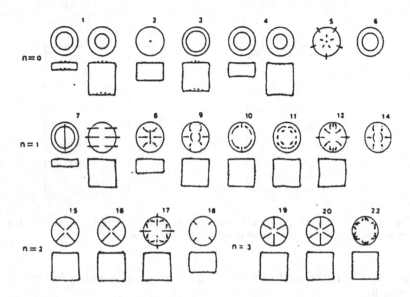

Figure 7.6. Approximate forms of vibration modes in cylindrical sonotrodes presented as cross sections and nodal lines on the surfaces. The circumferential order n indicates symmetry in the radial direction. Modes with even numbers are symmetric, whereas those with odd numbers are antisymmetric about the median plane of the cylinder [12].

tions. The radial displacement is proportional to $\cos(n\theta)$, while the circumferential order n indicates the symmetry with respect to rotation around the cylinder axis. Modes were longitudinally symmetric or antisymmetric, depending on the radial and tangential displacement symmetry $\xi(z) = -\xi(-z)$ with respect to the median plane of cylinder.

Figure 7.7 presents the frequency spectra of modes of different circumferential orders n. All modes except no. 2 show at least one nodal line at the output surface. At small length-to-diameter ratios $(l/d < 0.2)$, the mode corresponds to radial vibrations of a thin disc $(kd = 4.4)$. At $l/d > 1.5$, the mode corresponds to half-wave longitudinal vibrations of a slender rod (see expression 7.11).

Velocity c of the longitudinal wave propagating in a tapered sonotrode decreases with increasing diameter of the sonotrode.

Formulas (7.11) take no account of the influence of the lateral dimension of a waveguide on its resonant length. In the case of a cylindrical waveguide with diameter d, this influence can be accounted for by substituting velocity c'_l for c_l, which are related as follows

$$c_l' = c_l \left[1 - \nu^2 \pi^2 \left(\frac{d}{2\lambda} \right)^2 \right],$$ (7.12)

where ν is the Poisson coefficient.

If the condition $0.15 < d/2\lambda < 0.2$ is fulfilled, error in the resonant length determined by formula (7.11) with allowance for correction (7.12) is less than 3%.

This correction was suggested by Rayleigh. Another refinement, which allows still more correct estimation of the resonant frequency of cylindrical sonotrodes with different length-to-diameter ratios, was suggested by Mori [14], who assumed that the actual vibration mode of the longitudinal wave in cylinder with the length-to-diameter ratio close to unity can be considered as a result of interaction of two orthogonal waves. One of these waves is the longitudinal wave in a slender rod, the other being radial extensional wave in a thin disc. Mori derived an equation relating cylinder length l and diameter d

$$\left(\frac{kl}{\pi} \right)^2 = \frac{1 - (1 - \nu^2) \left(\dfrac{kd}{2a} \right)^2}{1 - \left(\dfrac{kd}{2a} \right)^2 (1 - 3\nu^2 - 2\nu^3)},$$ (7.13)

where $a = 1.84 + 0.68\nu$.

In Figure 7.8, the results of calculations of the resonant length of sonotrodes, made with account for the Rayleigh and Mori's corrections, are presented together with the experimental data of Mc-Magon. Figure 7.9 shows the influence of Poisson's coefficient, frequency, and sound velocity on the cylinder length calculated by formula (7.13)

Of great practical interest is the problem of the displacement amplitude distribution over the emitting surfaces of various cylindrical sonotrodes. Derks [16] analysed the amplitudes of axial and radial vibrations of aluminum cylinders as functions of their diameter (Figure 7.10 and Table 7.1) and found that with increasing diameter, the output amplitude strongly decays in the direction from the cylinder axis toward lateral surface. The experiments were performed at a frequency of about 20 kHz.

The distributions of vibration parameters in nonuniform and uniform waveguides are different. The resonant length of a nonuniform waveguide differs from $(\lambda/2)n$. Thus, the resonant length of a sonotrode

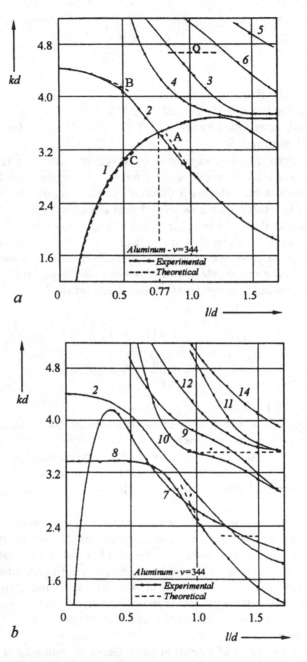

Figure 7.7. Frequency spectra of modes of the circumferential order n: (a) 0; (b) 1; (c) 2; (d) 3. Mode numbers correspond to those in Figure 7.6 [12].

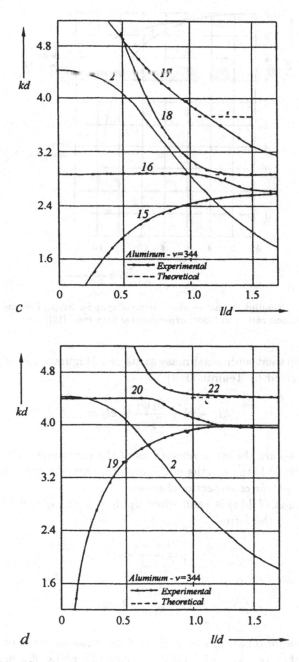

Figure 7.7. Continued.

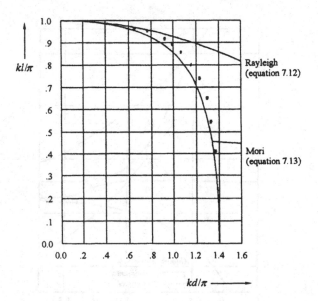

Figure 7.8. Nondimensional representation of cylinder length l versus its diameter d (the Poisson ratio $\nu = 0.344$; experimental data from [12]).

with an apparent additional mass at its end (Figure 7.2,b) is given by formula derived by Teumin [3, 4]

$$(kl_1 - \pi) + \frac{S_1 w_1}{S_2 w_2}\, \text{tg}\, kl_2 = 0, \qquad (7.14a)$$

where w_1, w_2 are the wave resistances of the sonotrode and the mass, respectively; l_1, l_2 the lengths of the uniform part of sonotrode and the mass; S_1, S_2 their cross-sectional areas.

Expression (7.14a) is valid when $d_2/d_1 \leq 2$ and $l_2 \leq 0.1\lambda$ and can be reduced to the form

$$l_1 = \frac{\pi - \varphi}{\omega} c_l, \qquad (7.14b)$$

where

$$\text{tg}\varphi = \frac{S_2 w_2}{S_1 w_1}\, \text{tg}\, kl_2.$$

If the uniform part of sonotrode and the mass are made of the same material, then $\text{tg}\,\varphi = (S_2/S_1)\,\text{tg}\,kl_2$. The nodal plane of a nonuniform waveguide is located at distance $c_n = c_l/4f_0$ from its end.

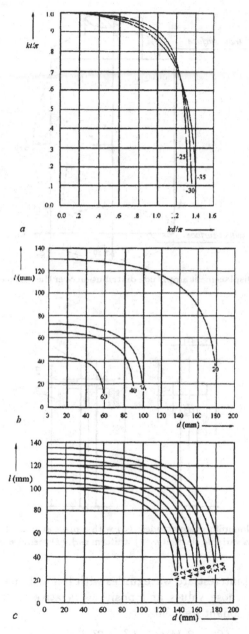

Figure 7.9. Influence of (a) the Poisson ratio ($\nu = 0.25$, 0.30, and 0.35), (b) frequency ($f = 20$, 30, 40, and 60 kHz; $\nu = 0.335$; $c = 5.2 \times 10$ m/s), (c) sound velocity ($c \times 10$ m/s = 4.0, 4.2, 4.4, 4.6, 4.8, 5.0, 5.2, and 5.4; $f = 20$ kHz; $\nu = 0.335$) on the cylinder length and diameter (longitudinal vibration mode) [14].

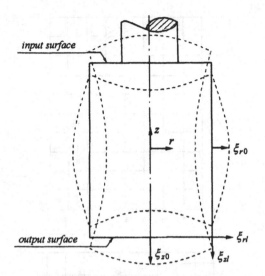

Figure 7.10. Displacement amplitude distribution over the surface of a cylindrical sonotrode [15].

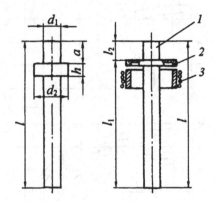

Figure 7.11. Nonuniform sonotrodes: (*a*) with a mass at some distance from the end and (*b*) with a variable elasticity. (*1*) uniform rod, (*2*) membrane, (*3*) solenoid.

Of practical interest are waveguides with lumped mass or elasticity (Figure 7.11), whose values and positions relative to the waveguide determine its resonant frequency

$$f_l \cong \frac{c_b}{l} \left[1 - \frac{(S_2 - S_1)^2}{S_1 S_2} \frac{h}{l} + \frac{S_2^2 - S_1^2}{2 S_1 S_2} \frac{h}{l} \cos k(l - 2a) \right], \qquad (7.15)$$

where l, h, a are the sizes of the waveguide parts (Figure 7.11,*b*).

Table 7.1   Typical amplitude ratios as measured for the longitudinal mode
in a cylindrical sonotrode [16].

Diameter (mm)	80	100	130	160	165
$\dfrac{\xi_{zl}}{\xi_{z0}}$	0.83	0.76	0.40	0.25	0.18
$\dfrac{\xi_{r0}}{\xi_{z0}}$	0.36	0.44	0.54	0.65	0.90
$\dfrac{\xi_{rl}}{\xi_{r0}}$	0	0.1	0.20	0.61	0.66

Expression (7.15) is valid under the following conditions: (a) $S_2/S_1 \leq 1/kh + 1$ and (b) relative change in the resonant frequency of a uniform waveguide in response to the attachment of an additional mass must be much less than unity

$$\frac{\Delta f}{f_0} = \frac{f_0 - f_l}{f_0} \ll 1,$$

here, f_0 is the resonant frequency.

A waveguide with a reactive elastic load can be used for the in-line frequency tuning of an ultrasonic stack with a varying load. Figure 7.11,*b* schematically shows a waveguide with a gradually varying elasticity. Elastic membrane *2* with a ferromagnetic ring is mounted on uniform rod *1* above the ferromagnetic core of solenoid *3*. Membrane elasticity is a function of urrent flowing across the solenoid. The input resistance and the resonant frequency of the waveguide depend on membrane tension, the following equation being valid

$$qx - \text{arctg} \frac{w_0}{w_0 \, \text{tg} \, kl_1(1-q) - \dfrac{m\omega_l}{y^2 - 1}} - \frac{\pi}{2} = 0, \qquad (7.16)$$

where $q = l_1/l$; $y = \omega_l/\omega$; $\omega = \sqrt{1/md}$; D, m is the membrane elasticity and mass, respectively; l_1, l are the waveguide part lengths (Figure 7.11,*c*); ω_l is the resonant frequency of the waveguide.

Of special interest is a waveguide with a varying load [5]. For instance, Figure 7.12,*a* shows a sonotrode with an increasing inertial load of length l_x. At $l_x = 0$, waveguide system with length $l = l_b$ is at resonance. Maximum mistuning of this system will be at $l_x = \lambda_0/4$.

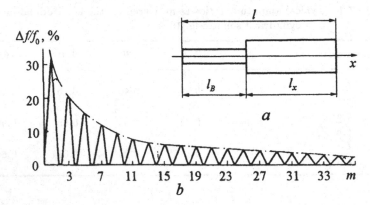

Figure 7.12. To the calculation of resonant frequency: (a) waveguide with a longitudinally varying load; (b) frequency curve.

With a further increase in the vibratory system length, frequency difference will decrease by the law (Figure 7.12,b)

$$\frac{\Delta f}{f_0} = \frac{f_m - f_0}{f_0} = \frac{1}{2m + 1}, \qquad (7.17)$$

where f_0 is the natural frequency of the waveguide with length $2m(\lambda_0/4)$; f_m the resonant frequency of the waveguide with length $(2m + 1)\lambda_0/4$; m the number of half-waves that can be accommodated in the vibratory system length.

Resonant frequency and dimensions of compound ultrasonic stacks can be determined from expression for the input impedance of vibratory system. This expression can be obtained by step-by-step analysis of the waveguide parts from input to output under the assumption that each next waveguide section serves as a load for the next but one section.

In the above consideration, energy losses in waveguides were ignored. Accounting for losses results in that vibration parameters at nodes differ from zero, whereas at antinodes they are not maximum even with the waveguide loads being $Z_l = 0$ or $Z_l = \infty$.

Kitaigorodskii and Yakhimovich [11] showed that losses in a sonotrode can be estimated by the loss coefficient related to quality factor as

$$\epsilon \cong \frac{1}{Q}. \qquad (7.18)$$

At resonance, $\cos kl \sim 1$ and $\sin kl \sim \varepsilon/2 \cdot \pi/2$, from which it follows that

$$Z_{\text{in}} = \frac{Z_l + w_0 \varepsilon \dfrac{\pi}{4}}{1 + \dfrac{Z_l}{w_0} \varepsilon \dfrac{\pi}{4}}. \tag{7.19}$$

The term $w_0 \varepsilon \pi/4$ can be considered as an internal resistance of the sonotrode, R_{int}. Then,

$$Z_{\text{in}} = \frac{Z_l + R_{\text{int}}}{1 + \dfrac{Z_l}{w_0} \dfrac{R_{\text{int}}}{w_0}}. \tag{7.20a}$$

If $Z_l \ll w_0$ and $R_{\text{int}} \ll w_0$, which is typical of the majority of technological ultrasonic stacks, then

$$Z_{in} = Z_l + R_{\text{int}}. \tag{7.20b}$$

The efficiency of the sonotrode, considered as a device transferring energy from the oscillator to the load, can be calculated by formula

$$\eta_w = \frac{N_l}{N_0} = \frac{R_l}{R_l + R_{\text{int}}}, \tag{7.21}$$

where N_0 and N_l are, respectively, the vibration powers consumed from the oscillator and supplied to the load.

Given the displacement amplitude and the coefficient of losses, e, one can determine the efficiency of a uniform waveguide

$$\eta_w = \frac{N_l}{N_0} = \frac{1}{\operatorname{ch} 2\epsilon l + 1/2(k_t + 1/k_t)\operatorname{sh} 2\epsilon l}, \tag{7.22a}$$

where k_t is the traveling wave coefficient.

If losses are small, then

$$\eta_w = \frac{1}{1 + 1/2(k_t + 1/k_t)\epsilon l}. \tag{7.22b}$$

For the traveling wave mode

$$\eta_w = \frac{1}{1 + \epsilon l}. \tag{7.22c}$$

Thus, the performance of ultrasonic stacks depends largely on internal losses. Therefore, waveguides should be fabricated from materials with a small coefficient of losses (absorption). Furthermore, from the

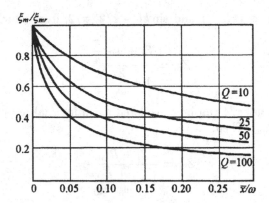

Figure 7.13. Vibration amplitude – load dependences of vibratory systems with various quality factors.

viewpoint of ultimate vibration power that can be delivered to any part of the waveguide without its failure or residual strains, waveguide materials should possess the maximum endurance limit. Such materials as titanium and its alloys, beryllium bronze, aluminum alloys, steels 40Cr, 45Cr, 35CrMnSi, etc. are beneficial. Some properties of materials utilized for fabrication of sonotrodes are summarized in Table 7.2. It should be noted that the coefficient of losses depends on the frequency and amplitude of cyclic stresses (these problems were detailed in chapters 1 and 3).

Quality factor is an important characteristic of ultrasonic stacks. With the frequency deviating from the resonant value, the amplitude of vibrations will fall. For unloaded system, amplitude decrement caused by frequency deviation Δf from the resonant value f_r is given by [11]

$$\frac{\xi_m}{\xi_{mr}} = \frac{1}{\sqrt{4Q_0 \Delta f / f_r + 1}}. \tag{7.23a}$$

When sonotrode is loaded, its quality factor decreases, and its amplitude-frequency characteristic can be described by formula

$$\frac{\xi_m}{\xi_{mr}} = \frac{f_r/f}{\sqrt{(\overline{Z}_l^2 + 1)\left(2Q_0\frac{\Delta f}{f_r}\right)^2 + \left(\frac{f}{f_r}\right)^4 + \frac{4Q}{\pi}\left[2Q_0\overline{r}\frac{\Delta f}{f_r} + \overline{x}\left(\frac{f}{f_r}\right)^2\right]}},$$

$$\tag{7.23b}$$

where $f = f_0 + \Delta f$ is the driving frequency; $\overline{Z}_l = Z_l/\omega_0 = \overline{r} + j\overline{x}$ is the reduced load impedance.

Figure 7.13 is given to illustrate this situation.

Table 7.2 Properties of some materials utilized in fabrication of sonotrodes [11].

Alloy	Treatment	Density $\rho \times 10^{-3}$, kg/m^3	Young's modulus $E \times 10^{-3}$, MPa	Sound velocity c_t, m/s	Wave resistance $\rho c_t \times 10^{-6}$, kg/m^2 s	Strength σ_B, MPa	Durability σ_{-1}, MPa	Coefficient of loss $\epsilon \times 10^4$
Steel 20	As-supplied	7.82	210.0	5174	40.60	400–560	170–270	5.0
Steel 45	As-supplied	7.81	209.2	5169	40.47	600–750	250–340	4.0
	Hardening, tempering at 150°C for 2 h	7.81	203.6	5100	39.93			2.1
	Hardening, tempering at 525°C for 1 h	7.81	208.0	5157	40.34			2.0
Steel 20Cr	As-supplied	7.90	209.8	5155	40.78			4.0
Steel 12CrNi3	As-supplied	7.88	208.1	5136	40.52	950–1400	420–640	4.5
	Hardening, tempering at 150°C for 2 h	7.88	203.6	5082	40.00			2.0
Steel 30CrMnSi	As-supplied	7.70	208.6	5206	40.10	1100–1700	480–700	3.0
	Hardening, tempering at 120°C for 2 h	7.70	204.1	5148	39.64			1.4
	Hardening, tempering at 630°C for 1 h	7.70	208.2	5200	40.04			1.5
Steel 1Cr18Ni9Ti	As-supplied	7.96	202.4	5039	40.17			4.4
Titanium VT-1	As-supplied	4.52	116.2	5072	22.91			1.4
Titanium alloy VT-3-1	As-supplied	4.50	120.7	5178	23.29			1.5
Copper M2	As-supplied	8.93	131.0	3842	34.00			5.0
Brass	As-supplied	8.50	100.0	3450	29.20			2.4
Aluminium alloy D16	As-supplied	2.66	72.0	5200	19.20			3.0

7.2 Design Principles of More Complex Sonotrodes

As mentioned above, not only uniform cylindrical rods but bodies of more complex geometry can be used as sonotrodes (Figure 7.1). In recent years, ultrasonic liquid-phase technology begins to use sonotrodes made in the form of a hollow cylinder (tube), in which excited longitudinal vibrations transform into radial vibrations to be then emitted to a load [17, 18]. In these devices, the electroacoustic transducer can be mounted either from one side of the sonotrode [17] or from both sides (the so-called push-pull system) [19].

In [20], an emitting sonotrode (known as the mode transformation system, MTS) was designed in the form of a hollow metal cylinder (Figure 7.1,c) with length L along the z-axis, internal radius R_1, and external radius R_0 (this corresponds to the wall thickness $\delta = R_0 - R_1$ and mean radius $R = (R_0 + R_1)/2$). The design principles of such vibratory systems are considered below.

First of all, it is necessary to find the eigen frequencies of elastic vibrations in a linear approximation and to calculate the resonant size of the system. To do that, one should find the spatial elastic displacement distribution that would satisfy the elastic equations for transverse and longitudinal waves with velocities c_t and c_l, respectively, under the boundary conditions of free surfaces at $r = R_1$, $r = R_0$ ($0 \leq z \leq L$), and $z = L$ ($R_1 \leq r \leq R_0$). The basic equation for elastic displacements characterized by the vectors ξ_l, ξ_t, and frequency ω can be written as

$$\Delta \xi_{l,t} + k_{l,t}^2 \, \xi_{l,t} = 0, \tag{7.24}$$

where

$$k_{l,t} = \frac{\omega}{c_{l,t}},$$

$$c_l^2 = E \frac{1 - \nu}{\rho(1 + \nu)(1 - 2\nu)},$$

$$c_t^2 = c_l^2 \frac{1 - 2\nu}{2(1 - \nu)}.$$

The boundary conditions are as follows [24]

$$\sigma_{rr}|_{r=R_1 R_0} = 0 \quad \sigma_{r\theta}|_{r=R_1 R_0} = 0 \quad \sigma_{rz}|_{r=R_1 R_0} = 0, \tag{7.25a}$$

$$\sigma_{zz}|_{z=L} = 0 \quad \sigma_{z\theta}|_{z=L} = 0 \quad \sigma_{rz}|_{z=L} = 0, \tag{7.25b}$$

where σ_{ij} is the elastic stress tensor.

Equations (7.24), (7.25) represent a complete set for calculating the resonant sizes of MTS. Boundary conditions at the point $z = 0$ determine the amplitude and the lengthwise distribution of elastic vibrations. Distribution patterns correspond to two possible types of oscillations, namely, penetrating and nonpenetrating waves. The former waves exhibit the following spatial distribution of displacements

$$\xi_r = \Xi_r^{(n)}(r) \cos n\theta \cos k(z-L) \sin \omega t,$$
$$\xi_z = \Xi_z^{(n)}(r) \cos n\theta \sin k(z-L) \sin \omega t. \qquad (7.26)$$

Here, the suffixes l and t are omitted because they are included in terms $\Xi_r^{(n)}$ and $\Xi_z^{(n)}$; the azimuthal dependence on θ is used in the general form for any n ($n = 0, 1, 2, \ldots$).

Nonpenetrating waves are described by formulas

$$\xi_r = \Xi_r^{(n)}(r) \cos n\theta \operatorname{ch} k(z-L) \sin \omega t,$$
$$\xi_z = \Xi_z^{(n)}(r) \cos n\theta \operatorname{sh} k(z-L) \sin \omega t. \qquad (7.27)$$

Radial dependence of penetrating waves (7.26) allows their further classification into the following waves:

(i) volume waves

$$\Xi_r^{(n)} = A_1 J_{n+1}(\kappa_l r) + A_2 N_{n+1}(\kappa_l r) + B_1 J_{n+1}(\kappa_t r) + B_2 N_{n+1}(\kappa_t r),$$
$$\Xi_z^{(n)} = -\frac{\nu}{k(1-\nu)} \left[A_1 \kappa_l J_n(\kappa_l r) + A_2 \kappa_l N_n(\kappa_l r) + B_1 \kappa_t J_n(\kappa_t r) + \right.$$
$$\left. + B_2 \kappa_t N_n(\kappa_t r) \right], \qquad (7.28)$$
$$\kappa_l^2 = k_l^2 - k^2; \quad \kappa_t^2 = k_t^2 - k^2; \quad k_{l,t} > k;$$

(ii) surface waves

$$\Xi_r^{(n)} = A_1 I_{n+1}(\kappa_l r) + A_2 K_{n+1}(\kappa_l r) + B_1 I_{n+1}(\kappa_t r) + B_2 K_{n+1}(\kappa_t r),$$
$$\Xi_z^{(n)} = -\frac{\nu}{k(1-\nu)} \left[A_1 \kappa_l I_n(\kappa_l r) + A_2 \kappa_l K_n(\kappa_l r) + B_1 \kappa_t I_n(\kappa_t r) + \right.$$
$$\left. + B_2 \kappa_t K_n(\kappa_t r) \right], \qquad (7.29)$$
$$\kappa_{l,t}^2 = k^2 - k_{l,t}^2; \quad k > k_{l,t};$$

(iii) mixed waves

$$\Xi_r^{(n)} = A_1 I_{n+1}(\kappa_l r) + A_2 K_{n+1}(\kappa_l r) + B_1 J_{n+1}(\kappa_t r) + B_2 N_{n+1}(\kappa_t r),$$
$$\Xi_z^{(n)} = -\frac{\nu}{k(1-\nu)} \left[A_1 \kappa_l I_n(\kappa_l r) + A_2 \kappa_l K_n(\kappa_l r) + B_1 \kappa_t J_n(\kappa_t r) + \right.$$
$$\left. + B_2 \kappa_t N_n(\kappa_t r) \right], \qquad (7.30)$$
$$\kappa_l^2 = k^2 - k_l^2; \quad \kappa_t^2 = k_t^2 - k^2; \quad k_l > k > k_t.$$

Here, $J_n(z)$ is the Bessel function, $N_n(z)$ is the Neyman function, $I_n(z)$ is the Infeld function, $K_n(z)$ is the MacDonald function, $A_{1,2}$ and $B_{1,2}$ are constants determined by the boundary conditions at $z = 0$.

For nonpenetrating waves, the functions $U_{r,z}^{(n)}(r)$ have the form of (7.28), where $\kappa_{l,t}^2 = k_{l,t}^2 + k^2$.

The substitution of above expressions into the boundary conditions (7.25) gives three equations for each wave type, which define the set of resonant values R_0, R_1, and L:

$$\alpha(a_0)\beta(a,b)+\eta(a_0)\left[C(\eta)-D(a,b)C(\mu)\right]-\mu(b_0)D(a,b)+\eta(b_0)=0,$$

$$\varepsilon(a)\beta(a,b)+\gamma(a)\left[C(\eta)-D(a,b)C(\mu)\right]-\varepsilon(b)D(a,b)\frac{\kappa_t}{\kappa_l}+\frac{\kappa_t}{\kappa_l}\gamma(b)=0,$$

$$\varepsilon(a_0)\beta(a,b)+\gamma(a_0)\left[C(\eta)-D(a,b)C(\mu)\right]-\varepsilon(b_0)D(a,b)\frac{\kappa_t}{\kappa_l}+\frac{\kappa_t}{\kappa_l}\gamma(b_0)=0.$$

Using the notations

$$A'(x) = \frac{dA(x)}{dx}, \quad a = \kappa_l R_I, \quad a_0 = \kappa_l R_0,$$

$$b = \kappa_t R_I, \quad b_0 = \kappa_t R_0 = \frac{b}{a} = a_0$$

one can relate terms, entering into (7.31) and referred to volume waves (7.28), by the following formulas

$$\alpha(c) = J_{n+1}(c) - \frac{\nu^2 a}{(n+1)(1-\nu)}J_n(c), \quad (c = a, b, a_0, b_0),$$

$$\gamma(c) = N_{n+1}(c) - \frac{\nu^2 a}{(n+1)(1-\nu)}N_n(c),$$

$$\varepsilon(c) = J'_{n+1}(c) - \frac{n+1}{c}J_{n+1}(c),$$

$$\gamma(c) = N'_{n+1}(c) - \frac{n+1}{c}N_{n+1}(c),$$

$$\mu(\tilde{a}) = J_{n+1}(\tilde{a}) + \frac{\nu(1-\kappa^2)}{\kappa^2(1-\nu)}J'_n(\tilde{a}), \quad (\tilde{a}=a, a_0); \quad \kappa^2 = \frac{k^2}{k_l^2},$$

$$\eta(\tilde{a}) = N_{n+1}(\tilde{a}) + \frac{\nu(1-\kappa^2)}{\kappa^2(1-\nu)}N'_n(\tilde{a}),$$

$$\mu(\tilde{b}) = J_{n+1}(\tilde{b}) + \frac{\nu(\theta-\kappa^2)}{\kappa^2(1-\nu)}J'_n(\tilde{b}), \quad \theta = \frac{2(1-\nu)}{1-2\nu}, \quad (\tilde{b}=b, b_0),$$

$$\eta(\tilde{b}) = N_{n+1}(\tilde{b}) + \frac{\nu(\theta-\kappa^2)}{\kappa^2(1-\nu)}N'_n(\tilde{b}),$$

$$C(\chi) = \left\{ \frac{\mu(a_0)}{\mu(a)} \chi(b) - \chi(b_0) \right\} \left\{ \frac{\mu(a_0)}{\mu(a)} \eta(a) - \eta(a_0) \right\}^{-1},$$

$$D(a,b) = \frac{n(b)\mu(a) - \alpha(a)[\eta(b) + \eta(a)C(\eta)] + n(a)C(\eta)\mu(a)}{a(b)\mu(a) - a(a)[\mu(b) + \eta(a)C(\mu)] + n(a)C(\mu)\mu(a)},$$

$$\beta(a,b) = \frac{D(a,b)[\mu(b) + \eta(a)\Delta(\mu)] - \eta(b) + \eta(a)\Delta(\eta)}{\mu(a)}. \tag{7.31}$$

For surface waves (7.29), the following changes should be made in formulas (7.31): $J_n \to I_n$ and $N_n \to K_n$.

For mixed waves (7.30), the following changes should be made in the terms of equations (7.31) that have the argument $\tilde{a}$: $J_n \to I_n$ and $N_n \to -K_n$.

Nonpenetrating waves differ from volume waves only in having the sign minus before the second terms in formulas for μ and η.

Basically, with parameters ω and $c_{l,t}$ being given, a set of resonant parameters R_0, R_I, and L can be derived from the transcendental equations (7.31). However, as the analytical solution to equations (7.31) presents a problem, they are usually solved by computing. A formal solution to equations (7.31) is a sequential set of quantities a_j, a_{0j}, b_j (where $j = 1, 2, 3, \ldots$ is the numerical count of the resonance). In the case of volume waves, the set of resonant parameters can be written as

$$R_I = \frac{p_j}{k_l}, \quad p_j = \sqrt{\frac{b_j^2 - a_j^2}{c^2 - 1}}, \tag{7.32a}$$

$$R_0 = \frac{a_{0j}}{a_j} R_I, \tag{7.32b}$$

where $c^2 = c_l^2/c_t^2$, $k = k_l/g_j$, $g_j = \sqrt{(b_j^2 - a_j^2)/(b_j^2 - a_j^2 c^2)}$. By introducing the longitudinal wavelength $\lambda_l = 2\pi/k_l$, one can rewrite equation (7.32a) as

$$2\pi R_I = \lambda_l p_j. \tag{7.32c}$$

The equations (7.32b) and (7.32c) relate the mean radius R and wavelength λ_l by the formula that is well known from experiment

$$2\pi R = \lambda_l p_{0,j}, \tag{7.33a}$$

where $p_{0,j} = p_j(1 + a_{0j}/a_j)/2$.

For small j, $p_{0,j} \sim 1$; therefore, R is directly proportional to λ. A similar relation can be derived for δ. The value for the cylinder length l may be derived from the boundary conditions at $z = 0$

$$L = \frac{\lambda_l}{2} mg_j, \qquad (7.33b)$$

where $m = 1, 2, 3, \ldots$.

In design practice, it is expedient first determine the ultimate resonator dimensions and the type of desirable oscillations in sonotrodes. For instance, one should find conditions for the ultrasonic resonance of volume waves in a metal MTS with given parameters c_l, c, and ω, when the sonotrode resonant sizes are in the limits $R_{0l} \leq R_0 \leq R_{0m}$, $R_{1l} \leq R_I \leq R_{1m}$, $L_l \leq L \leq L_m$ and when the MTS length L can accommodate $n_0 = 1, 2, 3, \ldots$ half-waves. Based on equations (7.31), the quantities a_0, a, and b can be written as

$$a_0 = \sqrt{k_l^2 - \left(\frac{\pi n_0}{L}\right)^2} R_0; \quad a = \sqrt{k_l^2 - \left(\frac{\pi n_0}{L}\right)^2} R_I;$$

$$b = \sqrt{k_t^2 - \left(\frac{\pi n_0}{L}\right)^2} R_I; \qquad (7.34)$$

Their upper and lower limits can be found by using the values $R_{0l}, R_{0m}, R_{1l}, R_{1m}, L_l, L_m, n_0$. To do this, one should derive a_{0l}, a_{0m}, a_l, a_m, b_l, b_m such that $a_{0l} \leq a_0 \leq a_{0m}$, $a_l \leq a \leq a_m$, $b_l \leq b \leq b_m$. The solution to equations (7.31) represents a set of three values a_{0s}, a_s, b_s such that $a_0 = a_{0s}$, $a = a_s$, and $b = b_s$. The set of quantities a_0, a, and b must satisfy all equations (7.31). To find them, it is necessary to solve every equation (7.31) independently with respect to the functions z_i ($i = 1, 2, 3$) of arguments x and y introduced instead of a_0 and b (arguments x and y have the limits $a_{0l} \leq x \leq a_{0m}$ and $b_l \leq y \leq b_m$, respectively). With allowance for this and based on (7.31), one can write for $z_i(x, y)$

$$\alpha(x)\beta(z_1, y) + \eta(x)\left[C(\eta) - D(z_1, y)C(\mu)\right] - \mu(b_0)D(z_1, y) + \eta(b_1) = 0,$$

$$\varepsilon(z_2)\beta(z_2, y) + \gamma(z_2)\left[C(\eta) - D(z_2, y)C(\mu)\right]$$
$$-q(b)D(z_2, y)\frac{\kappa_t}{\kappa_l} + \frac{\kappa_t}{\kappa_l}\gamma(y) = 0, \qquad (7.35)$$

$$\varepsilon(x)\beta(z_3, b) + \gamma(z_3)\left[C(\eta) - D(z_3, b)C(\mu)\right]$$
$$-q(b_3)D(z_3, b)\frac{\kappa_t}{\kappa_l} + \frac{\kappa_t}{\kappa_l}\gamma(b_3) = 0,$$

where $b_i = xy/z_i$ ($i = 1, 3$).

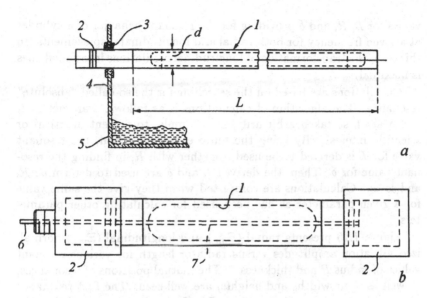

Figure 7.14. Ultrasonic radiators transforming longitudinal vibrations into radial: (a) radiator with one transducer; (b) push-pull system with two transducers. (1) tubular radiator, (2) electroacoustic transducer, (3) fixture, (4) tank wall, (5) load, (6) terminal for the power supply of first transducer, (7) internal connection for the power supply of second transducer.

From the first, second, and third equations of the set (7.35), one can get, respectively, three functions $z_1(x,y)$, $z_2(x,y)$, and $z_3(x,y)$ that are valid in their own areas x, y of three-dimensional space. These functions intersect at a point with coordinates $x = a_{0s}$ and $y = b_s$, representing a required solution, since $z_1(a_{0s},b_s) = z_2(a_{0s},b_s) = z_3(a_{0s},b_s) = a_s$. When there are no or several intersection points, the ranges for a_0, a, and b should be revised.

Thus, this method of ultrasonic resonance analysis is rather simple and effective to be used in finding the sizes of resonators.

Calculations of resonance curves and the relationship between radial and longitudinal displacement amplitudes can be easily performed by the finite element technique [21–23]. It should be emphasized that the comparison of thus derived results with those obtained from analysis of equations (7.31) allows the type of the resonance mode to be determined.

Calculations were made for steel with the elastic parameters $\rho = 7.8 \times 10^3$ kg/m^3, $E = 2.1 \times 10^{11}$ N/m^2, and $\nu = 0.28$ in the region $L \approx 3\lambda_l \approx 600$ nm for $f = 25$ kHz. The aim was to find optimum values for L, R, and δ providing for the resonant behavior of a cylinder at a given frequency for both radial and longitudinal displacements (in this case, the transformation of longitudinal oscillations into radial ones is maximal).

Calculations are based on the searching for the so-called "absolute" resonance. The algorithm of calculations is as follows. Parameters R and δ are first taken arbitrarily, for example, to present practical or scientific interest. By using the finite element method, the resonant value for L is derived to be used, together with R, in finding the resonant value for δ. Then, the derived L and δ are used to determine R, and so on. Calculations are completed when they give the same value for a given parameter as has been used for calculating other parameters.

Figure 7.15 presents radial (Ξ_{mr}) and longitudinal (Ξ_{mz}) normalized vibration amplitudes versus radiator length for various resonant values of radius R and thickness δ. The mutual positions of resonances, as well as their widths and heights, are well seen. The first resonance at $L \approx 515$ mm is somewhat above $5\lambda_l/2$, despite of the fact that the resonant values of R and δ have been chosen for the absolute resonance at $L = 600$ mm, which is close to $3\lambda_l$. It should be noted that the absolute resonance at $L \approx 515$ mm may actually have other resonant values of R, δ, and L, which may be responsible for its lower amplitude in comparison with that at $L = 600$ mm (Figure 7.15,a).

Analogous dependences for R and δ are shown in Figures 7.15,b and 7.15,c. The resonant value for R is less than $\lambda_l/2\pi$, which is probably because factor $p_{0,j}$ differs from unity.

The results of calculation of resonant L, R, and δ for steel and titanium MTS are summarized in Table 7.3.

In spite of different elastic parameters of steel and titanium, their resonant L, R, and δ are close.

The spatial distribution of longitudinal and radial displacements are presented in Figure 7.16. The anticipated phase angle of $\pi/2$ between radial and longitudinal oscillations is actually observed.

Table 7.3 The resonant dimensions of steel and titanium MTS at 25 kHz.

Material	R, mm	L, mm	δ, mm
Steel	31.5	600.0	9.0
Titanium	32.7	599.0	8.0

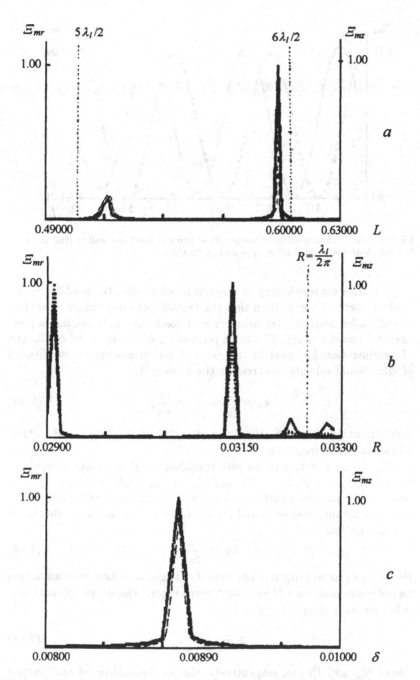

Figure 7.15. To the calculation of the resonant dimensions of steel MTS operated at 25 kHz: (a) length L, (b) radius R, and (c) thickness δ.

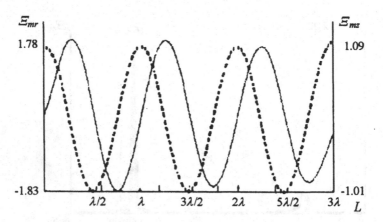

Figure 7.16. Distribution of longitudinal (dotted line) and radial (full line) vibrations lengthwise steel MTS operated at 25 kHz.

In ultrasonic technology, it is often required that the load be irradiated at intensity higher than that the transducer can produce. For this, as well as for matching transducers with loads, one may use horns (mechanical transformers). The basic parameter of a horn is the coefficient of amplification k_a equal to the ratio of the displacement amplitudes of vibrational velocity or stress at the horn ends

$$k_a = \frac{\xi_l}{\xi_0} = \frac{v_l}{v_0} = \frac{\sigma_l}{\sigma_0}, \qquad (7.36)$$

here subscripts 0 and l refer, respectively, to the input and output vibration parameters of horn.

The load, input, and output impedances of horn are related by expressions $Z_l = Z_{\text{out}} = z_l S_l$ and $Z_{\text{in}} = z_0 S_0$, where S_0 and S_l are, respectively, the input and output cross-sectional areas of the horn. The input-to-output cross-sectional area ratio is known as the coefficient of transformation

$$k_T = \frac{S_0}{S_l}. \qquad (7.37)$$

By choosing an appropriate coefficient of transformation one can obtain an optimum matching of load with transducer. The degree of matching is known as matching coefficient

$$k_l = \frac{R_l S_l}{R_{\text{in}} S_0}, \qquad (7.38)$$

where R_{in} and R_l are, respectively, the mechanical input and output (load) resistances of the horn.

It should be noted that transducer efficiency η_{ma} is governed by matching coefficient, which ranges from 10 to 15 in technological vibratory systems.

For horns operated in a traveling wave mode with a negligibly small loss, the coefficients of amplification and transformation are related as

$$k_a = \sqrt{k_T} = \sqrt{\frac{S_0}{S_l}}, \tag{7.39}$$

which implies that amplification coefficient depends on the input-to-output cross-sectional area ratio of the horn. Quantity $N = \sqrt{S_0/S_l}$ is known as area coefficient.

An important characteristic of horns is form (or stress) factor

$$\varphi = \frac{1}{c} \frac{v_m}{\varepsilon_m}. \tag{7.40}$$

Analysis of available vibratory systems [8] show that often the form factor $\varphi \sim 1$ and can hardly be more than 3. At large φ, horn cross section drastically changes lengthwise.

The design calculation of horns was described at length in works [1, 2, 4, 6, 24]. All of the authors of these works assumed that horns guide only longitudinal waves. For this condition to be fulfilled, it is necessary that the maximum cross-sectional dimension of horn be not more than half-wavelength. Otherwise, more complex wave modes traveling in the horn will reduce its efficiency.

Calculations can be based on equation (7.1a). Using a new variable

$$\xi = \frac{\xi'}{r}, \tag{7.41}$$

where

$$r = \sqrt{\frac{S}{\pi}},$$

the equation (7.1a) can be rewritten as

$$\frac{\partial^2 \xi'}{\partial x^2} + \left(k^2 - \frac{1}{r} \frac{\partial^2 r}{\partial x^2} \right) \xi' = 0. \tag{7.42}$$

Solution to this equation depends on the horn cross section function. The simplest case is when the concerned function $r(x)$ is a priori known. However, only some functions $r(x)$ lead to the differential equation (7.42) with known algebraic solutions. The substitution of chosen function $r(x)$ into (7.42) gives

$$\xi' = A\operatorname{ch}\left(\sqrt{k^2 - \beta^2}\,x\right) + jB\operatorname{sh}\left(\sqrt{k^2 - \beta^2}\,x\right), \qquad (7.43)$$

where A, B are constants defined by the boundary condition

$$\beta = \frac{1}{r}\frac{\partial^2 r}{\partial x^2} = \frac{1}{2\sqrt{S/\pi}}\frac{d^2 S}{\partial x^2}. \qquad (7.44)$$

Taking the output displacement amplitude of horn as ξ_{ml} and assuming that it is unloaded, one can get

$$\xi' = \xi_{ml}r_l\left[\operatorname{ch}\left(\sqrt{k^2 - \beta^2}\,x\right) + \varphi_0\operatorname{sh}\left(\sqrt{k^2 - \beta^2}\,x\right)\right],$$

where φ_0 is the value of function φ (given below) at point $x = 0$.

$$\varphi = \frac{1}{\sqrt{k^2 - \beta^2}}\frac{1}{r}\frac{dr}{dx}. \qquad (7.45)$$

Knowing φ, one can determine the displacement and stress distributions over horn.

The positions of displacement nodes are defined by expression

$$\operatorname{tg}\left(\sqrt{k^2 - \beta^2}\,x_{n\xi}\right) = -\frac{1}{\varphi_0}. \qquad (7.46)$$

The positions of stress antinodes and displacement nodes somewhat differ

$$\operatorname{tg}\left(\sqrt{k^2 - \beta^2}\,x_{l\sigma}\right) = \frac{2\varphi\varphi_0 + 1 + \Psi - 2\varphi^2}{2\varphi - \varphi_0(1 - \Psi - 2\varphi^2)}, \qquad (7.47)$$

where $\Psi = \beta^2/(k^2 - \beta^2)$.

The present-day ultrasonic technology employs various types of horns, which differ in the law according to which their cross sections change. The most common horns are tapered, exponential, catenoidal, stepped, and Gaussian (bell-shaped) (Figure 7.17) [6].

Table 7.4 lists formulas for calculating gains, resonance lengths, displacement node positions, and form factors for various horns.

Table 7.4 Formulas for calculating horns [11].

Type	Cross-section function	Displacement amplitude gain, k_a	Resonant length of a half-wave horn, l	Displacement node position, x_0	Form factor, φ
Tapered	$S_x = S_0(1-bx)^2$, where $b = \dfrac{N-1}{Nl}$	$k_a = N\left(\cos kl - \dfrac{N-1}{Nkl}\sin kl\right) \le 4.6$	$l = \dfrac{c_l}{2f_0}\dfrac{(kl)}{\pi} \le 0.72\lambda$, (kl) – roots of equation $\operatorname{tg}(kl) = \dfrac{(kl)}{\dfrac{(kl)^2 N}{(1-N)^2}+1}$	$x_0 = \dfrac{1}{k}\operatorname{arctg} k\dfrac{Nl}{N-1}$	2.29
Exponential	$S_x = S_0 l^{-bx}$, $b = \dfrac{2\ln N}{l}$	$k_a = N$	$l = \dfrac{c_l}{2f_0}\sqrt{\dfrac{\pi^2 + \ln^2 N}{\pi^2}}$	$x_0 = \dfrac{l}{\pi}\operatorname{arcctg}\left(\dfrac{1}{\pi}\ln N\right)$	2.5
Catenoidal	$S_x = S_0\operatorname{ch}^2\gamma(l-x)$, $\gamma = \dfrac{\operatorname{arcch} N}{l}$	$k_a = \dfrac{N}{\cos k'l}$, $k' = \sqrt{k^2 - \gamma^2}$	$l = \dfrac{c_l}{2f_0}\sqrt{\dfrac{(k'l)^2 + (\operatorname{arcch}N)^2}{\pi^2}}$, $(k'l)$ – roots of equation $\operatorname{tg} k'l = -\sqrt{1-\dfrac{1}{N^2}\operatorname{arcch} N}$	$x_0 = \dfrac{1}{k'}\operatorname{arctg}\left(\dfrac{k'}{\gamma}\operatorname{ctg}\gamma l\right)$	—
Stepped	$S_x = S_0$ at $0 \le x \le l/2$, $S_x = S_l$ at $l/2 \le x \le l$	$k_a = N^2$	$l = \dfrac{c_l}{2f_0}$	$x_0 = \dfrac{l}{2}$	0.8
Gaussian (bell-shaped)	—	$k_a = \varphi$	$l = \left[\dfrac{1}{2\pi}\left(\varphi+\dfrac{2}{\varphi}\right)+\dfrac{1}{4}\right]\lambda$	—	$\approx 2(b - 0.027 + 0.4376)^{1/2}\sim 5$, $b = \ln N$

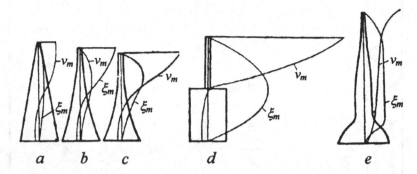

Figure 7.17. Longitudinal cross sections of some horns: (a) tapered, (b) exponential, (c) catenoidal, (d) stepped, and (e) Gaussian (bell-shaped). Curves show the lengthwise distribution of the amplitudes of vibration velocity v_m and deformation ξ_m.

In calculations, one may use the nomograms presented in Figure 7.18. Depending on applications, horns may differ in their cross sections (Figure 7.19).

Exponential hollow horns are calculated by the same formulas as solid ones. At $d_2/d_1 < 1/2$, the length of a half-wave horn can be accurately calculated by formula

$$l = \frac{c_l}{2f}\sqrt{1 + 0.37 \lg \frac{d_1^2 - d_2^2}{d_0^2 - d_2^2}}.$$ (7.48)

If the maximum cross-section size of a horn is smaller than half-wavelength and comparable with the horn length, calculations must involve the correction Δ accounting for radial waves due to the Poisson effect. The resonant length l_Δ of an exponential horn with allowance for this correction is given by

$$l_\Delta = l - \Delta = l\left[1 - \frac{\pi^2 \nu^2}{8}\left(\frac{D_0}{2l}\right)^2 \frac{N^2 - 1}{N^2 \ln N}\right],$$ (7.49)

where ν is the Poisson coefficient.

One should bear in mind that elastic waves would not propagate through the horn whose amplification coefficient is too high. For an exponential horn, this is the case if $(bc_l)/\omega)^2 > 1$.

As follows from analysis of tabulated formulas (Table 7.4), stepped horn possesses the maximum coefficient of amplification. On the other hand, it is characterized by a very sharp resonance and, hence, must be very sensitive to load. That is why ultrasonic stacks with a stepped

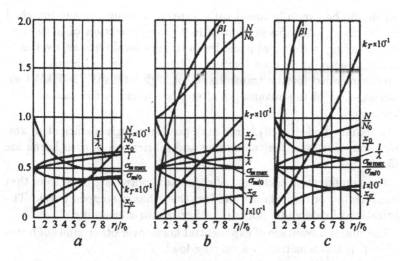

Figure 7.18. Nomograms for designing horns [11]: (*a*) tapered, (*b*) exponential, and (*c*) catenoidal. N is the dissipated power; r_0/r_l is the ratio of reduced input and output horn radii; $\sigma_{m\,\max}$ is the antinodal stress amplitude; σ_{ml0} and N_0 correspond to a uniform-section rod with radius r_l, vibrating with the same amplitude; x_σ is the coordinate of stress antinode.

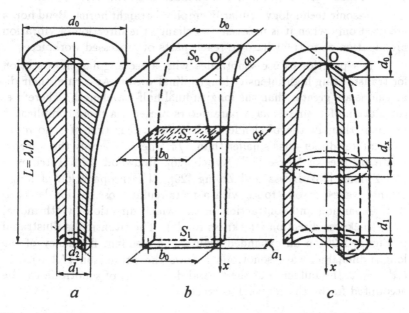

Figure 7.19. Some feasible forms of exponential horns: (*a*) hollow, (*b*) blade-shaped, and (*c*) trepan-shaped.

horn should be carefully tuned with respect to frequency and matched with the load. An abrupt change in the cross section of a stepped horn occurring in its nodal plane leads to high local stresses, overheating, and eventually to horn fracture. Therefore, stepped horns cannot be recommended for the transmission of high-intensity (> 1 kW) vibrations. One of the advantages of stepped horns is the ease of their fabrication.

Gaussian (bell-shaped) horns may possess a high coefficient of amplification. However, the calculation and fabrication of such horns are difficult.

The amplification coefficient of catenoidal horns is greater than that of exponential or conical ones and less than that of stepped horns. The fabrication of catenoidal horns is rather labour-consuming.

The major advantage of exponential horns over other kinds of horns is that it is less sensitive to a varying load.

Tapered horns, although presenting no fabrication problems, possess amplification coefficients smaller than those of exponential horns. Utilization of tapered horns is expedient at $N < 3$.

Half-wave horns are used most frequently. Stepped horns are typically composed of several half-wave sonotrodes connected in series.

Ultrasonic technology primarily employs straight horns. Bend horns are used only when it is necessary to change the direction of vibration propagation or to treat hard-to-reach parts of processed workpieces.

It is a practice to use sonotrodes with a rectangular cross section for transmitting high-intensity longitudinal waves. As a rule, their dimensions are greater than the longitudinal half-wavelength. Therefore, calculation of their resonant parameters presents a more complicated problem than in the of slender rods. For calculation of sonotrodes with lateral dimensions smaller than $\lambda/4$, one may emply the approximate theories of Morse [25], Kynch [26], Redwood [27], Hutchinson and Zilliner [28], Leissa and Zhang [29]. The propagation of longitudinal waves in sonotrodes with a rectangular cross section leads to their expansion and contraction in the width direction (width-mode) or in thickness direction (thickness-mode). The situation is illustrated in Figure 7.20 on an expanded scale. The resonant frequency of the longitudinal half-wave sonotrode follows the expression (7.11a). At $b, d > \lambda/2$, the influence of the lateral dimensions of sonotrode can be accounted for by the Rayleigh correction

$$c' = c \left(1 - \nu^2 k^2 \frac{bd}{4\pi}\right) \tag{7.50}$$

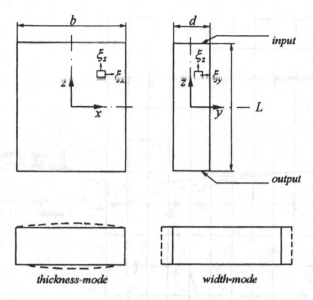

Figure 7.20. Solid rectangular resonator of length *l*, width *b*, and thickness *d*: two propagation modes for longitudinal wave are shown. Thickness-mode has much higher propagation velocity than width-mode (modes are shown as cross sections in the midplane $z = 0$.

or by using the expression

$$\frac{kl}{\pi} = 1 - \frac{1}{24}\pi^2\nu^2\left(\frac{kb}{\pi}\right)^2\left[1 + \left(\frac{d}{b}\right)^2\right].\qquad(7.51)$$

For high ratios kb/π, these corrections are insufficient (Figure 7.21). Better results could be obtained by using expression derived in terms of the modified Mori theory [14]

$$\frac{kl}{\pi}\left(1 - \frac{\nu^2}{1 - \left(\frac{\pi}{kb}\right)^2}\right)^{-1/2}.\qquad(7.52)$$

Figure 7.22 illustrates the influence of width on the resonant sizes of sonotrode.

To reduce the influence of the lateral dimensions of sonotrode on its resonant length and to obtain a more uniform distribution of displacement amplitude over radiating surface, blade-like and block-like sonotrodes are slotted (Figure 7.2,*e,f*).

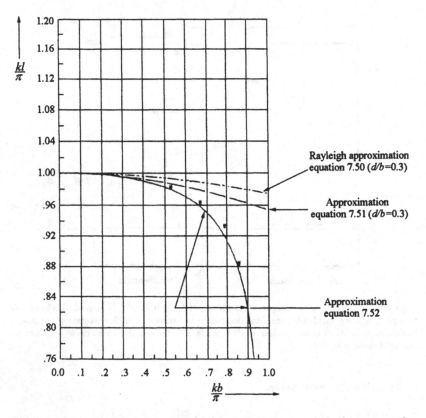

Figure 7.21. Non-dimensional representation of the length l of a rectangular resonator versus its width b: comparison of various theories ($\nu = 0.335$).

An element of vibratory systems, which is in a direct contact with a processed medium, is radiator. The present-day radiators can be divided into two groups:

(a) membrane radiators for processing liquid media;

(b) rod radiators for processing both liquid and solid media.

The latter involve also instruments for ultrasonic welding, processing of hard and brittle materials, and so on.

In each case, the method for transmitting vibrations and the type of radiator are chosen by taking into account the character of the medium (load) to be treated, irradiated area, and the necessary intensity of vibrations in load.

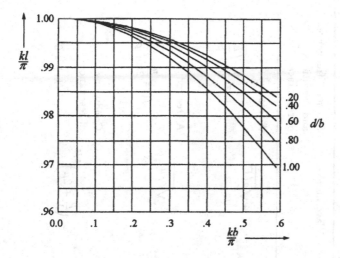

Figure 7.22. Resonant length l of a rectangular resonator versus its width b for various thickness-to-width ratios d/b.

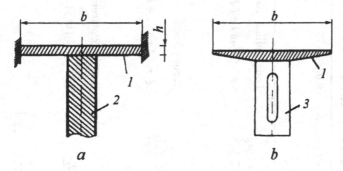

Figure 7.23. Membrane radiators: (a) peripherally fixed, (b) with a nonuniform cross section. (1) radiator, (2) waveguide, (3) transducer.

Membrane radiators are used to match load with the output impedance of transducer as well as to provide a desired efficiency and direction of treatment. Radiator of this type represents a free or peripherally fixed plate, whose thickness h is typically much less than $\lambda/2$, and lateral size b is comparable with $\lambda/2$ (Figure 7.23). Vibrations of the membrane radiator include longitudinal vibrations, which prevail in the central part of the membrane, and bending vibrations on its periphery. To raise the amplitude of bending vibrations, membrane is often made with radially diminishing thickness, which impaires its bending rigidity.

Table 7.5 Equations of vibrations and their solutions for finite-size bodies.

Type of vibrations	Equation of vibrations	Solution	Wave number	Resonant frequences and conditions of resonance
Longitudinal vibrations in rods	$\dfrac{\partial^2 \xi}{\partial t^2} = \dfrac{E}{\rho}\,\dfrac{\partial^2 \xi}{\partial x^2}$	Longitudinal displacement $\xi = (A\sin kx + B\cos kx)e^{j\omega t}$	$\omega\sqrt{\rho/E}$	A free rod $f = \dfrac{n}{2l}\sqrt{\dfrac{E}{\rho}}$
Torsional vibrations in rods	$\dfrac{\partial^2 \xi_t}{\partial t^2} = \dfrac{\mu}{\rho}\,\dfrac{\partial^2 \xi_t}{\partial x^2}$	Torque angle $\xi_t = (A\sin kx + B\cos kx)e^{j\omega t}$	$\omega\sqrt{\rho/\mu}$	A free rod $f_t = \dfrac{n}{2l}\sqrt{\dfrac{\mu}{\rho}}$
Radial vibrations in discs, rings, rods, and tubes	$\dfrac{\partial^2 \xi_r}{\partial t^2} = \dfrac{\lambda+2\mu}{\rho} \times$ $\times\left(\dfrac{\partial^2 \xi_r}{\partial r^2} + \dfrac{1}{r}\dfrac{\partial \xi_r}{\partial r} + \dfrac{\xi}{r^2}\right)$	Radial displacement $\xi_r = [A_0 J_0(kr) + B N_0(kr)]e^{j\omega t}$	$\omega\left(\dfrac{\rho}{\lambda+2\mu}\right)^{1/4}$	Conditions of resonance $J_0(kr) = 0$

		Resonance conditions of a free rod
Bending vibrations in slender rods	Transverse displacement $$\frac{\partial^2 \xi_b}{\partial t^2} = \frac{E}{\rho}\frac{I}{S}\frac{\partial^4 \xi_b}{\partial x^4}$$ $$\xi_b = [A\,\mathrm{ch}\,kx + B\,\mathrm{sh}\,kx + C\cos kx + D\sin kz]e^{j\omega t}$$ $$k = \sqrt{\omega}\left(\frac{\rho S}{EI}\right)^{1/4}$$	$$\mathrm{ch}\,kl\cos kl = 0;$$ rod of radius r, $$f_b = \frac{kl}{4\pi}\frac{r}{l^2}\sqrt{\frac{E}{\rho}}$$ rod with a rectangular cross section $a\times b$ (a is the dimension in neutral plane), $$f_b = \frac{(kl)^2}{4\pi}\frac{\sqrt{3}}{3}\frac{b}{l^2}\sqrt{\frac{E}{\rho}}$$
Axial bending vibrations of thin discs	Transverse displacement $$\frac{\partial^2 \xi_b}{\partial t^2} = \frac{Dh}{\rho}\times$$ $$\times\left[\frac{\partial^4 \xi_b}{\partial r^4} + \frac{2}{r}\frac{\partial^3 \xi_b}{\partial r^3} + \frac{1}{r^2}\frac{\partial^2 \xi_b}{\partial r^2}\right]$$ $$\xi_b = [AJ_0(kr) + BN_0(kr)]e^{j\omega t}$$ $$k = \sqrt{\omega}\left(\frac{\rho}{D}\right)^{1/4}$$	

Footnote: A, B, C, and D are coefficients determined from the boundary conditions; n is the integer; l is the rod length; r is the rod radius; $D = Eh^3/[12(1-\nu)^2]$ is the cylindrical rigidity; h is the disc thickness; λ and μ are Lamé's constants; ν is the Poisson coefficient; J_0 and N_0 are zero-order Bessel functions of the first and second kind, respectively; I is the inertia momentum of cross section S.

Membrane radiators have found an extensive application in ultrasonic cleaning, where large radiating surfaces are required.

Any sonotrode contacting a liquid or solid medium can serve as a radiator. An enlarged radiating surface improves the matching of ultrasonic stack with load and makes radiation more directed. However, too large radiating surface may give rise to bending vibrations and thus reduce the efficiency of elastic vibration feeding into the load. The use of radiators with the amplification coefficient less than unity (the so-called wide-band radiators) is expedient when load varies with time. In such radiators, varying load affects but little their vibrational parameters.

Radiators used for excitation of intense vibrations in corrosive and high-temperature media (for instance, molten steels) must be resistant to their destructive action. None of the presently known materials is appropriate for fabrication of such radiators; therefore, cooled radiators are recently coming into use for processing metal melts.

Apart from longitudinal vibrations, other vibrations can propagate in vibratory systems. Table 7.5 presents some data concerned with such vibrations.

7.3 Designing of Ultrasonic Stacks

As mentioned above, vibratory systems may be of various design. Some constructions are schematically shown in Figure 7.24.

Half-wave vibratory systems are typically used for irradiation of liquids, if there is no need for an enhanced intensity of vibrations or for special matching the transducer and the load (diagram *a*). If transducer has a power insufficient to produce a desired effect, several transduces can be accommodated on the radiator, usually of membrane type. Fixing points of the transducers should be spaced by a distance multiple to the bending wave length (diagram *b*).

Two- and three-half-wave systems (diagrams *c* and *d*, respectively) are commonly used to process melts and solids. In some cases, to concentrate the energy emitted by several transducers, one may use a system such as shown in diagram *e*. Alternatively, diagram *f* shows a single transducer loaded by two radiators.

In the system presented in Figure 7.24,*g*, ultrasound is radiated by a thick part of the transducer vibrating radially as the Poisson effect prescribes.

In some cases, to provide a more efficient transmission of vibrations into processed media, ultrasonic stacks may incorporate sections guiding longitudinal, bending, and torsional vibrations. For example, a

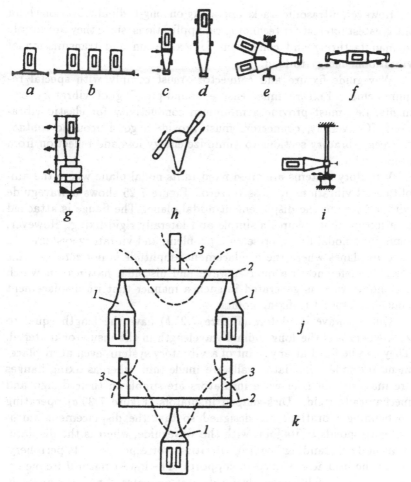

Figure 7.24. Some feasible ultrasonic stack arrangements: (*1*) transducer, (*2*) waveguide, (*3*) radiator, ↔ direction of vibrations; Δ fixture; → connector.

combination of bending and longitudinal resonators enables the transfer of vibrations in a desired direction. Simultaneously, such a system makes it possible to concentrate vibrations from several transducers (Figure 7.24,*h*) or, conversely, to distribute ultrasonic energy from one transducer among several radiators (Figure 7.24,*i*).

Vibratory systems operating on bending vibrations (Figure 7.24,*j*) are used, for instance, in ultrasonic welding.

Torsional vibrations can be generated by a vibratory system such as shown in Figure 7.24,*k*.

However, ultrasonic stacks operating on longitudinal vibrations hold
the greatest interest for technological applications, since they are simple
in manufacturing and convenient for excitation and transmission of
vibrational energy to load.

Waveguide fixture and connectors must comply with special requirements. Fixture must ensure sound-proofing of vibratory elements, i.e., must provide a minimum conductivity for elastic vibrations. Conversely, connectors must provide a good acoustic contact
between vibratory sections to minimize energy loss and reflection from
joints.

Vibratory systems are often fixed in its nodal plane where the amplitude of vibrations is close to zero. Figure 7.25 shows a waveguide
with a flange in the displacement nodal plane. The flange is attached
to a fixture that ensures a simple and laterally rigid fixing. However,
sometimes nodal fixing presents a problem, and vibratory systems are
fixed in planes where the displacement amplitude is not zero. In this
case, of preference are quarter-wave and discoidal fixtures, in which
a standing wave is generated in such a manner that its displacement
nodes correspond to fixing points.

Quarter-wave insulators (Figure 7.25,*b*) have the length equal to
$\lambda/4$, where λ is the longitudinal wavelength in the insulator material.
They can be fixed at any point of a vibratory system, even at displacement antinodes. Insulator walls are made thin, whereas fixing flanges
are massive. Quarter-wave insulators are simple in their design and
mechanically rigid. Disk-shaped insulators (Figure 7.25,*c*) operating
on bending vibrations are designed so that the displacement antinode corresponds to its joint with the waveguide, whereas the displacement node of standing bending vibrations corresponds to its periphery,
where the insulator is fixed to supports. The lowest natural frequency
f_b of a peripherally fixed elastic disc of diameter d and thickness h,
excited in its center, can be calculated with a sufficient accuracy by
formula

$$f_b = 1.88\, \frac{h}{d^2} \sqrt{\frac{E}{\rho\,(1-\nu)^2}}, \tag{7.53}$$

which can also be used to calculate the outer diameter of disc insulator. This formula is valid if the lateral dimension of the waveguide
is significantly smaller than the insulator diameter, otherwise the actual distribution of displacements over the insulator would essentially
differ from calculated distribution. Discoidal insulators can be expediently utilized only for mounting waveguides operated in a standing

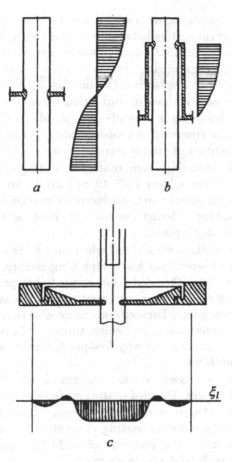

Figure 7.25. Fixation of waveguide sections with the use of: (a) nodal flange, (b) quarter-wave fixture, and (c) disk-shaped insulator. Vibration displacement distributions are shown nearby.

wave mode, when the displacement amplitudes at nodal points are close to zero. In some cases, vibratory system can be fixed by screws at some peripheral points, as well as by crimps or elastic pads. The rigidity of fixing can be enhanced by using two fixtures $\lambda/2$ apart.

To reduce the ultrasonic energy loss brought about by bending, shear, and other waves, fixture thickness must be as small as possible. Nevertheless, real fixtures always cause an additional dissipation of transmitted vibrational energy, which is partially due to losses in fixtures themselves and partially to their inaccurate design and imperfect choice of fixing points.

The efficiency of ultrasonic stacks is also largely dependent on the kind and quality of connectors between waveguide sections and between them and transducers. To introduce minimal energy losses, waveguide sections should be connected at points where stresses are minimum (i.e., at strain nodes or displacement antinodes).

The most common methods of connection are soldering, tightening, and via threads. Soldering is a simple and reliable method. Moreover, it is perfect from the viewpoint of a lossless energy transfer, as providing a good acoustic contact of mating parts.

For ultrasonic soldering, such materials as tin, brass, and silver are often used. Silver solders PSR-40 or PSR-45 are recommended for connecting magnetostrictive transducers to waveguides, whereas tin solders can be used for soldering low-power elements or when waveguide sections are frequently replaced.

Threaded connections are utilized when waveguide sections are frequently replaced or when load has a high temperature, as in the case of molten metals. Since threaded connections fail to provide a perfect acoustic contact of mating parts, propagating elastic waves will be partially reflected from joints. Furthermore, noticeable tangential stresses, arising in these connections, may deform thread and cause unscrewing of waveguide sections. That is why low-pinch threads should be used in threaded connections.

In high-power ultrasonic stacks, sonotrodes are often mated by tightening nodal flanges. Grinded mating surfaces may provide a good acoustic contact between connected elements. The contact can be further improved by stressing mating elements. The shortcoming of screwed connections is the presence of additional passive elements (screws, flanges) with relatively large mass.

Waveguide sections can also be connected by means of an electromagnetic clutch, representing an embracing solenoid with mating waveguide sections inside [3].

References

1. L. G. Merkulov, *Soviet Physics – Acoustics*, **3**, 246 (1957).
2. L. G. Merkulov and A. V. Kharitonov, *Akust. Zh.*, **5**, 183 (1959).
3. I. I. Teumin, *Ultrasonic Vibrational Systems* (in Russian), Mashgiz, Moscow, (1959).
4. I. I. Teumin, *Ultrasonic Stacks* (in Russian), Moscow, (1963).
5. I. I. Teumin, In: *High-Power Ultrasonic Generators* (in Russian), Nauka, Moscow, p. 207 (1967).

6. E. A. Neppiras, *Acustica*, **13**, 368 (1963).
7. E. A. Neppiras, *Ultrasonics International Conference Proceedings*, p. 96 (1977).
8. E. Eisner, In: *Methods and Instrumentation for Ultrasonic Research* (Russian translation), Mir, Moscow, p. 339 (1967).
9. A. E. Crawford, *Industrial Ultrasonics*, Loughborough University of Technology, p. 1 (1969).
10. V. F. Kazantsev, *Designing of Ultrasonic Industrial Transducers* (in Russian), Mashinostroenie, Moscow, (1980).
11. Yu. I. Kitaigorodskii and D. F. Yakhimovich, *Engineering Design of Ultrasonic Stacks* (in Russian), Mashinostroenie, Moscow (1982).
12. G. W. McMahon, *J. Acoust. Soc. Am.*, **36**, 85 (1964).
13. J. Rayleigh and Lord, *Theory of Sound*, MacMillan & Co., London, **1**, 252 et seq (1926).
14. E. Mori, K. Ithoh, and A. Imanuza, *Ultrasonics International Conf. Papers*, p. 262 (1977).
15. J. Hutchinson, *J. Acoust. Soc. Am.*, **51**, 233 (1972).
16. P. Derks, *The Design of Ultrasonic Resonators with Wide Output Cross-Sections*, (1984).
17. *Telsonic Ultrasonic Tube Resonator*, European Patent, no. 0.044.80 (1992).
18. M. Walter and D. Weber, *Ultrasonic Transducer*, United States Patent, no. 5 200 666 (1993).
19. V. O. Abramov, *Ultrasonic Transducer*, German Patent, (1996).
20. O. V. Abramov and O. M. Gradov, *Ultrasonics* (in press).
21. J. J. Connor and C. A. Brebbia, *Finite Element Techniques for Fluid Flow*, Newnes-Butterworth, London-Boston (1977).
22. D. H. Norrie and G. de Vries, *An Introduction to Finite Element Analysis*, Academic Press, New York (1978).
23. J. T. Oden, *Finite Elements of Non-Linear Continia*, McGraw-Hill, New York (1972).
24. L. O. Makarov, In: *Industrial Application of Ultrasound* (in Russian), MDNTP, Moscow, p. 112 (1959).
25. R. W. Morse, *J. Acoust. Soc. Am.*, **22**, 219 (1950).
26. G. J. Kynch, *British Journal of Applied Physics*, **8**, 64 (1957).
27. M. Redwood, *Mechanical Waveguides*, Pergamon Press, (1960).
28. J. Hutchinson and S. D. Zillmer, *Journal of Applied Mechanics*, **50**, 123 (1983).
29. A. Leissa and Z. Zhang, *J. Acoust. Soc. Am*, **73**, 2013 (1983).

Chapter 8

Measurement of Acoustic Parameters

Experiments with high-intensity ultrasound and its industrial applications call for the development of respective metrological methods. To approach this problem, one should take into account effects related to a high frequency of ultrasound and, as a rule, a nonlinear and nonstationary character of load, which is due to various physicochemical processes induced in irradiated media by ultrasound. In some industrial applications, one should also give due consideration for high temperature and corrosive activity of media, such as molten metals.

Acoustic metrological methods can be divided into three groups:

(1) Methods for measuring the input energy (or power) of a transducer and estimating the conditions of its coupling to other elements of a vibratory system;

(2) Methods for measuring vibration parameters (displacement, velocity, acceleration, and strain amplitudes) and determining their distribution over the vibratory system;

(3) Methods for measuring vibration parameters (acoustic pressure amplitude and acoustic energy density) and determining their distribution in processed media.

Below, these three groups of methods are considered with a special emphasis on their physical essence. Great contributions to ultrasonic metrology were made by Yu. I. Kitaigorodskii [11], V. F. Kazantsev [6], L. O. Makarov [9], and others [5, 7, 8].

8.1 Measurement of Energy Exchange between Oscillator and Transducer

In the case of electroacoustic transducers, metrological problems are reduced to measuring the input power of a transducer and its efficiency.

To date, a lot of electron wattmeters have been developed, which can be applied to measure electric power consumed by an ultrasonic transducer. All of these gauges are based on measuring voltage U and current I that, after being converted into signals U_v and U_c, are processed in summators, subtractors, and squarers. As a result, one can derive a signal proportional to the product of the instantaneous current and voltage across the transducer. In particular, the input power of an electroacoustic transducer can be measured by a wattmeter with squarers (Figure 8.1) [1].

If constant output voltages E_1 and E_2 of squarers 1 and 2 are related to their alternating input voltages U_1 and U_2 by relations

$$E_1 = \alpha_0 + \alpha_1 U_1^2,$$
$$E_2 = \alpha_0 + \alpha_1 U_2^2, \tag{8.1}$$

then, under the condition

$$U_1 = U_v + U_c,$$
$$U_2 = U_v - U_c \tag{8.2}$$

constant output voltage E will be equal to

$$E = E_1 - E_2 = \alpha_1(U_1^2 - U_2^2) = \alpha_1 k_v k_c U I \cos\varphi = k N_e, \tag{8.3}$$

where α_0, α_1, k are the coefficients of proportionality; k_v and k_c are the transformation coefficients for transformers T_v and T_c, respectively; U_v is the instantaneous voltage proportional to sypply line voltage U; U_c is the instantaneous voltage proportional to current I; φ is the phase angle between voltage and current.

Thus, the output voltage of the network is proportional to electric power consumed by the transducer.

Thermoelectric converters, vacuum-tube and semiconductor devices based on the Hall effect, etc. are used as squarers.

Calculating electron wattmeters to be used for measuring the electric power of transducers, one should take into account diversity of their working frequencies, presence of elements with sharp phase characteristics, and high intensity of signals. Thus, Mashonis described a device with a ferrite transformer for obtaining current signals [4]. The core

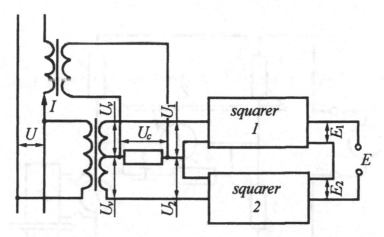

Figure 8.1. Circuit diagram of a wattmeter with squarers.

of the transformer, cut into halves, was mounted on a spring clamp. Such a construction enables the accommodation of the transformer on a current-carrying wire between the oscillator and transducer, which will prevent the flowing of current through the wattmeter and thus will greatly simplify its design.

To estimate acoustic power emitted by a transducer, one should know its input electric power and efficiency.

The efficiency of magnetostrictive transducers can be determined by the procedure well known from the electric and radio-engineering metrology, i.e., by using circle diagrams of input impedance [2, 3, 9]. Measurements usually employ a bridge scheme (Figure 8.2), the transducer under study and a resistance-reactance box being bridge arms. The bridge is balanced with the aid of a zero-indicator connected into bridge diagonal, which enables measuring the input resistance and reactance of the transducer. Both components of impedance are determined as functions of current frequency. The results are plotted in coordinates R_{in} and X_{in}. Near resonance, the line drawn through experimental points is close to a circle (Figure 8.3).

From experimental circle diagrams, one can calculate the efficiency and Q-factor of magnetostrictive transducer. To this end, the parameters of equivalent circuit diagram and the resonant frequencies of transducer should be determined. In calculations, energy dissipation is usually neglected.

To find the transducer impedance, one has to draw a straight line through the coordinate origin and the end point A of the horizon-

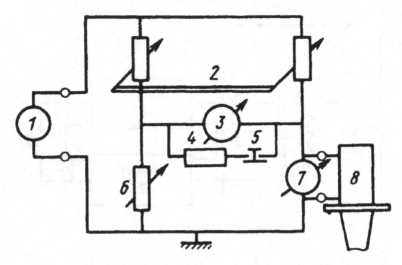

Figure 8.2. Block diagram of a bridge for obtaining circular diagrams: (*1*) electric generator, (*2*) regime regulator, (*3*) null indicator, (*4*) instrument shunt, (*5*) knob, (*6*) impedance box, (*7*) regime indicator, (*8*) magnetostrictive transducer.

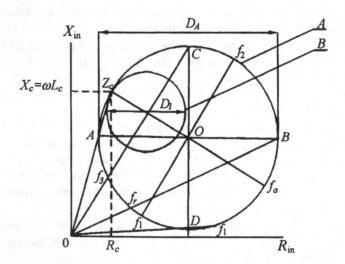

Figure 8.3. Circle diagrams of the input impedance of a magnetostrictive transducer: (*a*) idle-operated; (*b*) operated on a linear load. Figure illustrates the derivation of transducer parameters.

tal diameter until it intersects the circle at some point. The coordinates of this point, Z_c, represent the real and imaginary components of impedance, i.e., resistance R_c and reactance X_c.

The end point of the diameter drawn through point Z_c represents antiresonant frequency f_a. The end points of the diameter perpendicular to diameter $Z_c 0 f_a$ give frequencies f_1 and f_2 that are necessary for the assessment of Q-factor.

Electric quality (i.e., the quality of idle transducer) is given by

$$Q_e = \frac{2\pi f_a L_c}{R_c} \tag{8.4}$$

Whereas its Q-factor near electromechanical resonance is

$$Q_{ea} = \frac{f_a}{f_1 - f_2}. \tag{8.5}$$

Mechanical resonance frequency f_r can be found from the point of circle intersection with the straight line drawn through the coordinate origin and the end B of horizontal diameter. Frequencies f_3 and f_4, necessary for the determination of mechanical quality, can be found from the points of circle intersection with the straight lines drawn from the coordinate origin through the ends C and D of vertical diameter. Then, mechanical quality is given by

$$Q_m = \frac{f_r}{f_3 - f_4}, \tag{8.6}$$

and the magnetomechanical coefficient

$$k = \left[1 - \left(\frac{f_a}{f_r} \right)^2 \right]^{1/2} \tag{8.7}$$

The components of mechanical impedance are

$$R_m = \frac{Z_c^2}{D_A} - R_c, \tag{8.8a}$$

$$L_m = \frac{Q_m R_m}{\omega_r}, \tag{8.8b}$$

$$C_m = \frac{1}{\omega_r L_m}, \tag{8.8c}$$

where D_A is the circle diameter.

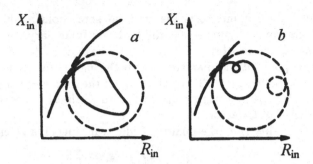

Figure 8.4.   Circular diagrams of the input impedance of a magnetostrictive transducer loaded with: (a) cavitation liquid; (b) load with additional resonances. Circle diagrams of idle transducer are dashed.

Electromechanical efficiency can be found from the relation

$$\eta_{em} = 1 - 2(k^2 Q_e Q_m)^{-1/2}. \tag{8.9}$$

It should be noted that magnetostrictive transducer has the maximum electromechanical efficiency at the frequency of mechanical resonance.

In the presence of a load, the equivalent circuit has to be supplemented with the load resistance R_l.

The circle diagram of a loaded transducer always lies inside the circle diagram of unloaded transducer (Figure 8.3,b), since energy loss in the former case is always greater. Circle diagram depends, but little, on the load properties. Analysis of circle diagrams allows one to get information on the mechanoacoustic efficiency of transducer (it is the higher, the greater is difference in the diameters of paired circles) as well as on the occurrence of natural resonances or non-linear effects in the load (Figure 8.4).

Comparing between the circle diagrams of loaded and idle transducer makes it possible to determine its electroacoustic efficiency at resonant frequency

$$\eta_{ea} = \frac{D_l}{R_l}\left(\frac{D_A - D_l}{D_A}\right), \tag{8.10}$$

where D_l is the diameter of the circle diagram of loaded transducer. The load resistance R_l is given by

$$R_l = \left(\frac{Z_c^2}{D_l} - R_c - R_m\right). \tag{8.11}$$

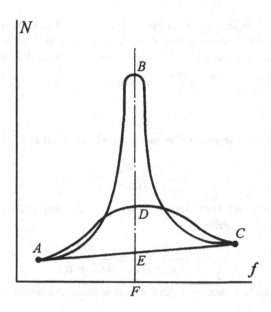

Figure 8.5. Frequency dependence of the input power of (*1*) idle and (*2*) loaded transducer.

For a correct estimation of transducer efficiency, it is necessary to provide a perfect acoustic contact between the radiator and a medium, as well as the absence of standing waves and cavitation in the medium.

An alternative method for determining transducer efficiency was proposed by Perkins [3]. With this method, the efficiency is deduced from the frequency dependence of the input power of idle and loaded transducer (Figure 8.5).

The electromechanical efficiency of idle transducer is given by

$$\eta'_{em} = \frac{N'_m}{N'_m - N'_e},$$
(8.12)

where N'_m is the idle mechanical power equal to

$$N'_m = \frac{U_m^2}{2R_m}$$
(8.13)

and

$$N'_e = \frac{U_m^2 R_c}{R_c^2 + (\omega L_c)^2}$$
(8.14)

is the idle power loss.

Electric power loss of transducer can be found graphically. To this end, nonresonant points A and C of the curve ABC have to be connected with a straight line. The intercepts EF and BE characterize the electric power loss of transducer and its idle mechanical power, respectively. Then,

$$\eta'_{em} = \frac{BE}{BE + EF}. \tag{8.15}$$

Intercept DE characterizes the mechanical power of loaded transducer

$$N_m = \frac{U_m^2}{R_m + R_l}. \tag{8.16}$$

Taking into account that electric powers of idle and loaded transducer are equal, one can write

$$\eta_{em} = \frac{N_m}{N_M + N_e} = \frac{DE}{DE + EF}. \tag{8.17}$$

Then, the mechanoacoustic and electroacoustic efficiencies of transducer are

$$\eta_{ma} = \frac{R_l}{R_l + R_m} = \frac{\eta'_{em} - \eta_{em}}{\eta'_{em} - \eta_{em}\,\eta'_{em}}, \tag{8.18}$$

$$\eta_{ea} = \eta_{ma}\eta_{em}\frac{R_l}{R_l + R_m}\frac{1}{1 + \dfrac{R_e(R_l + R_m)}{R_e + \omega L_e}} = \frac{1 - \dfrac{DE}{BE}}{1 - \dfrac{EF}{DE}}. \tag{8.19}$$

The maximum efficiency of transducer (the so-called optimum efficiency) is attained when load is optimal

$$R_{\text{opt}} = \frac{R_m Z_c^2 + R_c R_m^2}{R_c}. \tag{8.20}$$

It should be noted that the electromechanical efficiencies of loaded and idle transducer are related as

$$\eta_{em,\text{opt}} = 1 - \sqrt{1 - \eta'_{em}} = 1 - \alpha. \tag{8.21}$$

Since mechanoacoustic efficiency is

$$\eta_{ma,\text{opt}} = \frac{1}{1 + \alpha}, \tag{8.22}$$

then,

$$\eta_{ea,\text{opt}} = \frac{1 - \alpha}{1 + \alpha}. \tag{8.23}$$

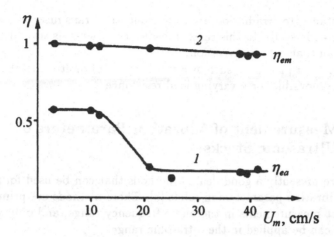

Figure 8.6. (*1*) Electroacoustic and (*2*) electromechanical efficiencies of transducer versus vibration velocity [21].

These methods of calculation of transducer efficiency are valid only for low-amplitude driving magnetostrictive forces, otherwise non-linear properties of transducer have to be taken into account.

Generally, load impedance has both real and imaginary components. Load is purely resistive only when it and ultrasonic stack are at resonance.

If load is nonlinear, the construction of circle diagrams is more sophisticated. By varying the frequency of exciting oscillator, its output voltage amplitude can be made such that the vibration velocity amplitude at the radiating surface of the transducer remains constant. In this case, the magnetostrictive transducer coil has a constant voltage-to-frequency ratio, whereas piezoceramic transducer has a constant voltage across its plates.

In the absence of non-linear effects, R_l depends neither on amplitude nor time. As soon as non-linear effects develop, the load resistance changes. For instance, cavitation reduces wave resistance, as a result of which emitted power increases with the amplitude of vibrations slower than it does in the absence of cavitation. This effect can be accounted for by introducing the effective resistance of emission, $\overline{R_l} = 2N_a/\xi_m^2$.

Varying load resistance and output acoustic power will change transducer efficiency, which is well seen in Figure 8.6. As soon as vibration velocity reaches the cavitation threshold, the electroacoustic efficiency of transducer significantly falls. Upon developed cavitation, the resistance to irradiation and, consequently, transducer efficiency η_{ea} are steady-state. In this case, transducer efficiency is several times lower than that before cavitation.

Some other processes, such as crystallization of molten metals, may also be responsible for a varying load resistance.

8.2   Measurement of Vibration Parameters of Ultrasonic Stacks

There are presently a good deal of methods that can be used for measuring vibration parameters in solids [8]. These methods are primarily for direct measurements in the audio-frequency range, and only a few of them can be applied in the ultrasonic range.

During direct measurements, a probe is brought into a contact with a vibratory system. The contact will certainly affect the measured parameters; however, errors may be neglected if the probe dimensions and mass are significantly less than respectively, wavelength of measured vibrations and mass of any component of the ultrasonic stack, for instance, waveguide or radiator.

This largely restricts the feasibility of employment of conventional vibrometric facilities in ultrasonics and calls for the development of contactless methods for measuring the amplitudes of displacements, velocities, accelerations, and strains.

Displacement amplitudes can be most conveniently measured with an optical microscope. This method is based on the observation of the lateral surface of a vibratory element under a microscope whose optical axis is perpendicular to the axis of an axial element or parallel to the axis of an annular element. A distinct point on the lateral surface becomes 'blurred' after the onset of vibrations. The extent of this 'blurring' corresponds to the peak-to-peak value (double amplitude) of vibrations. The microscope magnification is chosen based on the amplitude of vibrations, necessary fosuc depth, errors caused by transverse vibrations, and thermal strains in the vibratory element.

The advantages of this method are its relative simplicity, possibility to take measurements under both idle and loaded operation conditions, high linearity during the measurement of high-amplitude displacements (up to 100 μm), as well as the independence of sensitivity on frequency. However, this method makes difficult the measurement of small (below 1 μm) amplitudes or conversion of measured values into an analog electric signal. Nor it allows the study of the distribution of displacement

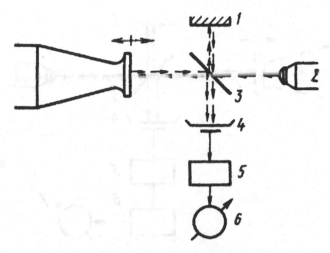

Figure 8.7. Schematic representation of laser interferometer: (*1*) mirror, (*2*) laser, (*3*) half-silvered mirror, (*4*) photocell, (*5*) detector, (*6*) indicator, (*7*) vibratory system.

amplitudes over radiating surface and vibratory elements positioned normally to the stack axis (for instance, suspension flanges).

It should be noted that this method can be conveniently used for calibrating probes and gauges that otherwise are incapable of measuring the absolute values of measured parameters.

An interesting approach was suggested by Teumin: a plate with a piece of abrasive paper is pressed to the polished surface of a cylindrical gauging waveguide [10]. During measurements, the paper is made to rapidly move along the gauging waveguide, scratching its surface. Due to the superposition of the translational motion of the paper and the vibrational motion of the waveguide, the polished surface appears to be scratched with numerous sinusoidal curves with the amplitudes being the displacement amplitudes of given points of the waveguide. Examining then the waveguide surface under a microscope, one can obtain the amplitude of vibrations and their distribution over the gauging waveguide.

Laser engineering also provides an opportunity to measure the amplitude and velocity of vibration displacements. By now, some modifications of laser interferometers and anemometers have been designed. In laser interferometer (Figure 8.7), the primary beam is splitted by half-silvered mirror *3* into two beams, one of which is incident onto a stable mirror, whereas the other falls onto the reflecting region of vi-

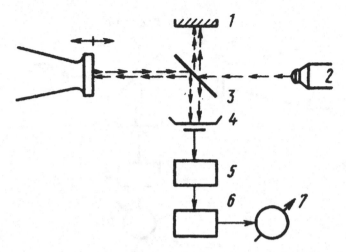

Figure 8.8. Schematic representation of laser anemometer: (*1*) mirror, (*2*) laser, (*3*) half-silvered mirror, (*4*) photocell, (*5*) Doppler effect gauge, (*6*) detector, (*7*) indicator, (*8*) vibratory system.

brating surface *7*. The reflected beams interfere in photocell *4* whose output signal depends on the phase difference between the light beams. The phase of the beam reflected from the vibrating surface is determined by its distance from the half-silvered mirror, i.e., this beam is modulated by surface vibrations. Accordingly, the output signal of the photocell is also modulated, which enables the displacement amplitude of the vibrating surface to be determined by demodulating the output signal.

Theoretically, laser interferometer is capable of operating in a wide range of ultrasonic frequencies; however, if the vibration amplitude exceeds the half-wavelength of the laser, it fails to provide an unambiguous correlation between the measured values and the output signal. As a result, the upper limit of measured displacement amplitudes is $\sim 0.2 \; \mu$m.

Laser anemometer (Figure 8.8) analyses the frequency of a measuring beam modulated due to the Doppler effect. Special processing makes the output signal proportional to the amplitude of vibration velocity of vibrating surface. Unlike laser interferometer, laser anemometer is unable to analyse vibrations of very low amplitude.

A considerable limitation of both laser instruments is that they must be mounted only on an optical bench, which makes them bulky and expensive. Practically, they can conveniently be used for calibrating other types of gauges.

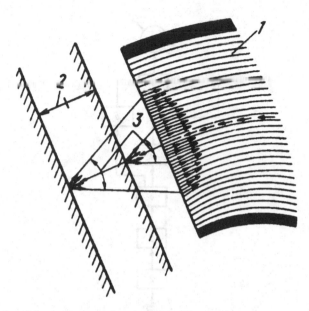

Figure 8.9. Schematic representation of the fiber-optic measurement of vibration amplitude: (*1*) fiber-optic bundle, (*2*) ultimate positions of vibrating surface, (*3*) angle of scattering.

A recent trend in ultrasonic metrology is toward the use of the methods of holographic interferometry.

Fiber-optic methods also hold great promise for this purpose. Figure 8.9 gives a schematic representation of one of the suggested fiber-optic arrangements [9]. Light delivered to a vibrating surface by one fiber of the bundle is scattered from it to be incident on the bundle face. The area of the face region (or the number of fibers), onto which the scattered light is incident, will be proportional to a gap between the fiber-optic bundle and the scattering vibrating surface.

The illuminated area of the bundle face (the number of illuminated fibers) can be determined by using a photocell whose analog signal will be proportional to the gap width and, in the final analysis, to displacement amplitude.

The amplitude of vibrations can be measured not only optically, but with the aid of a contactless parametric vibrometer of UBV type [11]. The vibrometer uses the varying magnetic reluctance of the gap between a probe and a vibrating surface to control the frequency of self-oscillator whose current is measured by an indicator. The vibrometer probe is made as a flat induction coil with the axis perpendic-

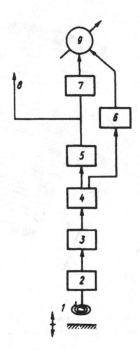

Figure 8.10. Block diagram of a UBV-2 contactless vibrometer: (*1*) vibrating surface, (*2*) inductive pickup, (*3*) heterodyne, (*4*) amplitude limiter, (*5*) frequency detector, (*6*) ultrasonic amplifier, (*7*) detector, (*8*) output to oscillograph and other devices, (*9*) indicator.

ular to vibrating surface (Figure 8.10). The magnetic circuit of this coil includes internal magnetic reluctances connected in series, a dissipation field, a thin conducting layer on the emitting surface (this type of vibrometers is used for analysis of vibrations of only metallic surfaces), and a gap. Gauging induction coil is a constituent of the oscillatory circuit of heterodyne oscillator *3* whose frequency is much higher than the upper ultrasonic frequency to be measured. Waveguide vibrations modify the gap width and, consequently, the reluctance of the coil, which finally leads to the frequency modulation of heterodyne oscillations. The modulation frequency and amplitude correspond, respectively, to the frequency and displacement amplitude of ultrasonic vibrations.

Further processing of data is performed by conventional methods. It should be emphasized that the sensitivity of vibrometer largely depends on the mean gap width; therefore, it must always be set to the mean width determined during vibrometer's calibration (for this pur-

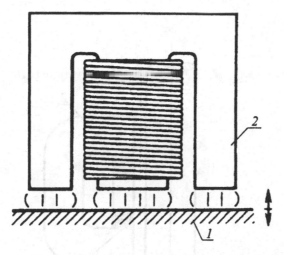

Figure 8.11. Electromagnetic pickup of displacement amplitude: (*1*) vibrating surface, (*2*) pickup.

pose, the device is equipped with a mechanical handler and an arrow indicator of the gap width). The vibrometer offers a relative accuracy of 2.5%, its sensitivity being linear in a frequency range of 10 to 50 kHz. The shortcoming of the vibrometer is that it measures only the normal component of surface vibrations. Because of this, the vibrometer can hardly be used for measuring displacement amplitudes of axial loaded vibratory systems.

Among contactless gauges those employing electromagnetic probes should be mentioned (Figure 8.11). The magnetic system of such probes is fixed over the surface of a ferromagnetic ultrasonic radiator. In the presence of vibrations, the gap between the probe and the vibrating surface changes, giving rise to a magnetomotive force in the ferromagnetic core of the probe and to an electromotive force in its coil. Emf corresponds to analysed ultrasonic vibrations with respect to phase, frequency, and amplitude. To enhance the spatial resolution of the method, the electromagnetic probe may be equipped with conical poles.

In electrodynamic probes, which are also often used, a strong permanent magnet is coupled, through the pole tip and thin gap, to the conducting layer of a sonotrode (Figure 8.12). The induction of vibrations in this system is accompanied by eddy currents whose intensity depends on the velocity of elementary conductors moving in the magnetic field, i.e., on vibration velocity. Gauging coil detects these eddy

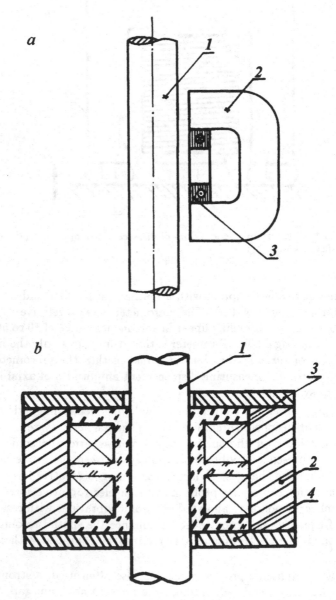

Figure 8.12.   (a) Superposable and (b) through-type electrodynamic probes: (*1*) waveguide, (*2*) permanent magnet, (*3*) coil, (*4*) discoidal ferromagnetic cores.

currents, yielding respective electric signals. Depending on the metrological problem to be solved, electrodynamic probe may be either superposable (Figure 8.12,*a*) [12] or of the through-type (Figure 8.12,*b*) [13]. In the latter case, magnetic circuit includes a circular permanent magnet and two discoidal magnetic coils pressed to the magnet. Sonotrode of the vibratory system to be monitored is placed inside the probe with a spacing of 0.05–0.1 mm. The use of two identical gauging coils instead of one is because in-phase vibrations near the two opposite poles of the permanent magnet might induce antiphase eddy currents whose overall action on a single gauging coil will be virtually nullified. At the same time, two antiphase magnetic fluxes of two identical gauging coils connected oppositely will induce in-phase electromotive forces.

Probe of this type has a specific feature: if gauging coils are connected aiding, their signal will reflect difference in the vibrations of both zones of transformation, i.e., the signal is proportional to the total vibration strains (stresses) in the monitored part of the system.

Probe sensitivity depends largely on the material, of which the monitored part of the vibratory system is manufactured. For instance, probe sensitivity for sonotrode from 40CrNi2MoA steel is almost 45 times higher than that for sonotrode from titanium alloy VTZ-1 [9].

Sometimes, one has to use contact metrological methods instead of more convenient contactless ones. This is because the sensitivity of contactless methods depends on the gap width, which may vary in the course of a long-term operation.

The simplest gauging probe is a standard pickup with a contact needle and a thin piezoelectric plate.

It should, however, be noted that standard pickups can be used for measuring displacement amplitude only in the range 0.1–5.0 μm at frequencies below 22 kHz. Moreover, such a pickup provides no particular information on the spatial polarization of displacements, since the electric signal of this probe is generated under the action of the needle tip vibrations, both transverse and longitudinal with respect to its axis.

Of the presently available contact probes, piezoelectric and magnetostrictive accelerometers [14] are worthy mentioning. The former represents a piezoceramic plate glued to some element of a vibratory system, and the latter consists of a winding on a miniature U-shaped laminated core made of a magnetostrictive alloy and closed with a permanent magnet (typically, ferrite). The magnetostrictive accelerome-

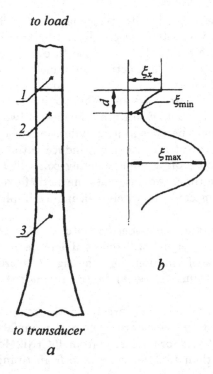

Figure 8.13. Measurement of acoustic power transmitted through an ultrasonic stack. (*a*) Part of the ultrasonic stack with a gauging waveguide: (*1*) sonotrode, (*2*) gauging waveguide, (*3*) horn. (*b*) Distribution of displacement amplitude over the gauging waveguide.

ter is soldered to its own metal waveguide possessing a screw butt for mounting the accelerometer on a vibrating surface. Such a probe is tolerant to electromagnetic induction, can detect polarized vibrations, and possesses a stable sensitivity.

Displacement amplitudes can also be measured with a clock-type mechanical indicator of small displacements [15].

Thus, the available technical facilities for measuring vibrational motion in ultrasonic stacks are diverse and their number can undoubtedly be considerably enlarged.

If load is almost linear, the emitted acoustic power can be estimated by measuring the vibration parameters of ultrasonic stack. In particular, this power can be found from the displacement amplitude distribution over the built-in gauging waveguide [16] (Figure 8.13) representing a uniform sonotrode made of a low-loss material (e.g., aluminum or titanium) and having the length $\lambda_m/2$ (λ_m is the wavelength in the gauging waveguide material).

Expression for the load resistance can be obtained from analysis of ultrasonic system operation mode

$$R_l = w_0 \frac{k_\delta}{k_\delta^2 \cos^2 \varphi + \sin^2 \varphi}, \qquad (8.24)$$

where $k_\delta = \xi_{min}/\xi_{max}$ is the traveling coefficient; $\varphi = 2\pi d/\lambda_m$; d is the distance from the gauging waveguide end to its first displacement node.

The emitted acoustic power is given by

$$N_a = \frac{1}{2}\xi_k^2\omega^2 R_l, \qquad (8.25)$$

where ξ_k is the displacement amplitude at the gauging waveguide end.

Quantity N_a represents the input power of the primary waveguide (radiator). In most cases, the above expression, which takes no account for the energy loss in the waveguide, is sufficient to estimate radiated power.

8.3 Measurement of Vibration Parameters in Loads

Methods that can be employed for measuring acoustic parameters in loads depend on their state (gas, liquid, or solid).

It is a common practice to measure acoustic pressures in gases with electrostatic or electrodynamical microphones, whereas piezoelectric and other types of microphones are rarely used for this purpose [7, 9]. The choice of microphone type is dictated by measured frequencies, intensity of vibrations, specificities of ultrasound radiation, as well as the level and character of noise.

In particular, condenser microphones have found a wide use due to their relatively high sensitivity, uniform frequency response, and miniature dimensions.

Consider some facilities for measuring acoustic parameters in liquids. Such a consideration seems to be necessary because of the commensurability of linear sizes of load and ultrasonic wavelength, complex geometry of load, its indefinite reflecting properties, occurrence of acoustic streamings and cavitation, etc., which often makes the theoretical calculation of acoustic parameters impossible.

On the other hand, it is real acoustic parameters and the character of their distribution in load that largely determine the physicochemical effects of ultrasonic treatment.

Local acoustic pressure in liquid load is often measured with hydrophones (typically, piezoceramic). In this case, hydrophone dimensions are of great metrological importance. Indeed, large hydrophones immersed in liquid may considerably disturb acoustic field, promote cavitation, and modify acoustic streaming. Additionally, they exhibit lower resonant frequencies and narrower operation frequency ranges than small hydrophones.

On the other hand, miniature hydrophones, in which the working substance volume and electric capacitance are reduced, have a more sophisticated design and lower sensitivity

It was empirically established that linear dimensions of a piezoelement should be less than 0.1–0.25 ultrasonic wavelength at the highest frequency of a desired operation range.

Figure 8.14 schematically shows some currently employed acoustic hydrophones. Depending on the design and piezoceramic properties, hydrophones exhibit various directional characteristics and sensitivities. Thus, hydrophone described in [17] contains a 2.5 mm thick piezoceramic disk 6 mm in diameter embedded in a fluoroplastic damper (Figure 8.14,c). The hydrophone shows a uniform amplitude–frequency response up to 200 kHz, whereas its directional characteristic is almost spherical in front and has a well pronounced dip from behind.

A wide-band waveguide-type hydrophone (Figure 8.14,d) [9] responds to acoustic pressure via a two-conductor screened high-frequency cable protruded from the sealed end of a metal casing. One conductor of the cable is ended with a pressed-on piezoceramic cylinder. Electric signal, proportional to the amplitude of acoustic pressure, is picked up from the surface of this cylinder by the other conductor of the cable. The extent of signal damping necessary for operating the system in a traveling wave mode and smoothing its amplitude–frequency characteristic results from the lateral emission of vibrations into the elastic insulator of the cable.

Hydrophones of this type with 2–3 mm piezoelements have uniform amplitude–frequency characteristics up to 1–3 MHz and a sensitivity of about 0.2 μV/Pa.

Complex receiver, representing the combination of a spherical hydrophone and two piezovibrometers, was described in [18]. Piezovibrometers are connected differentially and separated from the hydrophone. The hydrophone signal depends on acoustic pressure, whereas vibrometer signals depend on the projection of vibration velocity vec-

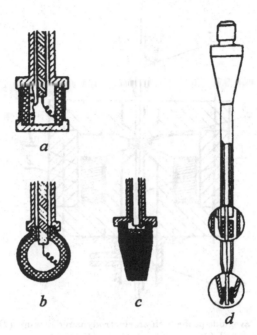

Figure 8.14. Cross-sectional views of various hydrophones: (a) cylindrical, (b) spherical, (c) for measurement of cavitation, (d) waveguide-type.

tor on the axis through the centers of vibrometers. The amplitudes of these signals and the cosine of the phase angle between them are multiplied together electronically, the product being the projection of ultrasound intensity on the axis through vibrometers. By adjusting the receiver to the position at which the product is maximum, one can find the intensity of radiation and the direction of its irreversible transmission.

Hydrophones can be used to study cavitation phenomena in liquids. Some types of cavitometers are described in detail by Makarov [9].

The above hydrophones are not applicable for measuring acoustic pressures in molten metals. For this, one can use an electrodynamic waveguide pickup [19] schematically shown in Figure 8.15. Coil *1* on ferromagnetic core *2* generates a constant magnetic field. Gauging coil *3* is placed in a core gap. Tungsten rod *4* serves as a sonotrode. To avoid standing waves, the free end of the rod carries absorber *5* made of tungsten powder and epoxy resin.

Facilities for measuring acoustic parameters in solid-state loads are virtually identical to those employed for concerned measurements in

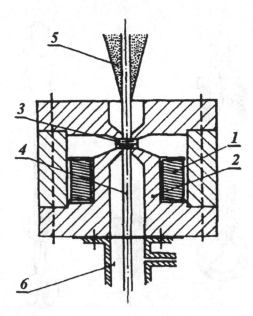

Figure 8.15. Waveguide probe with an electrodynamic pickup: (*1*) exciting coil, (*2*) core, (*3*) gauging coil, (*4*) tungsten sonotrode, (*5*) sound absorber, (*6*) sonotrode cooler.

ultrasonic stacks (see section 8.2). It should only be emphasized that strains and stresses are measured with tensometers, and temperatures are measured with calked thermocouples and other thermometric devices.

Acoustic power transmitted into a liquid or solid load can be evaluated calorimetrically. This method is based on the assumption that the entire acoustic energy fed into a load will eventually transform into heat.

During the calorimetry of liquid loads, they are made circulating through a measuring cell. The absorption of acoustic energy leads to the heating of liquid and thus to temperature difference ΔT across the measuring cell.

In this case, acoustic power transmitted into the load can be calculated by formula

$$N_a = kV\Delta T, \tag{8.26}$$

where V is the flow velocity; k is a constant.

To determine k, one has to perform liquid heating experiments at various ΔT and V.

Calorimetric method for measuring acoustic power output is described in more detail in [9, 20]. It should be noted that precise calorimetric measurements need rather expensive equipment. Besides, it is difficult to experimentally obtain necessary corrections for energy loss in vibratory system and measuring cell tubings. On the other hand, calorimetric method enables the measuring of acoustic power with no account for the processes (including nonlinear ones) occurring in the load.

References

1. I. Mataushek, *Ultrasonic Technique* (in Russia), Metallurgizdat, Moscow (1962).

2. H. Nodvedt, *Acoustica*, 4, 462 (1954).

3. A. Perkins, *Ultrasonic*, 4, 193 (1964).

4. A. Mashonis, *Ultrazvuk*, 12, 37 (1950).

5. D. A. Gershgal and V. M. Fridman, *Industrial Ultrasonic Equipment* (in Russian), Energiya, Moscow (1976).

6. V. F. Kazantsev and A. P. Panov, In: *Design and Applications of Ultrasonic Equipment in Machine Building*, part II, Izd. NTO Mashprom, Moscow (1978).

7. A. E. Kolesnikov, *Ultrasonic Measurements*, Izd. Standartov, Moscow (1970).

8. S. P. Alekseev, A. M. Kazakov, and N. N. Kolotilov, *Noice and Vibration Measurements in Mechanical Engineering*, Mashinostroenie, Moscow (1970).

9. L. O. Makarov, *Acoustic Measurements in Ultrasonic Technology*, Mashinostroenie, Moscow (1983).

10. I. I. Teumin, in: *Sources of High-Intensity Ultrasound*, Nauka, Moscow, p. 207 (1967).

11. Yu. I. Kitaigorodskii and A. V. Statov-Vitkovskii, *Akust.*, 16, 152 (1970).

12. J. Herbert, *Ultrasonic*, 3, 239 (1967).

13. Yu. S. Astashkin and O. V. Abramov, In: *Electrophysical and Electrochemical Machining*, Izd. NTO Mashprom, Moscow, no. 9 (27), 13 (1972).

14. Yu. I. Kitaigorodskii, In: *Ultrasonic Technology*, Metallurgiya, Moscow, p. 220 (1974).

15. S. Timoshenko, *Vibrations in Engineering*, Fizmatgiz, Moscow (1970).

16. I. I. Teumin, *Akust. Zh.*, 8, 372 (1962).

17. V. I. Domarkas, S. I. Sayauskas, and R. P. Patashyus, *Bull. Izobret.*, no. 37, A.S. 531076 (1976).

18. A. V. Kortnev and L. A. Davidenko, In: *Designing and Application of Ultrasonic Devices in Mechanical Engineering*, Part 2 (in Russian), Mashprom, Moscow, p. 161 (1978).

19. G. Grisshammer, *Ultrasonics*, **10**, 229 (1987).

20. V. N. Shchepetov, In: *Ultrasonic Gauges* (in Russian), TsINTI, Moscow, p. 165 (1960).

21. L. D. Rozenberg and M. G. Sirotyuk, In: *Ultrasonic Gauges* (in Russian), TsINTI, Moscow, p. 157 (1960).

Part III

Industrial Applications of High-Intensity Ultrasound

From the foregoing (Part 1) it follows that intense ultrasound is a potent tool to affect the processes of heat and mass transfer as well as the structure and interaction of solids. This makes ultrasound widely applicable in many fields of industry, agriculture, and medicine. Below is given a diagram illustrating the most potential applications of intense ultrasound. The employment of ultrasound in the production and processing of materials makes it possible to reduce cost prices, to obtain new products or to improve the quality of existing ones, to intensify conventional manufacturing processes or to develop new, ecologically advanced technologies.

HIGH-INTENSITY ULTRASOUND: MAJOR FIELDS OF APPLICATIONS

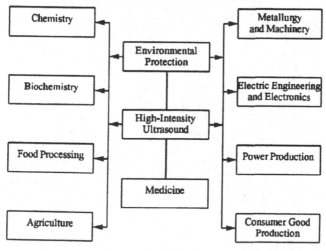

In the chapters that follow we shall treat the problems associated with the application of high-intensity ultrasound in raw minerals processing (chapter 9), metallurgy (chapter 10), and machinery (chapter 11).

Chapter 9
Ultrasound in Raw Mineral Dressing

Dressing of raw minerals (primarily, ores) involves, typically, ore dressing and recovery of valuable components. The recovery of valuable components from nonferrous, rare, and radioactive metal ores and ore concentrates is accomplished by hydrometallurgical methods.

The important circumstance that determines the state of the art in dressing and hydrometallurgical technology is that raw mineral materials to be processed are now often of low-grade and have a complex composition. To make the processing of such materials profitable, one has to use advanced technologies employing novel processes, reagents, and technical facilities.

This chapter considers the feasibility of intense ultrasound in basic dressing (flotation) and hydrometallurgical processes (leaching, grouting, sorption, and extraction). The basic feature of these processes is that they occur in heterogeneous systems with a mass transfer between liquid and solid (or second liquid) phases. Although proceeding by different mechanisms, these processes share a common feature: their rate and completeness depend largely on the area and structure of the interface, i.e., on the extent of interfacial interaction.

The mechanism of ultrasonic dressing and hydrometallurgical processes is associated with the development of nonlinear effects (cavitation and acoustic streaming) in heterogeneous systems and lies in the stimulation of emulsification, disintegration of solids and surface films, blending of liquids, promotion of interfacial diffusion and liquid penetration into mineral's pores and cracks.

9.1   Flotation and Emulsification of Flotation Reagents

Ore dressing is an essential stage preceding hydrometallurgical processes. There is ample evidence indicating that ultrasonic vibrations are effective in intensifying some dressing processes such as flotation and emulsification of flotation reagents, task-oriented modification of the surface layers of minerals in aqueous ore pulps, the breakdown of flotation reagent films adsorbed on mineral particles, defrothing, separation of heterogeneous systems, desliming, etc. [1–4].

Flotation is a dressing process lying in the separation of small solid particles due to their different wettability. During flotation, hydrophobic (i.e., poorly wettable) particles adhere to gas bubbles on their upward movement through a pulp (an aqueous slurry of mineral particles) to be collected in a froth, whereas waste particles stay in the main body of the pulp to be then discarded.*

The possibility of the formation of a floating particle–bubble complex depends on the physicochemical surface properties of mineral particles and liquid. To enhance the difference in the wettability of minerals to be separated, one has to add flotation agents to the pulp. These additives can be divided into

   (a) collectors, capable of impairing the wettability of minerals and thus enhancing the attachment of air bubbles to mineral particles, which improves their flotability;

   (b) frothers, providing a fine disintegration of air fed into a flotator and thereby the formation of a stable froth;

   (c) modifiers, capable of controlling the effect of collectors by either enhancing or diminishing it, which allows a particular mineral or a group of minerals to flotate;

   (d) suppressors, augmenting the wettability of mineral particles, which is accompanied by a decline in their flotability;

   (e) activators, stimulating the interaction of collectors with mineral particles.

   It should be noted that flotation additives are valuable chemicals (salts, acids, alkali, heteropolar and nonpolar compounds, tannins, etc.). Moreover, many of them (oils, fatty acids, amines, etc.) are

---

*An alternative is also possible when waste materials are carried to the froth, whereas valuable products stay in the main body of the pulp.

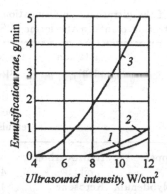

Figure 9.1. Effect of frequency and intensity of ultrasonic vibrations on the rate of mineral oil emulsification [5]: *1*, 800 kHz; *2*, 400 kHz; *3*, 22 kHz.

poorly soluble in water and their usage in excess amounts is not only wasteful, but may also impair the process efficiency. The use of water-insoluble flotation agents in the form of emulsions may reduce their consumption and even enhance the recovery of valuable minerals.

Ultrasound can be efficiently used to emulsify insoluble flotators. The emulsions thus produced are granulometrically uniform, fine, dense, and stable. The dynamics of ultrasonic emulsification depends on the ultrasonic field parameters such as frequency and intensity of vibrations, the nature of emulsified liquids, and the process temperature (some problems related to the mechanism of emulsification were considered in chapter 4). For illustration, Figure 9.1 presents the rate of mineral oil emulsification as a function of ultrasonic frequency and intensity.

The emulsion quality depends on the time of ultrasonic irradiation. A short-term irradiation gives rise to coarse, unstable emulsions, but a long-term irradiation may also reduce the stability of emulsions by inducing their coagulation.

Ultrasonic irradiation can raise the rate of emulsification, improve the quality of resultant emulsions, and, in the final analysis, reduce their consumption (Table 9.1.) [3]. At present, some ore concentration plants of the former USSR use ultrasound for the emulsification of nonpolar hydrocarbons, kerosene, heavy pyridine, cresol, phenol, a mixture of oleic acid with tall oil, and other flotators.

Emulsification is often conducted in the presence of stabilizers (surfactants) that are expected to reduce interfacial surface tension. Surfactant molecules are adsorbed on the emulsion drops thus preventing their coalescence. Ultrasonic emulsification can provide a high stability

Table 9.1   Efficiency of the ultrasonic emulsification of flotation agents at ore concentration plants in the former USSR.

Plant	Ore	Flotation agent	Agent consumption, g/ton		Agent saving, %
			without being emulsified	as emulsion	
Dzhezkazgan	Copper	Hydrocarbon oil	450	150	66.7
Balkhash	Molybdenum	Kerosene	450	300	33.3
Belousovskaya	Complex	Cresol	150	100	33.3
Berezovskaya	The same	The same	90	60	33.3
Karagandinskaya	Coal	Kerosene	1000	600	40.0
Leninogorskaya	Complex	Cresol and phenol	120	90	25.0
Tekeliiskaya	The same	Aerofloat	450	400	11.1
		Cresol	300	250	16.7
Almalykskaya	Copper	Pyridine	170	140	17.7
Zolotushenskaya	Complex	Flotation oil	80	65	18.8

of emulsions without adding stabilizers. A combined use of ultrasonic irradiation and surfactants results in a synergistic effect that gives rise to more stable and finer emulsions than those obtained with ultrasound or surfactant alone.

This method is used for the ultrasonic emulsification of nonpolar flotation agents in aqueous solutions of nonionic surfactants, sulfates, sulfonates, fatty acid salts, dithiophosphates, xanthates, and other heteropolar collectors. Collector molecules are adsorbed on the drops of a nonpolar agent, their polar groups being oriented to a polar aqueous phase. During the contact of such drops with mineral particles, the polar groups of the adsorbed collector molecules may interact with the crystal lattice ions of minerals and thus promote the spreading of nonpolar flotation agent drops over the surface of mineral particles.

Flotation agents are commonly emulsified in ultrasonic apparatus equipped with vortex, slot, or rotary hydrodynamic radiators. Such an equipment is quite efficient, low-cost, easy-to-fabricate, and easy-to-use [2]. Thus, the industrial application of oleic acid emulsions obtained ultrasonically made it possible to considerably reduce oleic acid consumption without impairing flotation efficiency. The use of ultrasonically emulsified grade L nonpolar hydrocarbon oil reduced the consumption of the oil threefold and allowed one to omit the stage of its pretreatment in the flotator, to enhance the flotability of concretions, to lower the effect of oil on the flotation of fine particles of copper sulfides, as well as to improve the quality of flotation froth and filtrability of the concentrate (as a result of diminishing oil content). Ultrasonic emulsification of kerosene increased the molybdenite content of the concentrate by 1.7% and reduced kerosene consumption by 30 g/ton. Ultrasonic treatment was found to be also efficient with respect to tall oil, bottoms of synthetic fatty acids, naphthenic acids, and other poorly soluble collector agents.

High-intensity ultrasound appears to be an effective tool for intensifying various flotation stages, as capable of giving rise to cavitation and acoustic streaming in pulp [4]. In this case, ultrasonic vibrations affect the basic events of flotation process (gas bubble movement, particle wetting, etc.).

Let us consider in more detail how ultrasound may influence the formation of the bubbles of the air fed into the pulp [6]. Ultrasonic vibrations begin to affect air bubbles once they are formed at the orifices of macrocapillaries used for air feeding. Being subjected to a cyclic acoustic pressure, a growing bubble will oscillate. Analysis of the bubble behavior in a gas flow moving through a capillary in an ultrasonic field showed that the bubble will detach from the capillary being much smaller in size than in the absence of ultrasound [7]. It was found that the higher the amplitude and frequency of oscillations, the smaller the bubbles at the instant of their detachment from the capillary. In addition to affecting the size of detaching bubbles, ultrasound considerably reduces the gas rate, at which there occurs a continuous formation of bubbles separated by only a thin layer of the liquid (the so-called chain mode of bubble formation).

Thus, intense ultrasound allows the formation of a large contact area in the gas–liquid–solid system. It should also be noted that cavitation developed in the liquid under the action of ultrasound gives rise to additional small gas-filled bubbles, which pulsate with a high displacement amplitude and contribute to flotation.

Ultrasound was found to alter the properties of liquid. Thus, solidophobic and hydrophobic effects induced by a 0.7–10 MHz ultrasound [8] were related to changes in the physicochemical properties of liquid, such as pH, electric conductivity, and redox potential (Figure 9.2). Ultrasonic vibrations applied to a liquid phase accelerate redox processes and favor the formation of hydrogen peroxide, nitric and nitrous acids, and other oxidants [9, 10. In flotation practice, vibrations can be used

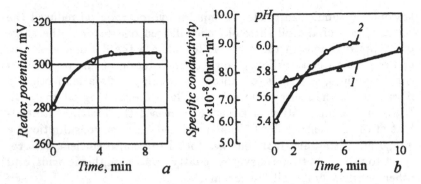

Figure 9.2. Effect of ultrasonic treatment on (a) redox potential and (b) water conductivity and pH [2]: 1, pH; 2, conductivity.

to break up xanthate ions, cyanocompounds, and other flotation agents in the pulp. Ultrasonic pretreatment affects the electrode potentials of sulfide minerals.

The use of ultrasonically pretreated liquids for flotation enhances the recovery of valuable minerals and reduces the flotation time and reagent consumption [4]. For instance, the recovery of chalcopyrite and limonite increases by 12 and 13%, respectively (22 kHz; ultrasound intensity 10 W/cm^2; exposure time from 30 s to 10 min), and that of bismuth by 4% (20 kHz; ultrasound intensity 2–4 W/cm^2; exposure time 60 min). Similar results were obtained for fluorite, calcite, and other minerals. At the same time, the flotability of molybdenite and quartz in ultrasonically treated water was noticeably decreased. Every mineral has its own optimum exposure time and ultrasound intensity at which ultrasonic flotation is most efficient.

Ultrasonic irradiation of a pulp before flotation contributes to the disintegration of mineral particles and to the removal of oxide films and dirt from mineral's surface.*

This was first pointed by Petersen [11], who studied the ultrasonic breakdown of the surface layers of sulfide minerals to be subjected to flotation. Pulps containing flotation agents were irradiated with ultrasound either before or during flotation. It was found that ultrasonic irradiation enhances the recovery of copper and lead minerals and reduces the consumption of collectors.

Ultrasound decreases the thickness of a hydrated coat formed by water molecules around mineral particles and air bubbles and thereby

---

*A more detailed consideration of ultrasonic grinding is given in the next section of the book.

increases the probability of particle attachment to bubbles. The hydrated coat destruction is most intense when ultrasonic devices operate at an elevated static pressure.

During ultrasonic flotation, acoustic streamings together with mechanical stirring provide a vigorous blending of the pulp and keep mineral particles suspended. Ultrasound breaks down mineral concretions, selectively crushes the constituting grains, and thus detaches various minerals present in raw materials. The selectivity of mineral disintegration is of particular importance. Thus, during the ultrasonic treatment of bauxite ore composed of kaolinite and boehmite, finely ground kaolinite particles are removed from a flotator, whereas more coarse boehmite particles stay in the pulp to increase its alumina content.

Ultrasound is effective in separating the flotation bulk concentrate into almost monomineral products, which is commonly a problem with other methods. This can be illustrated by the experiments of Akopova [12] who investigated the ultrasonic separation of a mixture of titanozirconium sand minerals into monomineral products. Ultrasound was generated by a magnetostrictive transducer operated at 22 kHz (input power of about 2 kW). Minerals (zircon, rutile, ilmenite, and staurolite) were powdered to a mean particle size of 0.15+0.052 mm (100–270 mesh). The mixtures of powders were treated ultrasonically for 1 min at various consumptions of tall oil that served as a collector.

It was found that there exists an optimum consumption of collector for each of the minerals studied. For instance, ultrasound raised the flotation recovery of zircon from 7.7% to 55.7% at the tall oil rate 20 g/ton. The ultrasonic flotation of ilmenite and staurolite was most effective with the tall oil consumptions 35 and 100 g/ton, respectively. In this case, the recovery of ilmenite made up 74.3%, whereas under the same conditions but in the absence of ultrasound ilmenite was not flotated at all (Figure 9.3).

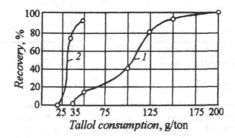

Figure 9.3.   Flotation of ilmenite: *1*, without ultrasound; *2*, in an ultrasonic field.

As the collector consumption increased from 35 to 50 g/ton, the difference in the flotation properties of minerals, except staurolite, decreased, but their recovery increased to 98.8 (zircon), 97.4 (rutile), and 94.0% (ilmenite). For comparison, the recovery of these minerals without ultrasound comprised 79.7, 14.2, and 14.7%, respectively. Staurolite failed to be flotated at the indicated tall oil consumptions.

The effect of ultrasound on mineral recovery was maximum when the exposure time varied from 1 to 5 min. More prolonged exposure times insignificantly enhanced the recovery of minerals and even decreased the recovery of ilmenite. Ultrasound greatly enhanced the selectivity of zircon and titanic mineral flotation.

Panteleeva *et al.* [13] investigated the effect of a preliminary ultrasonic treatment of pulps composed of pure minerals, viz. jarosite, limonite, and quartz, with various particle sizes. The oleic acid-kerosene emulsion was used as a collector. The experiments showed that a 1-min ultrasonic treatment improved the flotation recovery of jarosite and limonite (Figure 9.4), but was ineffective with respect to quartz (data not shown). Moreover, the application of ultrasound made it possible to reduce the collector consumption. For instance, in the case of conventional flotation, a 90% recovery of jarosite of any particle size was attained at a collector rate of 1000–2000 g/ton in comparison with only 250 g/ton in the case of ultrasonic flotation.

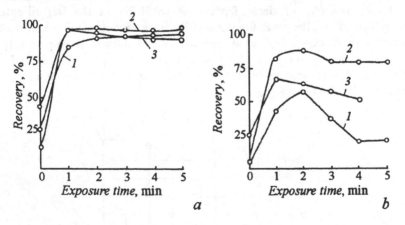

Figure 9.4. Dependence of the froth recovery of (*a*) jarosite and (*b*) limonite on the time of ultrasonic treatment at various mean sizes of particles: *1*, 0.5 + 0.2 mm (class 1); *2*, 0.2 + 0.044 mm (class 2); *3*, 0.044 mm (class 3).

9.2   Ultrasonic Disintegration of Minerals and Adhered Surface Films

Disintegration of solids can be accomplished by different methods, each of which has its own merits and demerits and its field of application [14]. Ultrasonic methods occupy a certain place. Ultrasonic disintegration allows the production of fine powders that are uniform in size and composition and have no foreign inclusions.

Ultrasonic disintegration is a subject of all basic effects developing in strong ultrasonic fields. The physical aspects of ultrasonic disintegration were considered, to some extent or another, in chapter 5. In this section we shall touch briefly on experimental results and the technological aspects of this problem.

As noted above, ultrasonic disintegration will take place if stresses acting upon a particle are above the strength of the material to be ground. Grinding can be expediently intensified by raising static pressure or by applying vibrations of two different frequencies.

Several setups were developed in Russia for disintegration processes at a high static pressure [14].

A UZVD-6 ultrasonic apparatus (Figure 9.5) embodies, as a source of vibrations, a 5-kW PMS-15A-18 magnetostrictive transducer 5 driven by a UZG-10 generator or its analog. Ultrasonic vibrations generated by the transducer are fed into a 800-cm^3 working chamber 2 through horn 4 and stack 3. Gas fed through lid 1 is used to maintain in chamber 2 an excessive static pressure of up to 1 MPa (10 atm.). The setup can be also used for deburring.

The main construction unit of a UPKhA apparatus (Figure 9.6), which is used for powder production and for making suspensions, is column 4 composed of four magnetostrictive cylindrical transducers 3 possessing a common cooling jacket 2. The working pressure in the column is maintained within 1 MPA (10 atm.). The magnetostrictive transducers operate at about 16 kHz, power requirement is from 4 to 15 kW.

Disintegration techniques available to date allow the grinding of a diversity of brittle metallic and semiconducting materials, oxides, carbides, borides, etc. In developing such techniques, it is of much importance to choose an appropriate working liquid, liquid-to-solid phase ratio, ultrasound intensity, static pressure, and exposure time.

For illustration, Tables 9.2 and 9.3 present the results of ultrasonic disintegration of Al_2O_3 and MoS_2 particles.

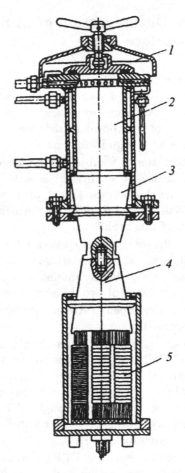

Figure 9.5. Cross-sectional view of a UZVD-6 ultrasonic apparatus: *1*, lid; *2.* working chamber; *3*, radiator, *4*, horn; *5*, magnetostrictive transducer.

Disintegration of Al_2O_3 particles was performed at an excessive pressure of 3 atm. The solid-to-liquid phase ratio was 1:4. It can be seen that ultrasonic treatment resulted in a significant size-reduction of particles. Similar results were obtained in MoS_2 disintegration experiments performed at an excessive pressure of 5 atm. Various alcohols, gasoline, benzene, and other liquids were used as working media.

Analysis performed in [14] showed that the time of treatment necessary to attain a desired particle size decreases with increasing cavitation index (Figure 9.7) and solid phase density and diminishing liquid viscosity.

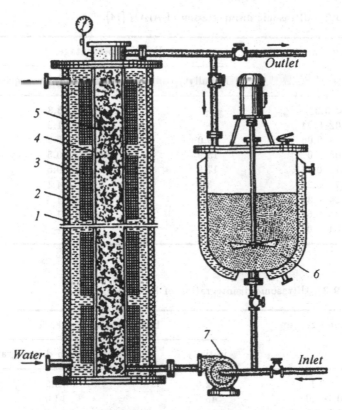

Figure 9.6. Cross-sectional view of a UPKhA apparatus: *1*, acoustic insulation; *2*, cooling jacket; *3*, magnetostrictive transducer; *4*, working column; *5*, concentrate; *6*, reservoir with a stirrer; *7*, pump.

Typically, the efficiency of ultrasonic disintegration is higher than that of ordinary mechanical grinding. Thus, the disintegration of 500 g of magnesium oxide from the initial particle size 4.2 μm to 0.7 μm in a UPKh-8M apparatus took 40–60 min, the content of < 0.5-μm particles being 60%. To compare, the amount of particles smaller than 1 μm in size comprised 16% after a 20-h milling of magnesium oxide in a ball mill [14].

Similarly, milling of 100- to 700-mkm zirconium carbide particles to a final size of 0.5–3.0 μm in a ball mill took 25 h, as compared to 2 h in the case of disintegration in the UZVD-6 apparatus.

Ultrasound can be used for the opening of fibrous materials. It was experimentally shown [14] that ultrasound can promote the chrysotile asbestos opening into elementary fibers without their shortening.

Table 9.2   Ultrasonic disintegration of Al_2O_3 [14].

Particle size, μm	Fraction percentage, %	
	initially	after a 15-min disintegration
< 0.02	–	13.3
0.02–0.04	–	21.2
0.04–0.06	29.0	28.9
0.06–1.0	13.9	20.5
0.10–0.16	17.8	9.5
0.16–0.24	3.8	4.1
0.24–0.36	6.3	1.5
0.36–0.64	10.1	1.0
0.64–1.0	19.1	–

Table 9.3   Ultrasonic disintegration of MoS_2 [14].

Particle size, μm	Fraction percentage, %	
	initially	after a 30-min disintegration
< 0.1	32.6	76.5
0.1–1.0	45.5	13.0
1.0–5.0	15.4	10.5
> 5.0	6.5	–

The disintegration of some high-strength materials can be effected by using a combination of kilo- and megahertz frequences or kilohertz and hertz frequencies [14].

Ultrasonic methods make it possible to degrade minerals to a desired size and to obtain products with a minimum content of foreign inclusions. Ultrasonic disintegration of minerals is expedient in the case of expensive products when it is necessary to obtain a small amount of powdered material.

Some high-strength materials were successfully powdered by alternating the ultrasonic dispersion of suspension with its mechanical milling in colloidal or vibratory mills. Such a method reduces the process time 2–3-fold, energy consumption 1.5–2 times, and minimal size of particles 4–5 times.

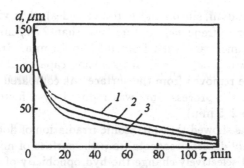

Figure 9.7. Mean size of gypsum particles as a function of ultrasonic treatment time at various cavitation indices K: *1*, 1.2×10^{-4}; *2*, 1.4×10^{-4}; *3*, 1.7×10^{-4}.

Ultrasonic treatment makes it possible to separate calcite and feldspar concretions from mica, montmorillonite, and ferric hydroxides; to separate fermorite grains from sphalerite and pyrite; to purify akkilite from the inclusions of baryte and other accompanying minerals; to remove ferric hydroxide and slime films from the grains of tourmaline, feldspar, quartz, and ore minerals; to remove manganic and ferric hydroxides from alumosilicates and phosphates; etc. Ultrasound allows the disintegration of fibrous minerals, such as graphite and molybdenite, to the highest fineness. For instance, ultrasonic vibrations of 18–22 kHz and 7–8 W/cm^2 made it possible to obtain a 20% molybdenite suspension with the main fraction particles 1 μm in size.

Ultrasonic removal of the films of secondary minerals, oxides, and reagents from the surface of valuable minerals, primarily occurring through cavitation, is of much practical importance during the processing of complex, oxidized, chatty, and other hard-dressing ores, as well as during the selective flotation of bulk concentrates. The removal of surface films from mineral particles improves the contact between the flotation agents and the primary mineral particles, reduces the consumption of reagents, and impart specific flotation properties to minerals. Ultrasound promotes the surface oxidation of minerals that are commonly difficult to oxidize. This will enhance the chemisorption of reagents, their attachment to the surface, as well as the rate and selectivity of flotation.

Relevant investigations are presently quite numerous. In particular, Agranat *et al.* [14], estimated the influence of ultrasonic vibrations on the removal of various coverings and secondary mineral films from the surface of feldspar, quartz sand, and artificial abrasive materials (synthetic corundum, silicon and boron carbides). The vibrations used were 15–25 kHz in frequency and not less than 2 W/cm^2 in intensity, the exposure time was varied from 0.25 to 12 min. Irradiation was performed in water or a surfactant solution capable of dissolving the substance to be removed from the surface. As compared with mechanical treatment, the process time was reduced by a factor of 15 to 30 (from 30–60 to 2–5 min).

Experiments showed that ultrasonic irradiation of flotation concentrates selectively desorbs reagents from the surface of minerals, which, in turn, should selectively change the hydrophobicity of mineral particles and their flotability. Thus, a 3–5-min treatment with ultrasonic vibrations (20 kHz; 2 W/cm^2) drastically reduces the flotability of galenite flotated at optimal consumptions of ethyl potassium xanthate and pine oil. It should be, however, noted that after ultrasonic irradiation of galenite concentrate, the pulp loses its collecting properties as a result of ultrasonic decay of xanthate.

Effect of ultrasonic vibrations on the surface films of reagents does not depend on mineral hardness. For instance, although the hardness of pyrite is greater than that of sphalerite and chalcopyrite, ultrasound completely breaks down the adsorbed layers of flotation reagents on the pyrite surface, but fails to remove them from sphalerite and chalcopyrite.

When the collector rate is above an optimum level or the size of mineral grains is too small, the efficiency of ultrasonic disintegration of hydrophobic films falls.

Raising the erosive activity of cavitation may contribute to the removal of adsorbed collecting reagents, such as sodium olefin, oleic acid, some unpolar oils (e.g., transformer and solar oils), and kerosene, from scheelite, baryte, quartz, antimonite, and other minerals. This can be achieved by creating an excessive static pressure in a working chamber or by a combined action of ultrasonic vibrations of different frequencies [15]. For instance, the ultrasonic treatment of pulp under an elevated static pressure for 15 s reduced the flotability of chalcopyrite particles by almost 90%, while did not virtually affect it at the atmospheric pressure even if the treatment lasted for 30–40 min.

Experiments on the disintegration of xanthate films adsorbed on the surface of gold particles and on the intensification of cyanidation were described in [16].

Thus the literature data indicate that ultrasound is a potent tool to disintegrate minerals and remove various films from their surface.

9.3 Application of Ultrasound for Defrothing and Dehydration of Ore Concentrates

Flotation froths are, as a rule, three-phase systems consisting of gas-filled bubbles separated by aqueous films and laden with mineral particles. Froth stability mainly depends on the nature and concentration of frothing reagents and inorganic salts in an aqueous phase, as well as on the size, concentration, and surface properties of mineral particles. In the case of a strongly mineralized, very stable froth, its separation and transportation may present difficulties with a concurrent loss of valuable components during dehydration processes (e.g., as a result of coagulant drainage).

An efficient ultrasonic breaking of mineral froths occurs even if vibrations are fed into the gas phase or directly to the froth. Optimal ultrasonic parameters for defrothing are 6–20 kHz and acoustic pressures not less than 140–150 dB. At concentration plants, ultrasonic defrothing is most commonly accomplished by aerodynamic and pulsation-rotary hydrodynamic transducers.

In aerodynamic transducers (sirens), the intensity of acoustic vibrations builds up as a result of their concentration by a deflector or a horn. Figure 9.8 shows the arrangement of an aerodynamic antifrother in a flotator trough.

In pulsation-rotary hydrodynamic transducers, operating elements are an impeller with cross-sectionally variable slits and a stationary cylinder with special-shaped fins. During rotation, pressure on the impeller axis drops, due to which froth arrives at impeller blades to be thrown down to the stationary cylinder fins, which assist in creating a pulsating turbulent air field between the impeller and cylinder. With this construction, defrothing occurs mechanically under the action of centrifugal forces, shock waves, and vibration of the gaseous phase.

The mechanism of ultrasonic defrothing can be explained as follows. Acoustic vibrations cause the pulsation of a froth bubble, which, in the phase of expansion, possesses an increased interface area and decreased wall thickness. Because of acoustic turbulence, the bubble wall thickness is not uniform, which can lead to bubble rupture. Liquid from disrupted bubbles appears in the lower layers of the froth, thus hindering its dehydration and accelerating its breakup. Tables 9.4 and 9.5 compare the rates a spontaneous and ultrasonic defrothing during the flotation enrichment of zinc and lead concentrates (forming unstable froths) and coal concentrate (forming a stable froth) [6].

The application of ultrasonic vibrations allows one to considerably increase the technological parameters of ore dressing and its econom-

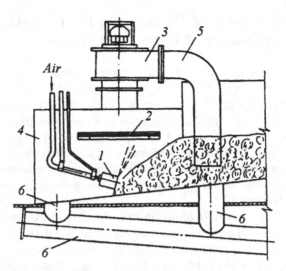

Figure 9.8. Arrangement of an aerodynamic antifrother in a flotation trough: *1*, antifrother; *2*, splash guard; *3*, fan; *4*, housing; *5*, fan outlet; *6*, concentrate collector.

ical efficiency. Drying of mineral concentrates is a final stage of their dehydration if coagulation and filtration cannot provide a necessary moisture content of ore concentrates. Typically, flotation concentrates consist of fine particles and are, therefore, porous, which explains their high moisture contents. Acoustic vibrations, excited in a gaseous phase, accelerate heat exchange and thus promote drying. Streamings induced by ultrasound reduce the thickness of boundary layers, thereby contributing to the removal of diffusional limitations.

A decisive factor of ultrasonic drying is the intensity of vibrations rather than their frequency. Nevertheless, the efficiency of drying is somewhat higher at lower frequencies, which can be due to frequency dependence of acoustic absorption.

There exist various types of ultrasonic drying installations, such as basket-type driers, driers with a vibrating conveyor, and "boiling layer" driers. Many installations combine the effect of acoustic vibrations with one of the conventional methods of drying. An essential advantage of acoustic drying is that it can be run at relatively low temperatures. Ultrasonic drying can reduce the moisture content of pyrite and lead oxide concentrates by 5.2 and 4.8%, respectively.*

---

*Ultrasound is also used for drying seed cotton, colloidal zirconium hydroxide, carboxymethylcellulose, silica gel, titanium dioxide, hormonal preparations, and other hygroscopic materials.

Table 9.4 Spontaneous and ultrasonic defrothing during the flotation enrichment of zinc and lead concentrates.

Spontaneous defrothing				Ultrasonic defrothing			
				Zinc concentrate		Lead concentrate	
Time, s	Zinc concentrate froth height	Lead concentrate froth height	Acoustic pressure, dB	Ultrasonic treatment duration, s	Froth height, cm	Ultrasonic treatment duration, s	Froth height, cm
0	12.7	12.7	155	1	4.7	1	4.0
5	9.0	9.0		2	3.0	2	2.0
10	6.0	5.0		5	1.5	5	1.0
15	4.0	2.5		9	0.0	8	0.0
20	2.0	2.0	160.5	1	2.0	1	1.5
600	1.6	1.0		2	1.0	2	1.0
1200	1.0	0.5		4	0.0	3	0.0
1800	0.5	0.0	165	1	0.0	1	0.0
3600	0.0	0.0	167	1	0.0	1	0.0

Table 9.5 Spontaneous and ultrasonic defrothing during the flotation enrichment of coal.

Spontaneous defrothing		Ultrasonic defrothing			
Time, min	Froth height, cm	Ultrasonic treatment duration, s	Acoustic pressure, dB		
			155	160.5	165
0	12.7	0	12.7	12.7	12.7
30	6.5	1	7.0	3.5	0.0
60	4.0	2	5.0	2.0	–
120	2.0	5	3.0	0.0	–
180	1.5	10	1.0	–	–
300	1.0	13	0.0	–	–

The products of flotation enrichment that are resistant to dehydration may be even dried by ultrasonic atomization, with which ore products are dried without their concentrating or filtrating. For instance, acoustic atomization of a 55–65% zinc concentrate led to the product with a residual moisture content of 9–10%, which corresponds to standard specifications.

9.4   Ultrasound in Hydrometallurgy

The most wide-spread hydrometallurgical processes are leaching, grouting, sorption, and extraction. Although having different mechanisms, these processes are similar in being dependent on the intensity of interfacial effects.

The effect of ultrasound on hydrometallurgical processes is associated with a cavitational disintegration of solids and surface films, agitation of liquids by acoustic streamings, and accelerated penetration of liquids into pores and cracks in minerals.

Ore and ore concentrate leaching lies in the preferential dissolving of valuable minerals. The kinetics of leaching is largely governed by the rate of interfacial diffusion of substances. Some mechanical processes, e.g. pulp stirring, act to reduce the thickness of diffusion layers, thus enhancing leaching. Ultrasonically driven acoustic streaming, cavitation, and other nonlinear phenomena may also affect the thickness of diffusion boundary layers (see chapter 2) and, hence, leaching kinetics.

Ultrasonic leaching is most efficient at low frequences of about 20 kHz and ultrasound intensities of several watts per square cm, which is primarily due to a well-developed cavitation under these conditions. Ultrasonic irradiation makes it possible to reduce the time of mineral leaching from several hours to 10–15 min and to avoid pulp heating. Moreover, the end product may contain as high as 95.5% of a particular mineral, which is far above the degree of enrichment attainable by conventional technology. Thus, Losev *et al.* [17, 18] studied the ultrasonic haloidation of coal, aimed at extracting germanium, and found that ultrasound accelerated bromation 160-fold. Similar results were obtained during the aqueous leaching of germanium from coal [19].

Andreev *et al.* [20] used ultrasound for uranium leaching from sandtuff with 5% soda. A 13-min ultrasonic leaching provided the recovery of 16% of uranium, while a 1.5-h conventional leaching led to a 8% recovery. The kinetics of these processes are shown in Figure 9.9.

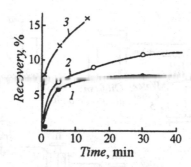

Figure 9.9. Dynamics of uranium leaching with a sodium carbonate solution: *1*, with a regular agitation; *2*, in a low-frequency ultrasonic field; *3*, in a high-frequency ultrasonic field.

Agranat *et al.* [1] described the ultrasonic ammonization of apatite with 50% nitric acid. The ground apatite in an autoclave was exposed to 20-kHz ultrasound with the intensity 2 W/cm^2, generated by a magnetostrictive transducer. The conventional ammonization of apatite took 3.5 to 4 h with a phosphorus recovery of 98–99%. At the same time, ultrasonic ammonization took as few as 10 min with the same phosphorus recovery.

Vasil'ev [21] used a 550-kHz ultrasound for leaching some lead- and copper-containing minerals and found that in most cases, ultrasound accelerated leaching 10- to 30-fold (Table 9.6). At the same time, the rate of copper leaching with 2% or 5% H_2SO_4 did not change under the action of ultrasound [22].

Ultrasound accelerated vanadium leaching from a titanium-magnetite concentrate 1.5–2 times and increased vanadium recovery to 0.92–

Table 9.6 Effect of ultrasound on mineral leaching.

Mineral	Solvent	Time of complete mineral dissolving		Process acceleration, times
		under ordinary stirring	under ultrasonic irradiation	
Cerussite $PbCO_3$	15% $NH_4C_2H_3O_2$	45 min	3 min	15
Crocoite $PbCrO_4$	2% NaOH	7 h	15 min	28
Galenite PbS	250 g/l NaCl, 60 g/l $FeCl_3$	12 h	30 min	24

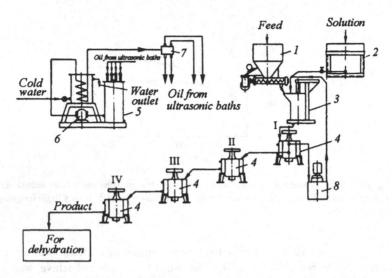

Figure 9.10. Arrangement for ultrasonic treatment of fine materials: *1*, feeder; *2*, solution tank; *3*, reactor; *4*, ultrasonic baths (I, II, III, and IV); *5*, heat exchanger; *6*, pump; *7*, manifold; *8*, pump.

0.96. A 18-kHz ultrasound with the iontensity 0.5 W/cm^2 accelerated the process of gold cyanidation fourfold. Agranat *et al.* [1] described a pilot setup for continuous ultrasonic treatment of fine mineral pulps with a particle size of no more than 50 μm. The setup incorporated feeder *1*, solvent tank *2*, reactor *3*, and four (I, II, III, and IV) ultrasonic baths *4* equipped with PMS6 magnetostrictive transducers (Figure 9.10).

The efficiency of a combined action of ultrasound and mechanical stirring can be seen from experiments on the leaching of some poorly soluble rare-metal minerals [12]. In this case, the role of mechanical stirring is to uniformly distribute particles in a liquid phase, whereas ultrasound acts to remove diffusional and capillary limitations.

Yakubovich [23] investigated the effect of ultrasound on copper dissolving in ammonium solutions. A UZG 10M generator, incorporating a PMS6 magnetostrictive transducer (20 kHz; 2.5 kW), was used as a source of ultrasonic vibrations. Ultrasound was able to accelerate copper dissolving in ammonium solutions 3–4 times (Figure 9.11).

Heterogeneous solid-liquid reactions, specific for some hydrometallurgical processes in production of nonferrous and rare metals, are often

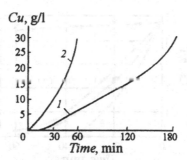

Figure 9.11. Dynamics of copper dissolving in 20% $(NH_4)_2CO_3$: *1*, without ultrasound; *2*, in an ultrasonic field.

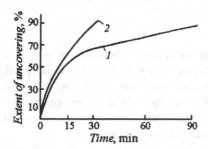

Figure 9.12. Dynamics of scheelite concentrate uncovering with nitric acid: *1*, without ultrasound; *2*, in an ultrasonic field.

strongly inhibited by the films of a newly formed solid phase coating the particles of an initial substance. For instance, a firm continuous film of tungstic acid is formed on the surface of scheelite particles during their dissolving in chloric acid. This film persists even under a vigorous stirring of pulp; however, ultrasound (20 kHz, 1.5 kW) is able to accelerate threefold the leaching of scheelite concentrate with 21% nitric acid at 100°C (Figure 9.12).

There is an increasing tendency in hydrometallurgy toward application of grouting, or a contact displacement of metals from solutions by more electronegative metals. Grouting is complex process involving multi-stage interfacial phenomena. The overall process rate depends on the particular rates of diffusion of reducing and grout ions, as well as on the rates of electrochemical reaction and crystal growth. Ultrasound intensifies all the stages of grouting process; for instance, a 19.5-kHz ultrasound with an intensity of 2.5 kW reduces 1.5-fold the time necessary to attain the same residual concentration of copper as in conventional copper grouting with a $ZnSO_4$ solution (Figure 9.13).

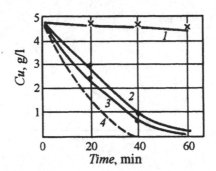

Figure 9.13. Time courses of copper concentration in grouting solution: *1*, without stirring; *2*, stirring at 800 rpm; *3*, stirring at 1100 rpm; *4*, in an ultrasonic field without stirring.

The intensity of mechanical blending, as well as the intensity of ultrasound above the cavitation threshold, weakly affects the rate and recovery of copper from solutions. This may be due to the existence of an extreme intensity of electrolyte exchange, above which concentrational polarization drastically decreases, and anodic or cathodic chemical reactions become limiting stages.

Depending on stirring conditions, the precipitate may have different structures. Without stirring, copper begins to precipitate in the form of needles oriented normally to the grouter wall. When the growing copper layer becomes rather thick, it is detached from the wall and sinks to the grouter bottom. A moderate mechanical stirring accelerates this process, but vigorous stirring suspends the precipitate. This is also the case under ultrasonic irradiation. Ultrasound removes in part concentrational limitations due to intense electrolyte exchange at the interface. The removal of the deposit from the grouter wall reduces the ohmic resistance of the system, due to which pores do not close and the rate of grouting does not drop. A very thin copper layer, which is always retained on the grouter wall, maintains the process at a sufficient rate, since the actual area of cathodic surface remains high and halvanic microelements are retained. Calculations show that ultrasound halves the activation energy of copper grouting. It should be emphasized that the efficiency of ultrasound depends on the nature of electrolytes used.

In many instances, the extraction of metals and their separation are due to sorption processes. The extraction of metals with ionites, viz. solid, water- and solvent-insoluble, beaded organic ion exchangers, occurs via the exchange of anions or cations of an exchanger for the

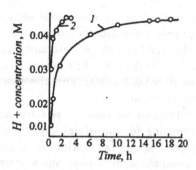

Figure 9.14. Sorption of sodium on a KU-2 cationite: *1*, without ultrasound; *2*, in an ultrasonic field.

equivalent amounts of cations or anions from solution. The application of ion-exchange resins for concentrating and refining considerably upgrades the technology of nonferrous, radioactive, and rare metals.

By now, the efficiency of ultrasound in ion-exchange metallurgical technology has been recognized. Under the action of ultrasound, bead micropores are cleaned, which enhances the sorption capacity of resins. Moreover, ultrasound facilitates the penetration of solutions into hard-to-reach microcapillaries and reduces the boundary layer thickness. Bronstein *et al.* [24] investigated the kinetics of sodium sorption on a KU-2 cationite from a NaCl solution under the action of ultrasound (16.4 kHz; 1 kW) and showed that ultrasound reduces the time of the steady-state attainment from 16 h to 2.5–3 h (Figure 9.14). The effect of ultrasound was maximum (a 3- to 5-fold increase in the rate of sorption), when the ion-exchange layer was less than 1 cm in thickness.

In a dynamic mode of operation, when the electrolyte flows through an ion-exchange column, the favorable effect of ultrasound is considerably lower (about 30%) [25], which can be attributed to the fact that now the mass transfer is no longer a rate-limiting factor as in a static mode of operation. In this case, ultrasound affects only diffusion in the solid phase.

Similar results were obtained during the study of heavy metal sorption [26]. In these studies, ultrasound also accelerated sorption processes and increased the sorption capacity of resin.

Extractive separation of substances is based on their distribution between two immiscible liquids, typically water and organic solvent. The prerequisite of the extractive separation of substances is their different solubilities in liquids constituting the separation system.

Two stages – proper extraction and phase separation – can be distinguished in the process of extractive separation. In a number of cases, ultrasound enhanced the process rate and increased the recovery of valuable ore components [27, 28]. The favorable effect of ultrasound is due to an increased mass transfer and the disruption of micellar structures in the extraction system.

Hydrodynamic ultrasonic radiators operating at 15–20 kHz appeared to be most promising for industrial applications. Ultrasonic extractors can be designed as special, upright or horizontal, counterflow columns, in which a combination of mechanical stirring and ultrasonic effects is realized.

References

1. B. A. Agranat, O. D. Kirillov, *et al.*, *Ultrasound in Metallurgy* (in Russian), Metallurgiya, Moscow (1969).
2. I. P. Golyamina (Ed.), *Ultrasound: A Small Encyclopedia* (in Russian), Sov. Entsiklopediya, Moscow (1979).
3. A. A. Baishulakov, Yu. V. Malakhov, and V. G. Varlamov, *Ultrasound in Ore Dressing and Hydrometallurgy of Kazakhstan* (in Russian), Alma-Ata (1979).
4. V. A. Glembotskii, M. A. Sokolov, *et al.*, *Ultrasound in Mineral Product Concentration* (in Russian), Alma-Ata (1972).
5. B. I. Pluzhnik and A. G. Shpakhler, *Izv. Vuzov, Gornyi Zhurnal*, no. 10, 174 (1964).
6. B. G. Novitskii, *Applications of Ultrasonic Vibrations in Chemical Technology* (in Russian), Khimiya, Moscow (1983).
7. I. S. Grachev, D. G. Kokorev, and V. F. Yudaev, *Inzh. Fiz. Zhurnal*, **30**, no. 4, 665 (1976).
8. Yu. P. Eremin, G. A. Denisov, *et al.*, *Obogashchenie Rud*, no. 3, 24 (1981).
9. K. Suslick (Ed.), *Ultrasound, Its Chemical, Physical, and Biological Effects*, VCH Publishers, New York (1988).
10. T. Mason (Ed.), *Sonochemistry, Ellis Horwood/John Wiley and Sons*, New York (1988).
11. B. Petersen, In: *Ore Flotation*, no. 129, 27 (1959).
12. K. S. Akopova, In: *Application of Ultrasound in Machine Building* (in Russian), TsINTIAM, Moscow, p. 123 (1963).
13. N. F. Panteleeva *et al.*, *Izv. Vuzov, Tsvetnaya Metallurgiya*, no. 3, 27 (1964).
14. B. A. Agranat *et al.*, *Fundamentals of Ultrasonic Physics and Technics* (in Russian), Vysshaya Shkola, Moscow (1987).
15. A. V. Fel'dman and N. N. Khavskii, In: *Dynamic Effects of*

High-Intensity Ultrasound (in Russian), Izhevsk, p. 9 (1981).

16. N. N. Khavskii (Ed.), In: *Application of Ultrasound in Metallurgy* (in Russian), Trudy MISiS, Moscow, no. 67, p. 115.

17. V. I. Losev *et al.*, *DAN SSSR*, **114**, no. 2, 372 (1957).

18. V. I. Losev and I. P. Lomashov, *Germanium in Coals* (in Russian), Izd. FN SSSR, Moscow (1962).

19. I. V. Vinogradov *et al.*, *Ukr. Khim. Zhurnal*, **26**, no. 3, 383 (1960).

20. Andreev *et al.*, *Zhurnal Fiz. Khim.*, **34**, no. 11, 2449 (1960).

21. V. V. Vasil'ev, *Application of Ultrasound in Analytical Chemistry* (in Russian), Izd. LDNTP, Leningrad (1965).

22. V. V. Vasil'ev *et al.*, *Vestnik Leningradskogo Universiteta, Ser. Fizika i Khimiya*, issue 4, no. 22, p. 146 (1959).

23. I. A. Yakubovich, In: *Ultrasonic Technics* (in Russian), Izd. NIIMASh, no. 2, p. 40 (1965).

24. I. K. Bronstein *et al.*, *Proceedings of Giredmet* (in Russian), Metallurgizdat, Moscow, **10**, p. 468 (1963).

25. G. N. Belova and I. K. Bronstein, *Proceedings of Giredmet* (in Russian), Metallurgizdat, Moscow, **15**, p. 189 (1965).

26. N. N. Khavskii *et al.*, In: *Ultrasonic Technics* (in Russian), Izd. NIIMASh, no. 3, p. 23 (1965).

27. V. Yu. Orlov and N. M. Zhavoronkov, *DAN SSSR*, **129**, no. 1, 161 (1959).

28. V. A. Nosov, *Ultrasound in Chemical Technology* (in Russian), Gostekhizdat USSR, Kiev (1963).

29. B. A. Agranat *et al.*, *Ultrasonic Technology* (in Russian), Vysshaya Shkola, Moscow (1974).

Chapter 10
High-Intensity Ultrasound in Pyrometallurgy

The regularities of propagation of high-intensity ultrasound in liquids and melts, as well as related nonlinear phenomena considered in part I of the book, secure an efficient implementation of ultrasonics in some technological processes dealing with molten metals.

This chapter is concerned with the employment of ultrasonics for metal refinement and crystallization, crystal growth, production of as-cast composites, and atomization of melts in the production of powders.

10.1   Ultrasonic Refinement of Melts

Exposure of a liquid to a high-intensity ultrasound reduces the dissolved gas content or, to put it otherwise, degasses the liquid. This effect can also be used for degassing metal melts.

The first papers, concerned with melt degassing, date back to the early 1930s, when a number of degassing methods employing piezo-electric and magnetostrictive transducers had been suggested [1, 2]. A decade later, Esmarch proposed a method for melt degassing that involved the excitation of vibrations by superimposing a permanent magnetic field on the alternating field of an induction furnace (the so-called electrodynamic excitation) [3]. The author experimented with molten aluminum alloy containing 5–7% of magnesium. The melt (8–10 kg) was treated by a 10-kHz ultrasound at 700°C for 30–60 min. The low efficiency of treatment could be accounted for by a low intensity of vibrations imparted to the melt.

The late 1950s saw rapid advances of Russian scientists in the ultrasonic degassing of molten aluminum- and magnesium-based alloys by vibrations directly fed into melts with the aid of sonotrodes [4].

Production of high-quality metals free of abscesses and cavities is a challenging problem in metallurgy. Abscesses and cavities are the result of a drastic decrease in gas solubility in a solidifying melt. It is the excess gas that forms abscesses and cavities. Degassing is expected to reduce the gas content of liquid metals to a level corresponding to gas solubility in solid metals at operation temperatures. Along with conventional methods based on chemical refinement and evacuation, ultrasonic degassing of melts is now coming into industrial use.

Investigations performed by G. I. Eskin on ultrasonic degassing of metal melts made it possible to establish the regularities of ultrasonic degassing of molten aluminum- and magnesium-based alloys, to design devices for ultrasonic treatment of melts simultaneously with shaping and continuous casting, and to develop theory of ultrasonic technology.

Mechanisms of ultrasonic degassing were studied with aluminum (grades A99 and A7), aluminum-magnesium alloys containing up to 8% of Mg, grade AL9 (Al–Si–Mg), AL19 (Al–Cu–Mn), and AL40 (Al–Si–Cu–Mg–Ni–Fe) industrial cast aluminum alloys, grade D1 and D16 (Al–Cu–Mg) wrought alloys, AK6 and AK8 (Al–Cu–Mg–Si) alloys, B95 and 1960 (Al–Zn–Mg–Cu) alloys, as well as grade AMg6 (Al–Mg–Mn) alloy. The use of such a diversity of alloys was dictated by the practical importance of their refinement.

Although being specific for every melt, ultrasonic degassing still obeys some general regularities of degassing. In particular, the efficiency of gas removal from melts parallels the intensity of vibrations. Isochrones of degassing are characterized by the occurrence of special regions (Figure 10.1), which can be attributed to the appearance and development of cavitation. Thus, region I, in which degassing is virtually absent, can be defined as the precavitation region. Region II, in which the efficiency of degassing drastically rises but then becomes steady-state, is the region of treatment conditions corresponding to cavitation.

Cavitation in a melt modifies the ratio of applied and absorbed acoustic powers, which is due to energy expenditure for cavitation and the related 8- to 10-fold drop in the wave resistance of cavitating liquid in response to the change of its state (the presence of cavities affects liquid density and the velocity of ultrasound propagation).

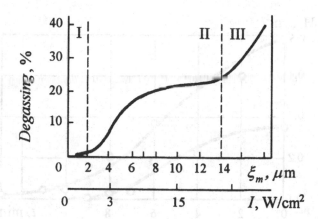

Figure 10.1. Efficiency of degassing of an aluminum-based alloy as a function of the amplitude and intensity of ultrasonic vibrations [5].

With developed cavitation, the efficiency of degassing again rises as the ultrasonic energy fed to the melt increases (region III), which is due to stabilization of the wave resistance of melt at a level corresponding to 0.1–0.2 of its initial value.

Raising the applied acoustic power still further may again lower the efficiency of treatment as the result of a drastic fall in the wave resistance, since cavities will virtually force liquid out of the cavitation region. However, this regime of ultrasonic treatment can hardly be attained, as it is very difficult to feed necessary ultrasonic powers to melts.

The above regularities were proved by numerous experiments with other aluminum- and magnesium-based alloys. The relationship between ultrasound intensity and the efficiency of degassing can be illustrated by the following data: degassing of molten AL9 alloy by vibrations with intensities of 4 and 24 W/cm^2 (all other experimental conditions being identical) requires 24 and 6 min, respectively.

The relationship between the efficiency of degassing and process time is of much practical importance. Figure 10.2 shows the dynamics of hydrogen content in molten AL9 alloy treated at a maximum attainable intensity of ultrasound. It can be seen that the dynamics of hydrogen liberation is virtually independent of its initial concentration. At both initial concentrations, ultrasonic treatment makes it possible to reduce hydrogen content to a minimum level of 0.1 cm^3/100 g alloy. The dynamics of hydrogen evolution allows one to speculate as to the

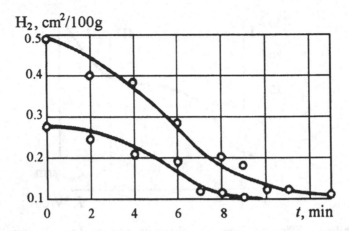

Figure 10.2.   Dynamics of hydrogen content in molten Al9 alloys with different initial concentrations of hydrogen [4].

mechanism of gas evolution in an ultrasonic field. Hydrogen concentration in molten AL9 alloy at 720°C is typically 0.45–0.75 cm^3/100 g metal; however, only part of this hydrogen is actually dissolved, the rest being contained in bubbles.

At the initial stage of ultrasonic irradiation, gas bubbles, always occurring in melts, coalesce and emerge to the surface. Further exposure to ultrasound gives rise to new tiny hydrogen bubbles that also coalesce to escape through the melt surface.

Analogous results were obtained in experiments on the degassing of other aluminum-based alloys. At the initial concentration of hydrogen 0.6 cm^3/100 g alloy, the exposure of molten AMg6 alloy (2 kg) to high-intensity vibrations for 20–30 s halves the initial hydrogen content, while irradiation for further 30–40 s reduces it to 0.1 cm^3/100 g.

Analysis of the temperature course of hydrogen evolution curves reveals the existence of optimum degassing temperatures lying, for Al9, between 720 and 760°C. At temperatures below 720°C, the high viscosity of melts prevents bubble coalescence. Above 760°C, the rate of hydrogen liberation becomes comparable with the rate of its absorption by the melt. This is because the solubility of moisture hydrogen in melts increases with temperature.

Radiator material may strongly influence ultrasonic degassing. In particular, titanium radiator was found to be more effective in degassing molten AL9 alloy than niobium or quartz radiators. This might be due to two reasons. First, titanium possesses a higher acoustic qual-

ity than niobium or quartz, since it has a smaller damping coefficient
and thus contributes to a lesser extent to energy loss in the vibratory
system. At the same input power of transducer, the power fed into
a load through niobium radiator is 25% lower than that fed through
titanium radiator [4]. Second, transition metals, to which titanium
and niobium belong, can efficiently chemisorb hydrogen, nitrogen, and
other gases. Disintegration of niobium or titanium radiators results in
that some amounts of these metals appear in the melts, where they
chemisorb dissolved hydrogen and thus decrease free hydrogen content
in melts.

G. I. Eskin assessed the efficiency of ultrasonic degassing with reference to other degassing methods – evacuation and refining of aluminum
melts with chlorides.

Reduced pressure may cause desorption of gas molecules from the
melt surface and diffusion of new portions of gas to both the melt surface and bubbles in the melt bulk. However, evacuation will not promote the formation of new bubbles. Therefore, for efficient degassing of
a bulk melt, it is expedient to combine evacuation with the stimulation
of bubble formation. This may be done by ultrasonic irradiation or
fluxing.

Figure 10.3 shows the dynamics of hydrogen evolution from molten
AL9 alloy caused by evacuation, refining with chlorides, and ultrasonic treatment. It is seen that degassing induced by the addition of
salts is insignificant. Thus, the use of salts is mainly owing to their
ability to remove solid nonmetallic inclusions from melts. Evacuation
is more efficient in degassing than fluxing, but ultrasound works still
better. A combined application of ultrasonic treatment and evacuation enhances degassing and diminishes residual concentrations of gases
in melts. Such a combined treatment is perhaps the most promising
method for production of high-quality ingots of aluminum alloys.

Experiments indicated the feasibility of efficient industrial ultrasonic refining of large-volume melts and melt streams during continuous
casting.

In continuous casting, when large melting furnaces and high-per-
formance casting machines are used, it is expedient to perform refining
operations in a casting zone. The feeding of ultrasonic vibrations directly to the melt placed in a crystallizer or on the way between a mixer
and the crystallizer is more efficient for degassing than their feeding to
the crystallizer walls.

The ultrasonic treatment of molten industrial aluminum alloys was
found to enhance their purity, castability, the density of as-cast and
wrought products, and the quality of ingots and castings.

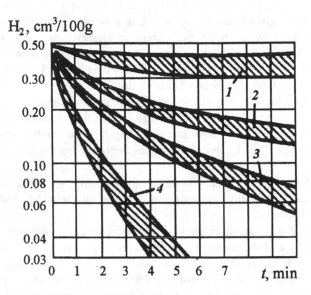

Figure 10.3. Degassing of molten A19 alloy by (*1*) fluxing with zinc chloride, (*2*) evacuation, (*3*) ultrasonic treatment, and (*4*) combined action of ultrasound and evacuation [4].

For illustration, Table 10.1 compares some industrial refining methods employed for a shaped casting of AL4 alloy.

Ultrasonically treated wrought aluminum alloys exhibit an increased density. Table 10.2 summarizes the results of rapid analysis (by the first-bubble method) of hydrogen content in aluminum melts during their continuous casting. It is seen that the efficiency of ultrasonic degassing largely depends on the chemical composition of melts, supplied acoustic power, and the rate of metal casting.

Ingots obtained under ultrasonic degassing exhibited improved mechanical properties at room and elevated temperatures. For instance, the elongation of an AK6 alloy ingot (460 mm in diameter) increased from 5.2 to 8%.

The ultrasonic degassing of molten AMg6 alloy in production of 3- to 10-mm thick sheets made it possible to reduce hydrogen content in rolled intermediate products to 0.25 cm^3/100 g, as well as to avoid their delamination and improve their weldability.

Along with investigation of aluminum-based alloys, attempts are now being made to study the degassing of magnesium-based and other alloys. Thus, analysis of hydrogen content in control and ultrasonically treated molten MA8 alloy showed that even a short-term

Table 10.1   Comparison of various industrial methods for refining grade
AL4 alloy [6].

Method	Hydrogen content, cm³/100 g	Density $\rho \times 10^3$, kg/m³	VIAM* porosity rating	Mechanical properties	
				σ_B, MPa	δ, %
Control melt	0.35	2.665	4	200	3.8
Refinement by fluxing	0.26	2.660	3–4	225	4.0
Treatment with hexachloroethane	0.30	2.663	2–3	212	4.5
Argon scavenging	0.26	2.667	2–3	233	4.0
Evacuation	0.18	2.681	1–2	228	4.2
Ultrasonic treatment	0.17	2.706	1–2	245	5.1

* Vserossiiskii Institut Aviatsionnykh Materialov (All-Russian Institute of
Aviation Materials).

Table 10.2   Efficiency of ultrasonic degassing of continuously cast aluminum alloys [6].

Alloy grade	Crystallizer dimensions, mm	Casting rate, kg/min	Location of ultra-sonic radiator	Acoustic power, kW	Hydrogen content, cm³/100 g		Dega-ssing effici-ency, %
					Before treatment	After treatment	
AK6	⌀250	9.3	In pouring spout	0.8	0.33	0.18	45
AMg2	⌀350	19.7	spout	0.8	0.42	0.30	27
AD1	⌀350	13.5	–"–	0.8	0.20	0.16	20
AK6	⌀460	15.7	–"–	0.8	0.40	0.21	46
AD1	1040×300	57.5	–"–	0.8	0.25	0.19	24
D16	1480×210	63.0	–"–	0.8	0.38	0.24	37
AMg6	1700×300	82.0	In tap box	6.0	0.60	0.45	25
AMg6	1700×300	82.0	–"–	9.0	0.60	0.40	33
AMg6	1700×300	82.0	–"–	11.0	0.60	0.30	50

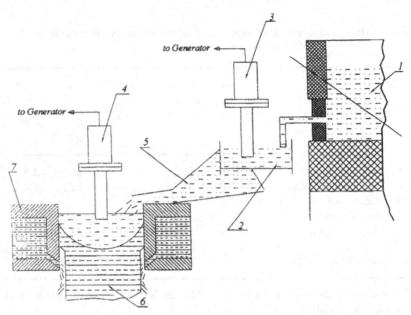

Figure 10.4. Schematic representation of ultrasonic treatment of melt stream [4]:
(*1*) furnace with melt, (*2*) intermediate tank, (*3*) ultrasonic transducer for filtration
and degassing, (*4*) ultrasonic transducer for structural refinement, (*5*) melt trough,
(*6*) crystallizer, (*7*) continuous casting mould.

exposure to ultrasound may reduce hydrogen content of alloys about
2.5-fold [6].

Apart from removing bubbles from melts, ultrasound induces flotation of solid nonmetallic particles and thus lowers their content in aluminum alloys by 10–20%.

The flotation effect of gas-filled bubbles pulsating in an ultrasonic
field was first observed by Rosenberg [9] during the ultrasonic refinement studies. Later, Novitskii showed that a pulsating bubble may at-
tract particles irrespective of their nature, density, and wettability [10].

Eskin suggested to use ultrasound not only for the degassing and
flotation of solid nonmetallic particles, but also for the fine filtration
of melts through multilayer gauze filters [7]. Figure 10.4 shows a combined ultrasonic installation for degassing, filtration (transducer 3), and
structural refinement (transducer 4) of continuously cast light alloys.
The efficiency of such combined treatment can be assessed by comparing the purity (with respect to nonmetallic inclusions) of large-sized
ingots of constructional aluminum-based alloys and the oxide content
in filter cake (Table 10.3).

Table 10.3 Efficiency of ultrasound for filtration of molten aluminum-based alloys [5].

Alloy grade	Ingot diameter, mm	Refining method	Number of filtering layers	Hydrogen content of ingot, cm^3/100 g	Oxygen content of filter cake, mass %	Ingots, mm^2/cm^2
D16ch	650	UF	5	0.13	0.09–0.12	0.005
		CF	1	0.16	–	0.025
V95pch	650	UF	2	0.14	0.06–0.10	0.008
		CF	1	0.16	–	0.021
V95pch	830	UF	3	0.18	0.12–0.40	0.005
		CF	1	0.25	–	0.035

Footnote. The mean concentration of oxygen in ingots is 0.01 mass %; UF stands for fine filtration through a glass cloth (0.6 × 0.6-mm mesh) in combination with ultrasonic irradiation; CF stands for conventional filtration through a glass cloth (1.3 × 1.3-mm mesh).

Thus, experimental data available to date indicate the feasibility of industrial application of ultrasound for melt refining during continuous and shaped castings of light alloys.

10.2 Structural Modification of Solidifying Metals

The vibration of solidifying melts leads to structural refinement of metals and improves their properties.

Sokolov [11, 12] was the first to use ultrasonic treatment of molten low-melting-point metals. However, technical difficulties, associated with the feeding of vibrations to melts, and the lack of suitable ultrasonic generators in the kilohertz range hindered the research along this line.

Since the 1950s, Teumin [13–18], Pogodin-Alekseev [19, 20], Polotskii [21–23], Eskin [4–8, 24, 25], and Abramov [26–32] in the USSR had been carrying out a broad range of research that culminated in the development of ultrasonic equipment for pyrometallurgical applications. Ultrasonic effects on the properties of as-cast materials can be summarized as follows:

(1) Reduction in the mean grain size;

(2) Control of columnar structure and formation of equiaxial grains;

(3) Variation in the distribution of phases in terms of their relative amounts, structural refinement, and mutual geometry;

(4) Improvement of material homogeneity and segregation control;

(5) Uniform distribution of nonmetallic inclusions.

At present, the method of ultrasonic treatment of molten and so-lidifying metals has been favorably proven by laboratory tests and is finding increasing use in large-scale applications. However, much of the currrent effort is directed toward the elucidation of the mecha-nism of action of ultrasonic vibrations on solidification (see section 5.1.4), the manufacturing of commercial ultrasonic equipment for met-allurgical applications, the development of ultrasonic treatment pro-cesses, the evaluation of final structure and properties of products, the assessment of technical and economic aspects of ultrasonic treat-ment.

This section attempts to address the reader to all these aspects of ultrasonic treatment of solidifying melts. The feeding of ultrasonic vibrations to a solidifying melt is not an easy problem and largely determines the potency of ultrasound in improving the quality of metal castings.

It is known that material tractability is strongly dependent on the ability of ultrasound to penetrate into the melt bulk, in particular, to the front of solidification. The penetrability of ultrasound depends on the method of its feeding and characteristics of casting processes, which should be adapted to ultrasonic technology. These are shaped casting, ingot casting, continuous casting and some refining processes, e.g., vacuum arc remelting (VAR), electroslag remelting (ESR), and electron beam remelting (EBR).

Continuous casting of metals has found a wide use in the past few decades. Continuously cast billets, especially of round or rectangular cross sections, often exhibit coarse-grained structure and axial porosity. During continuous casting, ultrasonic vibrations may be fed to a melt stream, to a liquid metal crater, to a mould, or to a secondary cooling zone.

Special steels, such as nickel-based superalloys and refractory met-als, are typically produced in VAR and EBR furnaces, although ESR is also a recommended method of their production. These methods have been developed to control segregation in heavy ingots and ensure their lower contamination with nonmetallic impurities and gases as compared to ordinary metals. However, all these methods suffer from a disadvantage associated with the formation of a coarse-crystalline columnar structure. This problem can, however, be solved by transmitting vibrations either through the bottom of a solidified ingot or via a consumable electrode.

Figure 10.5,a illustrates the transmission of elastic vibrations to a solidifying melt through hole 1 in the mould bottom. Radiating probe 6 is introduced through this hole with a spacing of tenths of a millimeter to avoid molten metal leakage. Figure 10.5,b illustrates the method of transmitting vibrations to the upper portion of the melt.

In this method, precautions should be undertaken as to avoid piping or the formation of an upper crust before the melt bulk is completely solidi fied. For this, the cooling of the mould top should be limited. In the top transmission method, ultrasound interacts with a liquid metal, whose volume continuously changes. The top transmission method has found a wide use for degassing aluminum alloys (see section 10.1).

The use of ultrasound in continuous casting is probably most promising. In this case, vibrations may have relatively low intensity, since the amount of metal that solidifies per unit time is constant and rather small. The transmission of vibrations directly to the melt crater is likely the best method of their supply (Figures 10.5,c and 10.4); in any event, neither the feeding of vibrations through a mould nor the irradiation of a melt in a tundish are very effective.

The difficulty of creating radiators that could effectively operate for long time periods in molten metals has turned the inventor's mind toward noncontact methods of vibration feeding. One of these is the transmission of vibrations through a billet crust in the secondary cooling zone.

Historically, Seeman and Menzel [36] were the first who applied ultrasonics in continuous casting. They used a 25-kW oscillator, operated at 40 kHz, for driving four magnetostrictive transducers. The system was tested in the casting of 290-mm diameter aluminum rounds. Ultrasonic irradiation was found to refine both macro- and microstructure of ingots and improved their mechanical properties.

Herman [39] described various proprietary designs of equipment used for transmitting ultrasound to the liquid metal crater. Unfortunately, no results of their tests have been reported.

In VAR and ESR, vibrations can be best transmitted from the magnetostrictive transducer through a sonotrode system set to the ingot

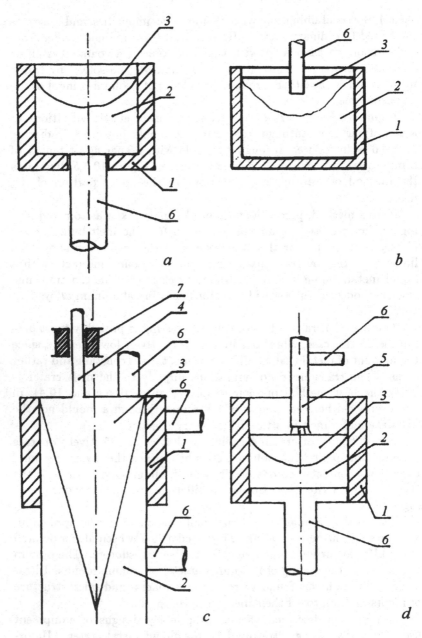

Figure 10.5.   Putative schemes of ultrasonic vibration feeding: (a) to mould from below, (b) to mould from above, (c) during continuous casting, (d) during remelting. (1) Mould (crystallizer), (2) solidifying ingot, (3) melt, (4) metal stream, (5) remelting electrode, (6) ultrasonic probe, (7) ultrasonic funnel.

bottom section and eventually to the solidification zone (Figure 10.5,d). Although offering a number of advantages, such as relatively low ultrasound intensity necessary for production of commercial-size ingots and the possibility of using conventional mild-temperature resonators, this method is not free of limitations. As the solidified portion of a solidifying ingot increases, the resonant frequency of the ultrasonic stacking-ingot system will vary to disturb the distribution of nodes and antinodes. This may interfere with the normal operation of the transducer and the whole vibratory system and lead to a nonuniform grain refinement in the direction of wave propagation. The result of this will be a gradual transition from fine-grained to coarse-grained structure at half-wave thickness intervals and increased vibrational energy loss during the transmission of vibrations to the solidification front.

Remelting furnaces equipped with ultrasonic stacks were used for processing a number of superalloys, as well as stainless and special steels.

The potentialities of ultrasonic installations with a direct acoustic contact between the melt and resonator are often crippled because of the lack of appropriate materials that would combine high fatigue strength and sufficient resistance to high temperatures, chemical attack, and cavitation.

Because of this, the method of contactless feeding of vibrations to the melt is of much practical interest. In terms of this method, a permanent magnetic field is imposed on the high-frequency field of induction furnace. The resultant electrodynamic forces produce vibrations in the melt [40–44].

In the mid-1930s, Sokolov [11, 12] investigated the effect of ultrasonic vibrations on solidification of pure metals, specifically, tin, zinc, and aluminum. By using piezoelectric transducers operated at frequencies from 600 to 4500 kHz, the author was able to reveal that elastic vibrations can produce a dendrite structure.

Later investigations [23, 45–49] were focused not only on Sn, Z, and Al, but also on Bi, Cd, Sb, and some refractory metals, such as Cr, Ti, Mo, and W. It was found that the grain structure of Cr, Mo, and W was refined as opposed to Ti (Figure 10.6).

Structural changes induced by the solidification of metals in an ultrasonic field can obviously modify their characteristics.

The majority of commercially available pure metals are actually alloys, because of an inevitable impurity contamination. The application of ultrasound to such metals modifies the distribution of phases and affects both grain refinement and zonal and dendritic segregation. As a result, the integral metal properties would change.

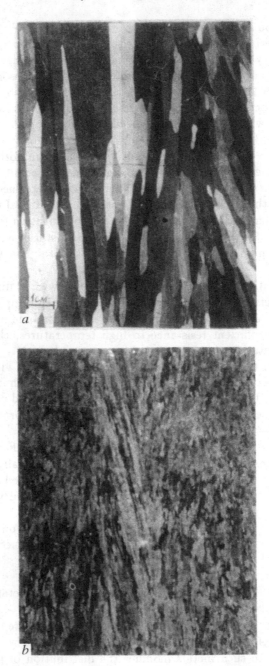

Figure 10.6. Molybdenum microstructure: (a) control ingot, (b) ultrasonically solidified ingot.

Therefore, to identify the effect of ultrasound on grain refinement, which is the most important metal characteristic, one has to use sufficiently pure metals exhibiting neither segregation nor second phase precipitation.

Several researchers, who approached this problem by using pure Al, Zn, Sn, and Sb, were able to reveal that the strength of as-cast materials can be improved through grain refinement. In particular, Eskin [4] demonstrated that ultrasonic treatment of high-purity Al substantially refined its microstructure (the number of grains per sq. mm increased from 3.1 to 37.5) and thus improved its mechanical properties (tensile strength increased from 5.4 to 7 kg mm^{-2}, and hardness from HB 17.2 to HB 19.7). The hardening was accompanied by an increase in elongation from 48 to 52%.

Seeman *et al.* [37] showed that an 8-fold increase in the grain size of Sn improved its hardness by 60%. For Zn and Al, the hardness gain was 63 and 80%, respectively.

Grain refinement in Sb was accompanied by a hardness reading rise from HB 34 to HB 52 [47].

Abramov and Gurevich [50] used a 20-kHz ultrasound to vibrate solidifying pure metals with differing crystalline structures: bcc (iron), fcc (nickel and aluminum), hcp (cobalt and zinc), tetragonal (tin), and rhombohedral (bismuth).

It was found that 1-kW power of ultrasound was sufficient to cause grain refinement of Sn and Al (Table 10.4), whereas a 2.5-kW ultrasound elicited the refinement of Fe, Co, Bi, and Zn. However, the results were inconclusive with regard to ultrasonic tractability of metals possessing different lattices.

Structural changes in metals were compared with their properties, viz. hardness (Table 10.4) and tensile strength at -196, 20, and 900°C for Fe, at -196, 20, and 700°C for Co, and at -196 and 20°C for Al (Figure 10.7).

Ultrasonic treatment increased the tensile strength of Fe 1.3-fold at -196°C and slightly improved its ductility. The effect of grain refinement on iron properties was found to be dependent on temperature; at 20°C, the tensile strength and hardness gains were 20 and 6%, respectively, whereas no difference was observed between ultrasonically treated and control specimens at 900°C. At the same time, the difference in ductility increased with temperature, in particular, ultrasonic treatment did not virtually affect elongation at -196°C, but increased it by 10% and 15% at 20 and 900°C, respectively.

Grain refinement augmented the tensile strength of Al at -196 and 20°C by about 50%. In this case, the hardness tested at room tempera-

Table 10.4 Effect of ultrasonic treatment on the grain size and hardness of metals.

Metal	Transducer input, kW	Grain size, no.	Mean grain diameter, mm	Brinell hardness
Fe	0	−1	0.50	100
	1	−1	0.50	−
	2.5	3	0.12	106
Al	0	1	0.25	16
	1	3	0.12	18
	2.5	4	0.09	19
Co	0	−1	0.50	143
	1	1	0.25	162
	2.5	3	0.12	170
Zn	0	1	0.25	32
	1	1	0.25	−
	2.5	3	0.12	40
Sn	0	−1	0.5	5
	1	2	0.18	6
	2.5	3	0.12	7
Bi	0	−1	0.5	9
	1	−1	0.5	−
	2.5	4	0.09	11

ture increased by 20%. At −196°C, the hardness gain was accompanied by a decrease in elongation. At 20°C, ductility was essentially the same for both ultrasonically irradiated and control aluminum ingots.

Grain refinement in Co improved its strength 2.1 to 2.4 times at all testing temperatures. At room temperature, Co hardness increased by 20%, but its ductility virtually remained unchanged. Ultrasonic treatment raised the hardness of Zn, Bi, and Sn by 22, 25, and 40%, respectively.

Thus, ultrasonically-induced grain refinement of metals leads to considerable changes in their mechanical properties irrespective of the crystal lattice type. Strength is most affected at low temperatures, especially in the case of fcc and hcp metals. With increasing testing temperature, the effect of grain refinement diminishes and virtually disappears in iron (bcc metal). Conversely, the effect of grain refinement on ductility is most profound for bcc metals and grows with increasing temperature.

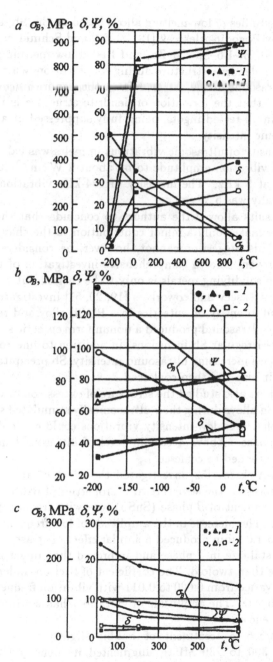

Figure 10.7. Effect of ultrasound on the mechanical properties of (a) iron, (b) aluminum, and (c) cobalt: (1) ultrasonically solidified ingots; (2) control ingots.

Detailed studies of low-melting alloys based on Sn, Sb, and Bi were performed by Pogodin-Alekseev [19, 20, 51] and Schmid *et al.* [46, 47]. Bi–Pb–Sn–Cd, Bi–Pb–Sn–Cd–Zn, and Bi–Cd low-melting-point alloys typically form dendrite structure during solidification, which allows the effect of ultrasound to be assessed from the dendrite arm reduction. It was found that the formation of dendrite structure is inhibited by ultrasound in all tested ingots, being fully suppressed at a sufficiently high ultrasonic intensity.

The intensity of ultrasonic vibrations in melts was calculated from the crucible vibration amplitude to be about 2 W cm^{-2} at 50 Hz and 39 W cm^{-2} at 9 kHz. The intensity of 284-kHz vibrations measured calorimetrically was 5 W cm^{-2}.

These results allowed the authors to conclude that the frequency of vibrations makes only a minor contribution to the changes in ingot structure. This suggestion cannot, however, be considered as unambiguously proven, since a comprehensive investigation of the effect of frequency on solidifying metals is only to be carried out.

Pogodin-Alekseev and coworkers [19, 20, 51] investigated the effect of ultrasound intensity on eutectics in a Pb–Sb alloy and revealed that low-intensity ultrasound produced a nonuniform eutectic structure containing large irregular Sb inclusions, in addition to fine round-shaped particles. With increasing ultrasound intensity, Sb precipitates initially diminished in size and then coalesced.

The authors also studied the structure of excess constituents of Pb–Sb and Zn–Sn alloys during their ultrasonically stimulated solidification and found that even low-intensity vibrations could cause dispersion of the excess phase. Above a certain level of ultrasound intensity, the excess phase tended to coalesce.

Pogodin-Alekseev [52] investigated the effect of audio- and ultrasonic frequences on the structure and properties of babbit and showed that the refinement of β-phase (SnSb) rose with frequency in a range of 20–60 Hz. The increase in the amplitude of low-frequency vibrations from 0.35 to 1.00 mm produced a similar effect. Ultrasonic irradiation reduced crystal size in β-phase and improved the impact properties of babbit more than twofold. The coefficient of friction under incomplete lubrication varied from 0.020 to 0.011 with vibration frequency increase from 20 to 60 Hz. The amplitude of vibrations had a strong effect only at low frequencies.

The ultrasonic treatment of babbit slightly reduced its hardness (from 300 to 230 MPa), augmented its density (from 7.34 to 7.39 g cm^{-3}), and increased tensile strength from 74 MPa (control specimen) to 90 MPa (ultrasonically treated specimen).

Abramov and Filonenko [53] investigated the effect of ultrasound on solidification of eutectic Sn–Cd, Sn–Pb, and Sb–Pb alloys prepared from high-purity components (max. 5 ppm impurity) and directionally solidified in a 10^{-5} torr vacuum. Ingots were tested for the temperature gradient at the solidification front, the crystal growth rate, overcooling during the growth of colonies, structural spacing, and phase thickness.

A correlation was observed between the structure spacing of regular Sn–Cd and Sn–Pb eutectics or the secondary arm spacing of irregular Sb–Pb eutectic and the crystal growth rate of control specimens (no ultrasound; a temperature gradient of 10 to 20 $\deg \mathrm{cm}^{-1}$).

When vibrational power was below a certain level specific for each alloy, the eutectics of Sn–Cd and Sn–Pb alloys retained their regularity under ultrasonic irradiation. Their structural spacing was increased by a factor of 2 to 2.5.

Ultrasonic vibrations with an intensity above the threshold value produced an irregular structure and increased the phase thickness. When the ratio of temperature gradient to crystal growth rate attained max. 10^4 $\deg \mathrm{s\,cm}^{-2}$, the crystals of the Pb-containing solid solution were concentrated at the ingot bottom.

In the hypereutectic Sb–Pb alloy, ultrasound augmented the thickness of precipitating phases (Figure 10.8). These results allowed Abramov and Filonenko to conclude that ultrasound increases the diffusion rate in melts by a factor of 5 to 7.

Pereyaslov and Sapozhnikov [54] examined the effect of ultrasound on the structure, mechanical properties, and corrodability of Pb–Sb alloys with Sb content varying from 0 to 12 wt.%. Solidifying melts were irradiated through the hole in the mould bottom. The displacement amplitude was 6 μm at 20 kHz. Ultrasonic treatment gave rise to a fine equiaxial structure. The addition of 0.1 to 1.0 wt.% Ga refined grains still further. In this case, mechanical properties varied insignificantly, but the resistance of Pb – max. 6% Sb alloys to corrosion in sulfuric acid considerably increased. The corrosion resistance especially increased in alloys that contained 0.2 wt.% Ga.

Eskin and coworkers [4–8, 24, 25] studied the effect of ultrasonic irradiation on aluminum and aluminum-based alloys. In particular, they investigated the solidification of aluminum and binary Al–Cu alloys with different constitution diagrams, as well as commercial castings and wrought alloys.

Ultrasonic irradiation was found to refine the macrostructure of castings. With the copper concentration increasing from 2 to 33%, the efficiency of ultrasonic treatment rose. In this case, the strength of

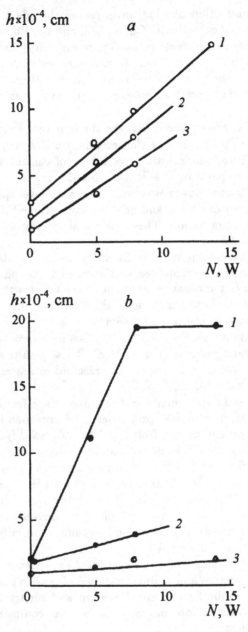

Figure 10.8. Dependence of the high-entropy phase (Sb) thickness on ultrasound intensity: (a) $G = 10$ deg/cm; $R = 5 \times 10^{-4}$ (*1*), 6.5×10^{-4} (*2*), and 8.9×10^{-4} cm/s (*3*). (b) $G = 20$ deg/cm; $R = 4.87 \times 10^{-4}$ (*1*), 6.87×10^{-4} (*2*), and 8.9×10^{-4} cm/s (*3*).

growing crystals diminished and the solidification temperature range widened.

Structural changes modified the strength and ductility of materials. When melts were cast in a metal mould, the tensile strength of Al alloys containing 2 to 8% Cu increased by 20 to 25% and that of Al alloys containing 12 to 33% Cu remained unchanged. As a result of casting in a ceramic mould, the tensile strength of single-phase and two-phase alloys increased by 50–70% and 20%, respectively. The elongation of all the alloys cast at a high rate remained unchanged and that of the alloys that solidified slowly in a gypsum mould increased by 30 to 100%.

Eskin also studied the effect of ultrasonic vibrations on the formation of intermetallics in Al–Mn, Al–Zr, and Al–Cr alloys.

Microstructure analysis showed that ultrasound, first, enhanced the solidification nuclei formation and, second, suppressed crystal precipitation.

Ultrasonic irradiation considerably improved the tensile properties of alloys, raising tensile strength by 24% (from 170 to 210 MPa), yield strength by 25% (from 140 to 175 MPa), and elongation by 40% (from 12.5 to 17.5%).

Ultrasonic irradiation of peritectically solidifying Al–Zr and Al–Cr alloys produced similar changes. These observations were later confirmed by other researchers [55].

Some investigators performed experiments with Al–Si melts. Thus, Abramov [27] examined a combined effect of ultrasonic irradiation and inoculation on the structure of Al–14% Si alloy. Sodium, as a microadditive, was introduced to the melt at a concentration of 0.1% immediately before pouring.

The macrostructure of untreated cast alloy showed the presence of the large segregates of excess Si crystals against the background of relatively coarse eutectics. By contrast, ultrasonic irradiation and microalloying with Na refined silicon crystals and uniformly distributed them over the ingot.

When inoculation was performed without ultrasonic irradiation, the shape of excess Si crystals changed, but the eutectics structure remained fairly coarse.

Eskin and coworkers [4–8, 24, 25] investigated a number of commercial Al alloys. Like in the case of model alloys, the ultrasonic irradiation of commercial alloys refined their structure and improved their mechanical properties (Table 10.5).

It was also found that ultrasound can substantially suppress zonal segregation of Zr, Ti, Cr, and Mn, i.e., the additives that strongly differ from aluminum in density.

Table 10.5 Effect of ultrasound on the mechanical properties of commercial aluminum-based alloys [4].

Alloy grade	Alloy system	Alloy condition	Ultrasonic treatment	UTS, MPa	Elongation, %	Area reduction, %	Brinell hardness
AL40	Al–Si–Cu–Mg–Ni–Fe	As-cast	No	160	5	–	64
		–"–	Yes	210	6	–	75
		Heat-treated	No	250	3	–	89
		–"–	Yes	330	5	–	96
AL19	Al–Cu–Mn	As-cast	No	270	3	–	–
		–"–	Yes	350	4	–	–
AL20		–"–	No	190	5	6	72
		–"–	Yes	210	11	12	76

Dobatkin and Eskin [5] determined the conditions of continuous casting, under which a dendrite-free grain structure was formed. This type of structure could be observed, if the grain size was smaller than or equal to the size of a dendrite cell produced at a given solidification rate. Cavitation in the melt appeared to be the necessary condition for an as-cast subdendrite structure to develop.

As a result, Dobatkin and Eskin could produce 850-mm diameter cast billets from most commercially available inoculated aluminum alloys with the subdendrite structure. Table 10.6 lists the linear grain sizes of subdendrite and dendrite structures averaged over 5 to 40 billets of several commercial aluminum alloys. In ultrasonically treated ingots, the as-cast structure exhibited a uniform refinement across the section, whereas untreated ingots had a fan-like dendrite structure.

Similar results were obtained for ultrasonically treated Mg alloys [57, 58].

Ultrasonically induced grain refinement and the dendrite-to-subdendrite structural changes increased the ductility of ingots at both room and high testing temperatures (Table 10.7).

Ultrasonic irradiation during a continuous casting of 550×160-mm ingots of grade MA2-1 magnesium alloy resulted in a marked grain refinement. Control ingot exhibited a nonuniform structure across its section: 1- to 3-mm equiaxial grains in the core and 5- to 7-mm columnar grains in the transverse direction near larger faces. The columnar zone was 45 mm long.

Table 10.6 Effect of ultrasonically induced cavitation in molten light alloys on the grain refinement of ingots [5].

Alloy system	Alloy grade	Ingot size, m	Inoculant concentration, %		Grain size, μm	
			Zr	Ti	Ultra-sound	Control
	1960, 1965, 1963, V96Ts-1	0.06–0.37	0.15	0.04	20–70*	300–1500
Al–Zn--Mg–Cu	V95pch, 7050, 7010, V934	0.06–0.98	0.15	0.04	20–140*	500–200
Al–Cu–Mg	D164, 2024	0.06–0.98	0.15	0.04	20–140*	300–200
Al–Cu–Mn	1201, D20-1	0.07–0.37	0.18	0.06	25–90*	300–120
Al–Mg–Mg	1561	0.37	0.16	0.06	90*	1500
Mg–Al–Zn	MA2-Ipch	0.12 0.55–0.165	0.003	–	50–500	120–400
Mg–Zn–Zr	MA14	0.17	0.1–0.4 0.03	–	60*	190
Mg–Y--Ce–Zr	MDZ-3	0.3×0.1	0.6 0.03	–	60*	300
Mg–Y–Nd--Zn–Zr	VMD-7 VMD-9	0.17	0.14 0.03	–	30*	200

* Subdendrite grain.

Table 10.7 Tensile strength of grade 1960 (Al–Zn–Mg–Cu) alloy under homogenizing conditions [6].

Ingot diameter, mm	Structure type	UTS, MPa	Elongation, %
	Testing at 20°C		
65	Dendrite	226	3.8
	Subdendrite	225	4.0
270	Dendrite	206	2.0
	Subdendrite	207	2.9
	Testing at 400°C		
65	Dendrite	35	98.0
	Subdendrite	37	132.0
270	Dendrite	39	108.0
	Subdendrite	40	122.0

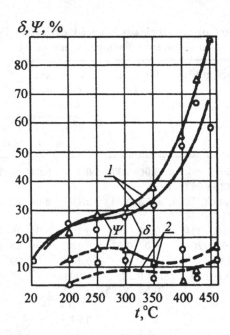

Figure 10.9. Ductility of homogenized grade MA2-1 alloy versus testing temperature [6]: (*1*) ultrasonically solidified ingot; (*2*) control ingot.

Ultrasonic treatment produced 0.3- to 0.5-mm equiaxial grains over the entire billet section and reduced the dendrite cell size. This appreciably improved material ductility, especially in the surface region. The ductility of control ingot was virtually independent of testing temperature, whereas in ultrasonically treated ingot both elongation and area reduction grew with temperature up to 450°C.

Ultrasonic treatment also raised the permissible strain level in metal working processes. The inherent effect of the refined as-cast structure on the quality of worked metals was retained even after relatively high strains. In fact, after the rolling of 150-mm thick cast slabs into 0.8–2.5-mm sheets, the ductility was greater for sheets produced from ultrasonically treated metal (Figure 10.9).

Pogodin-Alekseev [19, 59] studied the effect of elastic vibrations on copper-based alloys, viz. grade Br. OTsS-5-5-5, Br. OTsS-6-6-3, Br B2, Br B5, and BNT bronzes, and evaluated the contribution from displacement amplitude to the structure and properties of ingots produced by shaped and continuous castings.

When vibrations were transmitted directly to cast grade Br. OTsS-5-5-5 alloy, both low- and ultrasonic-frequency irradiation refined their macro- and microstructure and modified the mechanical properties of the alloy. Low and high frequences augmented the tensile strength of ingots by 20 and 40% and their elongation by 15 and 35%, respectively.

In the case of cast grade Br. OTs5-6-6-3 bronze, ultrasonic vibrations fed to shell moulds were found to modify the structure and properties of castings as well. In particular, tensile strength increased by 34 and 43% as a result of low-frequency and ultrasonic vibrations, respectively. It was also found that irradiated specimens offered better corrosion resistance.

Ultrasonic irradiation of grade BNT bronze increased its elongation and area reduction by 26 and 15%, respectively. Impact strength remained unchanged, while density was slightly increased.

It follows from the literature data that single-phase alloys possess fair ultrasonic tractability. This supposition was verified on a number of Cu–Al, Cu–Sn, and Cu–Zn alloys [29].

The Cu–Al alloys studied contained 1, 5, 7, 9, and 12% Al. The first three (α-alloys) and the last (β-alloy) were single-phase, whereas the alloy with 9% Al contained both α and β phases. All the alloys exhibited a narrow solidification temperature range.

The Cu–Sn alloys studied contained 2, 8.5, 13, and 15% Sn. The first three alloys (α-alloys) were single-phase, and the fourth was two-phase (α, β-alloy). The solidification temperature range was quite narrow for Cu–2% Sn alloy, but extended to 140°C for the alloys containing 8.5 and 15% Sn, and to 175°C for the alloy with 13% Sn.

In the Cu–Zn system, single-phase α-alloys contained 10, 20, and 30% Zn, two-phase α, β-alloy contained 35% Zn, and single-phase β-alloy contained 45% Zn. The solidification temperature range was from 10 to 20°C for all of these alloys.

Ultrasonic input power (1100 W) was kept constant in all tests. The macrograin refinement factor K, i.e. the ratio of the grain sizes of control and irradiated alloys, was taken as a tractability characteristic.

Experiments revealed no essential effect of ultrasound on the tractability of single-phase and two-phase alloys. However, ultrasound increased the tractability of the solid solution-type alloys, as follows from measurements of the grain refinement factor (see Table 10.8).

The micrograins were appreciably refined in solid solution-type alloys. The structure of two-phase alloys became more dispersed.

The Cu–Sn alloys exhibited a much better tractability than Cu–Al and Cu–Zn alloys did. This is probably due to a wider solidification temperature range of the former alloys.

Table 10.8   Ultrasonic tractability taken as the grain refinement factor
K of Cu–Al, Cu–Sn, and Cu–Zn alloys.

Al, %	Phases	K	Sn, %	Phases	K	Zn, %	Phases	K
1	α	7	2	α	9	10	α	11
5	α	13	8.5	α	26	20	α	6
7	α	11	13	α	20	30	α	6
9	$\alpha + \beta$	5	15	$\alpha + \beta$	14	35	$\alpha + \beta$	5
12	β	10				45	β	2

Vasin *et al.* [60] investigated the effect of ultrasonic irradiation
on the structure and properties of grade OTs10-2 bronze. Ultrasonic
vibrations were transmitted through a water-cooled funnel. Casting
flow rate and melt temperature were varied.

Under optimal conditions, ultrasonic irradiation led to structural
refinement, columnar zone suppression, and controlled porosity. As a
result, tensile strength was increased by 35%, yield strength by 20%,
elongation by 10%, and hardness by 8 to 12%.

There have been a few studies of the efficiency of ultrasonic treatment of various ferrite, austenite, and carbide steels, plain-carbon and
low-alloy steels, boron-containing steels, and cast irons. These studies
were performed with grade St20, St30, St40, St50, U8, U10, 17MnSi,
30CrMnSi, 35CrMo, 45Mn2, 55Si2 [27], and ShCr15 [61] plain-carbon
and low-alloy steels. The selection of the above metals was primarily
motivated by their practical importance and the necessity to improve
their mechanical properties and workability.

Vibrations were transmitted to a solidifying metal through the hole
in the mould bottom, or during continuous casting (grade ShCr15 ball
bearing steel), or during VAR (grade U10).

It was found that the efficiency of ultrasonic treatment of plain-
carbon steels increased with carbon content, so that hypoeutectoid
steels with carbon contents of less than 0.4% exhibited a relatively
low ultrasonic treatability. At the same time, ultrasound produced a
substantial grain refinement in grade U8 and U10 steels.

Such a dependence of ultrasonic treatment efficiency on the carbon
content of steels basically follows from the iron-carbon phase diagram,
i.e. from the solidification temperature range width, and the strength
of growing crystals.

Significant structural modifications obviously exerted effect on the
mechanical properties of steels (Table 10.9).

Table 10.9   Effect of ultrasound on the mechanical properties of plain-carbon and low-alloy steels.

Steel grade	Ultrasonic treatment	UTS, MPa	YS, MPa	Elongation, %	Reduction of area, %	U-notch Charpy toughness, MJ/m^2	Vickers hardness
St40	No	620	360	12	16	0.40	155
	No*	750	670	8	9	0.30	190
	Yes	710	450	8	10	0.30	175
	Yes*	810	700	7	8	0.30	246
St50	No	490	–	10	15	0.15	–
	Yes	630	–	18	28	0.28	–
U10	No	480	400	2	5	–	242
	Yes	850	410	3	9	–	254
40CrNi	No	880	560	9	14	–	–
	Yes	900	570	12	25	–	–
SiCrNiMo**	No	2380	2300	4	40	0.38	–
	Yes	2530	2430	3	50	0.57	–

* Additionally inoculated with 0.1% Ti.
** Thermomechanically treated, i.e. rolled at 525°C with a reduction of 60 to 78%, quenched, and tempered at 200°C for 1 h.

Ultrasonic irradiation of St50 steel improved its tensile performance at room temperature. The strength and ductility of this steel increased by 20–30 and 30%, respectively.

Ultrasonic irradiation of grade U10 steel increased its tensile strength, elongation, and area reduction by about 75, 30, and 60%, respectively; hardness and yield strength did not virtually change.

Experiments with grade ShCr15 ball bearing steel were aimed at elucidating the effect of ultrasound on ingot homogeneity, structure banding, dendrite segregation, and the shape and distribution of non-metallic inclusions. Comparative structural analysis showed that ultrasonically treated ingots possessed much lower carbide segregation, with a finer and more uniformly distributed carbide phase in comparison with control ingots (Figure 10.10).

Some works [62–64] were devoted to the effect of ultrasound on the structure and properties of ferrite steels and alloys with the bcc lattice, i.e. grades Cr13, Cr18, Cr25Ti, Cr27, Si3, Si6 (with carbon concentrations of 0.02 and 0.1%), Al14 and Al16.

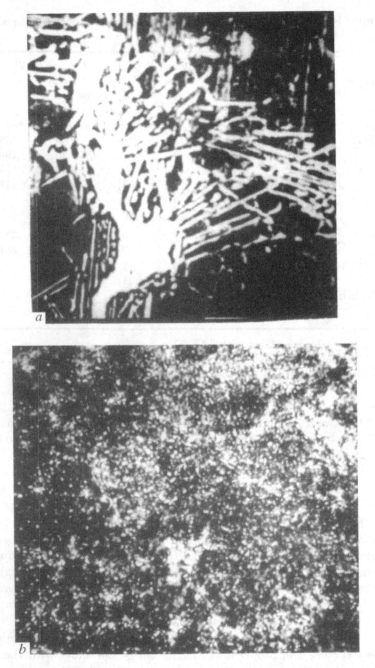

Figure 10.10. Carbide phase segregation in (*a*) control and (*b*) ultrasonically solidified ShCr15 steel ingots (×500 magnification).

Table 10.10 Effect of ultrasound on the mechanical properties of ferritic steels.

Steel grade	Condition	Testing tempe- rature, °C	Ultra- sonic treat- ment	UTS, MPa	YS 0.2%, MPa	Elong- ation, %	Area reduc- tion, %	Torsional shear
Cr13	As-cast	−196	No	230	–	1	4	–
		−196	Yes	290	–	2	6	–
		20	No	170	–	5	14	–
		20	Yes	20	–	8	30	–
Cr25Ti	As-cast	20	No	330	250	12	15	–
			Yes	460	360	14	18	–
		900	No	30	25	28	51	–
			Yes	35	25	68	90	–
		1100	No	6	–	53	77	25
			Yes	9	7	110	90	45
		1200	No	3	–	72	82	25
			Yes	6	3	80	90	40
Cr27	As-cast	20	No	450	320	6	6	50
		20	Yes	530	350	20	44	89
	As-heat treated	20	No	480	320	14	29	70

Ferrite steels and alloys showed a rather high ultrasonic treatability. The ultrasonic treatment of these materials suppressed their columnar structure, refined macro- and micrograins, and improved ingot homogeneity.

Changes in the ingot structure enhanced the strength and ductility of tensile specimens at room temperature and in metal-working temperature ranges (Table 10.10). Ultrasound also improved impact toughness and permissible reductions during the rolling of V-shaped specimens. Similar results were obtained for Fe–Si (grades Si3 and Si6) and Fe–Al (grades Al4 and Al16) alloys.

Ultrasonic irradiation increased the tensile strength, elongation, and area reduction of as-cast materials 1.2–1.6 times, 3–5 times, and 3–10 times, respectively. After subsequent thermal treatment and working, differences in the properties of control and ultrasonically treated ingots, even though being somewhat diminished, remained rather great.

Effect of ultrasound on the structure and properties of austenitic alloys was studied using Cr18Ni9, Cr25Ni20, Cr20Ni20Mo3, CrNi35WTiAl, and Mn12 steels [62–64].

Table 10.11   Effect of ultrasound on the mechanical properties of austenitic steels.

Steel grade	Testing temperature, °C	Ultrasonic treatment	UTS, MPa	Elongation, %	Area reduction, %	U-notch Charpy toughness, MJ/m^2
12Cr18Ni9	20	No	480	31	65	–
		Yes	520	45	75	–
Cr25Ni20	20	No	460	25	44	1.3
		Yes	560	32	71	1.8
	950	No	80	16	25	0.9
		Yes	70	26	32	1.2
Cr20Ni20Mo3	20	No	490	42	40	–
		Yes	550	54	68	–
	900	No	140	22	18	1.1
		Yes	170	28	26	1.7
	1100	No	50	12	7	0.3
		Yes	50	15	12	0.4

To be efficient in the treatment of austenitic steels, ultrasound should have higher intensity than in the case of ferritic steels.

Structural changes in austenitic steels were responsible for the differences in the mechanical properties of control and ultrasonically treated ingots (Table 10.11).

Grade W18 tool steel was used to elucidate the effect of ultrasound on segregation in carbide steels [30]. Ultrasonically treated ingots exhibited the refinement of macro- and fracture structures, as well as the modification of their microstructures. Moreover, carbide eutectics were also refined (Figure 10.11). Of particular interest is the fact that ultrasonically induced structural changes persisted after the complete processing cycle of W18 steel (forging at 1100°C, annealing at 800°C, oil quenching from 1280°C, and tempering at 560°C for 1 h). The extent of carbide segregation in ultrasonically treated steel (2.5) was substantially lower than that in control ingot (4.5).

The effect of ultrasound on the structure and mechanical properties of iron castings was studied by Gorev et al. [65–67] and Levi et al. [68–72]. Ultrasonic treatment drastically refined spherical graphite precipitates in iron castings irrespective of their carbon and silicon contents. As a result, the white iron structure of low-carbon, low-silicon alloys transformed into ferritic.

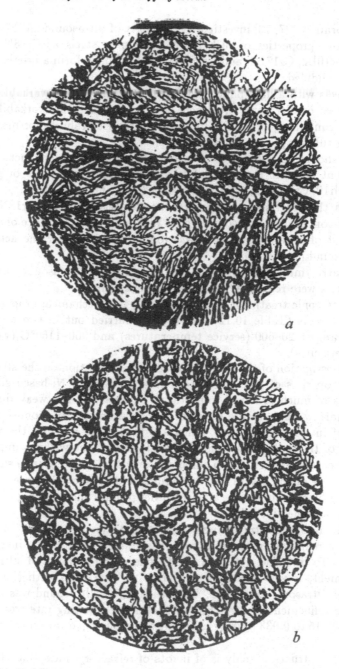

Figure 10.11. Microstructure of grade R18 steel (×100 magnification): (a) control ingot, (b) ultrasonically solidified ingot.

Abramov [27, 73] investigated the effect of ultrasound on the structure and properties of such boron-containing steels as Cr18Ni15B2, Cr18Ni10B2, Cr17B2, and Cr18Ni6Mn9B2, which form a brittle phase when solidified.

Steels with a high boron content are usually poorly workable. The objective of ultrasonic treatment was to improve the workability of boron-containing steels and to elucidate their suitability for manufacturing tubular products and strips.

It should be noted that ultrasonic treatment caused structural refinement of basic (i.e., free of boron) hypereutectic steels of grades Cr18Ni10, Cr18Ni15, Cr18Ni6Mn9, and Cr17 (Figure 10.12).

On the other hand, microstructure analysis of grade Cr18Ni15B3 boron-containing hypereutectic steel indicated the occurrence of a large amount of boride phase that was highly refined under the action of ultrasound (Figure 10.13).

Matrix microstructure was also modified by ultrasound, namely, eutectics were refined as compared to control ingot.

Ultrasonic treatment largely modified the mechanical properties of boride steels (Table 10.12). Tests were carried out in two temperature ranges: 20–600 (service temperatures) and 900–1150°C (working temperatures).

Investigation of the effect of ultrasonic irradiation on the structure and properties of precipitation-hardened refractory Ni-based superalloys is of much practical importance because of their weak ductility and narrow temperature range of deformation. In this connection, it was of interest to examine a method that could improve the workability of these alloys at technological temperatures without impairing their service properties. An enhanced ductility could ensure a substantial saving of the cost of rolled products and appreciably increase their quality by obtaining more efficient billet profiles and reducing the ratio of surface-defective billets.

Ultrasonic treatment was carried out in the process of vacuum arc remelting with vibrations transmitted through the ingot bottom. To find the optimum conditions of remelting and ultrasonic treatment, it was necessary to match melting rates with the intensity of ultrasonic vibrations. The efficiency of ultrasound was defined as the refinement of ingot microstructure. Remelting rate was varied from 0.015 to 0.033 kg/s by changing the electric parameters of remelting.

Macrostructure analysis of ingots of refractory, nickel-based alloys Cr20Ni80, CrNi77TiAlB, CrNi56WMoCoAl, and CrNi51WMoTiAlCoVB showed that they are ultrasonically workable. Indeed, ultrasonic treat-

Figure 10.12. Fracture of grade Cr18Ni15B3 boron steel (×8 magnification): (*a*) control ingot, (*b*) ultrasonically solidified ingot.

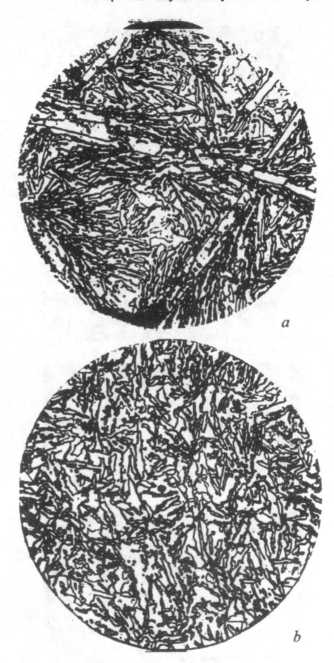

Figure 10.13.   Microstructure of Cr18Ni15B3 steel (×100 magnification): (*a*) control ingot, (*b*) ultrasonically solidified ingot.

Table 10.12 Effect of ultrasound on the mechanical properties of as-cast boron steels.

Steel grade	Boron content, %	Testing temperature, °C	Ultrasonic treatment	Tensile strength, MPa	Elongation, %	Area reduction, %	Max. rolling deformation, %
Cr18Ni15B2	2.4	550	No	250	–	–	–
			Yes	330	–	–	–
		1100	No	20	23	19	–
			Yes	20	23	32	–
Cr18Ni15B3	2.8	600	No	260	–	–	–
			Yes	400	–	–	–
		1100	No	70	14	14	16
			Yes	60	34	41	28
Cr18Ni10B3	3.2	350	No	100	0	0	–
			Yes	280	1	1.5	–
		1150	No	10	4	1	6
			Yes	25	8	3.6	37
Cr18Ni6Mn9B3	3.4	350	No	90	0	0	–
			Yes	260	0	0	–
		1150	No	19	3.5	0	4
			Yes	32	9.5	11	20

ment at optimal remelting rates and vibration intensities brought about the removal of the columnar structure of ingots and produced a sufficient refinement of their macrostructure over the entire ingot bulk.

The technological ductility of ingots, produced in a vacuum arc furnace conventionally and ultrasonically, was assessed from tensile and torsional strengths and toughness measured at temperatures of hot machining (1000 to 1200°C). The method of V-shaped specimen rolling was also used, which allowed a more complete estimation of metal behavior during hot plastic deformation (Table 10.13).

Analysis of experimental data indicated that the application of ultrasonic vibrations during remelting in a vacuum arc furnace improved the ductility of refractory alloys at temperatures of hot machining.

Ultrasonic treatment improved the extrusion workability of alloys CrNi51WMoTiAlCoVB and CrNi56WMoCoAl, so that they could be extruded in a wider temperature range.

Table 10.13   Maximum permissible rolling stress of V-shaped superalloy
              specimens.

Superalloy	Ultra-sonic treat-ment	Maximum rolling stress (MPa) at testing temperature (°C)					
		900	1000	1050	1100	1150	1200
CrNi77TiAlB	No	48	44	43	44	43	45
	Yes	51	48	50	51	57	47
CrNi56MoWCoAl	No	–	30	33	44	52	42
	Yes	–	34	37	54	56	57
CrNi51WMoTiAlCoVB	Yes	–	6	12	16	6	3
	No	–	11	14	19	14	9

Micrograins were found to be smaller in ultrasonically treated alloys, the degree of refinement being greater for CrNi51WMoTiAlCoVB.
For comparison, the micrograin counts were 0 and 2–3 for control and
ultrasonically treated CrNi51WMoTiAlCoVB ingots and 0–1 and 1–2
for control and irradiated CrNi56WMoCoAl ingots, respectively.

As-worked specimens of CrNi56WMoCoAl and CrNi51WMoTiAlCoVB
alloys were subjected to tensile test at room temperature, impact test
in the temperature range 1000–1200°C (Figure 10.14), and to stress–
rupture test under standard conditions (Table 10.14).

It can be seen that ultrasonic treatment induced no significant
changes in strength characteristics at room temperature and only slightly improved the stress-rupture strength of both alloys (Table 10.14). It

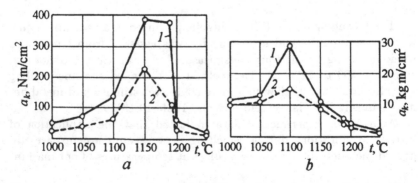

Figure 10.14.   Toughness of (*a*) CrNi56WMoCoAl and (*b*) CrNi51WMoTiAlCoVB
alloys at working temperatures: (*1*) control ingot, (*2*) ultrasonically solidified ingot.

Table 10.14 Effect of ultrasound on the mechanical properties of as-worked nickel-based superalloys.

Alloy grade	Ultrasonic treatment	Testing temperature, °C	UTS, MPa	YS 0.2 offset, MPa	Elong-ation, %	Area reduction, %	Stress-rupture strength, h		
							900°C, 270 MPa	900°C, 280 MPa	940°C, 220 MPa
CrNi56WMoCoAl	No	20	1290	910	14	14	–	–	–
	Yes	20	1300	920	14	16	–	–	–
	No	900	760	–	10	15	90	–	–
	Yes	900	800	–	10	15	111	–	–
CrNi51WMoTiAlCoVB	No	20	960	760	8	11	–	–	–
	Yes	20	1110	760	8	9	–	–	–
	No	950	600	–	10	11	–	93	75
	Yes	950	630	–	15	16	–	108	75

should be noted that these characteristics for control and ultrasonically treated ingots were close to the standard quality of these alloys produced industrially.

Ultrasonically treated CrNi51WMoTiAlCoVB alloy showed better ductility characteristics, elongation (1.5-fold), and area reduction (2-fold) than control specimen.

Ultrasound enhanced twofold the toughness of CrNi56WMoCoAl and CrNi51WMoTiAlCoVB alloys at temperatures of hot machining.

In work [74], noncorrodable Ni-based alloys Cr15Ni65Mo16Nb and Ni70Mo27W were studied. Ultrasonic irradiation was found to refine their structure and improve their ductility at both room and deformation temperatures.

10.3   Zone Refining Processes

The present-day technology often needs crystals of a specified chemical composition and crystalline perfection. This primarily pertains to the production of monocrystals.

Analysis of ultrasonically induced phenomena in molten and solidifying metals shows that ultrasonic vibrations modify the properties of melts, cause their mixing, and affect the form of crystallization front. All this may play a role in the processes of crystal growth and zonal refining.

The influence of ultrasound on the above processes was the subject of research in a number of works [75–84].

In particular, K. B. Yurkevich [76, 77] studied the zonal refining of zinc (the initial concentration of cadmium was 3.6×10^{-3} and 2.8×10^{-4} mass %) and tellurium (the initial concentrations of copper and selenium were 3×10^{-5} and 4×10^{-3} mass %, respectively).

The method of ultrasonic treatment used by the authors involved a direct transmission of vibrations to a molten zone. This method offers some advantages, such as a favorable influence on the crystallization front and boundary layer, vigorous melt mixing, etc. At the same time, special requirements are imposed, in this case, on the radiator material, which must be refractory and chemically inert with respect to metal melts.

Vibrations at 44 kHz were used. The ingots were subjected to zonal refining in an ultrasonic field of various intensities using a setup presented in Figure 10.15.

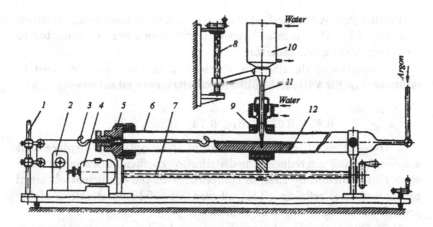

Figure 10.15. A setup for zonal refining of metals and semiconducting materials in an ultrasonic field [78]: (*1*) support, (*2*) gearbox, (*3*) stem, (*4*) motor, (*5*) vacuum seal, (*6*) quartz ampoule, (*7, 8*) feed screws, (*9*) heater, (*10*) transducer, (*11*) ultrasonic stack, (*12*) boat with a solidifying material.

Zonal refining of zinc with the initial concentration of cadmium 3.6×10^{-3} mass % was performed at a rate of 7×10^{-3} mm/s with the number of runs equal to 5 and molten zone width $\sim$40 mm. The results of spectral analysis of ten 15-cm long samples, cut sequentially lengthwise the ingots purified either conventionally or ultrasonically, were used to calculate effective impurity distribution coefficients:

ξ_m, μm	0	0.8	1.3	2.0	3.3
K_{eff}		0.62	0.53	0.48	0.54	1

It is seen that ultrasonic treatment reduced K_{eff} and intensified the refining, the effect being maximum (a 25% reduction in K_{eff}) at a vibrational amplitude of 1.3 μm.

In further experiments, the initial concentration of cadmium in zinc was 2.8×10^{-4} mass %; zonal refining was accomplished at a rate of 7×10^{-3} mm/s (10 runs) and a vibrational amplitude of 1.3 μm.

The ultrasonic zonal refining substantially enhanced the purity of metals. Thus, spectral analysis revealed no cadmium in three of four samples (the sensitivity of the method for cadmium was 5×10^{-5} mass %), the impurity concentrating at the ingot end.

Tellurium was used as an object for investigating the ultrasonic refining of semiconducting materials.

Purification by the method of directed crystallization was performed at a rate of 7×10^{-3} mm/s. Ultrasonic vibrations were transmitted to a melt via a tungsten radiator.

The lengthwise distribution of selenium in the ingot was used to estimate K_{eff} for various amplitudes of ultrasonic vibrations:

ξ_m, μm	0	0.2	0.3	0.6	1.0	1.7
K_{eff}		0.81	0.68	0.61	0.73	0.80	0.84

Ultrasonic zonal refining of tellurium at an optimum displacement amplitude of 0.3 μm reduced the distribution coefficient of selenium and copper in tellurium by 25%, as compared with the coefficient obtained by conventional refining. This allowed a twofold acceleration of the process at the same depth of refinement.

E. N. Ozerenskaya with coworkers carried out a number of aluminum zonal refining experiments [78, 79]. It was found that the maximum efficiency of ultrasonic refinement can be achieved with the intensity of vibrations less than the cavitation threshold in a melt, but sufficient for the development of acoustic streams near crystallization front. In this set of experiments, the number of runs was varied from 6 to 12, the ingot dimensions were 18×20×400 mm, vibrational frequency amounted to 44 kHz, the displacement amplitude of a radiator ranged from 0.2 to 2.2 μm. The degree of refinement was estimated from the residual electric resistance of samples at liquid helium temperature and by spectral analysis.

Ultrasound was found to intensify zonal refining and to enhance the degree of purification (Figure 10.16). The rate of zone displacement was 1 mm/s. Ultrasound appeared to be most efficient at a displacement amplitude of 0.2–0.5 μm (Figure 10.16, curves *2, 3*); at an amplitude of 2–2.2 μm, the arising cavitation prevented purification (curve *4*). With the optimum power of vibrations (corresponding, in a given case, to the displacement amplitude of the radiator 0.2–0.5 μm), the degree of purification per one run was greater than that attained in three runs of conventional zonal refining (curve *1*).

Closer examination of grade A99 aluminum subjected to zonal refining showed that ultrasonic treatment was rather efficient with respect to various eutectic-forming inoculants, both poorly soluble (Fe, Ni, Co, and Ba) and readily soluble (Mn, Mg, Ca, Si, and Zn), as well as those forming monotectics and peritectics (Ti and Cr). In ultrasonically purified ingot, hydrogen content was reduced from 0.13 cm^3/100 g (nonirradiated ingot) to 0.01 cm^3/100 g.

Analysis of the structure of aluminum ingot subjected to zonal refining revealed the formation of rather large (10–15 mm in size) crystals

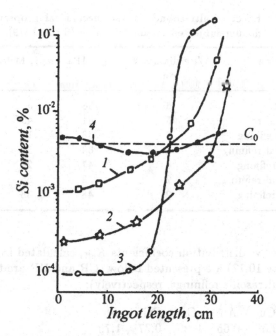

Figure 10.16. Lengthwise distribution of silicon in aluminum ingot: (*1*) control ingot (3 runs), (*2–4*) ingots treated by ultrasound at a vibration amplitude of 0.2, 0.5, and 2 μm, respectively (1 run).

extended in the direction of zone motion. At the same time, in ultrasonically refined ingots, the crystals were even larger, their number being sometimes 1–3 across the entire ingot.

Ultrasonic refining enhanced the ductility of aluminum ingots, as indicated by their cold drawing from a 18-mm to 2-mm diameter, other ductility characteristics being improved four- or twofold as compared with an untreated aluminum or that refined by conventional method (Table 10.15).

Aluminum subjected to ultrasonic zonal refining begins to recrystallize already at room temperature, the process being intensified as the degree of deformation increases. On the other hand, the recrystallization of aluminum subjected to conventional refining begins at higher temperatures and is completed only at annealing temperature of 300°C.

R. G. Sarukhanov *et al.* [80] investigated the effect of ultrasound on zonal refining of tin. The displacement velocity was 7×10^{-3} mm/s, the number of runs amounted to 10, the zone width varied from 5 to 8 mm. The amplitude of vibrations made up 1 μm at 44 kHz.

Table 10.15 Effect of ultrasound on the mechanical properties of aluminum subjected to zonal refining (6 runs) [79].

Crystallization conditions	Wire diameter, mm	σ_B, MPa	$\sigma_{0.2}$, MPa	δ, %
Initial ingot	6	108	55	9.7
Conventional refining	6	61	49	18.6
Ultrasonic refining	6	38	19	39.2
Conventional refining	4	63	55	10.0
Ultrasonic refining	4	41	24	16.5
Conventional refining	2	63	55	6.4
Ultrasonic refining	2	48	38	9.3

The effective distribution coefficients K_{eff}, calculated from spectral data, (Figure 10.17) are presented below (CR and UR are the conventional and ultrasonic refinings, respectively):

	Cu	Au	Cr	Ni	Sb
CR	0.58	0.65	0.20	0.37	1.70
UR	0.33	0.45	0.15	0.27	2.50

Comparison of these coefficients implies that ultrasonic zonal refining of tin substantially enhances its purity.

M. I. Dubrovin undertook a number of studies to elucidate the effect of ultrasound on the production and directed crystallization of semiconducting materials [81–83].

Ultrasound was found to substantially enhance the homogenization of HgTe–CdTe melts. The optimum conditions of ultrasonic treatment were chosen based on experimental estimates for the rate of cadmium telluride dissolution in mercury telluride. The derived dependence was described by a curve with two characteristic regions (Figure 10.18). In the first region, where the amplitude of vibrations was insufficient to induce cavitation, the rate of the dissolution was comparatively low; the second region, which begins from a vibration velocity amplitude of 0.4 m/s, was characterized by a drastic increase in the rate of dissolution induced by cavitation. As the amplitude of vibrations increased, this rate slowed down, which could be associated with a cavitation-dependent fall in the radiator load resistance.

Experiments indicate that the rate of cadmium telluride dissolving during ultrasonic homogenization of the melt rises almost by three orders of magnitude. It is noteworthy that ultrasonic treatment by

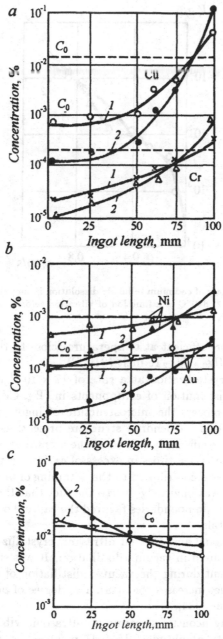

Figure 10.17. Lengthwise distribution of (a) copper and chromium, (b) gold and nickel, and (c) antimony in tin ingot: (1) control ingot, (2) ultrasonically refined ingot [80].

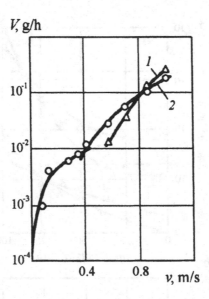

Figure 10.18. Rate of cadmium telluride dissolution in molten mercury telluride at (*1*) 810°C and (*2*) 840°C as a function of vibration velocity amplitude [81].

cavitation is more efficient at temperatures close to the liquidus temperature of the $Hg_{0.8}Cd_{0.2}Te$ melt (810°C).

Ultrasonic crystallization at a rate of 17×10^{-3} mm/s ensured a more uniform distribution of components in (Hg, Cd)Te ingots (Figure 10.19). Moreover, the microstructure of ingots was appreciably refined with no distinct dendrite structure present. Dendrites in (Hg, Cd)Te ingots were efficiently refined at the amplitude of vibrational velocity of ultrasound 5–6 times in excess of cavitation threshold. These dispersion processes contributed to the obtaining of more homogeneous material, which was obviously due to the fact that the increase in the number of intergrain boundaries favored the capture of matrix solution by growing crystals.

Apart from zonal and directed ultrasonic crystallization, vibrations can also be well used in vacuum distillation. It was shown that ultrasonic impingement during the vacuum distillation of molten tin from volatile impurities increases the attainable degree of purification by an order of magnitude [84].

By producing acoustic streamings, ultrasonic vibrations (25 kHz; vibration velocity amplitude 0.15–0.47 m/s) act to refresh the melt surface and thereby enhance impurity diffusion to the surface. Ultrasound was found to be most efficient in removing lead and bis-

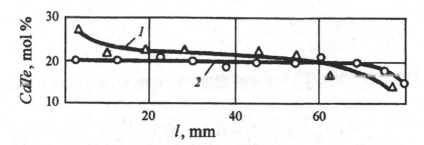

Figure 10.19. Lengthwise distribution of cadmium telluride in (Hg, Cd)Te ingot [82]: (*1*) control ingot, (*2*) ultrasonically solidified ingot.

muth impurities from tin melts and less efficient in removing antimony and indium. The degree of arsenic and zinc removal was insignificant.

Raising the amplitude of vibrations above a certain level (0.47 m/s) failed to improve the degree of purification still further.

10.4 Ultrasound in Production of Cast Composites

Ultrasound not only improves the structure and properties of metals and alloys but can also be used for producing cast composites from materials that are usually infusible (e.g., from immiscible melts of metals, oxides, borides, etc.).

In the early 1960s, G. I. Pogodin-Alekseev and coworkers published a number of papers [85–88], in which they demonstrated the feasibility of using ultrasound for production of materials that would not fuse under usual conditions. O. V. Abramov extended the investigations along this line [90–98].

This section presents the data pertaining to the production of antifriction Al–Pb composites, which, however, does not exclude the feasibility of ultrasonic production of other metal emulsions.

Al–Pb emulsions were produced by the method of ultrasonic vibration of a water-cooled funnel, through which the melt was cast into a water-cooled mould (crystallizer)*.

Before casting, the melt was superheated to 1100–1200°C to provide a 18–30% lead solubility in aluminum, whereas ultrasonic treatment with a simultaneous superheat removal led to a vigorous mixing of lead and aluminum and the formation of additional crystal nuclei,

*L. K. Vasin. Dissertation Theses (Cand. of Tech. Sci.).

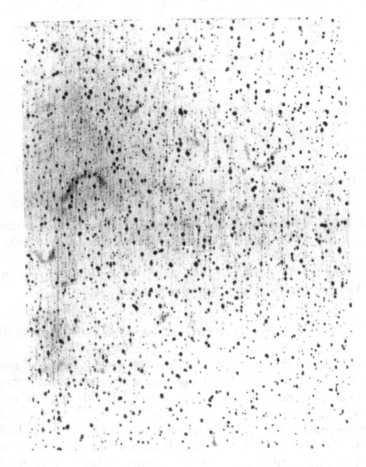

Figure 10.20.   Microstructure of Al–10% Pb–1% Cu–1% Sn composite (×70 magnification).

which stimulated the crystallization of metal emulsion and fixation of a uniform distribution of finely dispersed lead particles.

Such a technology was employed to produce the following Al–Pb-based composites*: Al–10% Pb–1% Cu–1% Sn, Al–10% Pb–10% Sn–1% Cu, and Al–10% Pb–10% Sn–4% Sb–0.5% Cu–0.5% Mg.

Analysis of the materials produced showed that the appropriately chosen conditions of casting and ultrasonic irradiation ensure high-quality ingots with a relatively uniform distribution of finely dispersed lead inclusions (Figure 10.20). The vertical lead distribution devia-

---

*The compositions are presented in mass %.

tion was 0.8–1%, the size of most lead particles ranged from 5 to 40 μm. Alloying the melts with Sn, Sb, and Cu did not substantially affect the homogeneity of Pb distribution over the ingot. Microstructural analysis showed that low-melting components occur as a fine continuous net over grain and dendrite boundaries. Moreover, 5–30-μm lead inclusions were also uniformly distributed in the matrix.

Local X-ray spectral analysis showed that the matrix, representing an aluminum-based solid solution, contains inclusions that involve alloying elements (Figure 10.21). In this case, large, irregularly shaped inclusions were enriched in tin, whereas small inclusions were enriched in lead.

Some physical parameters of materials produced, such as strength, ductility, coefficients of linear expansion and thermal conductivity, are listed in Table 10.16. The study of the casting properties of Al–10% Pb–1% Sn–1% Cu alloy indicated that its linear and volume shrinkages made up 1.75 and 5.25%, respectively. The adhesion strength of a steel plus Al–10% Pb–1% Sn–1% Cu alloy composite was 6.60–7.30 MPa, as assessed by shearing test.

To study the antifriction properties of aluminum–lead composites, the ingots were preannealed at 350°C for 1 h, followed by rolling and reannealing under the same conditions.

Antifriction properties of Al–Pb composites were tested at slip velocities 1.04, 2.08, and 5.34 m/s by the shaft–insert or roller–race methods using grade 45 steel, either unquenched (HB 196–200 units) or quenched (HRC 48–55 units).

Wear resistance was assessed by the dry friction test at a pressure of 1.4 MPa. Liquid friction test was performed at a pressure of 23 MPa and a maximum testing temperature of 250°C using diesel lubricating oil M14V and industrial lubricating oil 20 with viscosities corresponding to that of M20B oil. In these tests, the bearing capacity of the alloy (the specific load of seizing) and the critical temperature of oil layer discontinuity were determined.

To estimate the antifriction properties of Al–Pb composites, they were compared with the properties of B-83 babbit Sn–6% Cu–10% Sb, as well as AO9-1 (Al–9% Sn–1% Cu) and AO20-1 (Al–20% Sn–1% Cu) alloys.

Dry friction tests showed that the wear of race and roller is much greater for aluminum-tin than for aluminum–lead alloys (Figure 10.22).

Analysis of the specific load of seizing (P) at various slip velocities (V) revealed that PV for Al–10% Pb–1% Sn–1% Cu inserts working in

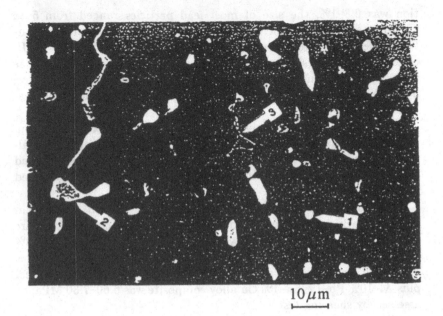

10 μm

Results of X-ray microprobe analysis

1.

Element	Atom. %
Al	15
Sn	16
Pb	68

2.

Element	Atom. %
Al	17
Pb	1
Sn	82

3.

Element	Atom. %
Al	60
Cu	40

Figure 10.21.   X-ray image of the section of Al–10% Pb–10% Sn–4% Sb–0.5% Cu–0.5% Mg composite (×500 magnification).

Table 10.16 Properties of Al–Pb composites.

Composite	Testing temperature, °C	Density ρ_S, g/cm³	Yield strength $\sigma_{0.2}$, MPa	Tensile strength σ_B, MPa	Relative elongation δ, %	Brinell hardness, MPa	Coefficient of linear expansion $\alpha \times 10^6$, deg⁻¹	Thermal conductivity, cal/cm deg s
Al–10% Pb–	20	2.79	53	136	22	264	24.7	0.36
–1% Sn–	60		53	133	21	262		
–1% Cu	100		50	129	21	253		

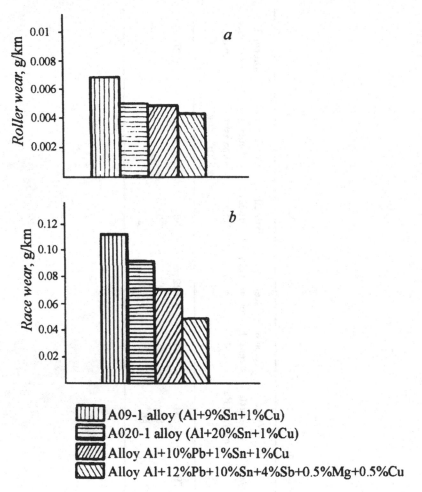

Figure 10.22. Wear of (*a*) grade 45 steel rollers and (*b*) races made of the alloys studied.

pair with a hardened steel shaft is 20–25 MPa m/s, and the coefficient of friction is 0.09 to 0.015. In the case of unhardened steel shaft, parameter PV and the coefficient of friction are 10 MPa m/s and 0.04–0.05, respectively. All of these values are close to those obtained for babbit B-83 under analogous testing conditions.

Figure 20.23 illustrates the dependences of friction momentum M_f and friction coefficient μ on the temperature of M14V oil. The ordinate scales were chosen so that the plots $M_f = f(t)$ and $\mu = f(t)$ coincide.

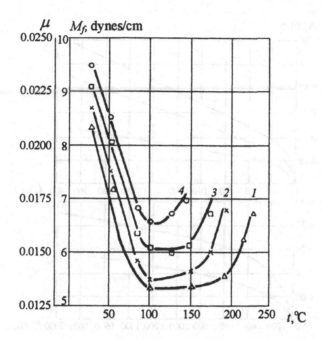

Figure 10.23. Friction momentum M_f and friction coefficient μ as functions of the temperature of diesel lubricating oil in tests of the pairs of grade 45 steel and one of the following materials: (*1*) AO9-1 alloy, (*2*) AO20-1 alloy, (*3*) Al–10% Pb–1% Sn–1% Cu composite, (*4*) Al–10% Pb–10% Sn–4% Sb–0.5% Cu–0.5% Mg composite.

Among the materials tested, Al–10% Pb–10% Sn–4% Sb–0.5% Cu–0.5% Mg composite possessed the highest critical temperature of boundary friction (180°C) and the lowest coefficient of friction.

In further tests, the wear resistance of materials was estimated at various slip velocities and specific loads. The materials studied were found to possess rather high wear resistance, close to that of babbit B-83 (Figure 10.24). Bench test of a 75-horsepower engine with inserts made of steel strip coated by Al–10%Pb–1%Sn–1% Cu brass composite or by B-83 babbit was performed at a slip velocity of 7 m/s with the use of DP-11 lubricating oil fed at a pressure of 0.35–0.37 MPa.

Test time was 250 h, but first indications of the fatigue failure of babbit inserts appeared as soon as within 8 h of starting the test at a maximum specific load of 21.2 MPa. Within further 8 h, almost the entire surface of inserts was covered with fatigue cracks.

At the same time, inserts with the antifriction layer made of Al–Pb composite operated for 250 h (226 of them at a maximum specific

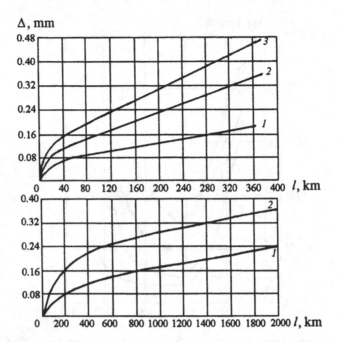

Figure 10.24.   Dependence of the wear of the insert made of Al–10% Pb–1% Sn–1% Cu composite on friction distance. (*a*) Slip velocity 1.04 m/s: (*1*) $P = 10$ MPa, shaft made of hardened 45 steel; (*2*) $P = 20$ MPa, the same shaft; (*3*) $P = 5$ MPa, shaft made of unhardened 45 steel. (*b*) Slip velocity 5.34 m/s: (*1*) $P = 4$ MPa, shaft made of hardened 45 steel; (*2*) $P = 2$ MPa, shaft made of unhardened 45 steel.

load of 42.5 MPa) without any indications of scratch, seizing, or fatigue failure. The overall wear of the insert surface in the region of maximum stress throughout the test time (250 h) amounted to 3–4 μm. According to the test data, the bearing capacity of the composite ranged from 25 to 30 MPa. Thus, high-intensity ultrasonic vibrations can be employed in the production of cast materials of the "frozen" emulsion type, such as antifriction Al–Pb alloys.

The use of dispersion-hardened metals with particles uniformly distributed in the matrix is a promising way to enhance the high-temperature strength of metals. Such materials show the necessary set of mechanical properties, unattainable in conventional alloys. Thus, dispersion-hardened metals will retain their strength at temperatures up to $0.9T_m$. Moreover, dispersion-hardened alloys are characterized by a high yield and stress-rupture strengths, very low creep rate, high resistance to restitution and recrystallization during annealing after cold deformation.

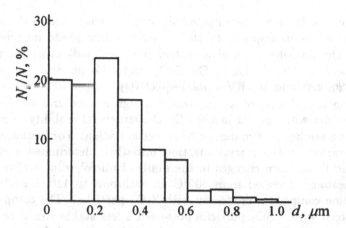

Figure 10.25. Size distribution histogram of ZrO_2 particles in CrNi60Co alloy.

Direct inoculation of a fine-grained oxide phase into a molten metal presents considerable difficulties, since the wettability of refractory oxides by liquid metal is not high, the more so that the surface of dispersed phase is usually somewhat contaminated.

Cast composites of the "frozen" suspension type were elaborated as applied to the strengthening of nickel-chromium matrices with a dispersed oxide phase [89–95], which underlies the production of almost all present-day refractory nickel alloys.

Grade Cr20Ni80 nichrome was initially reinforced with Al_2O_3, CeO_2, or ZrO_2 fine particles. However, after it was found that the best combination of service properties could be achieved with ZrO_2 particles, ensuing experiments to strengthen CrNi60Co and CrNi56VMoCoAl alloys were carried out only with these particles.

The oxide phase, preliminary refined to 1000-Å particles, was inoculated in an amount of 2 vol % of metal subjected to vacuum arc remelting. To improve the oxide phase distribution, the solidifying melt was imparted with ultrasonic vibrations. This was accomplished either by introducing interfacially active substances in an amount of 0.1 mass % or by precladding inoculated oxide particles with matrix metal.

Metallographic analysis showed that inoculated oxide particles were not preferably distributed over grain boundaries, but were uniformly distributed over the ingot. Electron microscopy of foils and carbon replicas revealed no large aggregates of particles in the ingot; their mean size was ~0.3 μm (Figure 10.25). Nor was there a substantial difference in the distribution pattern of reinforcing inclusions in relation to the chemical composition of the matrix.

Microhardness measurements indicated some increase in the hardness of alloys in response to the dispersed oxide phase inoculation. Thus, the microhardness of a control (uninoculated) nichrome ingot and those reinforced with Al_2O_3, CeO_2, and ZrO_2 particles comprised 210, 220, 220, and 256 HV units, respectively.

Dispersion-hardened alloys showed a higher structural stability in comparison with control ingots. By the structural stability is meant here the tendency of materials to recrystallization. For instance, the temperature of the recrystallization initiation, determined metallographically or from changes in the material microhardness after a 1-h annealing, increased from 600°C for nichrome to 1100–1200°C for nichrome composites reinforced with ZrO_2 particles. The composites reinforced with Al_2O_3 particles possessed a less stable structure than those strengthened with ZrO_2 particles. When the reinforcing phase was nickel-plated ZrO_2 particles, the composites displayed the least tendency to recrystallization.

The results of the measurement of grain size versus annealing temperature are in agreement with the microhardness data obtained for materials after their annealing at 1200°C for various times. These data show that dispersion-hardened alloys retain their improved hardness throughout the annealing period.

The mechanical properties and high-temperature strength of dispersion-hardened alloys were studied on prestrained specimens. For this, the ingots were subjected to hot deformation by technology commonly used for the deformation of rigid refractory alloys. The deformation of dispersion-hardened alloys occurred at 20 to 25% higher pressures than that of control ingots.

The study of the composite properties involved the examination of their mechanical properties at room and elevated temperatures, as well as analysis of their high-temperature strength (Figure 10.26) and resistance to oxidation (Table 10.17). Before preparing the test samples of CrNi56VMoCoAl–ZrO_2 composite, it was subjected to standard heat treatment, viz. air quenching from temperature 1220°C for 5 h followed by air ageing at 950°C for 8 h.

Analysis of the results of short-term mechanical tests showed that at all testing temperatures, the strength and ductility characteristics of dispersion-strengthened materials are close to those of control specimens with a small gain (about 10%) in the tensile and yield strengths of dispersion-hardened Cr20Ni80 nichrome and CrNi60Co alloy at both room and elevated temperatures.

The results of stress–rupture tests (Figure 10.26) are indicative of a substantial increase in the durability of dispersion-hardened chromium–

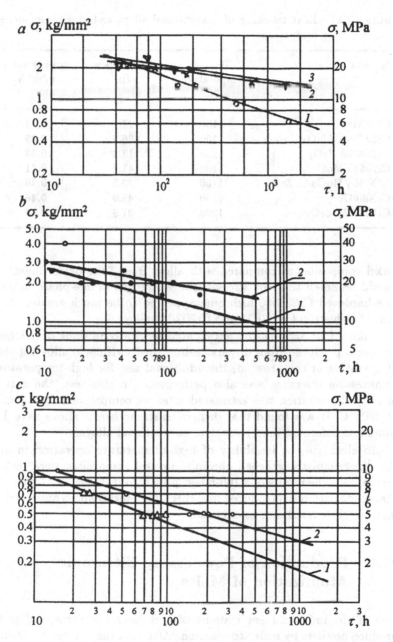

Figure 10.26. Stress–rupture strengths of composites at various temperatures. (a) 1000°C: (1) Cr20Ni80, (2) Cr20Ni80–Al₂O₃ composite, and (3) Cr20Ni80–ZrO₂ composite; (b) 1100°C: (1) CrNi60Co and (2) CrNi60Co–ZrO₂ composite; (c) 1200°C: (1) CrNi56VMoCoAl and (2) CrNi56VMoCoAl–ZrO₂ composite.

Table 10.17   Heat tolerance of conventional alloys and chromium–nickel composites.

Material	Testing temperature, °C	Weight gain, g/m^2	Scaling rate, g/m^2 h
Cr20Ni80	1100	27.6	0.28
Cr20Ni80–Al$_2$O$_3$	1100	26.3	0.26
Cr20Ni80–ZrO$_2$	1100	13.4	0.13
CrNi56VMoCoAl	1100	41.2	0.41
CrNi56VMoCoAl–ZrO$_2$	1100	30.3	0.30
CrNi60Co	1200	45.9	0.46
CrNi60Co–ZrO$_2$	1200	27.9	0.28

nickel composites as compared with alloys free of the oxide phase. It should be noted that the stress–rupture strength of one-phase dispersion-hardened Cr20Ni80 nichrome appeared to be much greater than that of industrial CrNi78Ti and CrNi70Al alloys.

Taking into account the anticipated service conditions of machinery parts produced from dispersion-hardened CrNi60Co alloy capable of operating in compression, the additional test for high-temperature compression tolerance was also performed. In this test, the extent of alloy deformation was estimated after its compression for 15 min at 1000°C. It was found that dispersion-strengthened alloys were 1.7 times less deformed than respective conventional alloys.

To elucidate the feasibility of high-temperature operation in air, the dispersion-strengthened chromium-nickel composites were tested for scaling. The results of a 100-h test presented in Table 10.17 indicate that oxide particles introduced into chromium–nickel alloys appreciably improve their resistance to oxidation.

10.5   Production of Powders by Ultrasonic Atomization of Melts

One of the industrial applications of high-intensity ultrasound is to produce powders by melt atomization. Although the studies along this line were initiated as early as in 1930–1940s, they are still at the stage of experiments. Recent advances are associated primarily with the employment of gas-jet generators for atomization.

Elastic vibrations can be fed to atomization zone through a liquid or gas. Accordingly, one can distinguish the methods of ultrasonic liquid and ultrasonic gas atomization.

The peculiarities of ultrasonic atomization of liquids, in particular molten metals, were considered at length in [99, 100].

In terms of ultrasonic liquid atomization in fountain, it is possible to obtain virtually uniform micron-size droplets, whose density can be widely varied. It is a practice to use for atomization focusing ceramic radiators operating at a natural frequency of several megahertz.

Since the efficiency of this method is low (not more than 1 l/h per atomizer), it can hardly be considered suitable for the production of metal powders.

The atomization of liquid in layer by kilohertz vibrations produces drops of about tens of micron in size. The efficiency of such atomization is two orders of magnitude greater than that in fountain. As in the case of liquid atomization in fountain, the atomizer operated by the principle of atomization in layer embodies an ultrasonic oscillator coupled to a transducer. This method can be used to produce powders of metals with a comparatively low melting point.

As shown in chapter 4, the employment of ultrasound for gas-jet atomization of melts can improve the uniformity of powders and reduce the atomizing gas discharge. Moreover, this method is applicable for the production of powders of metals and alloys with a comparatively high (more than 1000°C) melting point.

Generally, an atomization installation employing the principle of ultrasound feeding through liquid phase contains an electroacoustic transducer, a waveguide system, and a radiator. Figure 10.27 illustrates such an installation for the atomization of melts with a melting point below 1000°C [101].

The ultrasonic gas-jet atomizer operates on the principle of retardatio of a supersonic gas jet by an obstacle representing a hollow resonator.

Relatively few studies were concerned with the use of ultrasound in actual powder production processes. Thus, Pohlman described an ultrasonic liquid bed atomization system for producing powders from molten metals with melting points below 350°C [102, 103]. Atomization occurred under an atmosphere of inert gases and produced spherical powder particles with an extremely narrow size distribution. With a 20-kHz ultrasound applied to melts of low-melting-point metals, such as Pb, Zn, Cd, or Bi, the atomization rate was 1 $cm\,s^{-1}$. Pohlman emphasized a high efficiency of his ultrasonic atomization process. If energy requirement for preparing 1 ton of lead powder by conventional

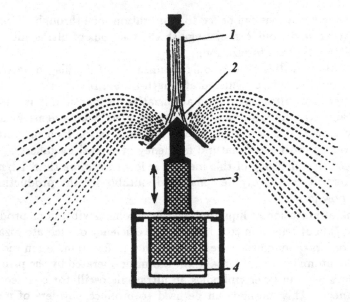

Figure 10.27.   Schematic representation of an installation for ultrasonic atomization of melts [101]: (*1*) nozzle, (*2*) radiator, (*3*) horn, (*4*) piezoceramic transducer.

ball mill method is 375 kWh with an output rate of 8 $\mathrm{kg\,h^{-1}}$, as little as 42 kWh are required to produce the same amount of lead powder by an ultrasonic installation with an output rate of 45 $\mathrm{kg\,h^{-1}}$.

Grant [104–106] outlined an ultrasonic gas-jet atomizer for producing powders of aluminum alloys and stainless steels. The use of ultrasound made it possible to substantially reduce the consumption of process gas. The size distribution of resultant powders was strongly dependent on gas pressure and extent of melt superheating. The dendrite cell size grew linearly with particle dimensions up to a value of 100 μm, above which it varied insignificantly (Figure 10.28). These data were used to calculate cooling rate, which was found to be 10^5 $\mathrm{^\circ C\,s^{-1}}$ for particles smaller than 100 mm in size.

Ro and Sunwoo [107] obtained quite different results. They compared the efficiency of ultrasonic and conventional gas-jet atomizers in producing aluminum alloy powders and were unable to find any reliable advantage of the ultrasonic atomizer.

Attempts to atomize steels were undertaken by Abramov *et al.* [108–110]. The objective was to assess the potentiality of ultrasonic gas-jet atomization as one of the technological processes in production of high-speed steels. The program involved the estimation of the

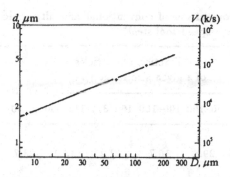

Figure 10.28. Relationship between the dendrite cell size d of aluminum alloy and the powder particle size D [105].

process parameters, powder quality, and the properties of compacted materials.

Grade 10W6Mo5 tool steel chosen for investigation contained 1% C, 6% W, 5% Mo, 4% Cr, and 2% V. Working gas was nitrogen. Ultrasonic atomizer operated at 14–15 kHz; the gas pressure ranged from 1.0 to 1.2 MPa; acoustic pressure was 176–180 dB. Conventional atomizer operated at the same gas pressure (1.0–1.2 MPa).

The parameters tested were pouring temperature, gas discharge, and steel stream diameter. The resultant powders were subjected to sieve analysis (for particles greater than 50 μm in size) as well as to sedimetric, metallographic, x-ray diffraction, and x-ray microprobe analyses. In addition, the bulk weight of powders and their oxygen content were measured.

The powder particles were 50 to 500 μm in size. The powders were compacted by hot extrusion at 1100–1150°C with a relative reduction of 80–90% and a deformation rate of 100 mm s^{-1}.

After annealing, the extruded blanks were tested for hardness, impact properties, and bending strength. The specimens were quenched from temperature 1200–1230°C followed by a triple tempering for 1 h at 560°C.

Investigation of ultrasonic atomization involved the elucidation of process conditions that could ensure its stable performance. These conditions were found to be close to those typical of conventional gas-jet atomization (a pouring temperature of 1670–1690°C, the gas rate 0.65–1.0 m^3 kg^{-1}.

Lower pouring temperatures reduced powder yield and could induce metal freezing in the teeming nozzle, while decreased gas rates drastically lowered the fineness of powders.

Table 10.18 Comparison of conventional and ultrasonic atomization of high-speed tool steel*.

Atomi-zation	Fraction of particles of a given size range (μm), %						Mean particle size,** μm	Gas rate, m^3kg^{-1}
	<50	50–100	100–160	160–315	315–500	>500		
Conven-tional	4.6	29.8	16.7	12.2	13.0	23.7	122	0.67
Conven-tional	9.5	30.5	20.0	12.5	10.0	15.7	107	1.03
Ultrasonic	18.6	39.9	14.2	14.8	5.0	7.5	91	0.65
Ultrasonic	27.2	52.7	7.8	6.3	0.0	6.0	81	0.90

* High-speed tool steel containing 1% C, 6% W, 5% Mo, 4% Cr, and 2% V.
** Mean particle size was estimated accounting for particles up to 500 mm in size.

It was found that ultrasound improved the powder yield by 10–15%, narrowed the particle size distribution, and reduced the process gas discharge (Table 10.18).

It was revealed that both atomization methods produce spherical powder particles free of oxide film on their surface (Figure 10.29). Virtually all powder particles had tiny attached particles on their surface, which were 5–20 μm in size, which was likely the result of collision and attachment of partially solidified particles during atomization.

Some particles showed surface "craters", which are believed to be due to the collision of already solidified particles, and occasionally some irregular formations. High-magnification observations revealed a dendritic structure of particle surfaces.

Metallographic analysis confirmed the dendritic structure of particle surface and indicated an increasing structure refinement with diminishing particle size. Estimation of cooling rate from the secondary arm spacing showed that it is 5×10^4 to 10^5 °C s^{-1} for the particles of 100 to 160 μm in size. The secondary arm spacing in such particles is about 1 μm.

Ultrasonically atomized powder particles possessed a finer structure than particles of the same size produced conventionally, which is indicative of a higher cooling rate during ultrasonic gas-jet atomization.

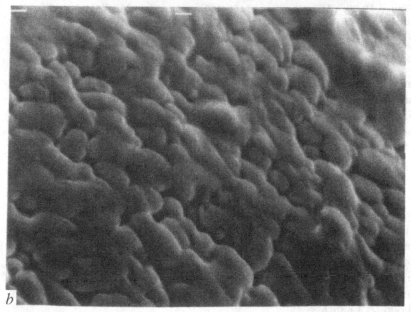

Figure 10.29. Micrographs of 100–160-μm particles of 10W6Mo5 steel powder obtained by ultrasonic gas-jet atomization: (a) ×500 magnification; b) ×5000 magnification.

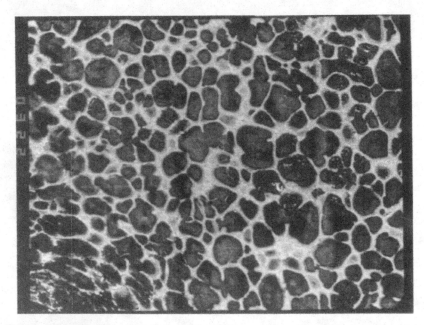

Figure 10.30.   Microstructure of 100–160-μm particles of 10W6Mo5 steel powder obtained by ultrasonic gas-jet atomization (×2000 magnification).

Ultrasonically atomized particles possessed some specific features: metallographic studies showed the occurrence of grains, i.e., dendrite arm sections, against a light background. The grains had three layers, viz. a gray-colored central region, a dark fringe, and a light-colored periphery (Figure 10.30). It can be surmised that these structure constituents are actually different phases. It should be also noted that the amount of central "grained" phase grew with decreasing particle size.

X-ray diffraction analysis revealed the presence of two phases, δ-ferrite and austenite, whose relative amounts depended on the powder particle size. As the particle size diminished (or cooling rate increased), the amount of δ-ferrite phase rose, while that of austenite phase proportionally decreased. The particle size was also related to the lattice parameter of δ-ferrite and austenite. The lattice parameter of δ-ferrite increased from 2.879 Å for 400–630-μm particles to 2.995 Å for 50–63-μm particles, whereas the lattice parameter of austenite increased from 3.606 to 3.635 Å.

X-ray diffraction data suggested that the grain core contained δ-ferrite; the dark fringe was δ-eutectoid and the light-colored area around the fringe was austenite.

Table 10.19   Mechanical properties of 10W6Mo5 high-speed tool steel (1% C–6% W–5% Mo–4% Cr–2% V) produced by various methods.

Method	Oxygen content, ppm	Mean austenite grain size, μm	Mean carbide grain size, μm	HRC hardness	Impact property $\times 10^{-2}$, $MJ\,m^{-2}$	Bending strength, MPa
Casting + working	–	–	–	–	300–500	3000–3500
Conventional gas-jet atomization	3×10^{-3}	7–10	0.8	64–65	200–300	2600–3600
Ultrasonic gas-jet atomization	3×10^{-4}	5–7	0.6	65–67	300–500	3400–5100

In the case of conventionally produced powders, the phase that appeared to be δ-ferrite was essentially lacking of coarse particles, whereas fine particles were presents (albeit in small amounts), once more indicating a higher cooling rate of ultrasonic gas-jet atomization in comparison with conventional gas-jet atomization.

Local X-ray microprobe analysis showed that the light, grain-containing background was most likely the eutectics of austenite grains surrounded by a network with occasional spherical carbide inclusions. This analysis also revealed a uniform distribution of alloying elements Cr, Mo, W, and V in powder particles.

Structural examination of extruded and annealed high-speed tool steel revealed that it contained sorbite-like pearlite and fine spherical carbide precipitates (Figure 10.31). When quenched and tempered, the steel showed the occurrence of martensite and carbide precipitates. The austenite grain size varied from 5 to 7 μm, which corresponds to nos. 11–12. There were no qualitative differences in the structure of specimens prepared from conventional and ultrasonically produced powders.

As for oxygen concentration, steel powder produced conventionally contained an order of magnitude more oxygen than powder produced ultrasonically.

The results of mechanical test of specimens prepared from powders produced conventionally and ultrasonically are presented in Table 10.19 alongside with the literature data concerning cast and worked steels.

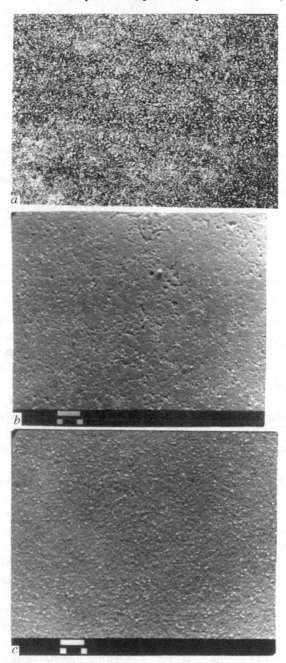

Figure 10.31. Microstructure of grade 10W6Mo5 steel (×1000 magnification) after (a) annealing, (b) quenching, and (c) tempering.

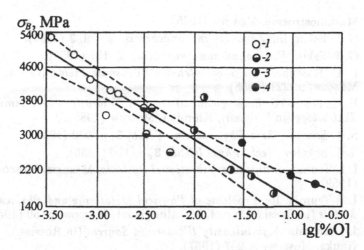

Figure 10.32. Dependence of bending strength on oxygen content: (*1*, *2*) 10W6Mo5 high-speed tool steel, (*3*) grade W18 steel, (*1*) ultrasonic gas-jet atomization, (*2*, *3*) conventional gas-jet atomization. The upper and lower dashed lines indicate confidence limits.

The mechanical properties of high-speed steels tend to be greatly impaired with increasing oxygen content, which was confirmed in our experiments. Ultrasonic gas-jet atomization reduced oxygen content of steels and thus improved their impact properties and bending strengths (Figure 10.32).

The powders prepared by us were used for producing turning tools for machining grade 40Cr steel with HB 229. The trials showed that tooling lifes were 1.2—1.4 times longer for tools produced from ultrasonically atomized powders than from conventional powders.

References

1. R. Bogle and J. Tayller, *Trans. Roy. Soc. Canada*, **20**, 25 (1926).
2. S. L. Hem, *Ultrasonic*, **4**, 202 (1967).
3. W. Esmarch, *Wissenschaftliche Verofentlichungen des Simens Werke*, Bd. 78 (1940).
4. G. I. Eskin, *Ultrasonic Treatment of Molten Aluminum* (in Russian), Metallurgiya, Moscow (1965).
5. G. I. Eskin, In: *Effect of high-intensity Ultrasound on Metal Interfaces* (in Russian), Nauka, Moscow, p. 6 (1986).
6. G. I. Eskin, *Ultrasonic Treatment of Melts in Shaped and Continuous Casting of Light Alloys* (in Russian),

Mashinostroenie, Moscow (1975).

7.   G. I. Eskin, In: *Advances in Sonochemistry*, vol. 3 (1993).

8.   G. I. Eskin, *Ultrasound and Sonochem.*, **2**, 137 (1995).

9.   L. D. Rozenberg, In: *high-intensity Ultrasonic Fields*, Nauka, Moscow, p. 221 (1968).

10.   B. G. Novitskii, *Employment of Acoustic Vibrations in Chemical Technology* (in Russian), Khimiya, Moscow (1983).

11.   S. J. Sokolov, *Acta Physicochim. URSS*, **3**, p. 939 (1935).

12.   S. J. Sokolov, *Tech. Phys. URSS*, **3**, p. 176 (1936).

13.   I. I. Teumin, *Ultrasonic Vibrational Systems*, Mashgiz, Moscow (1959).

14.   I. I. Teumin, In: *Problems of Physical Metallurgy and Physics of Metals* (in Russian), no. 4, Metallurgizdat, Moscow, p. 50 (1955).

15.   I. I. Teumin, *high-intensity Ultrasound Sources* (in Russian), Nauka, Moscow, p. 207 (1967).

16.   I. I. Teumin, In: *Problems of Physical Metallurgy and Physics of Metals* (in Russian), no. 7, Metallurgizdat, Moscow, p. 376 (1962).

17.   I. I. Teumin, *Izv. AN SSSR, Ser. Metallurgiya i Toplivo*, no. 1, 67 (1962).

18.   I. I. Teumin, In: *Industrial Application of Ultrasound* (in Russian), Mashgiz, Moscow, p. 163 (1959).

19.   G. I. Pogodin-Alekseev, *Ultrasound and Low-Frequency Vibration in Alloy Production* (in Russian), Izd. NTO Mashprom, p. 74 (1961).

20.   G. I. Pogodin-Alekseev, In: *Application of Ultrasound in Machine Building*, Izd. MDNTP, p. 72 (1963).

21.   I. G. Polotskii, *Application of Ultrasonic Vibrations for Investigation, Quality Control, and Treatment of Metals and Alloys* (in Russian), Izd. AN USSR, Kiev (1960).

22.   I. G. Polotskii, *Application of Ultrasound in Production and Heat Treatment of Alloys* (in Russian), Izd. NTO Mashprom, p. 12 (1961).

23.   I. G. Polotskii, In: *Proc. Inst. of Ferr. Metallurgy Acad. Sci. USSR*, Izd. AN USSR, Kiev, p. 91 (1963).

24.   G. I. Eskin and I. N. Fridlyander, *Metalloved. Termich. Obrab. Met.*, **4**, 320 (1962).

25.   G. I. Eskin, *Izv. AN SSSR, Ser. Metallurgiya i Toplivo*, **1**, 118 (1963).

26.   O. V. Abramov and I. I. Teumin, In: *Physical Fundamentals of Ultrasonic Technology* (in Russian), Nauka, Moscow (1970).

27.   O. V. Abramov, *Crystallization of Metals in Ultrasonic Field* (in Russian), Metallurgiya, Moscow (1972).

28. O. V. Abramov, *Hutnik*, no. 6, 12 (1970).

29. O. V. Abramov, In: *Application of Ultrasound in Machine Building* (in Russian), Nauka Tekhnika, Minsk, p. 10 (1964).

30. O. V. Abramov, E. N. Milenin, and V. E. Neimark, In: *Ultrasonic Methods of Intensification of Technological Processes* (in Russian), Mashinostroenie, Moscow, p. 349 (1970).

31. O. V. Abramov, I. G. Khorbenko, and Sh. Shvegla, *Ultrasonic Treatment of Materials* (in Russian), Mashinostroenie, Moscow (1984).

32. O. V. Abramov, In: *Effect of high-intensity Ultrasound on Metal Interfaces* (in Russian), Nauka, Moscow, p. 52 (1986).

33. H. Seeman and H. Menzel, *Zs. Metall.*, **9**, 869 (1955).

34. H. Seeman and H. Staats, *Acustica*, **6**, 325 (1956).

35. H. Seeman and H. Staats, *J. Acoust. Soc. Amer.*, **29**, 698 (1957).

36. H. Seeman and H. Menzel, *Zs. Metall.*, **1**, 318 (1947).

37. H. Seeman, H. Staats, and K. G. Preter, *Archiv. fur Eisenhuttenwessen*, **38**, 257 (1967).

38. H. Seeman and H. Staats, Patent BRD no. 933779, 31 (1935).

39. E. German, *Continuous Casting* (in Russian), Metallurgizdat, Moscow (1961).

40. M. Ya. Arkin, *Akust. Zh.*, **14**, 344 (1968).

41. Yu. A. Krikun, In: *Ultrasound in Machine Building* (in Russian), TsNIIPI, Moscow, p. 13 (1966).

42. Z. D. Chernyi, In: *Ultrasound in Machine Building* (in Russian), TsNIITSh, p. 9 (1966).

43. W. Esmarch, *Wissenschaftliche Verofentlichungen des Simens Werke*, **18**, Werkstoff sondere heft 8 (1940).

44. C. Eden, *Glasstech. Berichte*, **25**, 83 (1952).

45. V. I. Danilov and G. Kh. Chedzhemov, In: *Problems of Physical Metallurgy and Physics of Metals* (in Russian), Metallurgizdat, Moscow, p. 34 (1955).

46. G. Schmid and L. Ehret, *Zs. Electrochem.*, **43**, 269 (1937).

47. G. Schmid and A. Role, *Zs. Electrochem.*, **45**, 769 (1939).

48. G. Schmid and A. Roll, *Zs. f. Electrochem.*, **46**, 653 (1940).

49. P. D. Southgate, *J. Metals*, **9**, 514 (1957).

50. O. V. Abramov and Ya. B. Gurevich, *Fiz. Khim. Obrab. Mater.*, **3**, 18 (1972).

51. V. V. Zaboleev-Zotov and G. I. Pogodin-Alekseev, *Metalloved. Termich. Obrab. Metal.*, **1**, 2 (1958).

52. G. I. Pogodin-Alekseev, In: *Ultrasound in Machine Building* (in Russian), Izd. TsNIIPI, p. 114 (1966).

53. O. V. Abramov and V. A. Filonenko, *Fiz. Khim. Obrab. Mater.*,

1, 45 (1974).

54. P. P. Pereyaslov and A. V. Sapozhnikov, In: *Application of Novel Physical Methods for Intensification of Metallurgic Processes* (in Russian), Metallurgiya, Moscow, p. 185 (1974).

55. R. R. Malinovskii and N. N. Barbashin, In: *Proc. of Aviation Engineering Inst.* (in Russian), Mashinostroenie, Moscow, **67**, p. 79 (1966).

56. V. I. Dobatkin, G. I. Eskin, and S. I. Borovikova, In: *Processing of Light and Refractory Alloys* (in Russian), Nauka, Moscow, p. 151 (1976).

57. A. E. Aksyutina, I. I. Gur'ev, and G. I. Eskin, In: *Magnesium-Based Alloys* (in Russian), Nauka, Moscow, p. 136 (1973).

58. I. I. Gur'ev, G. I. Eskin, and A. E. Aksyutina, *Technology of Light Alloys* (in Russian), VILS, **12**, 25 (1975).

59. G. I. Pogodin-Alekseev and V. M. Gavrilov, *Application of Ultrasound in Production and Thermal Treatment of Alloys* (in Russian), Izd. NTO Mashprom, Moscow, **2**, p. 3 (1961).

60. L. K. Vasin, S. M. Potapov, and O. V. Abramov, In: *Ultrasonic Methods in Technological Processes* (in Russian), Metallurgiya, Moscow, p. 67 (1981).

61. O. V. Abramov and S. B. Maslenkov, In: *Novel Methods of Ultrasonic Treatment of Metals* (in Russian), Izd. NTO Mashprom, Moscow, p. 52 (1966).

62. O. V. Abramov, I. I. Teumin, and V. E. Neimark, In: *Novel Methods of Ultrasonic Treatment of Metals* (in Russian), Izd. NTO Mashprom, Moscow, p. 20 (1966).

63. O. V. Abramov, V. I. Lomakin, and V. E. Neimark, *Ultrazvuk. Tekhnika*, no. 2, 11 (1967).

64. O. V. Abramov and I. I. Teumin, *Proc. VIth All-Union Acoust. Conf.* (in Russian), Izd. NTO Mashprom, Moscow (1968).

65. K. V. Gorev, *Liteinoe Proizv.*, **4**, 35 (1961).

66. K. V. Gorev and L. A. Shevchuk, *Dokl. Akad. Nauk BSSR*, **3**, 298 (1959).

67. K. V. Gorev and L. N. Belozerskii, In: *Application of Ultrasound in Machine Building* (in Russian), Nauka Tekhnika, Minsk, p. 15 (1964).

68. L. I. Levi, *Izv. Vuzov, Ser. Chernaya Metallurgiya*, **1**, 155 (1969).

69. L. I. Levi and S. K. Kantenik, *Casting Alloys* (in Russian), Vysshaya Shkola, Moscow (1967).

70. L. I. Levi, E. V. Vygovskii, and V. I. Samsonov, *Izv. Vuzov, Ser. Chernaya Metallurgiya*, **5**, 150 (1968).

71. L. I. Levi, E. V. Vygovskii, and D. T. Karaseva, *Izv. Vuzov, Ser. Chernaya Metallurgiya*, no. 1, 155 (1969).

72. L. I. Levi, I. I. Tsypin, and E. V. Vygovskii, In: *Ultrasonic Methods of Intensification of Technological Processes* (in Russian), Metallurgiya, Moscow, p. 39 (1970).

73. O. V. Abramov, V. E. Neimark, and I. I. Teumin, In: *Application of Ultrasound in Machine Building* (in Russian), Nauka Tekhnika, Minsk, p. 57 (1964).

74. O. V. Abramov, N. N. Dmitriev, and T. V. Svistunova, *Fiz. Khim. Obrab. Mater.*, no. 2, 145 (1972).

75. O. V. Abramov, I. I. Teumin, V. A. Filonenko, and G. I. Eskin, *Akust. Zh.*, **13**, 161 (1967).

76. K. B. Yurkevich, E. N. Ozerenskaya, and S. A. Afanas'ev, In: *Application of Novel Physical Methods for Intensification of Metallurgic Processes* (in Russian), Metallurgiya, Moscow, p. 194 (1974).

77. K. B. Yurkevich, B. A. Agranat, and N. N. Dubrovin, In: *Application of Ultrasound in Metallurgic Processes* (in Russian), Metallurgiya, Moscow, p. 103 (1972).

78. G. I. Eskin, B. A. Agranat, and E. N. Ozerenskaya, In: *Application of Novel Physical Methods for Intensification of Metallurgic Processes* (in Russian), Metallurgiya, Moscow, p. 193 (1974).

79. E. N. Ozerenskaya, In: *Novel Physical Methods for Intensification of Technological Processes* (in Russian), Metallurgiya, Moscow, p. 82 (1977).

80. R. G. Sarukhanov, B. A. Agranat, and L. A. Firsanova In: *Application of Novel Physical Methods for Intensification of Metallurgic Processes* (in Russian), Metallurgiya, Moscow, p. 197 (1974).

81. B. A. Agranat, N. N. Khavskii, and M. N. Dubrovin, In: *Novel Physical Methods of Intensification of Technological Processes* (in Russian), Metallurgiya, Moscow, p. 67 (1977).

82. M. N. Dubrovin, B. A. Agranat, O. V. Pelevin, and A. M. Sokolov, In: *Novel Physical Methods of Intensification of Technological Processes* (in Russian), Metallurgiya, Moscow, p. 87 (1977).

83. M. N. Dubrovin, In: *Novel Physical Methods of Intensification of Technological Processes* (in Russian), Metallurgiya, Moscow, p. 94 (1977).

84. B. A. Agranat, L. A. Firsanova, and R. G. Sarukhanov, In: *Novel Physical Methods of Intensification of Technological Processes* (in Russian), Metallurgiya, Moscow, p. 91 (1977).

85. G. I. Pogodin-Alekseev and V. V. Zaboleev-Zotov, *Liteinoe Proizvodstvo*, no. 7, 35 (1958).

86. G. I. Pogodin-Alekseev, *Ultrasound and Low-Frequency Vibration in Alloy Production* (in Russian), Mashgiz, Moscow (1961).

87. G. I. Pogodin-Alekseev, In: *Application of Ultrasound in Machine Building* (in Russian), Izd. MD NTP, Moscow, p. 72 (1963).

88. G. I. Pogodin-Alekseev, V. I. Boiko, and V. V. Lenskii, In: *Application of Ultrasound in Machine Building* (in Russian), Izd. UP NTO Mashprom, Moscow, p. 37 (1963).

89. O. V. Abramov, S. A. Golovanenko, and I. V. Abramov, *Metally*, no. 6, p. 227 (1972).

90. I. V. Abramov, O. V. Abramov, and S. A. Golovanenko In: *Application of Ultrasound in Machine Building* (in Russian), Izd. NTO Mashprom, Moscow, p. 17 (1972).

91. O. V. Abramov, I. V. Abramov, and S. A. Golovanenko, In: *Physical and Chemical Mechanisms of Heat Tolerance of Metals* (in Russian), Metallurgiya, Moscow, p. 21 (1971).

92. I. V. Abramov, In: *Application of Ultrasound in Metallurgy* (in Russian), Metallurgiya, Moscow, p. 64 (1977).

93. O. V. Abramov, V. B. Kireev, and I. V. Abramov, In: *Experience of Industrial Application of Ultrasonic Technology* (in Russian), Izd. TO Mashprom, Moscow, p. 22 (1976).

94. O. V. Abramov, V. B. Kireev, and I. V. Abramov, *Metalloved. Termich. Obrab. Met.*, no. 12, 37 (1975).

95. O. V. Abramov, V. B. Kireev, and I. V. Abramov, *Fiz. Khim. Obrab. Metal.*, no. 1, 54 (1975).

96. I. V. Abramov, O. V. Abramov, and Zh. Yu. Chashechkina, *Proc. IVth Intern. Conf. on Composit. Mater.*, Smolenitse, Czech Republic, p. 39 (1980).

97. O. V. Abramov, V. O. Abramov, F. Sommer, and D. Orlov, *Mater. Lett.*, **23**, 17 (1995).

98. O. V. Abramov, V. O. Abramov, F. Sommer, and D. Orlov, *Mater. Lett.*, **29**, 67 (1996).

99. O. K. Eknadiosyants, In: *Physical Fundamentals of Ultrasonic Technology* (in Russian), Nauka, Moscow, p. 337 (1970).

100. K. Sollner, *Trans. Faraday Soc.*, **32**, 1532 (1936).

101. R. Ruthardt and E. G. Lierke, *Intern. Conf. on Metal Powders*, Washington, p. 105 (1980).

102. R. Pohlman and K. Stamm, *Forschungsber Landes Nordrhein – Westfallen*, no. 933, Koln und Oplanden, W. D. V. (1960).

103. R. Pohlman and E. G. Lierke, *5th Congress Internationale d'Acoustique Liege*, D35 (1965).

104. N. J. Grant, *J. Metals*, no. 1, 20 (1983).

105. N. J. Grant, *Rapid Solidification Processing*, 230 (1978).

106. V. Anaud, A. J. Kaufman, and N. J. Grant, *Proc. 2nd Intern. Conf. on Rapid Solidification*, p. 273 (1980).

107. D. H. Ro and H. Sunwoo, *US National P/M Conf.* (1983).

108. O. V. Abramov, Yu. Ya. Borisov, and R. A. Oganyan, *Akust. Zh.*, **27**, 801 (1981).

109. O. V. Abramov, R. A. Oganyan, and T. S. Shishkhanov, *Poroshkovaya Metallurgiya*, no. 6, 6 (1981).

110. O. V. Abramov, V. L. Gershov, and R. A. Oganyan, *Metalloved. Termich. Obrab. Met.*, no. 10, 37 (1982).

Chapter 11
Materials Processing

11.1 Ultrasound in Metalworking

It has been already shown that the transmission of high-intensity ultrasound to a solid metal increases the amount of structural dislocations and vacancies and thus may modify the properties of workpieces and their workability. Ultrasound can also affect the solid–solid interface and reduce boundary friction.

These physical phenomena pave the way for using ultrasound in metalworking. At present, ultrasound has proved its efficiency in metalworking processes such as wire and tube drawing, extrusion, forging, and rolling. The use of ultrasound in these processes reduces energy requirement, increases process rate and tool life, improves the quality of the workpiece surface, and extends metalworking technology to such materials that would fail to be processed by conventional methods.

In the 1950s, Blaha and Langenecker [1] investigated for the first time the effect of ultrasonic vibrations on the mechanical properties of metals and showed that superimposing an alternating stress on a workpiece greatly affected the process rate, the force necessary for workpiece deformation being considerably decreased (Figure 3.33, chapter 3).

This problem was extensively studied in the former USSR by Prof. Severdenko and his colleagues Dr. Klubovich and Dr. Stepanenko who carried out a wide range of research into a potential use of ultrasound in metalworking [2–6]. The ultrasound-assisted lowering of working forces can be estimated theoretically through changes in the workpiece material properties and processes occurring at the workpiece-tool contact surface. It should be noted that the efficiency of ultrasonic stressing strongly depends on the method of ultrasound feeding to deformation zone and process parameters, such as speed and strain. For example, the force necessary for ultrasound-assisted wire drawing is given by [7]

$$\sigma^U = (\overline{\sigma}_d - A\sigma_{mz})\left(1 + \frac{2}{\pi}\frac{v_i \cos\beta}{\omega\xi_{mz}}\cot\theta\right)\ln\frac{S_i}{S_f}, \qquad (11.1)$$

where σ_{mz} and ξ_{mz} are the maximum ultrasonic stress and displacement in deformation zone, respectively; σ_d is the mean deformation resistance; θ is the die half-angle; v_c is the velocity of deformation; β is the angle between the directions of deformation and ultrasound propagation; S_i and S_f are the initial and final cross-sectional areas of the wire, respectively. Similar relations were obtained by Severdenko *et al.* [2–6] for other metalworking processes.

Analysis of Eq. (11.1) suggests that drawforce depends on the method of ultrasound feeding. When the deformation zone is at the stress or displacement antinodes, the major contributing factor is the workpiece material properties or contact friction, respectively.

It should be noted that the ultrasound-assisted working of parts results in their heating due to ultrasonic energy absorption. Thus, the temperature in the deformation zone of aluminum, cold-extruded by ultrasound, increases to 60–120°C (in control experiments, the temperature was about 30°C) [8]. In aluminum heated to 60–120°C without imposing to ultrasouns, the extrusion force was 3–6% lower than that in aluminum extruded at 30°C. All this suggests that the effect of ultrasound on metalworking is due to a reduced contact friction, increased temperature in the deformation zone, and varying material strength and ductility.

Unfortunately, because of different experimental conditions and reported arrays of data, it is impossible to perform a rigorous analysis of the available literature data. Nevertheless, one can semiquantitatively analyze the relationship between some acoustic parameters (e.g., displacement or stress amplitude), process parameters (e.g., speed, strain, lubricant viscosity, temperature, workpiece geometry, and equipment design), and working forces or product quality for such main metalworking processes as setting, rolling, sheet formation, extrusion, wire and tube drawing and piercing.

It should be first noted that the efficiency of ultrasonic treatment increases with vibration amplitude, as follows from the analysis of data obtained for a wide range of materials worked by various methods of ultrasound feeding and metalworking processes.

Thus, Abramov *et al.* [8] evaluated the effect of ultrasound on extrusion as a function of vibrational displacement amplitude. For

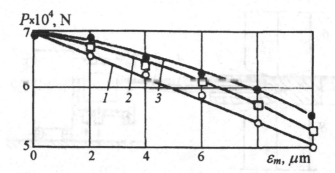

Figure 11.1. Effect of displacement amplitude on the extrusion force of aluminum upon feeding vibrations through (*1*) die, (*2*) container, and (*3*) punch [7].

this, aluminum specimens 15 mm in diameter and 30 mm in length were forward-extruded at a 75% strain, ultrasonic vibrations being fed through a female or male die or a container.

Figure 11.1 shows the process parameters versus vibrational amplitude. As the vibrational amplitude increases, extrusion force and deformation diminish nonuniformly regardless of the method of ultrasound feeding.

The relation between working forces and displacement amplitudes was found to depend on the deformation rate. As shown above, an ultrasound-induced decrease in contact friction occurs only when vibrational velocity is higher than tooling speed. In particular, this means that the decrease in deformation resistance due to a fall in contact friction should diminish with growing process speed.

This suggestion can be proved by the analysis of available extrusion data.

Ultrasonic vibrations with a 12-μm displacement amplitude (at 19.5 kHz, which corresponds to a vibrational velocity amplitude of 1.5 m s^{-1}) were fed to the deformation zone through a female die. If the displacement amplitude was constant, the increase in the extrusion speed would have affected the linear-to-vibration velocity ratio.

Experiments with aluminum showed that the effect of ultrasound on process dynamics slightly diminished with increasing extrusion speed, i.e., with decreasing v_m/v_1 (v_m and v_1 are the vibration and linear velocity amplitudes, respectively). Indeed, at extrusion speeds 4 and 33 mm s^{-1} (the velocity ratio v_m/v_1 is about 350 and 42), the ultrasound-induced drop in extrusion force made up 24 and 17 kN, respectively.

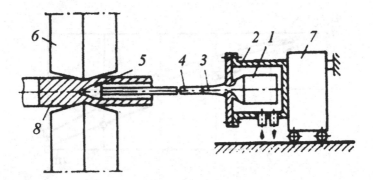

Figure 11.2. Cross-sectional view of an ultrasonic installation for tube piercing: (*1*) magnetostrictive transducer, (*2*) water-cooled jacket, (*3*) horn, (*4*) sonotrode, (*5*) piercer, (*6*) rollers, (*7*) axial bearing, (*8*) pierced workpiece.

Similar inference can be derived from the evaluation of working force – vibration amplitude relations: vibrational velocity diminishes with the displacement amplitude fall, thereby changing the v_m/v_1 ratio. The observations obtained during wire drawing are also in this line.

The limitation of drawing speed imposed by the application of vibrations makes them unsuitable for practical use in drawing of ductile metals that can be drawn at high speeds by conventional methods.

In this connection, of interest is the effect of ultrasound on the deformation processes usually proceeding at low rates.

Thus, Manegin *et al.* [12] investigated the effect of ultrasonic vibrations on the kinetic and dynamic parameters of the skew-roll mill piercing of solid tube rounds. Figure 11.2 shows an installation which includes a two-roll piercing mill and a built-in ultrasonic attachment consisting of magnetostrictive transducer *1* placed in water-cooled housing *2*, exponential booster *3*, horn *4*, and plug *5* secured at mill thrust bearing *7*.

The metals tested were lead, grade AD-1 aluminum alloy, OT-4 titanium alloy, grade St 20 mild steel, grade 0Cr18Mn8Ni2Ti and ShCr15 alloyed steels. The results are summarized in Table 1.1, from which it follows that application of ultrasound lowers the axial piercing force by 10 to 80%, while increases the process rate by 5 to 30% and the roll pressure by 5 to 15%. A slight ultrasound-induced increase in the roll pressure is due to increasing blank feed pitch.

Ultrasound was found to diminish piercing force and enhance process rate whether or not a lubricant was used.

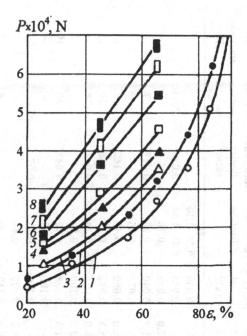

Figure 11.3. Strain dependence of the pressing force reduction in an ultrasonic field: (*1, 2*) aluminum, (*3, 4*) copper, (*5, 6*) iron, (*7, 8*) steel 20, (even numerals) ultrasonic pressing, (odd numerals) conventional pressing [2].

The ultrasonic effect on kinetic and dynamic parameters is governed by the vibrational displacement amplitude in the deformation zone, as evident from Table 11.1 showing that ultrasonic effect parallels the tool displacement amplitude in the deformation zone and the vibration-to-piercing velocity ratio.

In the extrusion tests with aluminum, copper, iron, and grade St 20 steel, the relative decrease in extrusion force was found to be essentially constant for deformations ranging from 19 to 90% [2]. Testing included both forward and backward extrusion runs, vibrations (8 to 10 μm in amplitude) being fed either to female or male dies, or simultaneously to both. Figure 11.3 illustrates the results of a forward extrusion test with vibrations fed through the female die. It is seen that the extrusion force decreased by 16–22% for aluminum, copper, and Armco iron, and by 10–14% for St 20 steel.

These results are similar to those obtained by Abramov *et al.* [8]. In his experiments, vibrations (12 μm in amplitude) were fed through a female die. The deformation resistance of aluminum was found to be independent of deformations varying from 26 to 75%.

Table 11.1 Effect of ultrasound on tube piercing.

Metal	Process	Displacement amplitude, μm	Axial plug resistance, kN	Roll load, kN	Axial piercing speed, m s^{-1}	Axial coefficient of friction	Piercing/vibrational velocity ratio	Plug resistance reduction, %	Process rate increase, %
Lead	Nonrotating plug	–	7.5–8.0	–	0.030	0.35–0.40	–	–	–
		16	5.5–6.2	–	0.035	0.48–0.58	0.020	20–40	–
	Rotating plug	–	6.0	27.0	0.028	0.16	–	–	–
		16	2.0	31.7	0.040	0.23	0.017	67	30
AD-1 aluminum alloy	Rotating plug, no lubrication	–	24.2	65.0	0.070	0.35	–	–	–
		3.5	22.5	66.0	0.080	0.40	0.160	7	13
		10	12.0	68.5	0.082	0.41	0.065	50	17
		13	10.0	68.0	0.074	0.38	0.045	58	3
	Rotating plug plus lubrication	–	19.7	72.0	0.100	0.50	–	–	–
		18	4.0	71.5	0.098	0.49	0.043	80	10
St 20 steel	Solid blank	–	24.0–24.5	–	0.10	0.68	–	–	–
		10	7.0–7.5	–	0.106	0.73	0.07	72–75	–
ShCr15 steel	Solid blank	–	28.2	78.5	0.105	0.48	–	–	–
		4	25.7	79.5	0.112	0.50	0.23	8	7
		18	7.2	105.0	0.21	0.84	0.09	70	12
0Cr18Mn8Ni2Ti steel	Solid blank	–	31.2	73.0	0.112	0.45	–	–	–
		18	5.5	75.0	0.122	0.49	0.05	83	9

Figure 11.4. Bismuth specimens extruded (*a*) conventionally and (*b*) with radial ultrasonic vibrations fed through a die [8].

It is of interest to consider the ultrasonic extrusion of a brittle metal such as bismuth. While conventional extrusion failed to produce a high-quality piece of this metal, radial vibrations applied to the female die enabled the production of rods with a sufficiently high quality (Figure 11.4).

The available data evidence that ultrasound can potentially be used in metalworking, although many studies have not yet passed the laboratory stage, and not every method can be recommended for practical application. This calls for continuation of relevant research.

As shown in chapter 3, ultrasound can modify mechanical properties of metals; therefore, it was of interest to evaluate the mechanical properties of pieces produced by ultrasonic technology.

The effect was found small but promising. For example, the microhardness of ultrasonically drawn deep steel cups in their wall-bottom transition was slightly greater than that of control cups. In other words, the product can be strengthened in its critical section [4]. Ultrasound increased the tensile strength of forward-extruded aluminum, copper, and Armco iron by 3 to 9% as compared with control specimens [8].

The ultrasound-assisted decrease in contact friction must diminish residual stresses in wrought products. Actually, analysis of copper strips produced at different reductions with or without ultrasonic irradiation showed that residual stresses were much lower and distributed more uniformly in the ultrasonically produced strips [4].

Ultrasonically extruded aluminum, copper, and Armco iron also displayed much lower residual stresses than control specimens did [7]. Exposure to ultrasound greatly improved the surface quality of rolled materials as well [4].

11.2   Ultrasound in Heat and Chemical Treatment Processes

Recent years have seen certain progress in the application of ultrasound for heat treatment of metals. In combination with chemical treatment, it appears to be a highly efficient approach to material processing along with thermomechanical, mechanothermal, and thermomagnetic methods.

Ultrasonic heat treatment may give rise to special time-saving effects that fail to be obtained by conventional methods.

This section considers ultrasound as a potential tool to be used in various thermal and thermochemical treatment processes. Valuable contributions to the research along this line have been made by Pogodin-Alekseev [10–12], Pogodina-Alekseeva [13–14], Aizentson [15–24] and Biront [25].

Feeding the vibrations to materials is the most difficult problem of ultrasonic treatment, which largely determines the efficiency of industrial applications of ultrasound.

Liquids are rather ineffective in transmitting the vibrations in almost all treatment processes except quenching. With a liquid-mediated transmission of ultrasound, high strains and stresses can hardly be achieved because of the great and uncontrolled reflection of vibrations from liquid-solid interfaces. Another reason for liquid inefficiency is related to cavitation and the screening effect of cavitation region. Moreover, the reproducibility of this method of ultrasound feeding is very low.

The effect of ultrasound on the kinetics of heat treatment and the resultant structure and properties of treated materials will be higher, if they are in a direct contact with the ultrasonic radiator. In this case, alternating stress and strain amplitudes may reach particular values necessary for efficient treatment.

As opposed to simple-shaped pieces (e.g., rods or wires), no suitable approaches are currently available to ensure the feeding of a specified ultrasonic intensity to all parts of an irregular workpiece. One should also bear in mind (during both the development of ultrasonic equipment

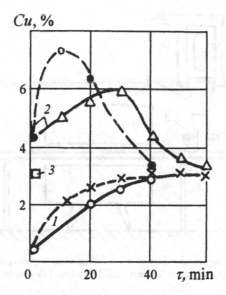

Figure 11.5. Effect of ultrasound on the durability of homogenizing annealing at 450°C of aluminum alloy containing 3% copper: (*1*) grain core, (*2*) grain periphery, (*3*) mean content of copper in the alloy. Solid and dashed lines show conventional and ultrasonic homogenizations, respectively [26].

and analysis of test data) the possibility of the arising of standing waves in treated materials, which may be responsible for a nonuniform ultrasound action on various parts of the workpiece.

Higher atomic mobility in sonicated materials must enhance phase transitions, whose kinetics is controlled by diffusion processes. This suggestion can be illustrated by data on the homogenization of aluminum alloys and phase transitions in steels.

Novik [26] studied the effect of ultrasound on the homogenization of cast aluminum alloys and revealed that ultrasound reduced the total time of homogenization 1.5 to 2 times. Figure 11.5 shows annealing-induced variations in the concentration of the second component (copper) in the bulk and surface of grains. Due to a nonuniform solidification, the grain bulk was initially depleted, while the grain periphery enriched in copper, which was because the excessive phases precipitating at the grain boundaries dissolved more readily than they could migrate to the grain bulk.

Ultrasound was found to enhance homogenization not only in metals, but also in thermoelectric, semiconducting alloys [27] and Cd–Hg–Te compounds [28].

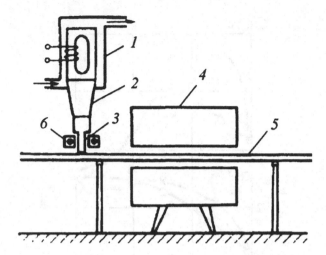

Figure 11.6.   Ultrasound feeding to a traveling wire: (*1*) magnetostrictive transducer, (*2*) horn, (*3*) radiator, (*4*) furnace, (*5*) wire, (*6*) displacement amplitude
probe.

Some authors studied the effect of ultrasound on phase transformations in steels. Thus, Kulemin and Nekrasova [29] proposed to use
ultrasound for spheroidizing a ball bearing steel wire. Their ultrasonic
assembly is shown in Figure 11.6. Bending vibrations were excited in
the wire as it was unreeled from the coil. The wire passed through the
nodes and antinodes of ultrasonic vibrations and was thus uniformly
treated lengthwise.

When ultrasonic strain amplitude is above some critical value ε_{mt},
the kinetics of austenite decomposition changes in both austenizing and
isothermal holding. In particular, ultrasound can augment the rate of
carbide agglomeration [29]. This effect was used for enhancing carbide
precipitation in grade ShCr5 ball bearing steel.

A prolonged annealing brings about carbide agglomeration and thus
decreases steel hardness (Figure 11.7). The favorable effect of ultrasonic
treatment lies in the lowering of annealing time by a factor of about
10. Apart from producing a spheroidizing effect, ultrasonic irradiation
promotes the dissolution of carbide network, thereby improving steel
quality.

The idea of ultrasonic quenching of steels is quite attractive [11,
12–14, 30–32]. Analysis of the available literature data shows that the
cooling of pieces can be considerably intensified by feeding vibrations
to quenching medium.

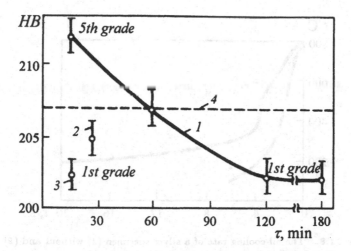

Figure 11.7. Hardness and carbide phase size in grade ShCr15 steel versus the time of annealing and ultrasonic treatment [29]: (*1*) heating to 780°C, interim cooling to 710–720°C, isothermal exposure, air cooling; (*2*) heating to 780°C, ultrasonic treatment for 24 min at $\varepsilon_m = 3 \times 10^{-4}$; (*3*) heating to 780°C, interim cooling to 710–720°C, ultrasonic treatment for 15 min at $\varepsilon_m = 4.3 \times 10^{-4}$; (*4*) the line below which hardness corresponds to the state specification GOST 801-60.

Figure 11.8 illustrates the effect of ultrasound on the cooling rate of a silver cylindrical specimen during its oil quenching [13].

Without ultrasound, the cooling curve exhibits the intervals of film and bubble boilings, as well as of convective heat exchange.

At the same time, ultrasonic irradiation rapidly degrades the vapor-oil film and thus prevents the occurrence of the film-boiling phase. In this case, the cooling rate peak shifts toward higher temperatures.

The absence of unstable film-boiling stage must obviously suppress the nonuniform cooling of various surface regions of pieces. An increase in the process time and intensity of ultrasonic vibrations leads to a rise in the cooling rate over all the temperature ranges studied. This is because of several factors, two of which are worthy mentioning:

(1) Acoustic streaming augments the rate of heat transfer through the solid surface;

(2) Cavitation breaks down a steam jacket around the material at the first stage of cooling, thus improving heat transfer.

It is generally believed that ultrasound is most potent in oil quenching, where it substantially increases the hardness and hardenability of

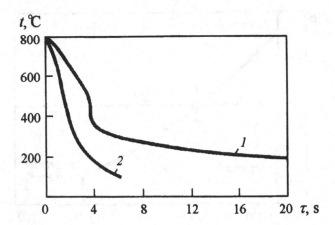

Figure 11.8. The oil-cooling rate of a silver specimen (*1*) without and (*2*) with ultrasound.

materials and improves their surface quality by preventing scale and oil burns.

Figure 11.9 shows the hardness pattern across the section of grade St 45 plain carbon steel quenched under various conditions [30]. It is seen that during ultrasonic oil quenching, cooling rate is above its critical value, which gives rise to a martensitic structure.

It should, however, be noted that ultrasonic oil quenching fails to reach the hardness levels attainable through water quenching. For instance, the hardness of steel containing 1% carbon quenched in oil, in oil plus ultrasound, and in water was HRC 30–38, 49–50, and 60–64, respectively [32].

Similar results were obtained by Pogodina-Alekseeva [13] who showed that the hardness of St 45 steel quenched in oil and water was HRC 50 and 54–56, respectively.

It should be emphasized that ultrasonic equipment used for quenching, i.e., power supplies, electroacoustic transducers, boosters, and horns, require additional operation expenditure and are not very efficient.

Hydrodynamic systems are more suited in this respect. Thus, Usatyi [31] used a hydrodynamic radiator for transmitting elastic vibrations to an oil bath and was able to find out that ultrasound enhanced the rate and uniformity of cooling, as well as hardness and hardenability of plain carbon steels (Figure 11.10). Hardness increased with ultrasound intensity to attain the level typical of water quenching. Usatyi established appropriate conditions for the oil quenching of ma-

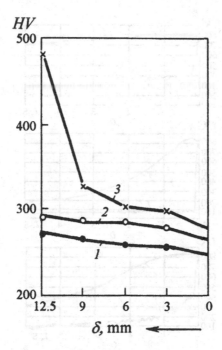

Figure 11.9. Hardness distribution over grade 45 steel specimen depending on cooling conditions [30]: (*1*) oil-quenching, (*2*) quenching in circulating oil, and (*3*) ultrasonic oil-quenching.

chine parts and metalworking dies made of plain carbon steels, which ensured the desired hardness and hardenability of parts without their cracking.

The mechanical properties of ultrasonically quenched aluminum alloys can be further improved by subsequent special heat treatment. In particular, ultrasonic quenching improved both the tensile strength and elongation of grade AL9 aluminum alloy.

It is noteworthy that ultrasound can be used to prepare novel quenching agents (e.g., emulsions) that can hardly be produced alternatively.

Aging pertains to heat treatment processes whose kinetics is governed by diffusion. Biront [25] overviewed the studies dealing with the aging of alloys placed in fluids vibrated at sonic and ultrasonic frequencies. Experiments were performed with various materials, such as duralumin, Pb- and Ni-based alloys, and steels. Radiation intensity was varied from 1.4 to 6.7 $W\,cm^{-2}$. In most cases, ultrasonic irradiation augmented the aging rate by a factor of 2 to several hundred

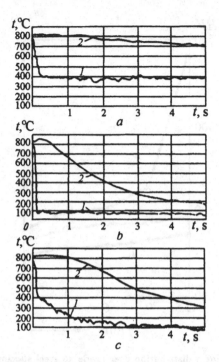

Figure 11.10.   Cooling of (*1*) surface and (*2*) bulk of grade U10 steel rod 10 × 50 in size in (*a*) oil, (*b*) water, and (*c*) oil excited with elastic vibrations [31].

and sometimes considerably affected the hardness of treated materials.

Some ambiguity in the relevant data is due to the method of ultrasonic radiation feeding to workpieces through a fluid, since it is extremely difficult, in this case, to correctly estimate the alternating stress amplitude in the workpiece to compare the results. Besides, the aging rate increase could be induced by an uncontrolled radiative heating.

These limitations can be avoided by using an ultrasonic arrangement presented in Figure 3.11 (chapter 3). A similar method of ultrasound feeding to specimens was used by Pogodina-Alekseeva [14]. Aluminum alloys of grades V95, AK6, AK4-1, and D16 were aged at temperatures optimal for the maximum effect without ultrasonic irradiation. The imposed ultrasonic irradiation improved the mechanical properties of aged D16 alloy and drastically shortened the process time.

Similar results were obtained for other alloys. As opposed to normal forced aging, the strength of ultrasonically aged D16 and binary Al-Cu model alloys increased with the alloy supersaturation which could be modified by varying quenching temperature.

Relatively few studies have been concerned with the effect of ultrasound on the precipitation hardening of Ni-based superalloys [10, 34, 35]. Experiments with grade CrNi77TiAlB alloy showed that ultrasound raised the tempering rate 2 to 4 times and enhanced hardness when tempering temperature was 750°C to 700°C [10]. It was also found [34, 35] that raising the cyclic stress amplitude increased still further the process rate (the aging rate increased 2 and 40–50 times at stress amplitudes 25 and 35 MPa, respectively). Ultrasonic irradiation modified the material structure as well [36]. In such a situation, precipitation occurred primarily in the form of solid solution grains, which was not the case in conventional tempering. Ultrasonic irradiation appeared most potent at the initial process stage and was not so effective after precipitation had started. However, increasing the time of ultrasonic tempering could intensify structural rearrangement. Ultrasonic tempering under optimum conditions greatly improved the performance of CrNi77TiAlB alloy at high temperatures [18].

Ultrasonic tempering of commercially pure iron, MSt3 plain carbon steel, and grade 12CrNi3 HSLA steel was extensively studied by Pogodina-Alekseeva [13]. The tempering rate was found growing whenever ultrasound intensity was sufficient for increasing the dislocation density. For instance, ultrasonic irradiation substantially enhanced the natural aging rate and hardness of grade MSt3 steel. The hardness increment produced by ultrasonic irradiation was retained during subsequent natural aging.

Al'ftan [37] tempered grade St 50 steel for 2 h at 200 to 450°C and revealed no effect of ultrasound of a given alternating stress amplitude on the properties of steel tempered at 200–250°C. At higher temperatures, however, ultrasound of the same stress amplitude improved both the hardness and impact toughness of this steel.

Similar results were obtained in experiments with grade 30CrMnSi steel ultrasonically tempered at 500°C. Lutsyak [38] studied the effect of ultrasonic irradiation on high-temperature tempering of plain carbon steels containing 0.57 and 0.73% C. The treatment temperature and time were varied between 600 and 700°C and between 0.02 and 30 h, respectively. Ultrasound augmented the rate of carbide particles agglomeration and reduced hardness. Aizentson [23] observed a faster agglomeration of carbide particles when the tempering of grade U12 spring-type plain carbon steel at 650°C was superimposed by ultrasonic irradiation. The ultrasonic effect was maximum at the initial stage of tempering. For example, ultrasonic irradiation acted to increase the mean particle size and spacing approximately twofold, if the tempering treatment was continued for 60 to 120 min. Extending the time of tempering reduced the efficiency of ultrasonic irradiation.

Tempering at a reduced temperature (300°C) produced a reverse effect, viz. increased the number of carbide particles and decreased their mean size with respective changes in steel properties [24]. Ultrasonic tempering at temperatures between 300 and 400°C enhanced HRC hardness by 1 to 2 units, while reduced it by 1 to 2 units at tempering temperatures above 500°C. In addition to modifying the martensite decomposition kinetics, ultrasound affected the retained austenite.

Of interest is the ability of ultrasonic irradiation to relieve internal stresses. Arkhangel'skii *et al.* [39] showed that ultrasonic vibrations can increase the rate of stress relaxation. In a Cr–W–Mn steel, stresses were reduced twofold when relaxation was carried out ultrasonically at 140–150°C and only by 10% when stresses were relieved conventionally at 150–160°C. Similar results were obtained with grade ShCr15 ball bearing steel and grade 40CrMnB HSLA steel [40].

At present, there is an adequate understanding of diffusion mechanisms involved in thermochemical treatment processes, most of which are controlled by diffusion of a solute from surface to bulk. Ultrasonic irradiation can increase diffusion rate and thus improve process efficiency [23].

Ultrasound was found to be efficient in such applications as carburizing, nitriding, and boriding [23].

Pogodin-Alekseev [10] investigated the effect of solid carburizers on Armco iron and grade 12Cr43 steel. In experiments, performed at 1000°C, specimens were fixed to a horn. Ultrasonic irradiation was found to increase the case thickness twofold and to reduce the process time manifold. The process efficiency improved with increasing vibrational amplitude.

Ultrasonic irradiation enhanced the rate of liquid or gas nitriding and carbonitriding of grade 30Cr10 and 30CrMnSi steels, which led to a heavier case and, therefore, to its higher hardness and wear resistance. The wear resistance of ultrasonically carbonitrided case was twice as great as in control.

Ultrasonic boriding at 900 to 1050°C [41] augmented the case depth 2 to 2.5 times (Figure 11.11), the process efficiency increasing with temperature. The case microhardness of ultrasonically treated steel increased to 2300 MPa, i.e., to the value typical of FeB.

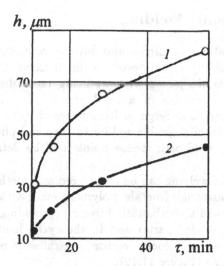

Figure 11.11. Borided layer depth versus the time of process (*1*) with and (*2*) without ultrasonic irradiation [41].

Ultrasonic chromizing of grade 30 CrMnTi HSLA steel and cast iron at temperatures from 1000 to 1100°C for 3 to 15 h augmented the case depth 1.5–2 times, which noticeably improved material wear and its resistance to heat and corrosion [23].

Pogodin-Alekseev [42] studied the ultrasonic silicification of grade Cr18Ni9 stainless steel. The treatment, carried out at 900–1200°C for 0.5 to 3 h, increased the case depth as much as sixfold (at 1000°C, from 0.05 to 0.3 mm).

In the same work, Pogodin-Alekseev studied molybdenum plating with molten Al at 1000°C for 5 to 60 min and found that aluminization would not occur without ultrasound. At the same time, ultrasonic treatment performed initially at a power of 4 kW (5 min) and then at a power gradually diminishing to 0.7 kW produced a fairly homogeneous aluminized case, whose thickness grew linearly with the process time to reach 60 μm after a 60-min treatment.

Similar observations were made during aluminization of Armco iron [31]. Ultrasonic irradiation appeared more efficient when a specimen was rigidly fixed to a horn but not when the aluminizer was used for coupling. It is noteworthy that ultrasonic treatment changed the chemical composition of the case from FeAl to Fe$_3$Al.

Thus, ultrasonic irradiation is an efficient tool to improve heat and chemical treatment processes.

11.3   Ultrasonic Welding

Potential applications of ultrasound involve cold welding and conventional hot welding superimposed with ultrasound. In the former case, materials are joined without being fused due to the intensification of mass transfer in a solid phase, whereas in the latter
case, welded materials undergo melting followed by their solidification
with the formation of a specific weld structure and phase transformations in the near-weld area (these points will be detailed in section
11.4).

Ultrasonic cold welding can be considered as a method to join similar or dissimilar materials (metals, polymers, etc.) with a simultaneous
exposure of the weld to both static force F_0 (providing for an acoustic
contact between welded parts) and to the cyclic loading with ultrasonic vibrations whose velocity vector is parallel or perpendicular to
the plane of joining (Figure 11.12).

Some effects produced by high-intensity ultrasound at the interface
of two solids were considered in section 5.2. These effects are associated
with changes in the surface state and contact zone temperature, as well
as with microplastic deformation processes, and can, therefore, promote
interdiffusion between welded solids.

Ultrasonic welding of polymeric materials is based on heat liberation as a result of ultrasound absorption and friction of two polymeric
surfaces [43-49]. Welding zone is highly heated due to the action of
both heat sources, the contribution from surface friction being dominant [43, 44].

The coefficient of ultrasound absorption by hard polymers is smaller
than by soft polymers, which enables a remote welding of hard polymers
when the welder and material are up to 25 cm apart. Soft polymers
absorb ultrasound so extensively that even at a distance as small as a
few millimeters from ultrasound source the acoustic power and, therefore, the temperature in welding zone would drop to the extent that
excludes the obtaining of a high-quality weld.

Polymeric materials can be welded either in a near field, when
the welder and material are less than 5 mm apart, or in a far
field, when they are more than 5 mm apart (Figure 11.13). Materials that can be well welded in near field involve soft polymers and
foam plastics, which are characterized by a significant dissipation
of acoustic energy (polyamide, polyethylene, polypropylene, polyvinyl chloride, etc.). Materials that can be welded by the far-field
method (which is most common) are polystyrene, acrylbutadienesty-
rene, etc.

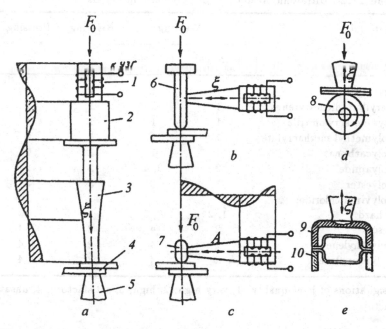

Figure 11.12. Schematic representation of (a) longitudinal spot welding, (b) transverse-longitudinal welding, (c) seam welding with the use of a resonant roller, (d) seam welding of films, (e) contour welding: (1) transducer, (2) primary horn, (3) removable horn, (4) welded parts, (5) anvil, (6) welding instrument, (7) resonant roller, (8) rotary bearing roller, (9) hollow instrument, (10) anvil.

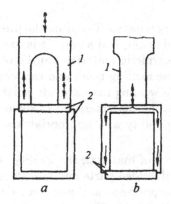

Figure 11.13. Ultrasonic welding in (a) a near field and (b) a far field: (1) welder, (2) welded parts.

Table 11.2   Ultrasonic weldability of polymeric materials.

Material	Welding		Riveting	Pressing
	near-field	far-field		
Polystyrene	1	1	1	1
Acrylbutadienestyrene	1	2	1	1
Styrene acrylonitrile	1	1	1	1
Polymethyl methacrylate	1	1	1	2
Polycarbonate	1	1	1	1
Polyamide	2	3, 4	2	2
Polyester	1	3	3	3
Polyvinyl chloride:				
hard	1, 2	2, 3	1	1
soft	2	4	2	3, 4
Polyethylene	2	4	2	3, 4
Polypropylene	2	4	2	3, 4

Designations of joint quality: 1, very high; 2, high; 3, satisfactory; 4, unsatisfactory.

For near-field welding, the working surface of a welder must have the form of parts to be welded. This method allows high-quality airtight welds to be obtained.

Far-field welding ensures the joining of materials when they are up to 200 mm from the welder, without imposing severe demands on the welder form.

The weldability of polymers depends on the presence of fillers, dyes, on the method of production, and so on. It is considered that a high-quality joint of two polymeric materials can be obtained if they have approximately the same melting point and chemical affinity. However, actually only tentative welding may provide reliable weldability data. Table 11.2 summarizes the characteristics of some polymeric materials in terms of their weldability and appropriateness for other types of ultrasonic joining.

Apart from the kind of material, the quality of ultrasonic welding depends on the form of welded parts and how their surfaces have been prepared. Of much importance are weld projections, also known as energy directors (Figure 11.14). The volume of these projections must be such that their material, after being fused, could cover the greatest possible area of welded surfaces without discharge. In this case,

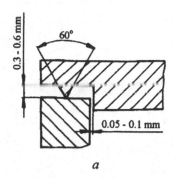
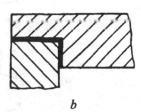

a   *b*

Figure 11.14. Geometry of a plastic material joint (*a*) before and (*b*) after welding.

acoustic energy is mainly concentrated in the welding zone, ensuring a high quality of the welded joint. Ultrasonic technology makes it possible to implant polymeric parts with metallic ones, such as nuts, pins, fingers, toes, etc. (Figure 11.15,*a*). During implantation, ultrasonic energy liberated in the contact zone fuses the polymer, and it embraces an implanted part. Upon cooling, the polymeric material is contracted to produce a tight joint with the implanted part. A detachable joint is produced by ultrasonic pressing of metal nuts or bolts into the hole in a polymeric material, whose diameter is equal to the mean diameter of a thread. The molten material fills the thread to produce, after cooling, a detachable joint with good mechanical properties [45].

Ultrasound can be used for riveting polymers with other materials. Vibrational energy softens a material in the zone of instrument contact with the rivet shank which, when being pressed, assumes the form of the instrument end to produce a mechanical joint of two parts (Figure 11.15,*b*). This method can be used for joining almost all thermoplastic polymers with metals.

A boosting production of synthetic thermoplastic textiles called for elaboration of methods for ultrasonic welding of clothes details [46].

Along with enhancing operation efficiency and saving fabrics, ultrasonic welding appears to be a promising innovation in textile industry, providing aesthetic joints, tailor modeling, etc.

The welding of thermoplastic fabrics differs from the welding of homogeneous materials. In the former case, the heat is liberated due to mechanical vibrations and friction between the fabric fibers themselves and between the surfaces of contacting details. Owing to such an ef-

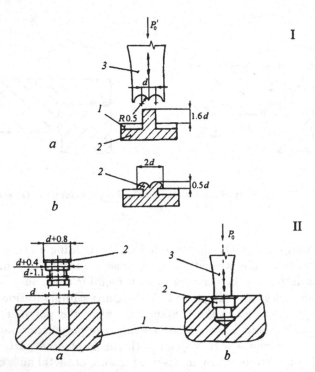

Figure 11.15. Principles of ultrasonic (I) pressing and (II) riveting: (*a*) parts before processing, (*b*) joint parts after processing, (*1*) polymeric part, (*2*) metal part, (*3*) ultrasonic instrument.

ficient process of heat production during ultrasonic welding of fabrics, the upper limit of their weldability approaches a thickness of 3–5 mm, while the lower limit is like that for homogeneous materials (0.1 mm). An imprint of the welder end on the welded fabrics is not a disadvantage of this method. Conversely, it may be an advantage, since an appropriately designed working surface of the welder may produce an aesthetic exterior of the weld.

A high coefficient of thermal expansion of polymers determines the requirement for weld cooling. Cooling time depends on the thickness of welded materials and may amount to 60% of a welding cycle. The strength of a resultant weld depends on the properties of welded materials, their thickness, weave type, the content of thermostable components, and weld structure.

The examples of ultrasonic welding are making of buttonholes, internal taping of trousers' cuffs, fixing of strips and plastic zippers. Ultrasonic welding can be well used for other auxiliary operations: making of holes, cutting of tapes and zippers to a required length, etc. Ultrasonic welding may substitute for stitching operations and thus reduce machining time by a factor of 1.5 to 12 and save sewing materials.

An installation for ultrasonic joining (welding, pressing, and reveting) of polymeric materials incorporates three main units: (1) pneumatic welding press with an ultrasonic welder involving a piezoceramic transducer, (2) working table, and (3) ultrasonic transistor generator. The installation can operate either individually under a manual control or as an automatic or auto-manual unit of a production line. If necessary, the ultrasonic welder driven by a generator can be made portable.

An example of ultrasonic welding installations is a 20-kHz (1.6-kW input) VUMA machine operated at controllable pressure (from 200 to 2000 N) and acoustic energy (from 10 to 9990 W.s). The machine is equipped with a rotary table providing a mechanized assembly of parts.

Most of the welders operate at 20 kHz and a power of up to 3 kW. Such an operation frequency is the result of a compromise between the cost of equipment, the noise it creates, and universality. Special modifications are also available for the welding of small and thin parts, which operate at 40 kHz and have power below 200 W. Such low-noise welders can be used for riveting and film welding when operation at 20 kHz is undesirable.

Welders are often furnished with rotary tables, which enables their incorporation into large mechanized lines performing various operations with workpieces, including their testing. This ensures a high quality of products.

Commercially produced welders range from portable manually operated instruments to large multipurpose automatic ultrasonic installations capable of making a few welds at a time. The latter can be used to weld large plane parts with many intermittent and continuous welds, e.g., plastic door panels. This equipment is expensive and can, therefore, be available upon request.

Recent investigations [51–54] have outlined the main principles of ultrasonic welding of metals. In particular, it was established that to obtain a strong weld, it is necessary that during ultrasonic welding the surfaces to be welded could slip with respect to each other. This condition is described by the following relation between the cyclic tangential force F_τ and the pressing force F_0

$$\frac{F_\tau}{F_0} > Kf_s, \tag{11.2}$$

where K is the coefficient of proportionality dependent on the presence of an oxide layer on contacting surfaces $(0 < K < 1)$; f_s is the coefficient of static dry friction.

The slip value x of welded surfaces (the upper and lower surfaces have $i = 1$ and 2, respectively) is related to their displacements ξ_i as

$$x = \xi_1 - \xi_2. \tag{11.3}$$

The slip value rises proportionally to contact pressure (shear stress), which is given by

$$x = \frac{\sigma_\tau}{Gh}, \tag{11.4}$$

where σ_τ is the shear stress that causes plastic deformation; G is the shear modulus of the metal welded; h is the total depth of the heat affected zone (HAZ).

Investigations have clarified the role of pressing force F_0, by which contacting surfaces are pressed. F_0 depends on specific conditions of ultrasonic welding and can be estimated to a sufficient accuracy from the following relation

$$F_0 = \sigma_{0.2}S, \tag{11.5}$$

where $\sigma_{0.2}$ is the yield strength of welded material; S is the working area of the welder.

During ultrasonic welding, plastic strains modify the microrelief of welded surfaces (see Figure 5.45), so that they approach to each other. This is accompanied by a change in the state of surface oxide films that degrade to expose bare metal areas. The following increase in the number and area of adhesive regions results in weld formation. For illustration, Table 11.3 lists the data on the kinetics of ultrasonic welding of 1-mm copper sheets at $F_0 = 4400$ H and welder power 4 kW [53].

It is seen that after as early as 1.2–1.5 s, the welding zone covers the entire area of metal contact with a welder. The strength pattern of the weld suggests the development of diffusion processes immediately after the onset of welding (within 0.3 s). Temperature in the weld zone does not exceed $0.6T_m$ (Table 11.4).

Table 11.3 Ultrasonic welding of copper sheets [53].

Process duration, s	Area of polished spot, mm^2	Welding area, mm^2	Mean shear rupture force, N
0.2	24.0	–	–
0.3	28.3	2.5	180
0.5	28.9	14.0	1100
0.9	38.4	26.5	2000
1.2	38.4	38.4	2600
1.5	38.4	38.4	2600

Table 11.4 Temperature in the ultrasonic welding zone [53].

Welded metal pairs	Thickness, mm	Welding parameters		Temperature, °C
		Time, s	Clamping force, N	
Al–Al	0.5 + 0.5	0.5		200–300
Mg–Mg	1.0 + 1.0	1.5	4400	300–350
Zn–Zn	0.85 + 0.85	0.6	2200	100–150
Fe–Fe	0.4 + 0.4	0.4	2600	800–900
Fe–Constantan	10.0 + 0.65	1.6	1900	730
Cu–Constantan	0.3 + 0.65	1.0	500	450

Although there remain some blanks in the physics of ultrasonic welding, this method has found an extensive industrial application. It provides the possibility of joining very thin films and wires to each other and even to thick sheets and parts. Materials can be ultrasonically welded without a preliminary cleaning of their surfaces, as well as without application of fluxes and other additives. Ultrasonic welding allows the retaining of the properties of welded materials in the near-weld region and a smaller deformation of the welding zone. Unlike other methods of welding, it is energy-saving.

Many modifications of ultrasonic welders are presently available, which are usable in machine building and electronics.

The most important element of any ultrasonic welder is a vibratory system consisting of an ultrasonic transducer, waveguide, radiator, and

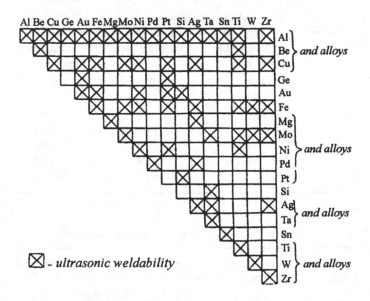

Figure 11.16.   Nomogram of ultrasonic weldability of metals and their alloys.

a welding instrument. Spot welding of materials employs longitudinal vibrations that are guided from the ultrasonic transducer to the welding instrument (Figure 11.12,*a*).

High-power welders have an arrangement (Figure 11.12,*b*), in which bending vibrations are generated in a resonant rod. Installations used for seam welding of metal parts are shown in Figure 11.12,*c, d*, and that for contour welding is shown in Figure 11.12,*e*.

Ultrasonic welding is used for joining many but not all metals. The nomogram in Figure 11.16 visualizes the weldability of a range of metals and alloys. Experiments show that the most appropriate thickness for welding of such metals as aluminum, copper, and steel is 3, 2, and 1.3 mm, respectively. Molybdenum, cobalt, tantalum, tungsten, and beryllium 0.5–0.25 mm thick parts can be welded only under certain conditions. Foils of gold, silver, platinum, and their alloys can be welded if their thickness is not less than 0.004 mm. The weldability of pure metals is usually greater than that of alloys. Ductile metals are typically weldable. Ultrasonic welding makes it possible to join soft metals to hard and brittle metals and even to glass, quart, and ceramics. It can also be used to join wires, especially microwires in electron circuits. A promising application of ultrasound is diffusion welding [55] (Figure 11.17).

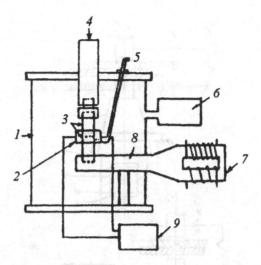

Figure 11.17. Setup for ultrasonic diffusion welding [55]: (*1*) chamber, (*2*) heater, (*3*) welded parts, (*4*) loading arrangement, (*5*) thermocouple, (*6*) vacuum pump, (*7*) transducer, (*8*) vibratory system, (*9*) power meter.

Thus, ultrasonics intensify the process of welding and improve the mechanical properties of welds.

11.4 Ultrasonic Contact and Arc Welding

As noted above, ultrasonic vibrations can be used to intensify conventional spot and arc welding [56].

During spot welding, ultrasonic vibrations contribute to the process efficiency by degrading oxide films on the surface of welded materials and increasing the area of microcontacts. In the following, they affect the formation of melted zone and eventually the weld, making its structure more fine-grained.

Figure 11.18 shows some typical diagrams for transmission of ultrasonic vibrations to the welding zone. The most appropriate feeding option is determined by material properties and geometry of parts to be welded, as well as by welding conditions and welder design.

As experiments show [56], the feeding of ultrasonic vibrations throughout the welding period is most advantageous, since it leads to a significant decrease in contact resistance R_k and its rapid stabilization, the effect being dependent on vibration amplitude. Ultrasound can reduce R_k of welded steels and aluminum alloys 3 to 10 times (Figure 11.19).

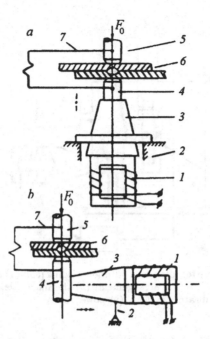

Figure 11.18. Diagrams of the welding zone feeding with (*a*) longitudinal and (*b*) bending vibrations [56]: (*1*) electroacoustic transducer, (*2*) fixture, (*3*) horn, (*4*) vibrated electrode, (*5*) reflecting electrode, (*6*) welded parts, (*7*) electric circuit of the welder.

Ultrasonics improve the mechanical properties of welded joint. Static shear tests testify to a decrease in strength data scatter and increase in shear force (Figure 11.20). Fatigue tests are indicative of the improvement of the fatigue endurance of welded joints (Figure 11.21). Ultrasonic vibrations give rise to compressive stresses at the weld spot and in the heat affected zone (HAZ) and thus reduce the cracking of welded materials. Similar results were obtained for ultrasonic arc welding.

An ultrasound-induced acceleration of austenite transformation can be used in welding practice. So far, the welding technology faces certain problems associated with the welding of alloyed medium-carbon (more than 0.2 mass % carbon) steels possessing an enhanced austenite stability at temperatures of ferrite-perlite and bainite transformations. After welding, such steels undergo, in the HAZ, a preferential martensitic transformation with the formation of a structure typical of quenched steels, i.e., with a high level of internal stresses and tendency to cracking and delayed fracture.

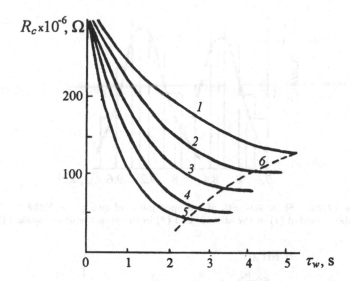

Figure 11.10. Dynamics of contact resistance during ultrasonic welding of two 1.5 mm thick plates of grade D16AT alloy at a vibration amplitude of (*1*) 4, (*2*) 6, (*3*) 8, (*4*) 10, and (*5*) 12 μm. Curve *6* shows stabilized resistance [56].

The risk of cold cracking can be reduced by employing a more sophisticated welding technology involving, for instance, ultrasonic treatment of welds at the stage of cooling of welded materials.

In works [57–60], attempts have been made to evaluate the feasibility and appropriateness of this method for welding the alloyed medium-carbon steels 30CrMnSiNi2A, 30Cr3, 35Cr, 40CrMnSiNi2Mo, and 10Cr3Mn2Mo. The authors established threshold strain amplitudes, above which there occurs an acceleration of austenite transformation at temperatures of austenization as well as of ferrite-perlite and bainite transformations.

As an example, Figure 11.22,*a* shows the lengthwise hardness distribution in grade 30CrMnSiNi2A steel subjected to ultrasonic irradiation during austenization and perlite transformation. The hardness diminished around the strain amplitude antinode, indicating austenite transformation. These results were used to plot the dependence of the amount of ferrite-perlite mixture (α%) and material hardness (HV) on the ultrasonic strain amplitude (Figure 11.22,*b*).

Ultrasound was found to induce austenite grain refinement, the effect being observed at cyclic stress amplitudes much greater than the threshold of accelerated austenization. For grade 30CrMnSiNi2A steel, the threshold amplitude made up 3.1×10^{-4}. At $\varepsilon_m = 1 \times 10^{-3}$, the

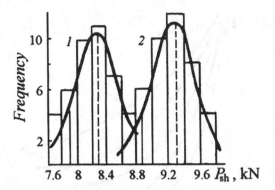

Figure 11.20. Shear strength of one-spot joints of two grade V95AT 1.5-mm thick plates welded (*1*) in the absence and (*2*) in the presence of ultrasound [56].

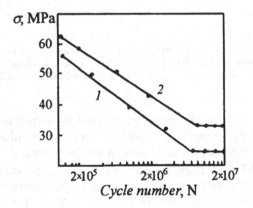

Figure 11.21. Fatigue strength of one-spot joints of two grade D16AT 1-mm thick plates welded (*1*) in the absence and (*2*) in the presence of ultrasound [56].

grain size was reduced almost twofold (from 30–32 to 15–17 μm) (Figure 11.23).

Ultrasonic irradiation of steel during isothermal austenite decompositon at the temperature of its minimum stability acted to accelerate ferrite-perlite transformation provided that ultrasonic strain amplitude was above a threshold value. For 30CrMnSiNi2A steel, this value (6.2×10^{-4}) is greater than the threshold value of accelerated austenite decomposition.

Analysis showed that transformations were proportional to strain amplitude and treatment time. The threshold amplitude ε_{mt} did not depend on the time of isothermal exposure.

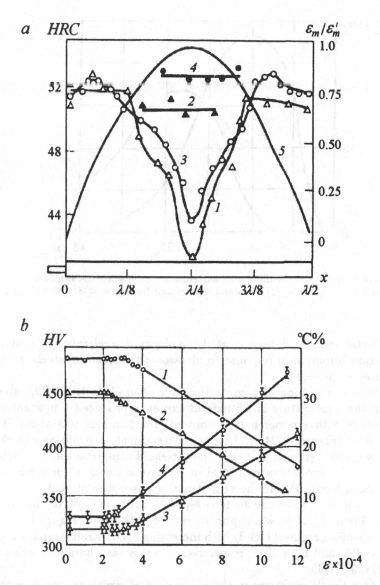

Figure 11.22. (a) Lengthwise distribution of (*1–4*) hardness and (*5*) relative strain amplitude and (b) dependence of (*1, 2*) hardness and (*3, 4*) perlite content on the vibration strain amplitude during the treatment of 30CrMnSiNi2A steel. Panel a: (*1, 3*) control samples, (*2*) 4-min isothermal ultrasonic treatment with $\xi_m = 37\ \mu m$, (*4*) 4-min ultrasonic treatment with $\xi_m = 10\ \mu m$ during austenization at 975°C. Panel b: (*1, 3*) hardness, (*2, 4*) perlite content α, (*1, 2*) 2-min ultrasonic treatment during austenization at 810°C, (*3, 4*) 2-min ultrasonic treatment during perlite transformation at 650°C in the region of the minimal stability of austenite.

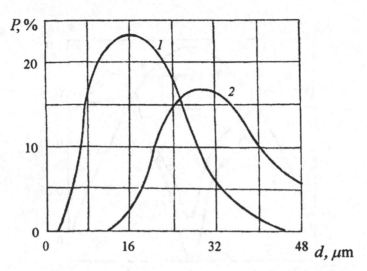

Figure 11.23. Size distribution of austenite grains in grade 30CrMnSiNi2A steel either (*2*) untreated or (*1*) ultrasonically treated for 2 min at 810°C with $\xi_m = 10^{-3}$.

Metallographic data supported the observed acceleration of austenite transformation in response to ultrasonic treatment of steels in the ferrite-perlite region.

Water quenching of a control sample of grade 30KhGSN2A steel from the temperature of isothermal exposure produced a martensitic structure with a minor bainite content (the hardness was about HV 460–510) (Figure 11.24,*a*). Ultrasonic treatment at strain amplitudes below 6×10^{-4} (i.e., smaller than the threshold value) did not essentially affect the microstructure of steel in comparison with a nonirradiated specimen (control). At the same time, ultrasonic treatment at $\varepsilon_m \sim$ 6.2–6.3 $\times 10^{-4}$ gave rise to ferrite-perlite areas (8–10% of the total area, Figure 11.24,*b*), whose portion increased to 30–40 (Figure 11.24,*c*) and 45–50% (Figure 11.24,*d*) with increasing cyclic strain amplitude to 7.6×10^{-4} and 1.2×10^{-3}, respectively. In this case, hardness declined to HV 340–360.

Along with variations in the proportion of structural components, one could observe some alterations in the perlite structure. In control samples exposed to 650°C for 3 min perlite occurred as large colonies with rather thick and extended cementite lamellae (Figure 11.24,*e*), whereas ultrasonic treatment at $\varepsilon_m \sim 6 \times 10^{-4}$ for 2 min during the isothermal exposure caused a reduction in the perlite interlamellar spacing and ferrite segregation in the form of large structures around perlite

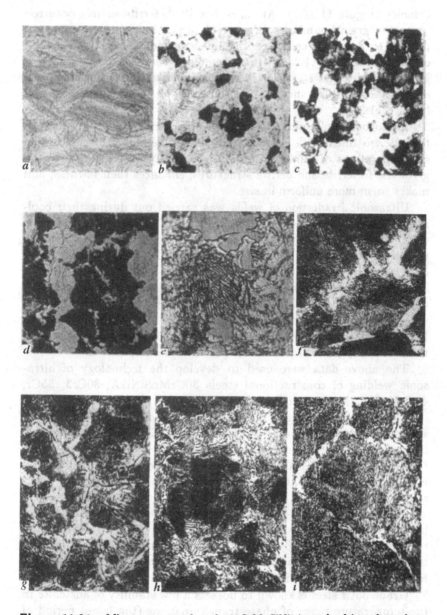

Figure 11.24. Microstructure of grade 30CrMnSiNi2A steel subjected to ultrasonic treatment at 650°C with strain amplitudes of (a, e) 0, (b) 6.3×10^{-4}, (c) 7.6×10^{-4}, (d) 1.2×10^{-3}, (f) 6×10^{-4}, (g) 7×10^{-4}, (h) 8×10^{-4}, (i) 10^{-3}. (a–d) ×500 magnification, (e–i) ×2000 magnification.

colonies (Figure 11.24,f). At $\varepsilon_m \sim 7 \times 10^{-4}$, ferrite mainly occurred along the grain boundaries (Figure 11.24,g). Raising the strain amplitude to 8×10^{-4} enhanced the fineness and uniformity of ferrite crystallites (Figure 11.24,h). In some areas, there was a segregation of globular perlite. At $\varepsilon_m \sim 1 \times 10^{-3}$, both ferrite and cementite showed a granular structure, indicating a tendency to perlite spheroidization (Figure 11.24,j).

Analogous results were obtained for other steels, suggesting that ultrasonic treatment with a strain amplitude exceeding a threshold value at temperatures of minimum austenite stability in the perlite region gives rise to ferrite and perlite structures, enhances their fineness, and makes them more uniform in size.

Ultrasonic irradiation of welds was carried out during their cooling at rates determined by the kinetics of austenite transformation. The mean rate of steel cooling in HAZ (HAZ was conventionally taken as an area heated to 650°C during welding) was 30 K/s.

The efficiency of ultrasonic irradiation was estimated from the metallographic data and mechanical properties of the welded joint.

Experiments showed that for the efficient ultrasonic treatment of HAZ, the strain amplitude should be 1.3 to 1.5 times greater than the threshold amplitude of isothermal austenite decomposition.

The above data were used to develop the technology of ultrasonic welding of constructional steels 30CrMnSiNi2A, 30Cr3, 35Cr, 40CrMnSiNi2Mo, and 10Cr3Mn2Mo.

Conventional welding of 30CrMnSiNi2A steel gave rise to martensitic structure with HV 530–540 in the HAZ. Exposure to ultrasound in the temperature range of austenization (1400–740°C), ferrite-perlite transformation (740–580°C), or bainite transformation (580–380°C) with cyclic strain amplitudes of $\sim 3.5 \times 10^{-4}$, 7.5×10^{-4}, and 10^{-3}, respectively, led to the formation of bainite structure with a noticeable (up to 20%) ferrite-perlite mixture content (Figure 11.25). Steel hardness in the HAZ decreased to HV 380–420 (Figure 11.26). In the temperature interval 740–380°C, the increase in the strain amplitude to 10^{-3} led to the formation in the HAZ of a fine-grained bainite plus ferrite-perlite mixture with HV 320–340.

Grade 35Cr steel is known to possess a low stability of austenite in the temperature range of phase transformations (the latent period of ferrite-perlite and bainite transformations during a welding thermodeformation cycle is 40–50 and 35–40 s, respectively). As a result, bainite structure with a certain content of martensite was found to be formed in the HAZ after conventional welding.

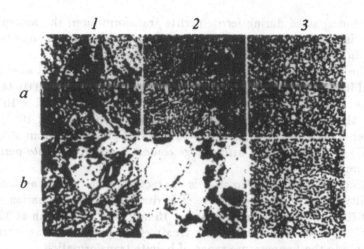

Figure 11.25. Structure of (a) control and (b) ultrasonically treated welded joints of grade 30CrMnSiNi2A steel (×500 magnification): (1) welded joint, (2) heat affected zone, (3) intact metal.

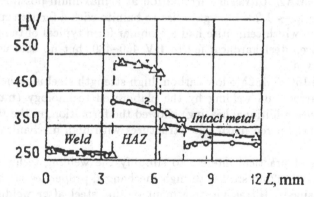

Figure 11.26. Hardness pattern in the HAZ of grade 30CrMnSiNi2A steel: (1) control, (2) ultrasonically treated specimen. HAZ, heat affected zone.

At the same time, ultrasonic irradiation in the temperature range of austenization, ferrite-perlite, and bainite transformations with strain amplitudes of 3.1×10^{-4}, 10^{-3}, and 10^{-3}, respectively, caused the formation of bainite structure with a large content (~40%) of ferrite-perlite mixture, in which ferrite segregates had a globular form. In this case, steel hardness diminished.

For ultrasonic treatment of grade 30Kh3 steel, the temperature ranges of austenization and bainite transformation were of particular importance, since during ferrite-perlite transformation, the austenite phase is stable for ~5 min; therefore, no essential effects may take place in the temperature range of austenization.

The cooling of the HAZ of conventionally welded 30Cr3 steel resulted in the formation of martensitic phase with some bainite content. Ultrasonic treatment at varying strain amplitude ($\varepsilon_m = 3.1 \times 10^{-4}$, when the HAZ was cooled from 1400 to 810°C; $\varepsilon_m = 10 \times 10^{-4}$ on further cooling to 580°C, and $\varepsilon_m = 12 \times 10^{-4}$ on cooling from 580 to 340°C) produced bainite with a large content (~30%) of ferrite-perlite mixture.

Grade 40CrMnSiNi2Mo steel is characterized by a high austenite stability in the temperature range of ferrite-perlite transformation. At 660–670°C, the austenite life is about 15 min, and only 3 min at 350–370°C. In view of this, ultrasonic treatment of this steel was carried out only in the temperature range of bainite transformation.

Martensite formed in the HAZ of the control sample of grade 40CrMnSiNi2A steel exhibited a considerable hardness (HV 600–620) and a high level of residual internal stresses, which led to cold cracking in the HAZ. Ultrasonic irradiation at a maximum possible strain amplitude $\varepsilon_m \sim 1.2 \times 10^{-3}$ gave rise to bainite with a ferrite-cementite mixture, in which cementite had a globular form typical of tempering. In this case, steel hardness fell to HV 460–480, but no cold cracking was observed.

Grade 10Cr3Mn2Mo low-carbon, high-strength steel is rather weldable; therefore, its welding by the elaborated technology (multi-run welding with additional heating) allowed the formation, in the HAZ, of bainite or tempered martensite structures with high mechanical properties.

It was of practical interest to simplify the welding technology of grade 10Cr3Mn2Mo steel with high mechanical properties of the weld being retained. Ultrasonic treatment of this steel after welding was carried out in the temperature range 550–200°C at constant strain amplitude $\varepsilon_m \sim 10^{-3}$.

The conventional cooling of welded 10Cr3Mn2Mo steel led to the formation of martensite (HV 340–360). At the same time, ultrasonic cooling gave rise to a fine-grained bainite structure with HV 270–310.

Structural changes induced by ultrasonic treatment modified the mechanical properties of steels. The results of mechanical testing of steel samples are summarized in Table 11.5.

Conventional welding of all the steels studied reduced their plasticity and raised their strength characteristics in the HAZ, the greatest reduction in plasticity being observed for grade 40CrMnSiNi2Mo steel.

Table 11.5 Mechanical properties of machinery steels subjected to ultrasonic welding.

Steel grade and conditions	Hardness, HV	Tensile strength σ_B, MPa	Yield strength $\sigma_{0.2}$, MPa	Relative elongation δ, %	U-notch charpy toughness, MJ m^{-2} 20° C	U-notch charpy toughness, MJ m^{-2} -40° C	Delayed fracture stress σ_i, MPa
30KhGSN2A, initial state	320–310	830–840	680	13	1.4	–	–
CW*	480–540	1800	1650	5	0.3–0.5	0.2	220
UW**	370–420	1200	910–950	12	0.9	0.4	320
30Kh3, initial state	400–420	870–910	720–780	14	0.9–1.0	–	–
CW	440–500	1200	980–1070	5	0.3–0.4	–	200
UW	350–380	810–830	690	11	0.8	–	300
35Kh, initial state	250–260	870	720	10	0.9	0.3	–
CW	470–530	1480	1250	6	0.3	0.1	180
UW	310–340	1060	910	9	1.2–1.4	0.5	340
40KhGSN2M, initial state	350	890–980	790	7.2	0.7–0.8	–	–
CW	600–620	1150	c.c.***	c.c.	c.c.	–	220
UW	480–640	1100	970	6	0.5	–	290
10Kh3G2M, initial state	260	730	580	18	2.2	0.3	–
CW	360	1300	1150	7	0.2–0.4	0.1	370
UW	310	1000	820	12	1.9	0.3	520

Footnote: * conventional welding; ** welding with a subsequent ultrasonic treatment; *** cold cracking.

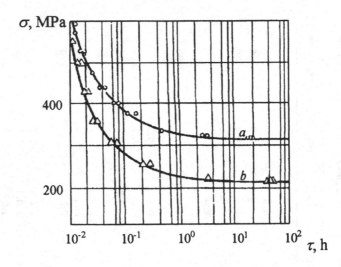

Figure 11.27. Delayed fracture testing of (*1*) ultrasonically treated and (*2*) control weld joints of grade 30CrMnSiNi2A steel.

Ultrasonic treatment somewhat diminished the strength, but considerably improved plasticity characteristics to levels typical of high-temperature tempered steels.

Delayed fracture testing of weld joints revealed a significant difference between control and ultrasonically treated welds. For grade 30CrMnSiNi2A steel, the delayed fracture stress of ultrasonically treated weld was 1.4 times greater than that for control samples (Figure 11.27). Analogous results were obtained for other steels.

Of interest are the results of fractographic analysis of fractured samples. For illustration, Figure 11.28 shows the photographs of the fracture of grade 35Cr steel specimens that failed at room temperature in U-notch Charpy test. The fracture of control sample has a streaming pattern typical of transcrystallite brittle rupture: distinguishable are steps between crack sections occurring on parallel fracture planes within a grain. In fractograms, one can see only some sections of tough fracture appearing as drawn crests along the fracture facet boundaries. Such a fracture pattern is associated with a tendency of metal to crack in the HAZ. At the same time, the fracture of ultrasonically treated sample is tough, cupped, and rather homogeneous. Such structure is highly resistant to cracking.

Fractographic analysis of samples with a very low impact toughness (< 0.1 MJ/m^2) indicated the emergence on the fracture surface

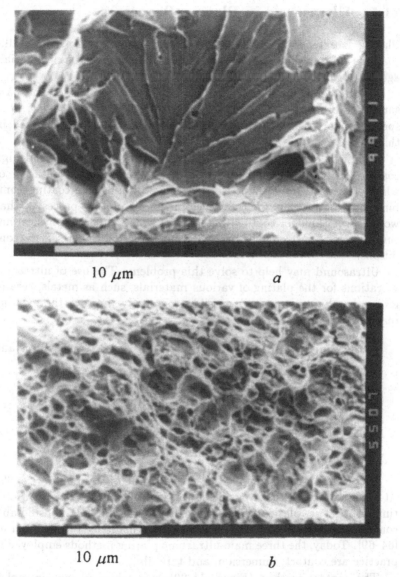

Figure 11.28. Fracture of (1) control and (2) ultrasonically treated 35Cr steel welded joints.

of cold crack nuclei, developing in planes perpendicular to the fracture surface.

Thus, experiments described evidence for the efficiency of ultrasonic treatment in minimizing the risk of cold cracking and in improving the quality of joint welds of medium-carbon, alloyed constructional steels.

11.5   Ultrasound in Plating Processes

Plating is extensively used in present-day manufacturing. Metallic coatings are multipurpose; in particular, they protect base materials against corrosion and wear.

Among coating methods, the hot dip process, in which the base surface is brought into a contact with the plating metal melt, is of special importance [61]. It is this process that will be considered in this section.

Although offering certain advantages, such as simplicity and high coating quality, this method suffers a number of disadvantages, of which the most grave is the limited range of substrate-coating combinations. This calls for the development of plating processes that would be suitable for coating materials of widely differing physical nature to ensure a strong substrate-coating adhesion under service conditions.

Ultrasound may help to solve this problem. The use of ultrasonic vibrations for the plating of various materials, such as metals, ceramics, and polymers makes it possible to greatly improve the existing manufacturing processes:

- ultrasonic plating can be carried out at lower temperatures and higher rates;

- there will be no need to employ expensive vacuum technology;

- the preparation of substrate surface may be simplified;

- the substrate-coating adhesion can be superior strong.

Ultrasound was first applied to the soldering of aluminum wire [62]. At the same time, Zakatov [63] published the results on ultrasonic tinning, that is, plating with a tin-containing solder. A considerable contribution to ultrasonic metal plating was made by Pugachev *et al.* [64–69]. Today, the three main ultrasonic plating methods employed in practice are contact, immersion, and thin film [69].

The contact method (Figure 11.29,*a*), in which ultrasonic tool is brought into a direct contact with a coated surface, is usually used for soldering. Ultrasonic solderers are generally of hand type where a waveguide heating rod is fixed to a gun handle to simplify operation. An opening is made in the waveguide tip to feed a melt to the process zone. Various types of ultrasonic solderers are presently available [68].

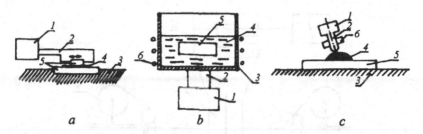

Figure 11.29. Some types of ultrasonic metallization: (a) with a contact, (b) with an immersion, (c) in thin layer. (1) Transducer, (2) waveguide, (3) anvil or tank, (4) melt or solder, (5) plated part, (6) heater.

In the immersion-type method (Figure 11.29,b), the processed material is immersed in a melt, the spacing between radiator and coated surface, as well as the tank size, being comparable with the ultrasonic wavelength in the melt.

In the thin-film process, a gap between the radiator and the processed surface is smaller than the longitudinal wavelength in the melt (Figure 11.29,c).

Comparison of process efficiencies and the quality of resultant coatings clearly shows that the thin-film method offers some advantages over other methods:

– the efficiency of wave energy conversion into the energy of acoustic streaming is higher;

– overheats, vibrational velocity gradients, and accelerations are greater;

– acoustic energy density in the process zone is higher;

– melt loss is lower.

The thin-film method was successfully used for coating Al, Ti, Cu, steels, glasses, ceramics, and polymers. The following discussion deals exclusively with this method as holding the greatest promise.

Pugachev and Semenova [68] described a semiautomatic installation designed for coating surfaces less than 500 mm in size. The installation incorporates a 4-kW ultrasonic generator operated at 22 kHz, portable acoustic system, and a control panel. During the plating, the ultrasonic system scans over the workpiece surface in two mutually perpendicular directions, which can be done either manually or automatically at

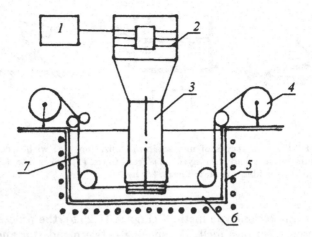

Figure 11.30. Schematic representation of an installation for continuous ultrasonic metallization [70]: (*1*) ultrasonic generator, (*2*) magnetostrictive transducer, (*3*) ultrasonic stack, (*4*) bobbins, (*5*) heated crucible, (*6*) melt, (*7*) band.

velocities in the range 1–50 $mm\,s^{-1}$. The vibratory system automatically operates at a resonant frequency. The process temperature can be varied from 20 to 500°C.

Dumitrash *et al.* [70] described an ultrasonic thin-film plating system operated in a continuous mode (Figure 11.30). The magnetostrictive transducer was excited by a 4-kW oscillator. The band speed with respect to the radiator could be varied from 4 to 40 $mm\,s^{-1}$.

These plating systems were employed to coat a diversity of materials, such as grade AMg6 aluminum alloy, grade VTi-5-1 titanium alloy, NbZr-1.5 niobium alloy, various steels, grade PTZB-3, PZT-19, and PZT-23 piezoceramics, grade MSN, SAM, UPS, and STAN polymers. The coating materials were Sn, In, Sn-Zn, and Cd–Bi–Pb alloys.

Some of the process parameters were optimized to produce coatings of desired thickness, uniformity, and strength. These are

- process time;

- velocity of substrate motion;

- melt temperature;

- vibrational amplitude.

Let us consider, as an example, the results of Dumitrash *et al.* [70], who optimized the ultrasonic plating of a niobium band and aluminum bar.

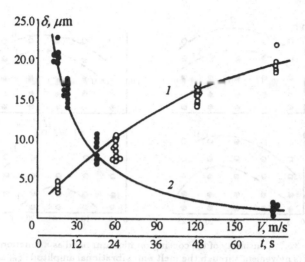

Figure 11.31. Dependence of the tin coating thickness on (*1*) band speed and (*2*) time of its occurrence in the melt.

Following cleaning, Nb band was passed through a tin melt. The band speed was varied from 4 to 40 mms^{-1}; the melt temperature was between 250 and 400°C.

Experiments revealed that the maximum coating thickness could be attained at a band speed somewhat above 30 mms^{-1} (Figure 11.31). Below a certain critical displacement amplitude, corresponding to the appearance of cavitation in the melt, no plating took place. The critical amplitude was not less than 4–5 μm for all coating materials studied (In, Sn, Sn–Zn, Bi–Cd–Pb).

By varying melt temperatures between 250 and 400°C, it was shown that the optimum temperature for producing a firm and uniform coating is 250°C (the region below curve *1* in Figure 11.32,*a*). Raising melt temperature narrowed the range of process parameters, at which high-quality coatings are obtainable, so that at 400°C, high-quality tin coatings could be produced only at band speeds lower than 4 mms^{-1} in a limited displacement amplitude range.

In the transition region between curves *1* and *2*, the coatings had discontinuities and sometimes could even peel from the substrate. Similar results were obtained in experiments involving indium plating of aluminum (the optimum process temperature was found to be about 200°C).

Prokhorenko *et al.* [69] conducted ultrasonic plating of Al and Ti alloys, piezoceramics, and polymers. They recommended that the eutec-

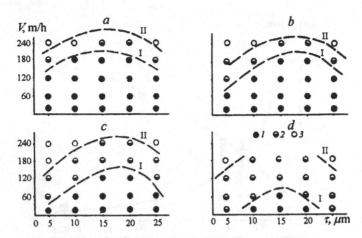

Figure 11.32. Continuity of tin coating on niobium band as a function of the velocity v of its movement through the melt and vibrational amplitude ξ_m at different temperatures: (a) 250, (b) 300, (c) 400, and (d) 450°C. Designations: *1*, continuous, tightly adhered coatings; *2*, defective, discontinuous coatings; *3*, no coating.

tic Sn–Zn alloy, exhibiting high ductility, strength, and corrosion resistance, should be used for the plating of Ti and Al alloys, or grade PZT and TB piezoceramic materials. It should be noted that the molten eutectic Sn–Zn alloy possessed low viscosity and surface tension and had wave impedance close to that of plated materials.

In addition to the above general requirements, materials for coating of polymers should satisfy the requirements imposed by the softening temperature of a polymer to be plated. For instance, styrene copolymers with a softening temperature of 100°C can be coated with a Bi-Cd–Pb alloy. Some useful information as to the conditions of ultrasonic thin-film plating is given in Table 11.6.

The efficiency of ultrasonic plating was expressed as adhesion strength of coatings. The process parameters were varied in the following ranges:

Ultrasound intensity I	20–30 W cm^{-2}
Displacement amplitude ξ_m*	2–6 μm
Substrate speed v	1–5 mm s^{-1}
Tool-substrate gap δ	0.1–0.3 mm
Tool-substrate angle α	20–40°

* This parameter unambiguously depends on the intensity of ultrasound.

Table 11.6 Some parameters of ultrasonic thin-film plating of various materials [69].

Substrate	Coating material	Substrate velocity $v \times 10^3$, m/s	Amplitude of vibrations ξ_m, μm	Coating thickness Δ, μm	Adhesion strength σ, MPa
Alloys:					
AMg6	Sn–Zn	1.5	5.65	15	94
VTi5-1	”–”	1.5	5.20	20	37*
”–”	”–”	”–”	”–”	”–”	32**
”–”	”–”	”–”	”–”	”–”	11***
Piezoceramics:					
TB	Sn–Zn	2.0	5.20	30	22
TBK-3	”–”	2.0	5.20	30	20
MBS-1	”–”	1.8	5.20	25	14
PZT-19	”–”	2.0	5.55	30	25
PZT-23	”–”	2.0	5.20	30	21
PZTB-3	”–”	1.8	5.55	25	20
Polymers:					
MSN	Bi–Cd–Pb	2.0	4.80	20	14
SAN	Bi–Cd–Pb	1.8	4.45	20	14
UPS	”–”	2.0	4.85	20	12
STAN	”–”	2.0	5.00	25	09

* Sand-water blasting and oxidation of the substrate surface.
** Sand-water blasting of the substrate surface.
*** Untreated substrate.

Since in some applications, plated materials are subject to cyclic stresses, they were tested for fatigue strength. The 10^7-cycle adhesion strengths σ_{-1} of the VTi5-1/Sn–Zn and AMg6/Sn–Zn joints were found to make up 5.5 and 11.2 MPa, respectively. The strength of the former joint could be additionally raised to 35 MPa by sand-water blasting of the substrate followed by its oxidation in air at 400°C for 1 h [69].

In 90% of cases, the fracture of piezoceramic-metal composites occurred in its piezoceramic component if it was of grades TB, TBK-3, PZT-19, and PZT-23, and in 70% of cases if it was of grade PZTB-3. At the same time, 55% of fracture events were observed in the joint if the substrate was of grade NBS-1.

The testing of ordinary commercially available silver-plated ceramics showed that the maximum tensile strength was 10.5 MPa, with the failure always occurring as the peeling of silver coating from ceramics.

The efficiency of metal plating of polymer substrates was tested by both thermal cycling and adhesion strength assessment. In the cycling test, each cycle involved 1-h exposure of a specimen to 80°C water followed by its cooling to −5°C. No failure was observed for ultrasonically coated materials after 10–12 testing cycles, whereas the products manufactured conventionally failed after five cycles.

The coating quality was also tested scleroscopically. To do this, a 0.5 × 0.5-mm mesh grid was applied to a plated sample before thermal cycling. In the case of UPS and STAN, two or three meshes peeled after eight to nine testing cycles, whereas the coating remained tightly adherent to NSP or SAN substrates. It should be noted that the quality of an electrolytic coating is considered to be satisfactory if it remains adhered after application of a 1.5 × 1.5-mm grid [69].

The strengths of metal coatings produced on polymers by electroplating technology ranged from 5.3 to 8.3 MPa [68].

In addition to the above coating–substrate combinations, later studies demonstrated the potentiality of ultrasonic plating of aluminum by Al–Zn alloys, and of polycrystalline superhard materials by Cu–Sn alloys containing Ti, Cr, Mn, Nb, and Ni additives, as well as by Al, Cu, and Ag [69, 71].

Equipment used for ultrasonic contact and immersion platings was described in detail by Prokhorenko *et al.* [69].

11.6   Fatigue Testing at Ultrasonic Frequencies

Most of the present-day equipment employs high-frequency vibrations of sonic and ultrasonic ranges; therefore, the problem of the strength of materials suffering high-frequency alternating loading is of much practical and scientific importance.

Like low-frequency deformation, high-frequency loading of materials is responsible for their fatigue, which is just the subject of this section.

To date, a considerable information has been accumulated as to the behavior of various materials loaded with high frequencies [72]. In particular, fatigue phenomena were investigated in carbon and alloyed steels, aluminum, nickel, titanium, molybdenum and other metals and their alloys, as well as in acoustically active materials, such as piezoceramics, magnetostrictive alloys, and ferrites. The studies were concerned with analysis of the influence of the loading frequency, temperature, structural parameters of tested samples, and the properties

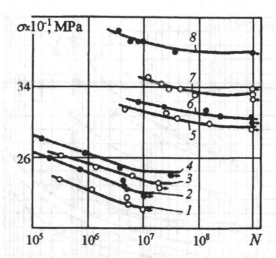

Figure 11.33. Fatigue strength curves of grade S10C low-carbon steel (0.08% C) obtained in the uniform dilatation-compression tests at various loading frequences (Hz): (*1*) 40, (*2*) 200, (*3*) 550, (*4*) 1800, (*5*) 13 000, (*6*) 22 000, (*7*) 50 000, (*8*) 100 000.

of test medium on the fatigue and durability characteristics of tested materials and the features of their cracking and failure.

Vast experimental data were obtained in studies of fatigue phenomena developing in Armco iron and carbon steels [72].

It has been established that fatigue strength and durability increase with the frequency of cyclic loading. For illustration, Figure 11.33 presents the results of fatigue testing of grade S10C low-carbon steel [73–75]. The trend of fatigue curves obtained upon high- and low-frequency loading is the same. These curves exhibit nearly horizontal sections that begin the earlier the higher is testing frequency. The slope of the curve section corresponding to a limited fatigue strength diminishes with increasing frequency of cyclic loading.

Similar results were obtained in tests employing various loading procedures – an asymmetrical cycle of uniform dilatation–compression or nonuniform bending dilatation–compression. Figure 11.34 shows the frequency dependence of cyclic strengths of iron and carbon steels. Curves *1–6* were obtained in nonuniform dilatation-compression tests, whereas curves *7* and *8* in uniform dilatation-compression tests.

Fatigue strength was estimated from 10^7–10^8 cycles. With increasing carbon content of steel subjected to the nonuniform dilatation–compression test, the fatigue strength difference grows with loading frequency. Thus, as loading frequency rose from 10^2 to 10^4 Hz, the

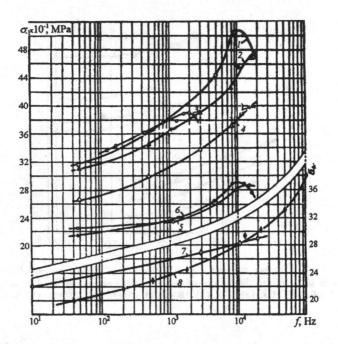

Figure 11.34. Frequency dependence of the cyclic strength of carbon steels with various carbon contents: (*1*) 0.85% C, (*2*) 0.35% C, (*3*) 0.42% C, (*4*) 0.2% C, (*5*) 0.1% C, (*6*) Armco iron, (*7*) 0.45% C, (*8*) 0.1% C.

fatigue strength of grade U8 steel increased by ~50% (from 330 to 500 MPa), and that of Armco iron by ~30% (from 220 to 290 MPa).

The fatigue strength of alloyed steels has also received a good deal of investigation. V. A. Kuz'menko *et al.* [72] gave a comprehensive review of the results of the testing of a wide range of chromium and chromium-nickel steels and showed that their fatigue strength increased with loading frequency, just as in the case of carbon steels.

The fatigue strengths of refractory nickel-based alloys were estimated at room and elevated temperatures at both low and high, loading frequencies. Fatigue strengths were found to increase with loading frequencies. Fatigue strength in air was lower than in a vacuum.

Fatigue testing of aluminum alloys was carried out mainly with grade D16T alloy (4.5% Cu, 0.5% Fe, 0.7% Mn, 0.4% Si, and 1.6% Mg) [72]. The loading frequency was varied from 16 to 20 000 Hz. The loading cycle was either symmetrical or asymmetrical; the mean cycle stress was either positive or negative. The results were analogous to those obtained for steels and nickel-based alloys.

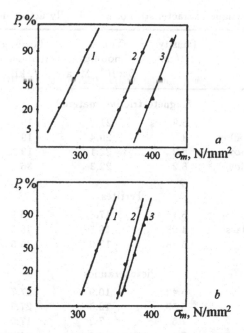

Figure 11.35. Fracture probability P of (1) molybdenum, (2) molybdenum with 0.8 mass % tungsten, and (3) grade TZM alloy as a function of the cyclic stress amplitude (5×10^7 cycles) at two different frequences (Hz): (a) 200 and (b) 20000 [80].

The fatigue testing of titanium alloys was conducted for a variety of materials with tensile strengths σ_B from ~690 MPa (lower-grade OT4-1 alloy) to ~1200 MPa (higher-grade VT22M alloy). Loading frequency was 16 or 20 000 Hz, temperature was varied from −196 to 20°C. The effects of frequency, loading procedure, and temperature on the fatigue strength of titanium alloys were found to be analogous to those observed for aluminum alloys.

B. Weiss and R. Stickler [76–79] undertook a detailed research into the fatigue failure of polycrystalline copper and arrived at a conclusion that high- and low-frequency loadings influence this material in a similar way.

The same authors carried out the fatigue testing of technical-grade polycrystalline molybdenum, molybdenum-based alloys with tungsten (0.8 and 1.5 mass %), and grade TZM molybdenum-based alloy (0.5% titanium, 0.06% zirconium). Loading frequencies were 200 and 20 000 Hz. The experimental data were processed to plot the probability of specimen failure versus the alternating stress amplitude (Figure 11.35). It was found that higher loading frequencies correspond

Table 11.7   Fatigue characteristics of acoustically active materials [72].

Material	Density ρ, g/cm^3	Young's modulus $E \times 10^4$, MPa	Loading frequency f, kHz	Fatigue strength σ_{-1}, MPa
	Magnetostrictive materials			
Nickel	8.9	21.8	19	100
Nickel-cobalt alloy	8.9	20.4	19	106
Iron-cobalt alloy	8.2	22.1	19	112
Iron-cobalt alloy	8.2	22.3	36	145
	Ferrites			
NF	5.14	17.8	36.5	35
NF + 1.5% glass	4.08	16.5	36.5	48
Vibrox	5.48	17.9	36.5	36
	Piezoceramics			
TB	5.43	10.9	27.7	25
TBK	5.53	12.3	27.5	31
PZT19	7.5	7.2	17.5	19
PZT23	7.6	8.2	24.7	23
PZTB3	7.2	7.2	27.0	17
PZTNB	7.3	6.3	18.0	14
PZTUN	7.6–7.9	6.2–8.5	17.8	10–28
PZTZNN	7.8	6.1–7.6	17.3	24–28
UTKSN	7.6–8.0	5–8	18.0	13–21
PZT8	7.55	10.8	28	38.5

to a greater peak alternating stress necessary for the specimen failure.

Ultrasonic loading frequencies were also employed for fatigue testing of tantalum, niobium, and some acoustically active (magnetostrictive, piezoceramic) materials.

V. A. Kuz'menko *et al.* [72] tested fatigue strengths of many acoustically active materials in both excited and nonexcited states under either uniform or non-uniform cyclic compression–dilatation loading. Tests were carried out over the frequency range 17–40 kHz. Magnetostrictive materials were tested as thin (0.1–0.4 mm) laminations, which enabled the strength estimation of their stacks with respect to the methods of lamination production and stacking. Table 11.7 summarizes fatigue strength values obtained for 5×10^8 cycles.

Thus, the reviewed experimental data on the fatigue strength of specimens at various loading frequencies indicate that fatigue strength monotonically increases with frequency. The type of loading does not influence the frequency dependence of fatigue strength.

11.7 Ultrasonic Machining

Current machine building deals with a wide range of novel structural materials possessing special physicochemical properties and updated methods of their machining.

Unusual properties of novel materials, complex shapes of workpieces, and rigid requirements imposed on their accuracy lead to a situation, when conventional methods of machining appear inefficient.

At present, new methods of machining are coming into use. Among them a particular place is occupied by ultrasonic machining methods proposed by Balamuth [89].

The three basic ultrasonic machining processes are

- machining of hard and brittle materials with a vibrated tool and free abrasive particles;

- final surface-hardening treatment;

- machining with vibrating tools.

Some characteristics of these processes will be considered below.

Ultrasonic machining of materials with abrasive particles lies in the feeding of abrasive slurry 3 into a gap between vibrated horn 2 and the workpiece surface (Figure 11.36). Pressing force F_p is applied either to the horn or to the workpiece. Abrasive particles subject to ultrasonic vibrations knock out tiny pieces of material from the workpiece surface, producing a depression which has a profile of the horn end.

Since abrasive particles gradually degrade (wear out); new portions of abrasive slurry should be fed into the process zone, which simultaneously assists in the removal of particles detached from the worked material.

For theoretical description of the process, it is necessary to relate the forces and stresses arising during ultrasonic machining to process parameters, such as the rate of hole growth, v_{ho}, and the process rate Q_V, or the volume of material, V, removed per unit time ($Q_V = V/\tau$).

The processes that develop when abrasive particles are pressed to the workpiece and tool surfaces have been considered by Kazantsev and

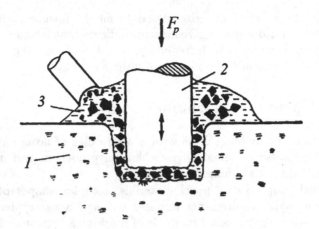

Figure 11.36. Schematic representation of ultrasonic abrasive machining: (*1*) processed material, (*2*) ultrasonic radiator, (*3*) abrasive slurry [52].

Markov [83–84], who established that the rate of hole growth $v_{ho} = Q_V/S$ (S is the area of hole) is given by

$$v_{ho} = Ac(\bar{\xi}) \left(\xi_m^2 F_p\right)^a f, \qquad (11.6)$$

where A, varying from 0.5 to 1, depends on the hardness of worked material and abrasive particles; $c(\bar{\xi})$ is the mean abrasive particle size function; f is the of frequency of vibrations. The relations are illustrated in Figure 11.37.

The physicochemical properties of the workpiece material can greatly affect the process rate: higher brittleness criterion Br corresponds to a better ultrasonic machinability of the material. According to their ultrasonic machinability, all materials can be divided into three categories:

(1) Br > 2. Materials of this group, such as glass, quartz, diamond, and ceramics, have the best ultrasonic machinability.

(2) 1 < Br < 2. This intermediate group involves quenched steels and hard metals.

(3) Br < 1. Ultrasonic machining of this group of materials (annealed steels, copper, lead, etc.) is ineffective.

Let us consider in more detail the relationship between the rate of ultrasonic machining, hole depth, and the contribution from the fresh abrasive slurry feeding.

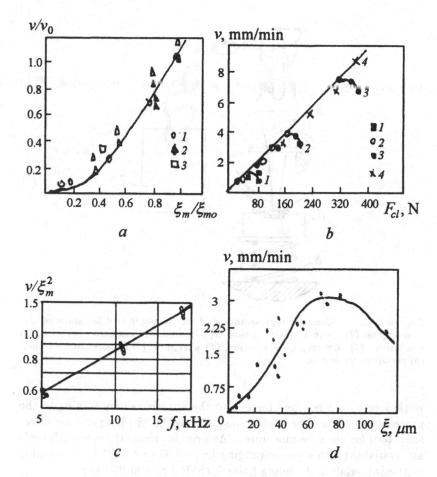

Figure 11.37. Dependence of the machining rate on (*a*) vibration amplitude, (*b*) pressing force at various rates of abrasive slurry feeding [83], (*c*) vibration frequency [85], and (*d*) mean abrasive particle size [83]. Panel a: (*1*) data from [85], (*2*) data from [86], (*3*) data from [87]. Panel b: (*1*) unforced abrasive slurry change, (*2*) suction at 0.1 MPa, (*3*) suction at 0.2 MPa, (*4*) suction at 0.3 MPa.

The rate of ultrasonic machining with abrasive slurry drops with increasing hole depth by the exponential law

$$v_h = v_{ho} \exp(-\alpha h), \qquad (11.7)$$

where α is the parameter that depends on pressing force and the method of abrasive slurry feeding.

For ultrasonic machining to be efficient, vibrations should be of a sufficient amplitude and frequency, the horn should be pressed to the

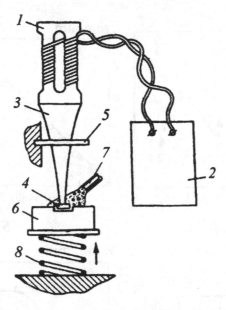

Figure 11.38. Schematic representation of an ultrasonic tool for abrasive machining [88]: (*1*) electroacoustic transducer, (*2*) ultrasonic generator, (*3*) horn, (*4*) instrument, (*5*) vibratory system fixture, (*6*) workpiece, (*7*) abrasive slurry feeder, (*8*) pressing arrangement.

workpiece with a required force, and the abrasive slurry feeding to the process zone should be continuous. Figure 11.38 illustrates an ultrasonic tool for abrasive machining. At present, many types of such tools are available (with a power ranging from 0.03 to 4 kW) for machining brittle materials and drilling holes 0.15–100 mm in diameter.

These machines may be either portable or stationary. The former are small and their ultrasonic assemblies can be handled during operation. They are employed for drilling holes of low depth and diameter, engraving, and marking. The rated power requirement of these tools is 30–50 W.

Stationary ultrasonic machines with a vertical arrangement of their vibratory system are most widespread. They may be low-power (0.03–0.2 kW), medium-power (0.25–1.6 kW), or high-power (1.6–4.0 kW). Besides, they may be designed as multipurpose or specialized tools.

Apart from ultrasonic machining with an abrasive grit, there is an alternative method, in which ultrasonic vibrations are fed directly to a cutting tool. The effect of ultrasound on machining depends on vibration amplitude, velocity, frequency, workpiece peripheral speed,

Table 11.8   Effect of ultrasound on hardness and specific deformation work upon a ball indentation [88].

Material	Vibrational frequency f, kHz	Vibrational amplitude ξ_m, μm	Indenter penetration depth h, mm	HB hardness, kg/mm^2	Specific deformation work A_{sp}, J/mm^3
D16 aluminum	–	0	0.2	14.8	13.8
alloy	7.9	4.5	0.2	2.4	2.9
	–	0	0.4	10.5	10.5
	7.9	4.5	0.4	5.4	4.9
	–	0	0.8	15.6	15.6
	7.9	4.5	0.8	5.3	5.2
Grade M1 copper	–	0	0.4	76.5	66.3
	7.9	4	0.4	31.9	29.6
	–	0	0.1	76.6	77.5
	7.9	5	0.1	35.0	34.8
Grade 45 steel	–	0	0.2	133.6	142.5
	7.9	3	0.2	101.0	66.6
	–	0	0.4	156.0	148.7
	7.9	3	0.4	115.0	81.5
VTi5 titanium	–	0	0.2	127.5	128.5
alloy	7.9	4.5	0.2	55.8	74.0
	–	0	0.05	181.0	162.5
	20.7	15	0.05	67.8	67.9
CrNi55WMoTiCoAl	–	0	0.15	298.0	331
nickel alloy	21.0	5	0.15	77.5	68

thickness of material to be cut, as well on physical properties of the workpiece and tool materials.

Ultrasonic vibrations imposed on the tool influence the speed and direction of effective cutting, the level and, sometimes, the sense of internal stresses. It may also modify friction conditions at the worked surface, and the efficiency of lubricating coolants. All these effects may reduce cutting forces and improve the surface quality and dimensional accuracy.

Experiments show that ultrasonic vibrations reduce both material hardness and specific deformation energy (Table 11.8). The actual improvement depends on indenter penetration and vibrational amplitude. Theory predicted a drastic reduction in the friction coefficient during

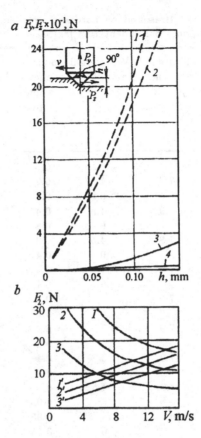

Figure 11.39. Dependence of forces $(1, 3)$ F_y and $(2, 4)$ F_z on (a) scratch depth and (b) velocity v of grade VTi5 titanium alloy tooling at different depths h (μm): $(1, 1')$ 20, $(2, 2')$ 15, $(3, 3')$ 10 [90]. Figures with apostrophes refer to experiments with ultrasonic treatment ($f = 20$ kHz, $\xi_m = 11$ μm). Panel a: $(1, 2)$ control experiments, $(3, 4)$ experiments with imposed ultrasonic vibrations.

ultrasonic indentation as compared to respective static process [88]. In addition to pressing the tool to the workpiece, machining also involves their relative motion.

Using an indenter cone, Markov [90] investigated the effect of ultrasound on the scratching of various materials and found that ultrasonic excitation of the cone along its axis reduced the degradation resistance of all the materials studied, viz. Pb, Cu, and grade VTi5 alloy.

Figure 11.39 shows variations in forces F_z and F_y caused by the penetration of grade W18 steel cone (the vertex angle 90°) into VTi5 alloy to depth h. Static testing was carried out at a maximum speed of

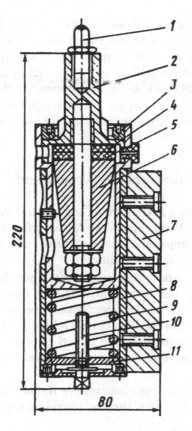

Figure 11.40. Cross-sectional view of a UZP-2 ultrasonic head for diamond smoothing [88]: (*1*) diamond tip, (*2*) radiator, (*3*) removable body, (*4*, *5*) piezoceramic transducer plates, (*6*) sleeve, (*7*) fixture, (*8*) stationary body, (*9*) spring, (*10*) screw, (*11*) nut.

10 mm min^{-1}; in ultrasonic tests, 20-kHz vibrations with an amplitude of 15–20-μm were used. The scratches produced in two test series differed in width, depth, and profile.

At scratching velocity $v > \omega \xi_m$, the efficiency of ultrasonic vibrations decreased. With increasing velocity v, forces F_y and F_z either grew (ultrasonic process) or diminished (conventional process) (Figure 11.39,*b*). The superimposing of ultrasonic vibrations on a cutting tool can be effective in threading, drilling, hole enlarging, hole trueing, reaming of holes of a small diameter, turning and gear shaping of ductile metals, grinding, honing, superfinishing, lapping, and other machining processes.

Special machines and ultrasonic systems have been elaborated for the ultrasonic cutting of brittle materials with the use of rotary diamond tools. The relevant acoustic systems, employing either magnetostrictive or piezoceramic transducers, can be mounted on conventional tools such as drills, mills, and borers. It should be noted that piezoceramic transducers require no water cooling.

Various ultrasonic heads are available for ultrasonic burnishing (Figure 11.40).

11.8   Ultrasonic Surface Treatment

Surface deformation is a widespread and effective method for hardening metallic materials. With this method, the surface layer of a material is a subject of high compressive stresses, which results in a better product strength, durability, and reliability.

By now, many surface deformation techniques have been elaborated and involve rolling, ball treatment, and shot blasting. Ultrasonic surface hardening, as one of the recent techniques in this series, was pioneered in the USSR in the late 1950s and early 1960s [91–93]. At that time, Soviet researchers evaluated the effect of ultrasonic surface hardening on residual welding stresses and proposed a hardening-finishing process.

Applied ultrasonic vibrations relieve residual stresses and improve surface finish and hardness, thereby providing a better wear resistance of products.

There are three major methods for ultrasonic surface deformation that differ with respect to working tools and bodies (Figure 11.41):

- single-tool process,

- multi-tool process,

- free-body process.

Ultrasonic single-tool process is similar to conventional ball deformation method, in which a ball (or a roll) is pressed with force F_N to the surface of a workpiece moving specifically with respect to the tool. The only difference between ultrasonic and conventional processes is that the tool is vibrated at a frequency and amplitude that are determined by operation conditions and the type of the electroacoustic transducer used. Tool *1* connected to horn *3* moves relative to the workpiece surface during a deformation cycle.

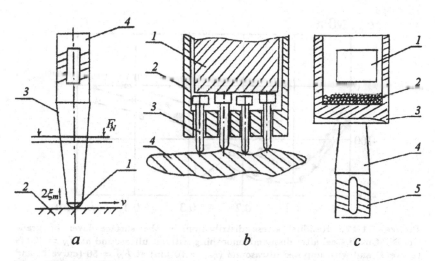

Figure 11.41. Schematic representation of ultrasonic surface treatment methods (*a*) with a single vibrating instrument [92], (*b*) with a multi-tool instrument [92], and (*c*) with free bodies [94]. Panel a: (*1*) tip, (*2*) workpiece, (*3*) ultrasonic horn, (*4*) electroacoustic transducer. Panel b: (*1*) horn, (*2*) cartridge, (*3*) tools, (*4*) workpiece. Panel c: (*1*) workpiece, (*2*) balls, (*3*) vibrated chamber, (*4*) ultrasonic horn, (*5*) electroacoustic transducer.

A multitool arrangement, which is generally similar to those used in conventional embossing, is illustrated in Figure 11.41,*b*. Tools *3*, possessing one degree of freedom along the system axis, are placed in cartridge *2* between workpiece *4* and horn *1*. The horn is pressed to the workpiece with a constant force. The tooling travels with respect to the workpiece during operation.

In many applications, it is convenient to use free bodies rather than fixed tools (Figure 11.41,*c*). Bodies may represent, for instance, ultrasonically excited balls. Workpiece *1* is placed in vibratory chamber *3* which contains steel balls *2* as working bodies. The vibratory chamber is connected to horn *4* and electroacoustic transducer *5*.

Ultrasonic treatment results in a drastic increase in dislocation density, crystal fragmentation, and variation in the stress–strain state of material's surface. All these effects depend on process conditions, including acoustic parameters, the properties and geometry of material, and stress pattern at its surface.

Ultrasonic hardening was found to raise the mean dislocation density in metals from 10^6–10^8 to 3×10^{11} cm^{-2}. To compare, surface rolling could augment mean dislocation density only to 6×10^{10} cm^{-2} [98].

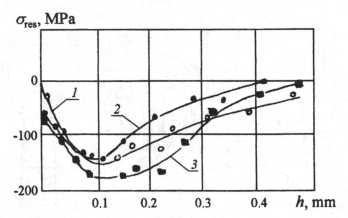

Figure 11.42. Residual stress distribution in the surface layer of grade Cr15Ni5Cu2Ti steel after diamond smoothing without ultrasound at $F_N = 400$ N (curve *1*) and with imposed ultrasound ($\xi_m = 10$ μm) at $F_N = 50$ (curve *2*) and 100 N (curve *3*) [96].

Ultrasonically induced structural changes modify the mechanical properties of product's surface and its bulk to the extent that the product performance becomes dependent on surface state.

Markov [95] and Mukhanov *et al.* [96, 97] investigated residual stresses in materials subjected to conventional and ultrasonic surface hardening. During diamond burnishing, residual compressive stresses were found to rise and their peak values to shift toward the material bulk with increasing pressing force F_N. Once a critical pressing force is achieved, the level and distribution pattern of residual stresses would not change in the surface layer. For grade Cr15Ni5Cu2Ti steel, this critical pressing force is 350–400 N (Figure 11.42). However, if burnishing process is intensified by ultrasound, the level and distribution pattern of residual stresses will behave differently: the stress level will be high even at a lower pressing force, whereas the stress–depth dependence will no longer be strong. Similar results were obtained in the studies of ultrasonic multitool process [101].

Ultrasonic hardening modifies the distribution of residual stresses in weldments as well. It was demonstrated [91] that the sag of butt-welded plates from grade St 3, 20CrMnSiNi, or Cr18Ni9Ti steels diminishes 1.5 to 2 times as a result of ultrasonic hardening with a pressing force of 50 N, displacement amplitude of 25 to 30 μm, frequency of 20 kHz, and a process speed of 30 cm min^{-1}. This finding suggests a variation in the distribution of stresses in the sample. More detailed analysis showed

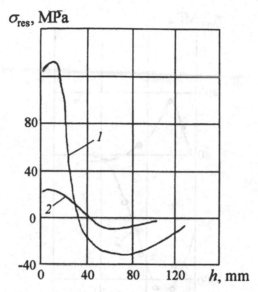

Figure 11.43. Depth distribution of residual stresses in a built-up grade 1915 aluminum alloy before (curve *1*) and after (curve *2*) ultrasonic treatment [100].

that ultrasonic hardening drastically reduces the level of tensile stresses in weldments.

Other researchers [99–101] obtained similar results in their studies of ultrasonically hardened grade St 3 mild steel, grade VAD-1 and 1915 aluminum alloys, and TiSi-5 alloy. As an example, Figure 11.43 shows the distribution of residual stresses in aluminum alloy weldments before and after ultrasonic hardening.

In free-body process, stresses are compressive. Figure 11.44 illustrates the distribution of residual stresses in grade 35CrNi2MoV steel that was subjected to grinding followed by ultrasonic hardening. The treatment resulted in a 100-μm hardened case with compressive stresses amounting to 70% of the yield strength. Experiments with grade 14Cr17Ni2 steel and grade VT8 titanium alloy gave similar results.

Ultrasonic deformation acted to improve material hardness, fatigue strength, and corrosion-fatigue resistance [103, 104], the extent of improvement being largely determined by process conditions. In diamond burnishing, the hardness grew to a particular level with incre- asing pressing force to be then essentially constant (Figure 11.45). In ultrasonically enhanced process, the same effect was achieved at considerably lower pressing forces. The optimum pressing force was found to

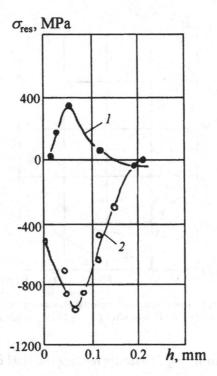

Figure 11.44. Depth distribution of residual stresses in polished grade 35CrNi2MoWA-Sh steel before (curve *1*) and after (curve *2*) ultrasonic treatment [102].

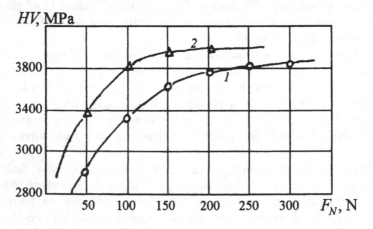

Figure 11.45. Microhardness *HV* of diamond-smoothed grade 12Cr18Ni9Ti steel as a function of pressing force F_N before (curve *1*) and after (curve *2*) ultrasonic treatment [105].

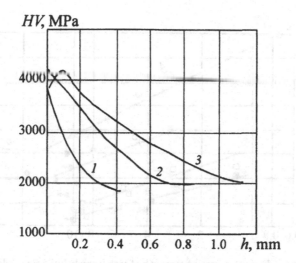

Figure 11.46. Depth distribution of microhardness HV in grade 12Cr18Ni9Ti steel diamond-smoothed at the displacement amplitude $\xi_m = 0$ (curve *1*), 15 (curve *2*), and 20 μm (curve *3*) [105].

increase with the strength of material. Thus, the optimum pressing force was 100 and 150 N for grade 2Cr18Ni9Ti austenitic steel and a stronger martensitic steel Cr15Ni5Cu2Ti, respectively.

The microhardness of materials diminished with depth, whereas the thickness of hardened layer increased with vibration amplitude and pressing force (Figure 11.46).

Ultrasonic plastic deformation and compressive loading of surface layers improved both the fatigue and fatigue-corrosion resistance of treated materials [103–106]. Figure 11.47 shows the effects of ultrasonic hardening on the corrosion-fatigue strength of steel samples that had transverse weld ridges on their surface. When welded samples were ultrasonically hardened on both welded and opposite sides, the fatigue strength improved from 45 to 120 MPa, which corresponded to more than a 10-fold increase in the fatigue strength of the joint at a constant level of cyclic stress.

Similar results were obtained in fatigue tests of a turbo-compressor blade hardened by ultrasonic free-body process [107, 108]. The efficiency of ultrasonic hardening increased with the process temperature. For instance, fatigue strength increased 1.3 times at room temperature and as much as 1.9 times at 600°C [106].

As mentioned above, ultrasonic hardening produced a regular pattern of microdepressions at the workpiece surface, which improved the

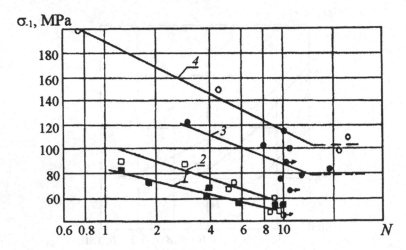

Figure 11.47. Fatigue-corrosion strength of transversely built-up grade AK8 steel [103]: (*1*) control specimens, (*2–4*) specimens ultrasonically treated from the side opposite to building-up (curve *2*), from the side of building-up (curve *3*), and from either side (curve *4*).

lubricant-holding conditions at the workpiece surface and, therefore, enhanced its wear resistance [95, 109].

11.9   Application of Ultrasound in Electrochemistry

Ultrasound is well suited to industrial applications associated with electrochemical coatings and improvement of their quality [50].

The role of ultrasound in electrodeposition of metals and alloys is to reduce hydrogen polarization and facilitate ion discharge, which is expected to improve cathode current density, the rate of metal deposition, and the firmness and structural properties of coatings. Once this is achieved, the porosity of coatings decreases, while their adhesion to the substrate rises. Sometimes, lustrous coats may be obtained without using glazing agents.

Ultrasound intensifies the stirring of an electrolyte in a galvanic bath due to acoustic streaming, which reduce the thickness of a near-cathode layer depleted in deposited metal ions.

As a rule, the use of ultrasonic vibrations does not require a substantial rearrangement of available galvanic baths. Ultrasound is commonly fed to the region near the surface of the parts to be coated.

A typical device for ultrasonic coating represents a tank with electrodes, power supply unit, and control equipment. The working volume of such tanks varies from 5 to 200 l, the radiating area from 150 to 1400 cm^2, the input from 0.3 to 4 kW, the operating frequency from 15 to 20 kHz.

The efficiency of ultrasound application depends on its intensity. Experiments showed that vibrations with an intensity of 0.3–0.5 W/cm^2 are most suitable, whereas higher intensities may reduce the quality of coatings by causing their erosion or discontinuity.

The present-day ultrasonic technology deals with copper, nickel, chromium, cadmium, brass, zinc, silver, and gold plating. The employment of ultrasound for coppering makes it possible to raise 5–6 times the maximum permissible current densities in cyanide and sulfate electrolytes, and 8–8.5 times in pyrophosphate electrolytes, which will increase the deposition rate 10 to 15 times and allow the production of vacuum-tight, fine-grained, well-adhered coatings. The required corrosion resistance of pieces can be achieved at lesser thicknesses of coatings.

The application of ultrasound in nickel plating appreciably widens the range of permissible current densities, enhances coating yield with respect to current, and allows light and virtually nonporous coatings to be obtained in common electrolytes. The permissible coat thickness can achieve as high as 25 μm with a hardness of 4500 MPa. In some electrolytes, such as those containing (g/l) nickel sulfate (200), boric acid (30), sodium chloride (10), sodium fluoride (4), disulfophthalic acid (2), and formalin (2), lustrous coats can be obtained on unpolished surfaces.

Ultrasound also enhances the rate of chromium plating and improves coating hardness. In the case of cadmium plating, ultrasound allows high-quality coatings to be obtained at high current densities ($\sim$10–15 A/dm^2), which accelerates metal deposition 8–10 times without drastically changing the electrolyte composition and anode passivation, but with some decrease in the scattering power of electrolytes. Ultrasonic cadmium plating appears most efficient with an electrolyte containing cadmium oxide (30–35 g/l), sodium cyanide (85–100 g/l), and sodium hydroxide (30–40 g/l).

Ultrasonic brass plating produces lustrous coatings in cyanide electrolytes at a current density of 2 A/dm^2. In the case of increased electrolyte concentration, current density should be raised. To obtain coating of a specified composition at a given current density, it is necessary to experimentally determine the appropriate copper-to-zinc ratio in the electrolyte.

Ultrasonic zinc plating in an electrolyte containing (g/l) zinc oxide (40–60), sodium hydroxide (100–300), and sodium cyanide (100–130) enables a 10–15-fold increase in cathodic current density.

The scattering power of silvering electrolytes is little affected by ultrasound. The greatest increase in current density is observed in electrolytes with a high concentration of silver. When the concentration of silver is about 40 g/l, the maximum permissible current density is 10–12 A/dm^2.

The use of ultrasound for gold plating in cyanide electrolytes allows the current density to be increased 17- to 20-fold (up to 3.5–4 A/dm^2).

11.10   Ultrasonic Impregnation

Nowadays, windings are commonly impregnated in autoclaves in the so-called conditioning regime, when electric devices with windings are submerged in impregnating varnish to be successively exposed to atmospheric, excessive, and vacuum pressures for 3–5 min at each, which is expected to provide a good filling of winding voids with the varnish. The process is run in two or three cycles (30 to 45 min long) with a drying stage between each impregnation cycle. The number of cycles depends on winding construction and the requirements imposed. The drying stage takes from 10 to 35 h, depending on varnish properties and winding size. The total impregnation process may take some days.

Ultrasound can significantly intensify the process and ensure a better impregnation over a shorter time period. Sometimes, only one impregnation cycle without an interstage drying will suffice. The optimum impregnation is achieved at $\xi_m = 1.0 \div 30$ μm, which corresponds to an acoustic intensity of 0.5–1.0 W/cm^2 at a vibrational frequency of 18–20 kHz. The varnish viscosity should be as in the case of vacuum-autoclave impregnation (measured with a VZ-4 viscosimeter), the process temperature being maintained at 50–60°C. Ultrasonic impregnation allows any of the varnishes and compounds tolerant to heating above 30°C to be used, e.g., those of grades ML-92, KO-835, and BT-987, otherwise ultrasonic impregnation is not suited, as it will require too extensive cooling to maintain the assigned temperature.

The overall time of ultrasonic impregnation is from 5 to 15 min, depending on winding density. The entire technological cycle involves

(1) a preliminary stage (removal of dust, drying, etc.),

(2) placing of devices in a basket or fixing them to a conveyer to submerge then in impregnating varnish in an ultrasonic tank,

(3) ultrasonic impregnation for a specified time period,

(4) withdrawal of impregnated devices from the tank and exposure them to air for 30–60 min to allow excessive varnish to flow off,

(5) cleaning of winding ends and drying.

Drozhalova and Artamonov [110] compared the windings of transformers, chokes, and tentative-type transformer coils impregnated ultrasonically or by the vacuum-autoclave method. The efficiency of impregnation was assessed by several parameters, such as the amount of varnish penetrated into windings, insulating resistance between winding layers (under normal ambient conditions, at 105°C above zero, and under high-humidity conditions), dielectric strength, open-circuit current, and load current.

Transformer coils were made from PEV-2 enameled wire or PELShO wire (both 0.2 mm in diam.) with an interlayer insulation made of LKS fiber glass fabric, PETF film, and grade KT-0.5 paper. Each coil had 8 windings. Choke coils with a toroidal winding and transformer coils with different-size windows were wound with PEV-2 wire.

Coils were ultrasonically impregnated by insulating varnishes of grades BT-987 (oil-bitumen), ML-92 (a solution of alkyd varnish and melamine formaldehyde resin in butanol), FA-97 (a solution of castor alkyd and parabutyl phenolformaldehyde resin in xylene), and KO-835 (a resin mixture solution in xylene and toluene). The impregnation time was 10 min; the drying time was as usual. A control autoclave impregnation with BT-987 varnish was performed in the conditioning regime with two impregnation stages, each 30–45 min long.

Presented below are the results of the testing of transformer coils that were weighed before and after impregnation. The mass increment for ultrasonically impregnated coils amounted to 28.72 g or 5.75%, and to 26.2 g (5.17%) for the coils impregnated in autoclave. Thus, the mean mass increments of coils impregnated conventionally and ultrasonically were almost the same.

To estimate the depth of varnish penetration into windings, some coils were cut. Microscopic examination of ultrasonically impregnated coils showed that winding voids were fully filled with varnish, thus providing a good cementation of wires. The degree of winding saturation with varnish can be estimated from breakdown voltage. The respective tests were carried out after a preliminary drying of coils at $+110°C$ for 3 h. It was found that the dielectric strengths of interlayer and interwinding insulations in the coils impregnated ultrasonically and in autoclave were virtually equal, indicating that the saturation of windings with varnish during a single ultrasonic and double conventional impregnations was similar.

The interlayer and interwinding insulation resistances measured at room temperature and at $+105°C$ after the preliminary coil drying at $+105°C$ for 3 h are listed in Table 11.9. It is seen that at room temperature, the insulation resistances of ultrasonically impregnated coils are 2–10 times higher than those of conventionally impregnated coils, whereas at elevated temperatures the resistance gain is still greater (10–100). After storing in a wet chamber (95–98% humidity) at $40 \pm 2°C$, the insulation resistance falls in a similar way for coils impregnated ultrasonically or conventionally.

Choke coils impregnated with BT-987 varnish were tested by by thermal shocks (a total of 10 thermal shocks at -60 and $+105°C$). After such a treatment, no visible damage to coils was observed. The open-circuit and load currents were also normal. The transformers and chokes fabricated with the use of ultrasonically impregnated coils corresponded to standard performance specifications.

Comparative analysis of transformer coils impregnated with EPK-4 epoxy compound for 15 min either ultrasonically or by the vacuum-autoclave method showed that in both cases the amount of impregnated EPK-4 compound was similar. All the transformers tested corresponded to standard specifications. In other words, a single-stage ultrasonic impregnation makes it possible to obtain electric devices with parameters similar to or even better than those attainable by a double vacuum-autoclave conditioning impregnation. Ultrasound reduces the time of impregnation 3- to 4-fold, requires no interim drying stage, and substantially simplifies technological process.

Analogous results were obtained in the impregnation experiments with the rotors, stators, and armatures of small-size electric machines. In this case, ultrasonic impregnation allowed the process time to be shortened 4 to 6 times. A promising application of ultrasound is also the impregnation of glass fiber fillers, which enables more firm materials to be obtained over a shorter time period.

Table 11.9 Effect of ultrasound on the insulation resistance of varnished windings.

Method of impregnation	Test temperature, °C	Insulation resistance, MΩ						
		Between windings					Between layers	
		Grade KT-0.5 paper	Glass cloth	Two layers of PETF film plus a layer of KT-0.5 paper	Two layers of glass cloth plus a layer of PETF film	Two layers of glass cloth	Grade PEV wire	Grade PELShO wire
Vacuum-autoclave impregnation	20	9.45×10^4	3.31×10^4	6.78×10^5	2.03×10^5	6.89×10^4	1.05×10^5	1.99×10^4
Ultrasonic impregnation	20	1.55×10^6	1.35×10^6	1.82×10^6	5.98×10^5	2×10^5	1.65×10^6	8.44×10^4
Vacuum-autoclave impregnation	+105	1.65×10^2	3.21×10^2	1.88×10^4	1.05×10^4	1.97×10^3	2.38×10^5	5.11×10^2
Ultrasonic impregnation	+105	2.16×10^5	2.28×10^4	4.08×10^5	4.9×10^4	8.8×10^3	1.25×10^5	3.15×10^3

11.11   Ultrasonic Cleaning

Experiments show that ultrasonic cleaning is the most efficient method for the removal of surface contaminants.

To compare, the residual surface contamination after different cleaning procedures comprises 80% (rinsing), 55% (vibratory cleaning), 20% (manual cleaning), and 0.5% (ultrasonic cleaning).

Technological processes and equipment for ultrasonic cleaning began to come into an extensive industrial use in the late 1950s and early 1960s. By that time, the physical fundamentals of ultrasonic liquid technology had been mainly established [111, 112]. Those studies formed the basis for a more close insight into the processes developing in liquids under the effect of ultrasound and allowed researchers to specify basic requirements for vibrational sources and technological equipment, which were considered in devising the ultrasonic cleaning technology.

The relevant physical investigations have shown that ultrasonic cleaning is based on certain nonlinear effects developing in a liquid under the action of high-intensity ultrasound. First of all, these are cavitation and cavitational erosion of surface, although acoustic streamings, sonocapillary effect, and radiation pressure may also contribute to surface cleaning.

The main and most developed technological lead at that time was based on low-intensity ultrasonic irradiation of liquids with a power density not exceeding 2–3 W/cm^2 [113], i.e., appropriate for cavitation to develop in quite a large volume.

In the works that followed, attempts were made to improve the efficiency of ultrasonic cleaning. For example, it was proposed to impose excessive static pressure on a cavitating liquid to increase its erosive activity [111, 114], or to use special freon composites to enhance the solubility of surface films [115], or to raise the intensity of ultrasonic irradiation of liquid to 12–15 W/cm^2 [116, 117]. These ultrasonic intensities suggest the vibrational displacement amplitudes more than 12–15 μm, due to which the respective technological process has become known as high-amplitude cleaning.

Let us consider in more detail the physical principles of ultrasonic cleaning.

Presumable mechanisms of ultrasonic breakdown of grease films were suggested by Agranat *et al.* [118] (Figure 11.48)*. Breakdown of films may occur via their separation, emulsification, erosion, hy-

---

*The original scheme was slightly modified by the author of this book.

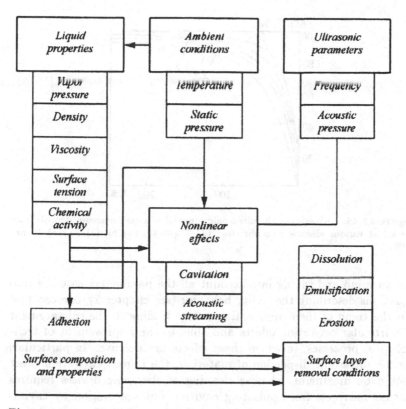

Figure 11.48.   Block-diagram of the mechanism of ultrasonic cleaning.

droabrasive decomposition, and dissolution. The arrows in the diagram show in which way each factor contributes to the breakdown of grease.

Some factors may directly influence cleaning process, while others do it through specific ultrasonic effects. Analysis of the diagram indicates that cavitation contributes to the removal of all types of contaminants. Investigations into the dynamics of a cavitation bubble show that cavitation can really improve technological processes, including cleaning. There are significant differences between the models describing a single bubble and an entire cavitation region, which lie in the effect of neighboring bubbles on the bubble considered, the presence of obstacles in an acoustic field, the possibility of nonspheric collapse with the formation of microjets, out-of-phase pulsations of cavities, etc. Nevertheless, there is a satisfactory qualitative agreement between the calculated dynamics of single bubble and cavitation region. With

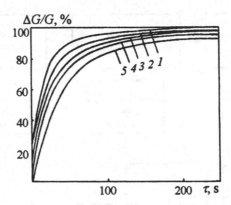

Figure 11.49.   Kinetics of the ultrasonic removal of a carbonaceous deposit from a steel at various vibration amplitudes: (*1*) 7, (*2*) 14, (*3*) 20, (*4*) 30, (*5*) 40 μm [119].

this in mind and taking into account all the parameters entering into equations describing the cavity behavior (see chapter 2), one can find out the trend of their preferential change leading to the enhancement of particular cavitation effects and thus to the improvement of technological processes, in which these effects are decisive. In particular, for the ultrasonic dispersion of materials, the impact cavitation effect should be maximum, whereas cleaning of electronic devices requires intense microjets from pulsating cavities, but not erosion by cavitation.

The kinetics of cleaning is described by expression [119]

$$\frac{\Delta G}{G} = 1 - \exp(-Qt^{0.5}), \tag{11.8}$$

where ΔG and G are the removed and initial amounts of contaminating material, respectively; Q is the energetic parameter of cleaning, depending on the intensity of vibrations and the extent to which cavitation is developed.

Figure 11.49 gives the kinetics of removal of a carbonaceous deposit from steel plates at various vibration amplitudes.

The selection of acoustic parameters for efficient cleaning lies in the creation of conditions for the maximum cavitation to develop. Such problems were considered in a general form in chapter 2. Here it seems expedient to appeal to some specific features of ultrasonic cleaning.

To understand how physicochemical properties of liquid may influence cleaning, one should first pay attention to such parameters as va-

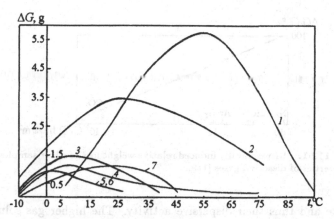

Figure 11.50. Temperature dependence of the weight loss of aluminum plates during their ultrasonic cleaning in various liquids: (*1*) water, (*2*) kerosene, (*3*) gasoline, (*4*) alcohol, (*5*) acetone, (*6*) carbon tetrachloride, and (*7*) trichloroethylene.

por pressure and surface tension. It has been already mentioned that with decreasing vapor pressure the erosive activity rises. In view of this, aqueous solutions must be more efficient in cleaning than organic solvents possessing a higher vapor pressure (Figure 11.50).

As for the surface tension of liquid, its effect may be dual. On the one hand, an increased surface tension impairs the wettability of parts to be cleaned, which prevents the penetration of the cleaning solution into small holes and other narrow spacings. This problem is usually surmounted by the addition of surfactants to cleaning liquids, which reduces their surface tension and thus improves the wettability of parts. In this case, very thin adlayers formed on grease films also contribute to their detachment from cleaned surfaces. On the other hand, with increasing surface tension the collapse of cavitation bubbles is more intense, although cavitation threshold increases as well.

With increasing viscosity, the erosive activity of liquid in weak ultrasonic fields falls, but increases in intense fields. As a result, the efficiency of high-intensity ultrasonic cleaning in viscous liquids rises. Such conditions can be recommended for the removal of contaminants tightly bound to the surface.

Small variations in density, typical of real liquids, leave the efficiency of cleaning almost unaffected.

There is a point to touch briefly the relationship between cleaning efficiency and the nature of gas dissolved in cleaning liquid. Penetrating into cavitation bubbles, the dissolved gas affects the rate of their

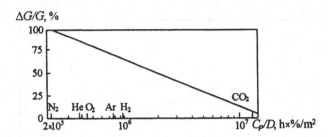

Figure 11.51. Ultrasonically induced relative weight loss of silver samples versus the properties of dissolved gases [119].

collapse and thus their dispersive activity. The higher gas solubility, the greater its amount in cavities. The rate of gas penetration into bubbles rises with the diffusion coefficient. Figure 11.51 plots the relative weight loss of samples versus the ratio of the gas solubility to its diffusion coefficient. It can be seen that smaller ratios correspond to higher dispersion rates.

As for the effect of ambient conditions on cleaning efficiency, of prime interest are here temperature and static pressure. Temperature influences erosion through a diversity of reasons. These may temperature-dependent changes in vapor pressure, surface tension, viscosity, liquid density, gas solubility, properties of cleaned material, as well as various thermodynamic phenomena determining the growth and collapse of bubbles.

Experiments showed the existence of an optimum temperature interval, over which the intensity of cavitation (and thus cleaning) is maximum for both aqueous cleaning solutions and organic solvents. Data on the erosion of a low-carbon steel in distilled water, buffer solution (pH 8), and 3% NaCl are shown in Figure 11.52, whereas the organic solvent erosion data are presented in Figure 11.50.

The insignificant erosive activity at the initial portions of curves is likely due to an enhanced solubility of gases (which inhibits the collapse of bubbles), as well as to a low chemical activity and high viscosity and surface tension of the cleaning solution.

The small degree of erosion at elevated temperatures is related to high vapor pressure so that near a boiling point the erosive activity falls to zero.

The role of static pressure in erosion was quite fully considered above. It should be noted that the formulated statements are valid not only for aqueous solutions, but for organic solvents as well (Figure 11.53).

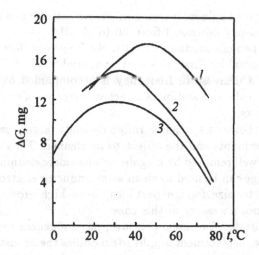

Figure 11.52. Temperature dependence of the mild steel erosion during ultrasonic cleaning in (1) 3% NaCl solution, (2) water at pH = 8, and (3) distilled water [120].

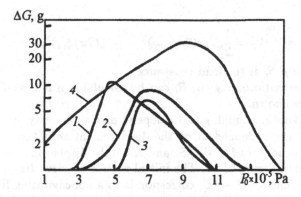

Figure 11.53. Weight loss of grade AK-6 alloy samples as a function of hydrostatic pressure in (1) carbon tetrachloride, (2) gasoline, (3) trichloroethylene, and (4) water.

The efficiency of ultrasonic cleaning must depend on vibration frequency and acoustic pressure amplitude. The interpretation of experimental results obtained in a wide frequency range presents difficulty, since it is not always possible to attain vibration amplitude sufficient to cause cavitation at any frequency. Nevertheless, the authors of works

[121, 122] observed a linear increase in erosive activity and strains when vibration frequency decreased from 30 to 10 kHz.

There are two circumstances that make frequency decrease to audible range undesirable. First, this may unfavorably affect the personnel. Second, such a decrease in frequency is accompanied by an undesirable increase in the size and mass of active elements of transducers and ultrasonic stacks.

A proper choice of frequency range depends largely on the kind of surface contaminants and the object to be cleaned. For instance, light grease can be well removed by megahertz ultrasonic cleaning. The same frequency range can be used to clean semiconductor electronic elements from micrometer-size foreign particles, since high erosive activity of cavitation is not necessary in this case.

The amplitude of acoustic pressure, p_m, influences cavitation. An increase in the displacement amplitude modifies the acoustic properties of liquid, thus changing the load resistance and sound pressure.

A close analysis of water cavitation over a wide range of displacement amplitudes was performed by Panov [117]. At small amplitudes, when there is no noticeable cavitation in liquid, acoustic power transmitted to the load is proportional to the square of the displacement amplitude:

$$W_a = \frac{1}{2} S_r \rho c (2\pi f \xi_m)^2 = \frac{1}{2} R_l (2\pi f \xi_m)^2, \qquad (11.9)$$

where $R_l = \rho_c S_r$ is the load resistance.

In a precavitation region, R_l can be calculated if ρ, c, and radiator area S_r are known.

In a cavitating liquid, ρ and c depend on the intensity of cavitation which, in turn, is dependent on the displacement amplitude.

The dependences $I = W_a / S_r$ and R_l on the displacement amplitude are plotted in Figure 11.54. The initial portion of curve $I(\xi_m)$, approximated by function $I \sim \xi_m^2$, corresponds to a noncavitating liquid. As the displacement amplitude increases to $\xi_m = 4$–$5\ \mu$m, the dependence becomes linear.

When cavitation begins to develop in liquid, its load resistance drops (curve *2*), as well as the liquid density and sound velocity. In this case, $R_l \sim 1/\xi_m$. Experiments show that with developed cavitation, the load resistance of medium may be tens of times lower than its wave resistance.

The extent of changes in liquid density produced by cavitation can be estimated from a change in the total volume occupied by the cavitation region. This volume is proportional to the number of cavitation

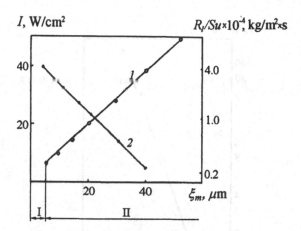

Figure 11.54. Acoustic intensity transmitted to a liquid load (curve *1*) and its impedance (curve *2*) versus displacement amplitude [117].

bubbles and can be estimated by the known procedure based on the transmission of sound radiation into a closed volume with an outward capillary.

Figure 11.55 shows the relative change in the total volume occupied by cavities with as a function of displacement amplitude. Measurements indicate that the volume of cavitation region is proportional to the cube of displacement amplitude. In other words, with increasing displacement amplitude there occurs a dramatic increase in the volume occupied by cavities and, consequently, a drastic decrease in liquid density.

These cavitation-induced changes in liquid properties are responsible for the complex behavior of sound pressure with increasing displacement amplitude. The experimental dependence $p_m(\xi_m, h)$ is given in Figure 11.56. Pressure was measured with a miniature hydrophone placed at distance h from the radiator along its axis.

The results show that initially the sound pressure rises with vibrational displacement amplitude. At $\xi_m = 7$–8 μm, when cavitation is developed, the sound pressure is maximum irrespective of the distance to the source.

Further increase in displacement amplitude causes a fall in the sound pressure, which is especially pronounced at $\xi_m = 10$–20 μm. The highest pressure level (0.35–0.40 MPa) is observed in the region next to the source of vibrations. The acoustic pressure drop in the rest of the liquid is related to a substantial absorption of acoustic energy in the near-radiator region that acts as a screen.

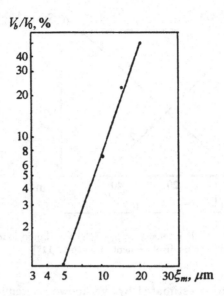

Figure 11.55. Relative volume of cavitation bubble versus displacement amplitude [117].

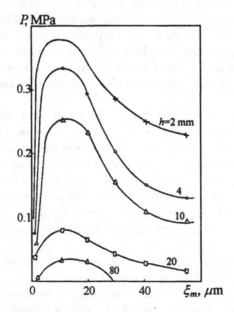

Figure 11.56. Acoustic pressure versus the vibration displacement amplitude at various distances from the ultrasonic source [117].

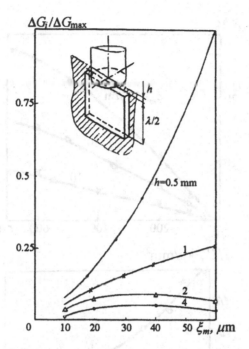

Figure 11.57. Relative erosion versus the vibration displacement amplitude at various distances from the ultrasonic source [117].

Spectral analysis of emitted radiation showed that, beginning from a displacement amplitude of 8–9 μm, the balance in energy spectrum shifts toward the prevalence of basic harmonics due to noise. In other words, there is a relative decrease in the intensity of cavitation in working volume with increasing displacement amplitude.

The relationship between erosion effects and displacement amplitude is also of practical importance. Figure 11.57 gives experimental layout and the results of investigation of erosion of a tin plate placed at different distances h from the radiator.

At $h = 0.5$ mm, erosive activity rises with displacement amplitude up to $\xi_m = 60$ μm. As distance h rises, there appears a maximum on the dependence $\Delta G(\xi_m)$, which is due to the formation of the zone of high sound absorption in the cavitation region. With increasing displacement amplitude, the contribution from the near-radiator region to erosion continues to increase to comprise 80–85% of total erosion at $\xi_m = 50$–60 μm.

Panov [117] revealed a zone, beyond which the erosive effect of cavitation is negligible. The solid surface subjected to a cavitating liquid

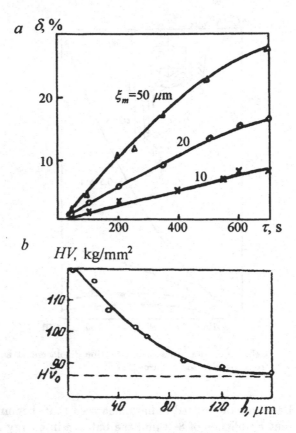

Figure 11.58. (a) Plastic deformation in the surface layer as a function of the time of ultrasonic treatment at various vibration displacement amplitudes, and (b) depth distribution of the hardness of grade AMg6 alloy ultrasonically treated for 180 s in water at a displacement amplitude of 30 μm [117].

experiences the action of high-pressure pulses produced by collapsing cavities or cumulative microjets. Pressure pulses give rise to stresses in the surface layer of the solid and, therefore, to its plastic deformation and strengthening.

Using X-ray methods, Panov [117] attempted to estimate the cavitation-stimulated plastic deformation of the surface layer of grade AMg6 aluminum alloy ultrasonically treated at various displacement amplitudes (Figure 11.58,a). The depth of hardened layer might reach 100 μm (Figure 11.58,b). It was also shown that the erosion fracture of the sample occurs when plastic deformation develops at its surface.

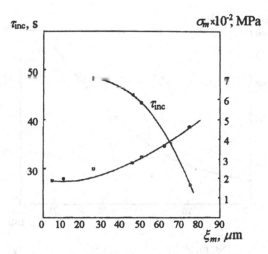

Figure 11.59. Effect of displacement amplitude on the latent time τ and cavitation stress σ_m in a nickel sample spaced by 1.14 mm from the horn [122].

Generally, erosion increases with displacement amplitude (Figure 11.59). It can be seen that ultrasonically induced stresses in the sample increase, and the latent period of erosion diminishes, as the vibration amplitude rises from 5 to 75 μm at 20 kHz [122]. This can be explained by an enhanced specific rate of erosion per unit area and an enlarged total area of eroded surface (Figure 11.60).

The role of acoustic streamings in ultrasonic cleaning is also worth noting. Some characteristics of streamings, such as their scale and velocity, were investigated by Panov [117] under various operation conditions of ultrasonic system. At a displacement amplitude of 8–10 μm, there appeared a continuous cavitation region near the radiator and directed hydrodynamic streamings that moved much faster than those formed under the same conditions, but without cavitation. The streamings contained many different-size bubbles that were translated with their own velocities differing from the streaming velocity. Bubble velocities were dependent on their diameters, distance to the source, and vibration displacement amplitude. The increase in the bubble diameter or vibration amplitude augmented the velocity of bubble translation, whereas the opposite took place with increasing distance from the source.

Figure 11.61 shows the dependences of the mean velocity of bubbles and streamings on the vibration displacement amplitude. It can be seen that the mean velocity of bubbles may reach 3–4 m/s.

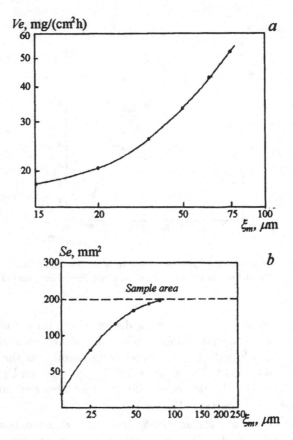

Figure 11.60. (a) Erosion rate and (b) eroded area of a brass sample as functions of displacement amplitude [121].

Apart from large-scale streamings covering most of the processed volume, the increase in displacement amplitude was accompanied by pulsations of the near-radiator cavitation region with frequency corresponding to that of ultrasonic vibrations. Estimations showed that the velocity of microstreamings associated with these pulsations and directed outward (during the phase of bubble expansion) or inward (during the phase of bubble shrinkage) the cavitation region was about 100 m/s.

Moreover, oscillations of single bubbles near the cavitation region occurred synchronously with oscillations of the entire region. The displacement amplitude of these oscillations was 0.5 mm, the velocity amplitude was estimated to be about 60 m/s.

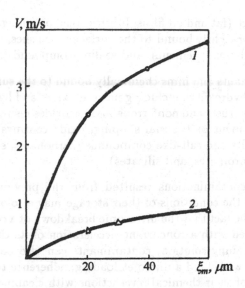

Figure 11.61. Dependences of (*1*) mean bubble velocity and (*2*) mean streaming velocity on the vibration displacement amplitude [117].

Panov [117] proposed a microstreaming cleaning model, in terms of which a translatory, pulsating, and vibrating bubble creates, when approaching the surface, microjets at the interface that cause the detachment of contaminants and enhance their emulsification and dissolution.

As for the technical aspects of ultrasonic cleaning, it seems reasonable to consider the problems associated with the classification of contaminants, the choice of cleaning solutions, design of relevant ultrasonic devices, and technology of ultrasonic cleaning.

A contaminant is characterized by a set of properties. According to the origin and the character of their binding to a surface, the following types of contaminants can be distinguished [123, 124]:

(1) Inorganic contaminants loosely be bound to the surface (dusts, chips, ashes, carbon black, sand, clay, etc.), or mechanically intruded into the surface (abrasive grains, mineral or metallic particles), or deposited on the surface (salt crusts after treatment in salt baths, scales, etc.), or fused with the surface, or crystallized from melts (fluxes, slags, burn-on sand, ceramic coatings, etc.).

(2) Organic contaminants and coatings loosely bound to the surface (dust, plastic chips, and wood shavings), or poorly adhered to

the surface (fat and oil films, lubrications, polishing and grinding pastes), or tightly bound to the surface (varnishes, resins, paints, enamels, latexes, cements, and sealing compounds).

(3) Contaminants and films chemically bound to the surface, i.e., oxides and hydroxides, including natural oxides and hydroxide films on ferrous (rust) and nonferrous metals, oxides resulted from thermal treatment or thermal shaping, oxide coatings (e.g., bluing films), salts and salt-like compounds (carbonates, sulfides, phosphates, chromates, and silicates).

Combined contaminations resulted from the previous operations with parts and the conditions of their storage may also occur.

In view of the fact that the ultrasonic breakdown of a contaminating film is associated with a concurrent acceleration of its chemical interaction with cleaning solution, contaminants can also be classified by their ability to withstand a microjet loading, adherence to the surface, and character of their chemical interactions with cleaning solution.

According to the first trait, contaminants can be either tolerant or sensitive to cavitation. By the second trait, they can be tightly or loosely bound to the surface. By the third trait, they are able or unable to chemically interact with a cleaning solution (chemical composition of the solution may, however, be modified to induce its interaction with contaminants).

Comparing the origin of contaminants and the peculiarities of their interaction with cleaning solution in an ultrasonic field, one can suggest the following combined classification of contaminants:

(1) Cavitation-sensitive, loosely bound to the surface to be cleaned, chemically inert with respect to cleaning solution (dusts, etching slimes);

(2) Cavitation-sensitive, tightly bound to the surface to be cleaned, chemically interacting with cleaning solution (products of corrosion);

(3) Cavitation-tolerant, loosely bound to the surface to be cleaned, chemically interacting with cleaning solution (grease films);

(4) Cavitation-tolerant, tightly bound to the surface to be cleaned, chemically interacting with cleaning solution (varnish films, paints, polishing pastes, scale, and oxide films);

(5) Cavitation-tolerant, tightly bound to the surface to be cleaned, chemically inert with respect to cleaning solution.

This classification is useful for elaborating a proper cleaning technology allowing a high degree of impurity removal. Clearly, the efficiency and quality of cleaning depend largely on the chemical properties of the cleaning solution.

Requirements that are usually imposed on cleaning solutions are a high chemical activity against contaminants, good wettability, a high capacity for detachment of contaminants from the surface and their dissolution, a resorption-preventing ability, inertness with respect to materials to be cleaned, ecological safety, and low cost.

Without going into a close consideration of these requirements, it is clear that they are contradictory and there may be no universal cleaning solution for all kinds of contaminants. It is a common practice to use water, water solutions of alkalies and acids, and organic solvents as cleaning solutions. Distilled water is known as a powerful cleaning agent. The efficiency of cleaning solutions can be raised by adding detergents. The most commonly used organic solvents are hydrocarbons, their chlorine and fluorine derivatives, alcohols, etc.

For a proper choice of ultrasonic cleaning technology, one should take into account the properties of a material to be cleaned and contaminants it contains. In this respect, technological objects can be grouped as follows:

(1) Objects that demand and admit severe regimes of cleaning based on the erosive effect of cavitating liquid;

(2) Objects that do not demand but admit severe regimes of cleaning;

(3) Objects that neither demand nor admit severe regimes of cleaning;

(4) Objects that demand but do not admit severe regimes of cleaning.

The first group includes varnish coats, scales, abrasive particles intruded into surfaces during their finishing, and other impurities and contaminating films formed on parts from structural materials (steels, alloys, nonferrous metals, etc.).

The second group involves precision parts and machine units that are mainly contaminated with grease films, lubricating or cooling liquid remainders, polishing pastes, chips, etc.

Various electronic devices, such as wiring or printed-circuit boards, can be referred to the third group.

The fourth group includes cavitation-sensitive objects with rather firm contaminating films. Their cleaning requires a pretreatment to

Band out   **Band in**

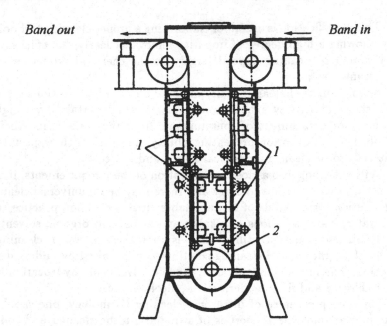

Figure 11.62. Converter arrangement with regard to a traveling steel band in an ultrasonic cleaning line [125]: (*1*) converters, (*2*) steel band.

loosen firm films, which allows their subsequent removal by mild cleaning procedures.

The present-day ultrasonic cleaning practice employs immersion-type, sequential, and contact methods.

With the first method, small parts placed in a basket or hanged on a suspension are immersed in a cleaning solution as close to the radiator as possible. To enhance the efficiency of cleaning, concerned devices are usually made to enable the displacement of cleaned parts relative to radiators.

Large-size or moving objects are cleaned by sequential methods (see, for instance, Figure 11.62).

Contact method is used to clean hard-to-reach interiors of thin-wall parts. In this method, ultrasonic vibrations are transmitted directly to the wall of a cleaned part, which acts then as a radiator.

The basic component of ultrasonic cleaners is a tank whose volume can range from 0.2 l to hundreds of liters at an input power of 50 W to tens of kilowatts. Ultrasonic vibrations, as a rule produced by piezoceramic transducers, can be fed into the tank either through its bottom or walls, or from above. Optimal cleaning conditions are often achieved by

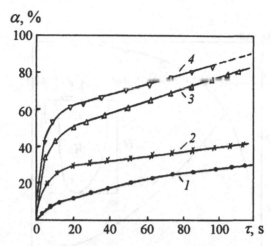

Figure 11.63. Kinetics of removal of greases and mechanical contaminants from steel by various methods [126]: (*1*) immersion, (*2*) immersion plus air bubbling, (*3*) immersion plus air bubbling plus ultrasonic irradiation, (*4*) immersion plus ultrasonic irradiation at an increased static pressure (pressurized ultrasonic cleaning).

filtering and regenerating the cleaning solutions. This method is suitable for the cleaning of a variety of workpieces. In particular, Agranat and Khavskii [125] described a device for the cleaning of a 750-mm wide transformer steel strip moving at a rate of 100–150 m/min (Figure 11.62).

To enhance cleaning efficiency, ultrasonic cleaners are often equipped with various attachments, such as those providing excess static pressure.

Of interest are investigations concerned with the cavitation-abrasive breakdown of solids in an ultrasonic field [129]. The efficiency of erosion is drastically increased by adding abrasive particles to the liquid subject to a high excessive static pressure. This method is especially well suited to deburring.

The technology of cleaning is dictated by particular conditions. For illustration, consider the problem of removal of grease films from the surface of steel samples with a cleaning solution containing 20 g/l trisodium phosphate, 10 g/l sodium carbonate, and 3 g/l surfactant [126].

Cleaning was carried out by various methods (Figure 11.63), viz. with immersion (curve *1*), with air bubbling through cleaning solution (curve *2*), with ultrasonic irradiation (curve *3*), and with ultrasonic irradiation at an elevated static pressure of 4 atm (curve *4*).

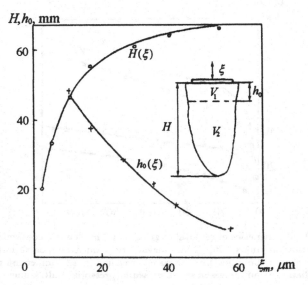

Figure 11.64.   Process zone in cavitation region [117].

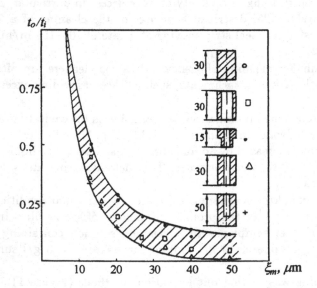

Figure 11.65.   Relative times of cleaning of various workpieces versus displacement amplitude [117].

Technological regimes of high-amplitude cleaning were analyzed in [117]. Figure 11.64 gives a schematic representation of the region, in which cavitation has the intensity sufficient for efficient cleaning.

This region is determined by the source dimension D and the depth of cavitation effects, H, proportional to displacement amplitude ($H = k\xi_m^n$). Curves in the figure refer to cleaning in water solutions. Cleaning takes place primarily in the neighboring zone (V_1), where erosion is most profound (zone of severe technological regime). In the erosion-free zone, whose boundary is at distance h_0 from the source, cleaning is due to microstreamings formed in the near-surface liquid layer as a result of bubble pulsations. Figure 11.65 shows the relative times of cleaning of a unit surface from loosely bound contaminants (polishing pastes, grease films, lubricants, cooling liquids, etc.) in the erosion-free zone.

Some authors [118, 127, 128] recommend a combined usage of low and high ultrasonic frequencies for cleaning. In this case, cavitation bubbles arising at low frequencies acquire additional oscillations in a high-frequency field, thereby enhancing the erosive activity of ultrasonic cavitation.

References

1. F. Blaha and B. Langenecker, *Z. Naturwiss.*, **20**, no. 9, 556 (1955).

2. V. P. Severdenko, K. V. Gorev, and E. G. Konovalov, *Ultrasonic Machining of Metals* (in Russian), Nauka i Tekhnika, Minsk (1966).

3. V. P. Severdenko, V. V. Klubovich, and A. V. Stepanenko, *Ultrasonic Rolling and Drawing* (in Russian), Nauka i Tekhnika, Minsk (1970).

4. V. P. Severdenko, V. V. Klubovich, and A. V. Stepanenko, *Ultrasonic Shaping of Metals* (in Russian), Nauka i Tekhnika, Minsk (1973).

5. V. P. Severdenko, V. V. Klubovich, and A. V. Stepanenko, *Ultrasound and Ductility* (in Russian), Nauka i Tekhnika, Minsk (1976).

6. V. V. Klubovich and A. V. Stepanenko, *Ultrasonic Treatment of Materials* (in Russian), Nauka i Tekhnika, Minsk (1970).

7. O. V. Abramov, *Ultrasonic Metal Shaping Experience* (in Russian), Mashinostroenie, Moscow (1980).

8. O. V. Abramov, A. V. Kulemin, Yu. V. Manegin, In: *Application of Novel Physical Methods for Intensification of Metallurgical Processes* (in Russian), Metallurgiya, Moscow, p. 203 (1974).

9. Yu. V. Manegin, O. V. Abramov, and Yu. F. Luzin, In: *Novel Physical Methods for Intensification of Technological Processes* (in Russian), Metallurgiya, Moscow, p. 44 (1977).

10. G. I. Pogodin-Alekseev, *Metallov. Termich. Obrab. Mater.*, 6, 14 (1958).

11. G. I. Pogodin-Alekseev and V. S. Mirotvorskii, *Metallov. Termich. Obrab. Mater.*, 9, 2 (1966).

12. G. I. Pogodin-Alekseev and O. B. Khavroshkin, *Ultrazvukovaya Tekhnika*, no. 2, 6 (1967).

13. K. M. Pogodina-Alekseeva, *Effect of Ultrasonic Vibrations on Diffusion Processes in Solid Metals and Alloys* (in Russian), NTO Mashprom, Moscow (1962).

14. K. M. Pogodina-Alekseeva, In: *Ultrasound in Machine Building*, TsNIIPI, 2, p. 115 (1969).

15. E. G. Aizentson and L. V. Spivak, *Izv. Akad. Nauk SSSR, Metallurgiya i Gornoe Delo*, 2, 124 (1963).

16. E. G. Aizentson and I. K. Utrobina, *Ultrazvukovaya Tekhnika*, no. 2, 24 (1967).

17. E. G. Aizentson and I. K. Utrobina, *Ultrazvukovaya Tekhnika*, no. 5, 24 (1967).

18. E. G. Aizentson and L. V. Spivak, *Ultrazvukovaya Tekhnika*, no. 3, p. 35 (1967).

19. E. G. Aizentson, L. V. Spivak, P. A. Malinin, and L. D. Pilikina, In: *Application of Ultrasound in Machine Building*, NTO Mashprom, Moscow, p. 80 (1963).

20. E. G. Aizentson and L. V. Spivak, *Izv. Vuzov, Ser. Chernaya Metallurgiya*, no. 6, 127 (1965).

21. E. G. Aizentson and L. V. Spivak, *Izv. Vuzov, Ser. Chernaya Metallurgiya*, no. 2, 128 (1967).

22. E. G. Aizentson and L. V. Spivak, *Izv. Vuzov, Ser. Chernaya Metallurgiya*, no. 4, 107 (1968).

23. E. G. Aizentson, *Fiz. Met. Metalloved.*, 17, no. 4, 624 (1964).

24. E. G. Aizentson, P. A. Malinin, and A. I. Uvarov, *Fiz. Mekh. Matem.*, 17, no. 5, 777 (1964).

25. V. S. Biront, *Ultrasound in Thermal Treatment of Metals* (in Russian), Metallurgiya, Moscow (1977).

26. S. F. Novik, In: *Ultrasound in Machine Building* (in Russian), TsNIIPI, Moscow, 2, p. 123 (1969).

27. A. A. Andreeva, In: *Novel Physical Methods for Intensification of Metallurgical Processes* (in Russian), Metallurgiya, Moscow, p. 256 (1974).

28. M. N. Dubrovin, In: *Ultrasonic Methods for Intensification of Technological Processes* (in Russian), Metallurgiya, Moscow, p. 107 (1975).

29. A. V. Kulemin and S. Z. Nekrasova, In: *Effects of High-Intensity Ultrasound on Metal Interfaces* (in Russian), Nauka, Moscow, p. 139 (1986).
30. V. N. Chachin and V. E. Eremin, *Metallov. Termich. Obrab. Mater.*, p. 9 (1966).
31. Yu. P. Usatyi, *Metallov. Termich. Obrab. Mater.*, **2**, 12 (1971).
32. R. K. Zhelokovtseva, *Temperature Control of Carbon Steel Oil Quenching using a Hydrodynamic Radiator*, Cand. Sci. (Engineering) Dissertation, MISiS (1982).
33. G. I. Eskin, *Ultrasound in Metallurgy* (in Russian), Metallurgiya, Moscow (1975).
34. E. A. Al'ftan and V. S. Ermakov, *Akusticheskii Zh.*, **4**, no. 4, 307 (1958).
35. V. S. Ermakov and E. A. Al'ftan, *Metallov. Termich. Obrab. Mater.*, **7**, 22 (1958).
36. K. V. Gorev and P. A. Parkhutik, In: *Ultrasound in Production of Alloys and Their Heat Treatment* (in Russian), NTO Mashprom, Moscow, **2**, p. 28 (1962).
37. E. A. Al'ftan, *Izv. Vuzov, Ser. Chernaya Metallurgiya*, **9**, 160 (1966).
38. B. G. Lutsyak, In: *Application of Ultrasound in Machine Building* (in Russian), TsINTIIMASh, Moscow, p. 21 (1963).
39. I. M. Arkhangel'skii, E. M. Kremlev, and K. M. Pogodina-Alekseeva, In: *Advanced Ultrasonic Methods of Metal Processing in Machine Building* (in Russian), NTO Mashprom, Moscow, p. 3 (1970).
40. I. M. Arkhangel'skii, B. M. Drankin, and K. M. Pogodina-Alekseeva, In: *Application of Ultrasound in Machine Building* (in Russian), NTO Mashprom, Moscow, p. 56 (1972).
41. A. I. Natchuk, In: *Application of Ultrasound in Production and Heat Treatment of Alloys* (in Russian), NTO Mashprom, **2**, p. 26 (1961).
42. G. I. Pogodin-Alekseev, In: *Application of Ultrasound in Production and Thermal Treatment of Alloys* (in Russian), NTO Mashprom, p. 33 (1962).
43. *Ultrasonic Technology* (in Russian), B. A. Agranat, ed., Metallurgiya, Moscow, p. 504 (1974).
44. *Ultrasound. Concise Encyclopedia*, I. P. Golyamin, ed., Sovetskaya Entsiklopediya, Moscow (1979).
45. S. S. Volkov, N. Orlov, and Chernyak, *Ultrasonic Welding of Plastic Materials*, Energiya, Moscow (1974).
46. S. Svegla, In: *Universal Electrotechnical Congress*, Moscow, section 4B, report no. 27 (1977).

47. J. R. Sherry, *Inserting Parts Ultrasonically*, Modern Plastics, February, p. 41 (1974).

48. A. Shoh, *Ultrasonics*, no. 5, 209 (1976).

49. S. Svegla, *Ultrazvukove zvaronie a jeho pouzitic v electrotechnike*, Trend-Vuma, **12**, 27 (1981).

50   O. V. Abramov, I. G. Khorbenko, and S. Svegla, *Ultrasonic Treatment of Materials* (in Russian), Mashinostroenie, Moscow (1984).

51. Yu. I. Kitaigorodskii, M. G. Kogan, V. A. Kuznetsov, N. N. Roskalin, and L. L. Silin, *Izv. Akad. Nauk SSSR*, OTN, no. 8, 88 (1958).

52. L. L. Silin, G. F. Balandin, and M. G. Kogan, *Ultrasonic Welding* (in Russian), Mashgiz, Moscow (1962).

53. Yu. V. Kholopov, *Ultrasonic Welding* (in Russian), Mashinostroenie, Leningrad (1972).

54. A. M. Mitskevich, In: *Physical Fundamentals of Ultrasonic Technology*, Nauka, Moscow (1970).

55. T. Enjo, K. Jkenchi, and H. Fujita, *J. of Japan Inst. of Light Metals*, **36**, no. 8, 498 (1986).

56. E. N. Parkhimovich, *Welding and Overlaying Welding in an Ultrasonic Field* (in Russian), Nauka i Tekhnika, Minsk (1988).

57. E. V. Kisterev, A. A. Shevchenok, and O. V. Abramov, *Izv. Vuzov, Ser. Chernaya Metallurgiya*, no. 1, 110 (1986).

58. E. V. Kisterev, O. V. Abramov, and R. I. Entin, *Fiz. Khim. Obrab. Mater.*, no. 5, 164 (1986).

59. O. V. Abramov, R. I. Entin, and E. V. Kisterev, *Problemy Prochnosti*, no. 6, 105 (1986).

60. E. V. Kisterev, O. V. Abramov, and A. V. Kulemin, In: *Physical and Physicochemical Methods in Technology* (in Russian), Metallurgiya, Moscow, p. 83 (1986).

61. V. V. Klubovich, M. D. Tyavlovskii, and V. A. Lapin, *Ultrasonic Soldering in Radioengineering and Instrument Production* (in Russian), Nauka i Tekhnika, Minsk (1985).

62. G. I. Apukhtin, *Ultrasonic Soldering of Aluminum and Its Alloys* (in Russian), Izd. Akad. Nauk SSSR, Moscow (1956).

63. M. G. Zakatov, *Recent Advances in Ultrasonic Soldering* (in Russian), NTO Mashprom, Moscow (1959).

64. S. I. Pugachev, *Tekhnologiya Sudostroeniya*, no. 5, 100 (1964).

65. E. V. Guseva, E. N. Dolgov, and S. I. Pugachev, *Tekhnologiya Sudostroeniya*, no. 6, 50 (1971).

66. E. N. Dolgov, S. I. Pugachev, and E. Ya. Tarat, *Tekhnologiya Sudostroeniya*, no. 4, 53 (1971).

67. A. M. Mitskevich and S. I. Pugachev, *Ultrasonic Welding and Plating* (in Russian), NTO Mashprom, Moscow (1979).

68. S. I. Pugachev and N. G. Semenova, In: *Effect of High-Intensity Ultrasound on Metal Interfaces* (in Russian), Nauka, Moscow, p. 7 2 (1986).
69. A. A. Prokhorenko, S. I. Pugachev, and N. G. Semenova, *Ultrasonic Plating of Metals* (in Russian), Nauka i Tekhnika, Minsk (1987).
70. P. G. Dumitrash, A. V. Rychagov, and O. V. Abramov, *Izv. Akad. Nauk Moldavian SSR, Ser. Fizikotechn. i Matem. Nauk*, no. 3, 50 (1978).
71. N. V. Delenkovskii, *Almazy i Sverkhtverdye Materialy*, no. 5, 4 (1981).
72. V. A. Kuz'menko, L. E. Matokhnyuk, and G. G. Pisarenko, *Fatigue Testing at High Load Frequencies* (in Russian), Naukova Dumka, Kiev (1979).
73. M. Kikukawa, K. Ohji, and K. Ogura, *Proc. 6th Jap. Congr. Test Mater.*, Kyoto, p. 4 (1963).
74. M. Kikukawa, K. Ohji, and K. Ogura, *Proc. 7th Jap. Congr. Test Mater.*, Kyoto, p. 7 (1964).
75. M. Kikukawa, K. Ohji, and K. Ogura, *Trans. ASME, D*, **87**, no. 4, 857 (1965).
76. W. Nessler, H. Mullner, B. Weiss, and R. Stickler, *Metal Science*, no. 5, 225 (1981).
77. H. Mullner and B. Weiss, *Materials Science and Engineering*, **48**, 1 (1981).
78. B. Weiss, R. Stickler, J. Fembock, and K. Pfaffinger, *Metals*, **34**, 636 (1980).
79. H. Mullner, B. Weiss, and R. Stickler, *1st Internat. Symp. on Fatigue*, Fhreshold, Stockholm (1981).
80. B. Weiss, R. Stickler, J. Fembock, and K. Pfaffinger, *Fatigue of Engineering Materials and Structures*, **2**, 73 (1979).
81. B. Weiss, R. Stickler, S. Schider, and H. Schmidt, *Proc. 1st Intern. Symp. on Fatigue and Corrosion Fatigue* (1984).
82. L. D. Rozenberg, V. F. Kazantsev, and L. D. Makarov, *Ultrasonic Machining* (in Russian), Izd. Akad. Nauk SSSR, Moscow (1962).
83. V. F. Kazantsev, In: *Physical Fundamentals of Ultrasonic Technology* (in Russian), Nauka, Moscow, p. 11 (1970).
84. A. I. Markov, *Ultrasonic Machining of Hard Materials* (in Russian), Mashinostroenie, Moscow (1968).
85. E. A. Neppiras and R. D. Fosket, *Philips Techn. Rundschau*, **19**, no. 2, 37 (1957).
86. E. A. Neppiras, *Metalwork Products*, **100**, no. 27, 1283 (1956).
87. G. Nishimura, K. Yanagishima, and T. Shima, *J. Fac. Eng. Univ. Tokyo*, **26**, no. 2, 129 (1959).

88. A. I. Markov, *Ultrasonic Treatment of Materials* (in Russian),
     Mashinostroenie, Moscow (1980).

89. L. Balamuth, *Method of Abroding*, English Patent no. IV602801.

90. A. I. Markov, In: *Ultrasound in Machine Building* (in Russian),
     TsNIIPI, Moscow, p. 45 (1966).

91. A. V. Mordvintseva, In: *Ultrasound in Welding* (in Russian),
     Proc. of Bauman MVTU, no. 35, p. 32 (1959).

92. I. I. Mukhanov, Yu. M. Golubev, and V. N. Filimonenko, In:
     Proc. of Novosibirsk Machine Building Conf., part 1, Novosibirsk
     NTO Mashprom, p. 21 (1964).

93. I. I. Mukhanov and Yu. M. Golubev, *Vestnik Mashinostroeniya*,
     no. 11, 52 (1966).

94. A. V. Kulemin, V. V. Kononov, and L. A. Stebel'kov, *Problemy
     Prochnosti*, no. 1, 70 (1981).

95. A. I. Markov, In: *Industrial Application of Ultrasound* (in
     Russian), Mashinostroenie, Moscow, p. 157 (1975).

96. I. I. Mukhanov, *Vestnik Mashinostroeniya*, no. 6, 51 (1968).

97. I. I. Mukhanov and Yu. M. Golubev, *Metalloved. i Termich.
     Obrab. Metallov*, no. 9, 141 (1969).

98. V. F. Kazantsev, *Effect of High-Intensity Ultrasound on Metal
     Interfaces* (in Russian), Nauka, Moscow, p. 186 (1986).

99. I. G. Polotskii, A. Ya. Nedoseka, and G. I. Prokopenko,
     Avtomaticheskaya Svarka, no. 5, p. 74 (1974).

100. I. G. Polotskii, *Vestnik Mashinostroeniya*, no. 4, 74 (1977).

101. V. G. Stepanov, E. Sh. Statnikov, M. I. Klestov, and
      E. M. Shevtsov, *Tekhnologiya Sudostroeniya*, no. 7, 32 (1974).

102. Ya. I. Blyashko, V. A. Volosatov, and D. M. Bavel'skii,
      Problemy Prochnosti, no. 7, 112 (1980).

103. V. G. Badalyan, V. F. Kazantsev, E. Sh. Statnikov, and
      E. M. Shevtsov, *Vestnik Mashinostroeniya*, no. 8, 56 (1979).

104. V. G. Stepanov, E. Sh. Statnikov, M. I. Klestov, and
      E. M. Shevtsov, *Tekhnologiya Sudostroeniya*, no. 1, 70 (1975).

105. A. I. Markov, N. D. Ustinov, M. A. Ozerova, *Vestnik
      Mashinostroeniya*, no. 9, 57 (1973).

106. V. L. Senyukov, V. S. Skvortsov, and E. Sh. Statnikov, In:
      Ultrasonic Methods in Instrument Production (in Russian),
      Metallurgiya, Moscow, p. 5 (1979).

107. I. A. Stebel'kov, USSR Inventor's Certificate no. 456 704.
      Method of Surface Hardening. Published in *Byull. Izobret.*, no. 2
      (1975).

108. A. A. Matalin, *Technological Methods for Enhancing Part
      Endurance* (in Russian), Tekhnika, Kiev (1971).

109. I. I. Mukhanov and Yu. B. Kuroedov, In: *Industrial Application*

of Ultrasonic Technology (in Russian), Moscow, p. 100 (1976).

110. V. I. Drozhalova and B. A. Artamonov, *Ultrasonic Impregnation of Parts* (in Russian), Mashinostroenie, Moscow (1980).

111. B. A. Agranat, V. I. Bashkirov, and Yu. I. Kitaigorodskii, In. *Physical Fundamentals of Ultrasonic Technology* (in Russian), Nauka, Moscow, p. 166 (1970).

112. L. O. Makarov and L. D. Rozenberg, *Akust. Zh.*, 3, no. 4, 374 (1957).

113. V. I. Volodarskaya, Yu. I. Kitaigorodskii, and V. F. Korolev, In: *Ultrasound in Machine Building* (in Russian), NTO Mashprom, Moscow, p. 64 (1960).

114. B. A. Agranat, V. I. Bashkirov, and Yu. I. Kitaigorodskii, *Ultrazvukovaya Tekhnika*, no. 3, 38 (1964).

115. F. A. Bronin and A. P. Chernov, In: *Ultrasonic Cleaning of Parts in Freon Mixtures* (in Russian), Mashinostroenie, Moscow (1978).

116. A. P. Panov and Yu. F. Piskunov, In: *High-Amplitude Ultrasonic Cleaning* (in Russian), Mashinostroenie, Moscow (1980).

117. A. P. Panov, In: *Effect of High-Intensity Ultrasound on Metal Interfaces* (in Russian), Nauka, Moscow, p. 217 (1986).

118. B. A. Agranat, M. N. Dubrovin, N. N. Khavskii, and G. I. Eskin, In: *Physical and Engineering Fundamentals of Ultrasonics* (in Russian), Vysshaya Shkola, Moscow (1987).

119. B. G. Novitskii, In: *Application of Acoustic Vibrations in Chemical Engineering* (in Russian), Khimiya, Moscow (1983).

120. M. S. Plesset, *Trans. ASME, Ser. D.*, 54, 559 (1972).

121. J. M. Hobbs and D. Rachman, In: *Erosion by Cavitation of Impingement*, ASTM & STP 408, p. 159 (1967).

122. *Erosion*, Academic Press (1979).

123. O. K. Keller, G. S. Kratysh, and G. D. Lubenitskii, In: *Ultrasonic Cleaning* (in Russian), Mashinostroenie, Moscow (1975).

124. O. V. Abramov, I. G. Khorbenko, and S. Svegla, In: *Ultrasonic Machining* (in Russian), Mashinostroenie, Moscow (1984).

125. B. A. Agranat and N. N. Khavskii, *Akust. Zh.*, 22, no. 1, 42 (1976).

126. M. V. Dolgov and B. S. Artomoshkin, In: *Ultrasound in Metallurgy* (in Russian), Metallurgiya, Moscow, p. 44 (1977).

127. B. A. Agranat, V. I. Bashkirov, Yu. I. Kitaigorodskii, and N. N. Khavskii, In: *Ultrasonic Technology* (in Russian), Metallurgiya, Moscow (1974).

128. N. N. Khavskii and A. V. Fel'dman, In: *Physical and*

Physicochemical Methods in Technology (in Russian),
Metallurgiya, Moscow, p. 17 (1986).

129.   B. A. Agranat and N. N. Khavskii, In: *Ultrasonic Methods of
Intensification of Technological Processes* (in Russian),
Metallurgiya, Moscow, p. 28 (1970).

Conclusion

The monograph is devoted to some problems of physics and technology of high-intensity ultrasound and its employment for materials processing. Applications of high-intensity ultrasound in chemical and biochemical technology have been reviewed at length in the literature and therefore were not considered in this book.

In agriculture, food and light industries, intense ultrasound is suitable for

- air humidification in textile industry;

- aspirated plant feeding in greenhouses;

- toxic chemicals spraying in horticulture;

- prevention and cure of diseases of domestic animals;

- presowing treatment of cereal, vegetable, and flower seeds;

- mayonnaise and margarine production;

- defoaming in fermentors;

- cleaning of glassware, trays, and molds;

- extraction of valuable components from agricultural and medicinal raw materials;

- corn drying;

- wool cleaning;

- sperm treatment for artificial fertilization.

In household equipment, ultrasound can be employed in inhalators, air saturators, washers, spotters, percolators, knives, etc.

In energetics, it can be used for atomizing burners, cryogenic devices, and antiscaling appliances.

In medicine, ultrasound is potential for dental, therapeutic, and surgical applications.

It is also suited for waste water treatment, tap water purification, and industrial exhaust disposal.

Undoubtedly, this list of ultrasonic applications is far from being complete, but in the scope of one monograph it is impossible to cover all aspects of feasibility of high-intensity ultrasound.

Subject Index

transducer, 466
Impregnation, 652
    general principles, 652
    results, 653
Impurities,
    and solidification, 344
Index,
    cavitation, 314
    reflection, 35
    refraction, 35
Inoculation, 346
Instrumentation, *see* also
  Technology of ultrasound
Interferometer, 473
Internal friction, 12
Interstitial diffusion, 224

Lamb's waves, 47
Lamé constants, 43
Lamé's equation, 44
Lasers,
    anemometer, 474
    measurement applications,
      473
Liquid technologies,
    atomisation, 570
    cleaning, 656
    crystal growth assistance,
      552
    degassing,
      of Al and Mg melts, 516
      water, 264
    electrochemistry, 650
    hydrometallurgy, 506
    plating, 526
    cast composite preparation,
      559
    solidification, 523
Liquids,
    acoustic streaming of, 124
      general principles, 124
      in solidifying melt, 140
    atomisation of, 237
    cavitation in, 80
    radiation pressure in, 150
    sonocapillary effect in, 288

Loss,
    angle, 64
    magnetic, 386
Low-melting metal solidification,
  532

Machining, 637
    application,
      process, 640
      results, 640
    equipment,
      for machining of hard
        materials, 637
      for machining with
        vibrating tools,637
      for final surface hardening,
        637
    general principles, 637
    history/review of literature,
      637
    mechanism, 638
Mach's number, 67
Magnesium alloys,
    degassing of, 516
    solidification, 536
Magnetostriction, 384
    static, 385
    hysteresis, 386
Martensite transformations, 617
Materials,
    magnetostrictive, 385
      characteristics, 387
    piezoceramic, 403
Measurement techniques,
    vibration parameter
      in stacks, 472
    by electrodynamic probes,
      478
    by electromagnetic
      pickup, 477
    by fiber optic device, 475
    by laser anemometer, 474
    by laser interferometer,
      473
    by vibrometer, 476

Printed in the United States
by Baker & Taylor Publisher Services